Geochemical Processes

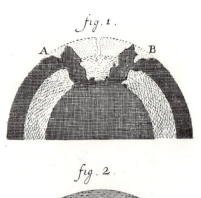

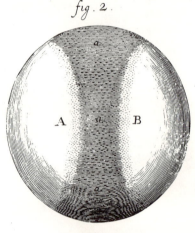

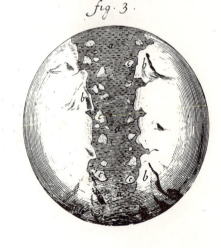

Formation of the oceans, according to Thomas Burnet, *The Theory of the Earth: Containing an Account of the Original of the Earth, and of All the General Changes Which It Hath Already Undergone, or Is to Undergo Till the Consummation of All Things* (2nd edition, London, 1691).

Geochemical Processes
Water and Sediment
Environments

A. Lerman

Northwestern University
Evanston, Illinois

A Wiley-Interscience Publication
JOHN WILEY & SONS
New York • Chichester • Brisbane • Toronto

Library of Congress Cataloging in Publication Data

Lerman, Abraham.
   Geochemical processes.

   "A Wiley-Interscience publication."
   Includes bibliographical references and indexes.
   1. Sedimentation and deposition. 2. Geochemistry.
3. Hydrology. I. Title.

QE571.L45   551.9   78-15039
ISBN 0-471-03263-8

Printed in the United States of America

10 9 8 7 6 5 4 3 2 1

Preface

The approach of this book to geochemistry can be summarized in the question: "What happens, and how, and how fast does it happen, when waters, solids, and gases interact in the earth's surface environment?" The environment of the earth's surface is made of solids and fluids, and the interactions among them are responsible for much of what is taking place in the physical world around us. The dissolved load of natural waters and the materials of which sediments are made are the products of reactions taking place practically everywhere on land, in the atmosphere, and in the hydrosphere. Thus the term *water and sediment environments* applies effectively to much of the surface environment of the earth, including the zone of up to a few kilometers above and below the land and ocean surface. Evolution of the environment, driven either by nature or man, or both, usually presents itself to us as a more or less complex variety of processes —geological, physical, chemical, and biological. To this end, the inclusive title *Geochemical Processes* was chosen for the book, to introduce a text that emphasizes processes and time-dependent phenomena. (A kinetic as compared to thermodynamic approach would have been an appropriate descriptive term, had not the word *kinetics* stood in a close and specific connotation to the mechanisms of chemical reactions.) To address myself to all, or even most, of the important geochemical processes operating in water and sediment environments would have been, at the very least, naive. The subjects presented in the book reflect personal interests and bias for which I bear the blame of omissions or commissions.

The book is primarily intended to be a graduate-level text, although it also contains additional material in the form of details, tabulated data, and literature references that should be useful to water-oriented geologists, chemists, limnologists, ocean-oriented scientists, and environmental engineers. It has been my experience in teaching graduate and undergraduate courses in geochemistry and environmental subjects that texts containing a certain volume of basic material make it considerably easier to deal with the bulk of their specific subject matter. This is particularly so if the subject matter falls in the domains of interest of several disciplines. With a view to the role of fundamentals in a teaching and research text, I assembled some basic background material in sections and chapters adjoining those that deal with the more specific topics. The fundamental material includes sections on diffusional and advective transport, chemical and physical characteristics of solutions and suspensions, and the behavior of settling particles. As much as possible, the narrative parts of the individual chapters are separated from the mathematical formulations and from the derivation of equations. To make the mathematical sections readable and easier to follow, explanatory text and examples are included.

The book begins with a chapter introducing the broad picture of the global geochemical cycles, and proceeds to treat in more detail the individual processes responsible for the major fluxes of materials on land, in waters, and, to a lesser degree, in the atmosphere. The subsequent chapters deal with the following subjects: transport of dissolved and suspended materials; controls of the chemical composition and acidity of rain; the physical and chemical weathering of the land surface; interactions between solids and waters in rivers, lakes, and oceans; regeneration of biologically formed materials; transport across the sediment water interface; and chemical reactions and diagenesis in sediments.

Most of the writing was done while on leave of absence granted by Northwestern University. My leave was supported by a 1976–1977 fellowship from the John Simon Guggenheim Memorial Foundation of New York, and by the hospitality and an appointment to the visiting faculty at the Institute of Aquatic Sciences and the EAWAG, both of the Swiss Federal Institute of Technology (ETHZ). My foremost thanks are due the Guggenheim Foundation, the Director of the EAWAG, Professor Werner Stumm, and all those of the administrative, scientific, and technical community of the ETH at Dübendorf and Zurich, whose word and deed helped me in my work.

A. LERMAN

Evanston, Illinois
September 1978

Contents

1 Geochemical Cycles   **1**

1.1   Geochemical Reservoirs   1
1.2   Residence Times   4
1.3   Some Mathematical Basics for Cycles   14
1.4   Global Cycle of Phosphorus   17
1.5   Carbon and Oxygen Cycles   23

2 Fluxes and Transport: Advection and Dispersal   **43**

2.1   Flux   43
2.2   Advection   44
2.3   Diffusion   56
2.4   Relative Effectiveness of Diffusion and Advection   58
2.5   Dispersal in Flow   60
2.6   Eddy Diffusion in Water   65

3 Fluxes and Transport: Molecular Diffusion   **73**

3.1   Basic Equations   73
3.2   Diffusion Coefficients in Solution   79
3.3   Effects of Environmental Variables on Diffusion   86
3.4   Diffusion of Gases in Water   94
3.5   Thermal Diffusion in Solutions   96
3.6   Diffusion in Gases   101
3.7   Diffusion in Solids   110

4 Atmospheric Processes   **122**

4.1   Water in the Atmosphere   122
4.2   Rain as a Transport Agent   126
4.3   Uptake of Gases and Solids by Rain   137
4.4   Acidity and Chemical Composition of Rain   148
4.5   Atmosphere-Ocean-Land Mercury Exchange   166

5 Physical and Chemical Weathering   **180**

   5.1 Beginnings of Weathering   180
   5.2 Particle-Size Distributions: Fundamentals   183
   5.3 Particle-Size Spectra of the Weathered Crust   210
   5.4 Transport of the Products of Weathering   217
   5.5 Solubilities of Minerals   225

6 Oceans and Lakes: Waters, Solids, and Solutes   **257**

   6.1 The Cycle of Water   257
   6.2 Sedimentation   260
   6.3 Regeneration of Suspended Materials   312

7 Sediment-Water Interface   **333**

   7.1 Up and Down Transport   333
   7.2 General Role of the Sediment-Water Interface   336
   7.3 Models of the Sediment-Water Interface   338
   7.4 Fluxes at the Sediment-Water Interface   340
   7.5 Uptake and Adsorption by Solids   342
   7.6 Silica in the Interface Zone   349
   7.7 Fluxes and Diagenetic Models   354
   7.8 Transport out of the Sediments   366

8 Migration, Reactions, and Diagenesis in Sediments   **369**

   8.1 Background Problems   369
   8.2 General Model of Sediment-Pore Water System   370
   8.3 Mathematical Models   373
   8.4 Nutrients and Related Species in Sediments   391
   8.5 Mineral-Pore Water Reactions   398

Appendices   **409**

   A   The Error Function   409
   B   Solution of Differential Equations   412
   C   Settling Velocities of Particles   420
   D   Physical Constants and Units   435

References   **437**

Symbols and Functions   **460**

Author Index   **465**

Subject Index   **473**

Geochemical Processes

CHAPTER 1

Geochemical Cycles

The upper part of the earth's crust, the hydrosphere, and the atmosphere are the biggest units into which the environment of the earth's surface can be subdivided. Flows of materials between these three major reservoirs bind them into one system, where each of the reservoirs affects the others to a greater or lesser degree. The geochemical cycles of chemical elements or species are conceptual models of their geochemical behavior within the different parts of the earth. Transport and transformations of materials near the earth's surface are called the *exogenic cycle*, to distinguish it from the slower and less well understood geochemical processes within the lower crust and mantle which represent the *endogenic cycle*.

1.1 GEOCHEMICAL RESERVOIRS

A division of the earth's surface environment into the major sections that are involved in the exogenic geochemical cycles of different elements is sketched in Figure 1.1. The nature and sizes of the geochemical reservoirs chosen for such a division depend on which particular cycle, or parts of a cycle, one wishes to consider. Such big units as the atmosphere, oceans, ocean-floor sediments, and land surface, drawn in Figure 1.1, may suffice for a broad picture of the geochemical cycle of an element, if no finer resolution of the scales is needed. The individual reservoirs can be made very small when, for example, one considers the geochemical cycles within systems of relatively small dimensions. The main transport paths between the individual reservoirs, indicated by arrows in Figure 1.1, are the flows or fluxes of materials. From the geochemical point of view, the fluxes of solid, dissolved, and gaseous species are driven by the thermodynamic and mechanical disequilibrium conditions that exist locally within the system.

Fluxes between the reservoirs carry materials in two forms: (1) in a state in which the material occurs in the reservoir (such as transport of solid

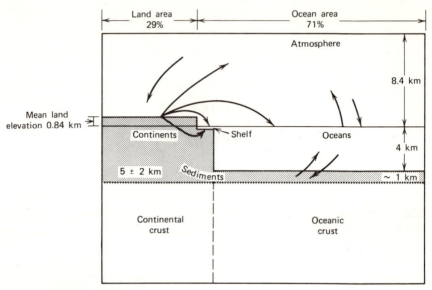

Figure 1.1 Major reservoirs and fluxes of the exogenic geochemical cycle.

detritus from land to the atmosphere and oceans, transport of dissolved species between different parts of the hydrosphere), and (2) in a state that has been altered by biogeochemical processes (such as transport of dissolved solids by rivers, precipitation or biological formation of solids in waters, fluxes of gases forming from dissolved or solid materials). In general, transport of materials between the big reservoirs, such as those shown in Figure 1.1, is carried out by forces of global nature, which do not depend on the chemical composition or mass content of the individual reservoirs. Wind, flow of water, and gravitational motions are the main transport agents on the global scale. The tendency of a chemical species to be transported either in an original or in a chemically modified state is a reflection of its geochemical behavior within a particular reservoir, and its behavior on exposure to the transport media.

The compartments of the earth's surface environment shown in Figure 1.1 serve as a guideline of the scale of the geochemical cycles that are considered in this chapter. The following features can be noted with regard to Figure 1.1. The mass of sediments and the mass of ocean water are the same, to an order of magnitude (10^{24} g). Both are much greater than the mass of the atmosphere (10^{21} g). The mean thickness of sediments on the continents is 5 ± 2 km, comparable to the mean depth of the ocean. The estimate of the thickness of the sedimentary layer on continents may vary by a factor of 2, because of a comparable difference between the estimates

Table 1.1 Major Chemical Constituents of the Earth's Crust, Sediments, Ocean Water, and Atmosphere

Element	Crystal ionic charge and radius[a]		Continental crust		Oceanic crust		Average sediments		Ocean water		Atmosphere	
	charge	r(Å)	(wt %[b])	(vol %)	(wt %[b])	(vol %)	(wt %[c])	(vol %)	(wt %[a])	(vol %)	(wt %)	(mol % or vol %[a])
O	−2	1.32	46.40	93.04	43.80	92.57	47.61	91.32	86.0	99.0 as H_2O	23.15	20.95 (O_2)
Si	+4	0.42	28.15	1.04	24.00	0.93	24.40	0.86				
Al	+3	0.51	8.23	0.56	8.76	0.63	6.03	0.40				
Fe	+3 / +2	0.64 / 0.74	5.63	0.46	8.56	0.74	3.79	0.30				
Ca	+2	0.99	4.15	1.40	6.72	2.39	7.86	2.54	0.04	0.025		
Na	+1	0.97	2.36	1.31	1.94	1.13	1.36	0.72	1.08	0.11		
Mg	+2	0.66	2.33	0.38	4.5	0.78	2.44	0.39	0.13	0.04		
K	+1	1.33	2.09	1.75	0.83	0.73	2.00	1.61	0.04	0.062		
Ti	+4	0.68	0.54	0.05	0.90	0.09						
Mn			0.095		0.15							
H			0.14		0.2				10.7	(see O)		
P	+5	0.35	0.105		0.14		0.16	0.003	0.09	0.0002		
S	+6	0.30	0.026		0.025		0.62	0.007	0.28	0.002		
C	+4	0.16					2.91[d]	0.013			0.046	0.03 (CO_2)
Cl	−1	1.81					0.83	1.85	1.94	0.833		
N											75.53	78.09 (N_2)
Ar											1.28	0.93 (Ar)

[a]Weast (1974).

[b]Taylor (1964).

[c]From Garrels et al. (1975, p. 61).

[d]Inorganic C, 2.4; organic, 0.5.

of the total mass of sediments—the lower estimates are 1.7×10^{24} and 1.8×10^{24} g (Poldervaart, 1955; Gregor, 1968), and a higher estimate is 3.2×10^{24} g (Garrels and Mackenzie, 1971).

The elemental chemical compositions of the continental crust, oceanic crust, average sediment, ocean water, and atmosphere are given in Table 1.1. Both the weight- and volume-percent abundances of the elements are listed for each of the five big geochemical reservoirs. For the gases in the atmosphere, the weight- and volume-percent figures are comparable. For solids in the crust and sediments, the differences are large because of the large differences between the ionic crystal radii of the elements. The abundances of elements listed in the table show effectively that the crust, sediments, and water are made of oxygen, with silicon or hydrogen as the second most important element by weight but insignificant by volume.

1.2   RESIDENCE TIMES

The flows through the geochemical reservoirs are controlled by the input and removal mechanisms, chemical reactions within the reservoirs, and the physical behavior of the reactants and reaction products. Input of a chemical species into a geochemical reservoir may include fluxes from other reservoirs and production of the species within the reservoir. For example, carbon is being transported from land to the oceans in solution and as undecomposed organic detritus; oxidation of the detritus in the ocean transfers carbon into solution, which amounts to an additional input source of carbon in the ocean. Similarly, removal from a reservoir can be accomplished by outward directed fluxes, as well as by internal removal reactions.

1.2.1   Residence Times: Total and Fractional

If the rates of input to a reservoir and removal from it are equal, then the amount of material in the reservoir does not change with time and a steady state is maintained. At the steady state, the mean residence time τ of material flowing through the reservoir is defined as

$$\tau = \frac{\text{amount in reservoir}}{\text{sum of all input } or \text{ removal rates}} \qquad (1.1)$$

or, in a shorter notation,

$$\tau = \frac{M}{\Sigma F_i} \qquad \text{(time)} \qquad (1.2)$$

such that whatever units are used for amount M, the fluxes F_i are in the same units per unit of time.

Each of the input or removal fluxes F_i corresponds to a fractional mean residence time τ_i, defined with respect to that particular process,

$$\tau_i \equiv \frac{1}{k_i} = \frac{M}{F_i} \qquad \text{(time)} \qquad (1.3)$$

where $k_i = 1/\tau_i$ is the rate constant for the process, in units of time^{-1}. For example, outflow of water F_1 (m^3 yr^{-1}) from a lake of constant volume V (m^3) corresponds to the rate constant $k_1 = F_1/V$ yr^{-1}, and evaporation flux of water F_2 (m^3 yr^{-1}) corresponds to another rate constant $k_2 = F_2/V$ yr^{-1}. Both fractional residence times, or their reciprocals, enter in the total mean residence time relationship 1.2, and a general form of the latter is either one of the two following relationships:

$$\tau = \frac{1}{k_1 + \cdots + k_n} \qquad (1.4)$$

$$\frac{1}{\tau} = \frac{1}{\tau_1} + \cdots + \frac{1}{\tau_n} \qquad (1.5)$$

Equation 1.4 shows that the stronger fluxes or, in other words, the rate constants k_i of higher values may control the total mean residence time, making the contributions of the weaker fluxes insignificant.

If input and output of a reservoir are not balanced, then the reservoir is not in a steady state and its contents are changing with time. In this case, equation 1.1 defines an "instantaneous" residence time, which is meaningful in comparison to the mean residence time τ, as defined for a steady state.

In Table 1.2 are tabulated some characteristic residence times for a number of geochemical reservoirs. The reservoirs in the table represent a finer subdivision of the earth's surface environment than those shown in Figure 1.1, although even the list in Table 1.2 holds only for a broad global picture. The tabulation of the residence times for the heterogeneous and homogeneous components of the reservoirs corresponds to the distinction made at the beginning of Section 1.1 between the chemically altered and original components of the reservoirs. For example, atmospheric precipitation consisting of liquid and solids is a heterogeneous component of the gaseous atmosphere; biological organic and skeletal materials in waters are heterogeneous components of the aqueous geochemical reservoirs; mechanical denudation removes material directly from the land reservoir, and this process is classified as a flux of homogeneous components;

Table 1.2 Some Characteristic Residence Times τ of the Major Geochemical Reservoirs

Reservoir	Homogenous components τ(yr)	Processes determining τ	Heterogenous components τ(yr)	Processes determining τ
Atmosphere	$10^{-2} - 10^{7}$	Gases, shortest τ for H_2O, longest for He	$10^{-1.5 \pm 0.5}$	Precipitation, washout of particles
Land (average thickness 60 cm, 2.5 g cm^{-3})	$10^{4.5 \pm 0.5}$	Mechanical denudation	$10^{5 \pm 0.5}$	Chemical weathering
Surface ocean (200 m)	10^{2}	Water exchange with deep	$10^{0.5 \pm 0.5}$	Settling of biologically fixed material
	10^{2}	Evaporation		
	10^{3}	River inflow of water		
Deep ocean (4000 m)	10^{3}	Water exchange with surface ocean	$10^{1 \pm 1}$	Settling of particulates
	$10^{2} - 10^{8}$	Removal of dissolved salts from river input		
Land biota	$10^{1 \pm 1}$	Mean lifetime of trees		
Ocean biota (plankton)	10^{-1}	Life cycle time		

similarly, the components of the biota, and water and dissolved solids in the ocean are homogeneous components of their respective reservoirs. The residence times for the individual reservoirs are order of magnitude figures that give an idea of the length of time involved in different processes. Some of the noticeable points of Table 1.2 are the following.

In the atmosphere, the range of the residence times of different gases is very broad, from a few days for water vapor to the estimated 10^7 yr for the nonreactive helium. Atmospheric solids are removed by settling and by washout, and therefore their residence times are comparable to the residence time of water in the atmosphere.

The land reservoir is taken as a layer 60 cm thick over the dry land surface. Mechanical denudation of soils and rocks is faster than the chemical weathering, and this is reflected in the shorter residence time of the components of land with respect to the mechanical weathering. Both the chemical and mechanical weathering of land materials are treated in more detail in Chapter 5.

For the surface ocean, represented by a 200-m-thick surface water layer, the fractional residence times of water vary within an order of magnitude. Evaporation and exchange by mixing with the deep ocean are faster than the renewal of water by river inflow.

The residence times of dissolved species in the deep ocean differ one from another by several orders of magnitude. The longest residence times are those of sodium and chloride ions, and the shorter residence times are for metals which form poorly soluble oxides and hydroxides.

The residence times of the chemical elements in the living land biota are represented by the longevity of trees. The residence times in grasses and annual crops are obviously much shorter. As a whole, the land plants live longer than the oceanic phytoplankton, the lifetime of which is a fraction of a year. The dead plankton resides for a relatively short time in the surface water layer of the ocean. In the bulk of the ocean, the residence times of suspended materials that settle on the bottom depend on the particle's size, which accounts for the plus or minus one order of magnitude in the residence time shown for the settling particulates in the deep ocean.

1.2.2   Residence Time and Mean Age

The material content of certain geochemical reservoirs has a well-pronounced age structure. Examples of such reservoirs include the global sedimentary reservoir with sediments of different ages, plant and animal populations made of individuals of different ages, and bodies of water with a range of residence times within them. Among the geochemical reservoirs listed in Table 1.2, the atmosphere and ocean are reservoirs with an age structure; the individual gases or dissolved components have different residence times within each reservoir. Two examples of reservoirs with an age distribution are shown in Figure 1.2: the age distribution of the global sediment mass and the age distribution of trees in a forest. In both cases, the mass decreases with an increasing age, although the rates of decrease are different. The exponential decrease in the mass of sediments with an increasing age is relatively slow, whereas a power-law decrease in the number of trees with an increasing age results in a steep decline of the old-age survivors.

For the geochemical reservoirs characterized by an age distribution of material within them, the concept of the mean residence time defined in equations 1.1 to 1.5 also applies, but the mean residence time of material may or may not be equal to its mean age within the reservoir. An analogy can be drawn between the geochemical reservoirs with age structure and living populations with respect to the mean age and mean residence time (Bolin and Rodhe, 1973): in industrialized societies, the mean age of human population is about 30 yr, but the mean residence time is about 70 yr. The mean age is the mean age of the living population or of the material present in the reservoir. The mean residence time is the mean age

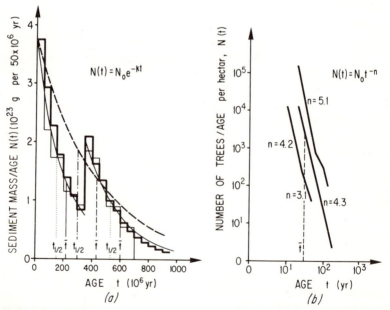

Figure 1.2 Age distributions. (*a*) Global mass of sediments (from Li, 1972; Garrels et al., 1976). (*b*) Trees in a forest (from Leak, 1975). $\bar{t}$ is mean age, $t_{1/2}$ is half-life.

of the individuals leaving the population or of the material being removed from the reservoir.

A distribution of ages within a reservoir can be steady if the input rate (equal to the number of births of age 0 per unit of time) is balanced by the removal rate of the material of age greater than 0 (equal to the total number of deaths of all ages per unit of time). A distribution of some quantity $N(t)$ per unit of time as a function of time (or age) can be written in the form

$$N(t) = N_0 f(t) \quad \text{(amount per unit of time)} \quad (1.6)$$

where N_0 is a dimensional constant and $f(t)$ represents the functional part of the relationship. Four age distributions corresponding to equation 1.6 are shown in Figure 1.3. The distribution in Figure 1.3a is an exponentially declining distribution that may be compared with the mass distribution of sediments (Figure 1.2). The distribution in Figure 1.3c is a negative power-law relationship resembling age distribution in animal populations with high mortality of the young, and distribution of sizes in a variety of natural materials. In Figure 1.3b, the curve is the one-half of the Gaussian distribution for $t \geqslant 0$. In Figure 1.3d, the age spectrum resembles an age distribution curve in a population with low mortality of the young, such as for a human population.

The mean age $\bar{t}$ of the distribution $N(t)$ is

$$\bar{t} = \frac{1}{N_T} \int_0^\infty t N(t)\, dt \quad \text{(yr)} \quad (1.7)$$

where N_T is the total amount of material in the reservoir given by

$$N_T = \int_0^\infty N(t)\, dt \quad \text{(amount)} \quad (1.8)$$

If the age distribution declines to zero at some finite age $t = T$ (Figure 1.3d), then the upper limit of integration in the preceding two equations is T instead of ∞. The definition of the mean age of a reservoir content, equations 1.7 and 1.8, is independent of whether the system is in the steady or nonsteady state.

The mean residence time τ is

$$\tau = \frac{\displaystyle\int_0^\infty t\, dN(t)}{\displaystyle\int_0^\infty dN(t)} \quad \text{(yr)} \quad (1.9)$$

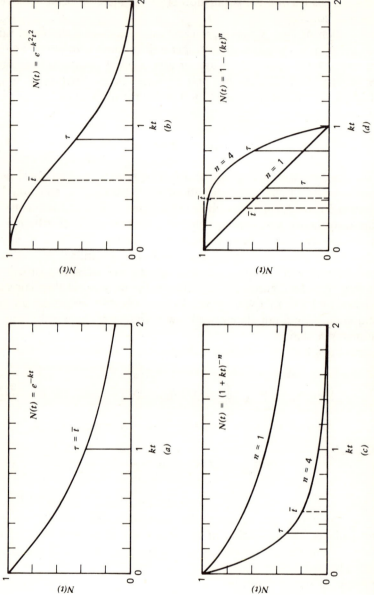

Figure 1.3 Age distribution functions. (*a*) Exponential age distribution. (*b*) Gaussian, for positive age *t*. (*c*) Power law, approaching zero asymptotically. (*d*) Age distribution attaining zero at finite time.

The residence times τ in equations 1.9 and 1.2 are identical, as is shown at the end of this section. If in a steady-state system there is either no mass or no individuals of very old age, then this corresponds to the condition of $N(t) \rightarrow 0$ as $t \rightarrow \infty$. Therefore, relationship 1.9 can be simplified to

$$\tau = \frac{1}{N(0)} \int_0^\infty N(t)\,dt = \frac{N_T}{N(0)} \qquad \text{(yr)} \qquad (1.10)$$

where $N(0)$ is the value of $N(t)$ at $t=0$.

The mean ages and residence times of the populations represented by the age spectra shown in Figure 1.3 are summarized in Table 1.3. The age spectra discussed so far belong to populations with simple birth and death processes—that is, all input to the population reservoir is of age $t=0$, whereas removal from the reservoir is spread over a range of ages. (In a biological population model, input of individuals of age $t>0$ may be attributable to immigration.)

The mean age $\bar{t}$ and residence time τ of material are the same if the reservoir is characterized by an exponential age spectrum $N(t) = N_0 e^{-kt}$ (Figure 1.3a). In other cases, the mean age and residence time differ from one another, and the difference depends on the spectrum shape and the parameter values in the time-dependent part of the relationship $f(t)$.

A measure of time that is convenient to read off such diagrams as the age distributions shown in Figure 1.2 is the half-life. Half-life $t_{1/2}$ is the time at which the value of $N(t)$ is one-half of its initial value, or

$$\frac{N(t_{1/2})}{N(0)} = \frac{1}{2}$$

For the exponential age distributions of sediments shown in Figure 1.2a, the half-lives are shorter than the mean ages.

For the populations of trees, the age distributions of which are plotted in Figure 1.2b, one of the distributions is $N(t) \propto t^{-4.3}$ and its initial age is $T \simeq 20$ yr. The mean age and half-life of this population can be determined using equation 5 of Table 1.3:

$$\text{mean age } \bar{t} = \frac{(4.3-1) \times 20}{4.3-2} = 29 \text{ yr}$$

$$\text{half-life } t_{1/2} = 2^{1/4.3} \times 20 = 23 \text{ yr}$$

Both the mean age and half-life are very close to the point of initial age. If the initial age were taken as $T=1$ yr, we would have $\bar{t} = 1.43$ yr and $t_{1/2} = 1.17$ yr. The very low mean age and half-life in such a population is

Table 1.3 Residence Time τ, Mean Age $\bar{t}$, and Half-Life $t_{1/2}$ of Some Age Distributions. Parameter k has Dimensions Time^{-1}, t is Age, and T is a Fixed Age Value

Age distribution function $N(t)$ (time^{-1})	Age range	Residence time τ	Mean age $\bar{t}$	Half-life $t_{1/2}$	
1. $N(t)=e^{-kt}$	$0 \le t < \infty$	$\dfrac{1}{k}$	$\dfrac{1}{k}$	$\dfrac{0.69}{k}$	$t_{1/2} < \tau = \bar{t}$
2. $N(t)=e^{-k^2 t^2}$	$0 \le t < \infty$	$\dfrac{\pi^{1/2}}{2k}$	$\dfrac{1}{\pi^{1/2}k}$	$\dfrac{0.83}{k}$	$\tau > t_{1/2} > \bar{t}$
3. $N(t)=(1+kt)^{-n}$	$n>2$	$\dfrac{1}{(n-1)k}$	$\dfrac{1}{(n-2)k}$	$\dfrac{2^{1/n}-1}{k}$	$t_{1/2} < \tau < \bar{t}$
	$0 \le t < \infty$				
	$n=1$	$\dfrac{(1/kT+1)\ln(1+kT)}{k} - \dfrac{1}{k}$	$\dfrac{T}{\ln(1+kT)} - \dfrac{1}{k}$	$\dfrac{1}{k}$	
	$0 \le t < T$				
4. $N(t)=1-(kt)^n$	$n>0$	$\dfrac{nT}{n+1}$	$\dfrac{T}{2}\dfrac{1-2(kT)^n/(n+2)}{1-(kT)^n/(n+1)}$	$\dfrac{1}{2^{1/n}k}$	$t_{1/2} > \tau > \bar{t}$
	$0 \le t \le T$ $(k=1/T)$				
5. $N(t)=(kt)^{-n}$	$n>2$	$\dfrac{T}{n-1}$	$\dfrac{(n-1)T}{n-2}$	$2^{1/n}T$	$\tau < t_{1/2} < \bar{t}$
	$T \le t < \infty$				

the result of the power-law decrease in the number of individuals with an increasing age, making the life expectancy in such a population very low. For all practical purposes, the upper age limit in populations of this type may be taken as $t = \infty$ instead of the real upper age ($t = 200$ yr for the population, the mean age and halflife of which were computed above); as long as the number of individuals at the upper age limit is about $\frac{1}{100}$ or less of the number at the starting age, the relationships for τ, t, and $t_{1/2}$ given in equation 5 of Table 1.3 are valid. The lower age limit that is greater than zero implies that there is no mortality between the ages of zero and some higher age. In some experimental woods, virtually no tree mortality has been recorded during the first 36 years, but significant decrease has been observed in the number of trees older than 36 years (Mitscherlich, 1970, p. 89).

Derivation of Residence Time.   A distribution of something per unit of time $N(t)$ is given in equation 1.6. The amount of material of age t decreases by some fractional amount $-\lambda N(t)$ before it matures to age $t + \Delta t$. The quantity $\lambda N(t)$ is the removal flux of material of age between t and $t + \Delta t$. The coefficient λ is a fractional decrease rate

$$\lambda(t) = -\frac{1}{N(t)} \frac{dN(t)}{dt} \qquad (\text{yr}^{-1}) \qquad (1.11)$$

that is, in general, a function of t.

The total flux out of the population or reservoir is, using the definition of $\lambda(t)$ from equation 1.11,

$$\int_0^\infty \lambda(t) N(t) \, dt = -\int_0^\infty dN(t) = N(0) - N(\infty) \qquad (1.12)$$

The result of equation 1.12 is identical to the total removal flux ΣF_i in equation 1.2.

The mean age of the material that is being removed is also its residence time in the reservoir, and it is

$$\tau = \frac{\displaystyle\int_0^\infty t \lambda(t) N(t) \, dt}{\displaystyle\int_0^\infty \lambda(t) N(t) \, dt} \qquad (1.13)$$

The definitions of the mean age of the material present in the reservoir, equations 1.7 and 1.8, and the mean age of the material leaving the reservoir, equation 1.13, are both based on the definition of the mean of a distribution. The numerator of equation 1.13 can be rewritten using equation 1.12 and integrated by parts, giving

$$\int_0^\infty t\lambda(t)N(t)\,dt = -\int_0^\infty t\,dN(t)$$

$$= -tN(t)\Big|_0^\infty + \int_0^\infty N(t)\,dt = \int_0^\infty N(t)\,dt \qquad (1.14)$$

The latter equality is true if the function $N(t)$ tends to 0 with an increasing t faster than the increase in t, as then the product $tN(t)$ is 0 at both limits. Substitution from equations 1.14 and 1.12 into 1.13 gives the same result as equation 1.2,

$$\tau = \frac{1}{N(0)}\int_0^\infty N(t)\,dt = \frac{M}{\Sigma F_i} \qquad \text{(yr)} \qquad (1.15)$$

1.3   SOME MATHEMATICAL BASICS FOR CYCLES

1.3.1   Material Balance Equations

A mathematical model of a geochemical cycle can be represented by a system of box-reservoirs and flows between them. Then the treatment reduces to a mass balance problem: each reservoir receives input from outside, from other reservoirs, and possibly from within, and it looses material to other reservoirs and possibly by chemical reactions within it. The mass balance condition for any (ith) reservoir can be written as

$$\frac{dM_i}{dt} = \sum_j F_{ji} - \sum_j F_{ij} \pm \sum_k F_{ki} \qquad \text{(mass time}^{-1}) \qquad (1.16)$$

where $i,j = 1,\ldots,n$ are reservoir numbers ($i \neq j$), M_i is mass concentration in the reservoir, t is time, F_{ji} are the fluxes from other reservoirs to the ith (input), F_{ij} are fluxes out of the ith to other reservoirs (removal), and F_{ki} are fluxes due to production or consumption reactions within the ith reservoir (input and/or removal). The flux terms as written are of dimensions mass time^{-1}. For a system of n reservoirs there may be n differential equations defining the rate of change in the reservoir content dM_i/dt as a function of external and internal fluxes. Systems of such simultaneous equations can portend considerable algebraic difficulties if the fluxes between the reservoirs F_{ji} and F_{ij} are concentration and time dependent, and the intrareservoir reactions are also some power functions of the reservoir concentrations. Many systems of simultaneous equations, of the type given in equation 1.16, with the physical and chemical transport terms of varying degrees of complexity, have been solved numerically in application to ecological, economic, engineering, physical, and chemical environmental problems. As important as are the results outlining evolution of reservoir systems in terms of the reservoir content through time, the procedures and intermediate steps of numerical computations are often lost on the readers who are not intimately involved in a particular problem and, in general, on those who cannot easily go into the details of computations. This is unfortunately the case with this type of mathematical modeling, but only the simplest three-reservoir or three-component systems with first-order chemical reactions are amenable to easy computation by hand.

The fluxes between such global reservoirs as land, ocean, and atmosphere (Figures 1.1 and 1.5) may, at least in the first approximation, be regarded as proportional to the reservoir masses or concentrations,

$$F_{ij} = k_{ij} M_i \quad \text{(mass time}^{-1}\text{)} \tag{1.17}$$

where k_{ij} (time^{-1}) is a rate constant as defined in equation 1.3. This definition of the flux is justified if the material being removed from a reservoir represents an average composition of a large exposure area or of a large well-mixed volume.

An explicit definition of the fluxes, such as in equation 1.17, can be used in the mass balance relationship 1.16 and the latter can be solved provided the rate constants k_{ij} for the individual reservoirs are known. Some examples of the use of equation 1.16 in models of the cycles are given in subsequent sections of this chapter.

If the reservoir concentrations do not change with time, then $dM_i/dt = 0$, and the system is in a steady state. Thus a model of a geochemical cycle at a steady state can be given by a system of simultaneous equations,

$$\sum_j F_{ji} - \sum_j F_{ij} \pm \sum_k F_{ki} = 0 \tag{1.18}$$

from which the reservoir masses M_i can be computed, provided the rate constants k_{ij} for the fluxes and chemical reactions within the reservoirs are known.

1.3.2 Time to Steady State

In an evolving geochemical cycle—that is, when $dM_i/dt \neq 0$—the manner in which the contents of the individual reservoirs approach the steady state depends on the transport mechanisms within the reservoirs. If a reservoir is considered as a well-mixed box with fluxes in and out, then the steady state is attained only after infinitely long time. A practical measure of the length of time required to attain the steady state can be taken as the time when the reservoir content has become sufficiently close to the final steady-state concentration. A concentration value M_t that is within 5% of a steady-state concentration M_{ss} is sufficiently close, and the time to steady state t_{ss} can be defined as the time required for the reservoir content to change by 95% of the difference between the initial $M_{t=0}$ and the new steady-state value M_{ss}. Thus the time to steady state t_{ss} is the time when the following condition is attained:

$$\frac{M_t - M_{t=0}}{M_{ss} - M_{t=0}} = 0.95 \qquad (1.19)$$

An opposite extreme of a well-mixed reservoir is a reservoir with no mixing at all, such that the material arriving with input fluxes does not mix with the reservoir content but travels through it as piston or plug flow. For the two types of transport through a reservoir of constant volume, changes in the mean reservoir concentration with time are shown in Figure 1.4. In the case of flow without mixing, the newly arriving material displaces the old within one residence time 1τ, and if the input and other conditions persist, then the time to steady state is $t_{ss} = 1\tau$.

In a well-mixed reservoir, the rise in concentration is much slower and the 95% level of the steady-state value is attained after about three residence times, such that the time to steady state is $t_{ss} \simeq 3\tau$.

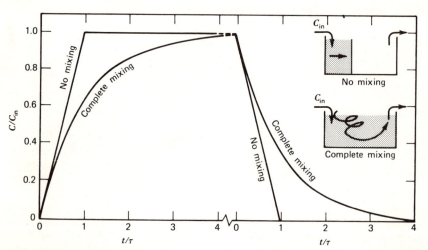

Figure 1.4 Increase in concentration due to input to a reservoir with mixing and without mixing. Decrease in concentration due to flushing.

In the no-mixing case, the arriving material at concentration C_{in} (see Figure 1.4) displaces the material in the reservoir at concentration zero. If the reservoir volume does not change with time, a mean concentration $\overline{C}$ within the reservoir is

$$\overline{C} = C_{in} \frac{t}{\tau} \qquad (1.20)$$

In effect, at times shorter than the residence time τ the reservoir contains a region with the input material (C_{in}) and a region free of the input material, the former increasing and the latter decreasing in volume. Output from the reservoir, indicated by arrows in the reservoir diagrams of Figure 1.4, is at concentration zero at times $0 < t < \tau$ and it becomes C_{in} at time $= \tau$. If, however, the newly arriving and old material in the reservoir are removed in proportion to their masses, then the mean concentration $\overline{C}$ is meaningful, despite the fact that the reservoir content is not homogeneous.

In the case of input mixing completely with the reservoir content, output concentration and concentration within the reservoir are the same,

$$\overline{C} = C_{in}(1 - e^{-t/\tau}) \qquad (1.21)$$

The ratio $\overline{C}/C_{in}$ increases with time, as shown in the left-hand part of Figure 1.4, and it decreases with time if the reservoir content is being replaced by material at concentration zero (flushing).

1.4   GLOBAL CYCLE OF PHOSPHORUS

1.4.1   Model of the Cycle

A diagram of the major reservoirs and fluxes for the global cycle of phosphorus is shown in Figure 1.5. The model deals with total concentrations of phosphorus within the individual reservoirs, irrespective of the chemical form of the phosphorus compounds in such different reservoirs as the land biota, ocean water, and sediments. The cycle shown in the figure is the exogenic cycle, as the flow of the element through the lower parts of the oceanic and continental crust is not considered. Excluding the reservoir of the mineable phosphorus deposits, the cycle is balanced and the total amount of phosphorus is conserved within the system. The key feature of the balanced cycle is the flux of phosphorus from deep ocean to the sediments, which is shown as equal (an assumption) to the input flux of dissolved phosphorus from land to the surface ocean. Some additional

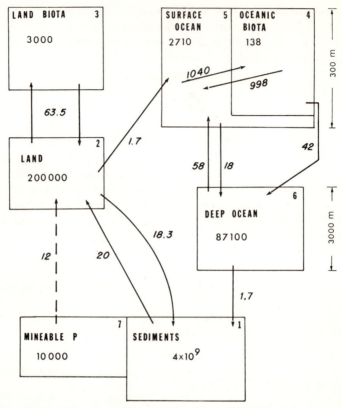

Figure 1.5 Geochemical reservoirs for the global cycle of phosphorus. Reservoir contents in units of 10^6 tons P, fluxes in units of 10^6 tons P yr^{-1} (Table 1.4). The cycle is at steady state, except for the transfer of mined phosphorus to land (from Lerman, Mackenzie, and Garrels, 1975). Printed by permission of the Geological Society of America, Inc. Copyrighted 1975.

features of the cycle diagram may be noted. The flux of phosphorus via river flow from land is approximately 10% of the phosphorus flux in solid sediment detritus. The values of the biological productivity and respiration fluxes of the oceanic biota correspond to a short residence of phosphorus in the oceanic biota pool, about 0.14 yr. Most of the phosphorus in the oceanic biota is regenerated in the surface ocean and only a small fraction, about 4%, settles to the deep ocean. The residence time of phosphorus in the land biota (mostly trees) is considerably longer, about 45 yr. The flux of phosphorus from the land biota to land is essentially the flux of dead vegetation that mixes with land. The land reservoir is a layer spread over the surface of the continents, 60 cm deep.

Table 1.4 Phosphorus Cycle. Contents of Reservoirs, Fluxes, and Residence Times[a]

Reservoir (number i or j)	Mass M_i (tons P)	Flux F_{ij} (10^6 tons P yr^{-1})	Residence time $\tau_i = M_i/F_{ij}$ (yr)
1. Sediments	4×10^{15}	$F_{12}=20$	$\tau_{12}=2 \times 10^8$
2. Land	2×10^{11}	$F_{21}=18.3$	$\tau_{21}=1.09 \times 10^4$
		$F_{23}=63.5$	$\tau_{23}=3.15 \times 10^3$
		$F_{25}=1.7$	$\tau_{25}=1.18 \times 10^5$
3. Land biota	3×10^9	$F_{32}=63.5$	$\tau_{32}=47$
4. Oceanic biota	1.38×10^8	$F_{45}=998$	$\tau_{45}=0.14$
		$F_{46}=42$	$\tau_{46}=3.3$
5. Surface ocean	2.71×10^9	$F_{54}=1040$	$\tau_{54}=2.6$
		$F_{56}=18$	$\tau_{56}=150$
6. Deep ocean	8.71×10^{10}	$F_{61}=1.7$	$\tau_{61}=5.12 \times 10^4$
7. Mineable phosphates	1×10^{10}	$F_{72}=12$	$\tau_{72}=830$
		or	or
		$F_{72}=12 \times e^{0.07t}$	$\tau_{72}=60$

[a]Lerman, Mackenzie, and Garrels (1975).

The phosphorus contents of the individual reservoirs (M_i) and the residence times of phosphorus with respect to the various removal fluxes ($\tau_{ij} = M_i/F_{ij}$ yr) are summarized in Table 1.4.

1.4.2 Functional Relationships

The portion of the phosphorus cycle model represented by reservoirs 1 through 6 in Figure 1.5 is at a steady state. Addition of a flux, such as the flux of phosphorus from mineable deposits to land, or interruption of any of the other fluxes, or sudden changes in some of the reservoir sizes—all these are perturbations of the cycle that may lead to development of new relationships between the reservoir masses.

In the presence of the flux of phosphorus from mineable deposits to land, the following relationship describes the rate of change in the phosphorus content of the land reservoir (the meaning of the subscripts is as in Figure 1.5 and Table 1.4):

$$\frac{dM_2}{dt} = \frac{M_7}{\tau_{72}} + \frac{M_3}{\tau_{32}} + \frac{M_1}{\tau_{12}} - M_2\left(\frac{1}{\tau_{23}} + \frac{1}{\tau_{25}} + \frac{1}{\tau_{21}}\right) \qquad (1.22)$$

where the first three positive terms on the right-hand side of the equation represent input fluxes to reservoir 2 and the negative term is the sum of all the removal fluxes. Comparison of equation 1.22 to 1.16 indicates that no fluxes due to chemical reactions are included in the phosphorus cycle model; this stems from the fact that only total phosphorus is considered as a chemical species flowing through the individual reservoirs.

Equations similar to 1.22 can be written for the other six reservoirs and the system of seven equations can be solved numerically, with the aid of a computer.

Another example of perturbation of the steady state of the phosphorus cycle may be cessation of the phosphorus fluxes from the land to land biota, and from surface ocean to oceanic biota. In Figure 1.5 this would correspond to omission of the arrows for the fluxes F_{23} and F_{54}. The physical significance of this perturbation is total cessation of photosynthesis and reproduction ("doomsday scenario"). Omission of these fluxes from the cycle diagram leads to the following relationship for the land reservoir:

$$\frac{dM_2}{dt} = \frac{M_3}{\tau_{32}} + \frac{M_1}{\tau_{12}} - M_2\left(\frac{1}{\tau_{25}} + \frac{1}{\tau_{21}}\right) \tag{1.23}$$

As in the previous example, similar equations can be written for the other five reservoirs (without the mineable phosphorus reservoir, which is not included in this scenario) and the system of six simultaneous linear differential equations is soluble numerically.

1.4.3   Perturbations of the Cycle

The picture of the phosphorus cycle and its residence times in different global reservoirs (Figure 1.5, Table 1.4) are sufficient to give approximate answers to the question, "What happens within the cycle after it has been perturbed?" Two perturbations and their possible consequences are considered below.

Industrial Perturbation.   An industrial perturbation is a result of the addition of phosphorus in fertilizers to land. The estimated mass of the mineable phosphorus reservoir is 10^{10} tons P. This is only 5% of the mass of phosphorus present in the land reservoir or, effectively, in soils. Thus if the mined phosphorus is being mixed with the contents of the land reservoir, as implicit in the mathematical model of equation 1.22, then the effects of the man-produced flux are damped by the large land reservoir and the long residence time of phosphorus in it. If, however, the phosphorus in fertilizers were fed to land plants and into rivers by a more direct route, then its effects on the reservoirs located downcycle from the land reservoir could be much more pronounced. The effect of the more direct input can be assessed from Figure 1.5, which shows that the masses of phosphorus in land plants and in the surface ocean are significantly smaller than the mass of mineable phosphorus reserves. In Figure 1.6 are shown the computed consequences of phosphorus flux from the mineable to land reservoir, when the fertilizer flux to land doubles in magnitude every 10 yr and the river flux of P also doubles every 10 yr. Under these conditions, the mineable phosphorus reservoir would be exhausted in about 60 yr, while the only reservoirs that would be significantly affected are the surface ocean (increase of 38%) and oceanic biota (increase of 30%).

The conclusions drawn from the industrial perturbation scenario of the phosphorus cycle depend on the nature and sizes of the reservoirs (Figure 1.5). For example, if a smaller reservoir of near-shore water and sediments were inserted between the land and surface ocean (compare Figure 1.1), then increase in the phosphorus content of the near-shore box could be very pronounced, whereas the surface ocean and oceanic biota might be affected little. In human terms, the effect of the increased use of fertilizers may be very noticeable on a "local scale," such as in a near-shore reservoir, although the global average of the system as a whole would not change significantly.

Doomsday Perturbation. A second possible perturbation is an instantaneous cessation of biological productivity on land and in the oceans. Under the conditions of the model, the standing crops of the oceanic and land biota would decay, and the decay rates would be controlled by the

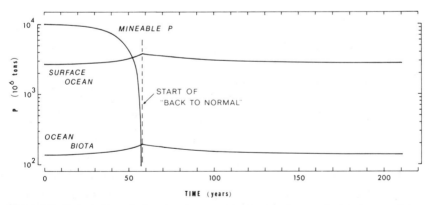

Figure 1.6 Perturbation of the phosphorus cycle. Increased rate of phosphorus mining, depletion of the mineable phosphorus reservoir, and its major consequences (from Lerman, Mackenzie, and Garrels, 1975). Printed by permission of the Geological Society of America, Inc. Copyrighted 1975.

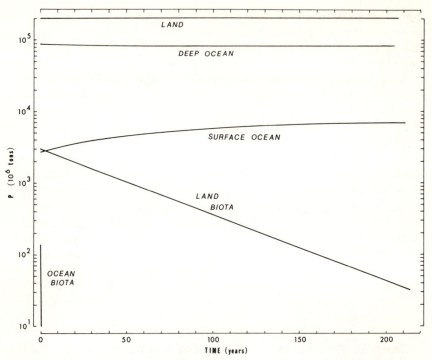

Figure 1.7 Perturbation of the phosphorus cycle. Doomsday or sterilization of all biota. Changes in the phosphorus contents of the main reservoirs (from Lerman, Mackenzie, and Garrels, 1975). Printed by permission of the Geological Society of America, Inc. Copyrighted 1975.

rate constants or residence times of phosphorus in the two reservoirs (Figure 1.7). A 5% level of the present amounts of phosphorus would be reached in $3\tau_{32} \simeq 150$ yr for the land biota and $3\tau_{45} \simeq 0.5$ yr for the oceanic biota. The decay of the oceanic biota might cause at the most a $\frac{138}{2710} = 5\%$ increase in the phosphorus content of the surface ocean, and this increase would be redistributed between the surface and the much bigger deep ocean reservoir. However, there are two removal fluxes of phosphorus from the surface ocean (Figure 1.5): flux $F_{56} = 18 \times 10^6$ tons P yr^{-1} owing to water exchange between the surface and the deep, and flux $F_{46} = 42 \times 10^6$ tons P yr^{-1} owing to removal of undecomposed plankton. Thus the total residence time of phosphorus in the surface layer including water and plankton ($M = 2710 + 138 = 2850$ million tons P) is, according to equation 1.5,

$$\tau = \frac{1}{18/2850 + 42/2850} \simeq 50 \quad \text{(yr)}$$

If no plankton were left then the removal of phosphorus from surface ocean water would be only by means of the water mixing mechanism, with respect to which the residence time in surface ocean layer is $\tau_{56} = 150$ yr. This indicates that as a result of the plankton elimination, the residence time of phosphorus in surface ocean water would increase threefold and would lead to a substantial increase in the phosphorus content of the surface ocean, to the value of approximately 9000×10^6 tons. The preceding discussion stresses the role of the plankton and its settling in controlling concentrations of dissolved substances in the ocean. This effect is discussed in more detail in connection with the production and regeneration of biological materials (Chapter 6).

1.5   CARBON AND OXYGEN CYCLES

The involvement of carbon dioxide and oxygen in life processes of plants and animals is, and has always been, an important link in the biogeochemical cycles of these elements, as long as there has been photosynthetic activity on earth. In addition to carbon and oxygen, biological material contains hydrogen, nitrogen, sulfur, and phosphorus. The average composition is nCH_2O with variable proportions of N, S, P, and metals. The proportions of the main constituents expressed as the mol ratios $C : N : S : P$ for the oceanic and land biota have the following representatative values:

Oceanic plankton
  Average $C : N : S : P = 106 : 16 : 1.7 : 1$ (Redfield et al., 1963)
  Plants $= 332 : 27 : 2.8 : 1$
  Plants + animals $= 286 : 27 : 1.7 : 1$ (Delwiche and Likens, 1977)

Land plants
  $C : N : S : P$ $= 510 : 4.2 : 0.76 : 1$ (Delwiche and Likens, 1977)
  $C : N : S : P$ $= 882 : 9 : 0.6 : 1$ (Deevey, 1973)

In this chapter the cycles of carbon and oxygen are dealt with on two very different time scales: first, on a geologic time scale, examining such processes as deposition and oxidation of organic matter in sediments, and their possible effects on the CO_2 and O_2 contents of the atmosphere; and second, on a time scale of human generations, where the CO_2 concentration in the atmosphere has been rising for the last two centuries owing to the burning of fuels.

1.5.1   CO$_2$ and O$_2$—Longer Time Scale

Model of the Cycle.   One of the possible models of the joint cycle of CO$_2$ and O$_2$ is shown in Figure 1.8. The diagram presents the more important geochemical reservoirs that have played a role in the CO$_2$–O$_2$ cycle, at least during Phanerozoic time (i.e., the last 600 million years). A brief discussion of the major reservoirs and fluxes operating between them will explain the cycle diagram.

Atmospheric CO$_2$ is taken up in photosynthesis, such that O$_2$ is released to the atmosphere and C goes into building the biota. As shown in the figure, only the oceanic biomass (reservoir 3) participates in this process. The terrestrial biomass, although it exceeds the standing crop of the oceanic biota by a factor of 100–400 (Bolin, 1970; Whittaker and Likens, 1975), is considered as a closed cycle on a geologic time scale: all the terrestrial organic matter that is being formed oxidizes and decays, returning CO$_2$ to the atmosphere. There are two possible sources of error in this representation: (1) the time spent by dead organic material in soils may be significantly long even for a geologic time scale; (2) some fraction of terrestrial organics may be transported by streams to the ocean, where it may be included in the sediments (reservoir 5).

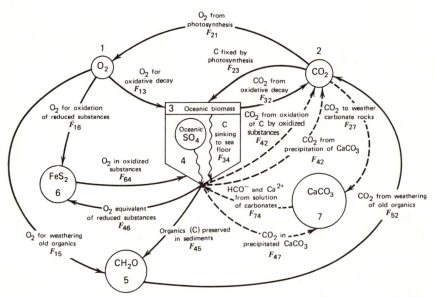

Figure 1.8   Major reservoirs and fluxes of the exogenic cycles of CO$_2$ and O$_2$. Dashed arrows indicate equilibria between sedimentary and dissolved carbonates, and atmospheric CO$_2$ (from Garrels et al., 1976). By permission of the *American Scientist*.

The major avenues of removal of CO_2 and O_2 from the atmosphere are the settling and burial of oceanic biota (fluxes F_{34} and F_{45}), and precipitation of $CaCO_3$ (flux F_{47}). The portion of the cycle consisting of reservoirs 2, 7, and 4 is the atmospheric CO_2–$CaCO_3$–ocean-water system. In the figure, an equilibrium is assumed to exist between the $CaCO_3$ in sediments and the dissolved calcium and CO_2 in the oceanic reservoir. In the real world, surface ocean water is supersaturated with respect to calcite, and deep ocean water is undersaturated. By assuming that the oceans (reservoir 4) are at an equilibrium with $CaCO_3$, the sedimentary reservoir $CaCO_3$ (reservoir 7) becomes an "instantaneous" buffer capable of adding and subtracting CO_2. According to the $CaCO_3$ precipitation reaction,

$$Ca^{2+} + 2HCO_3^- \rightleftarrows CaCO_3 + H_2O + CO_2$$

every mol of $CaCO_3$ precipitated (flux F_{47}) contributes one mol of free CO_2 (flux F_{42}). At equilibrium, the CO_2 fluxes owing to precipitation (F_{47} and F_{42}) and the fluxes owing to dissolution of $CaCO_3$ by atmospheric waters (fluxes F_{27} and F_{74}) represent the portion of the cycle controlled by thermodynamic relationships between solid $CaCO_3$ and an aqueous solution, such that no time constants are being considered for these fluxes (Table 1.5).

A major part of the biota (reservoir 3) is oxidized (flux F_{32}), returning CO_2 to the atmosphere, but a small fraction sinks through the deep ocean to the ocean floor (flux F_{34}). Some of the settling and settled material is oxidized through bacterially mediated reactions with sulfate in ocean water,

$$2CH_2O + SO_4^{2-} + 2H^+ \rightleftarrows 2CO_2 + 2H_2O + H_2S$$

such that CO_2 is formed in the process of oxidation, as shown by flux F_{42}. The remaining fraction of organic material becomes part of the sedimentary record (flux F_{45}).

Concentration of organic material and isotopic composition of carbon in it have been essentially constant during the last 600 million years: 0.5 ± 0.3 wt % of organic carbon in sediments deposited mostly through the ocean indicates that the conditions of photosynthesis, oxidation, settling, and inclusion in sediments have fluctuated only little. This feature supports the basic pictorial representation of the model in Figure 1.8.

Reduction of organic matter, as shown above, produces CO_2 and sulfide. The latter is viewed as an addition to the reservoir of reduced substances, shown as mineral pyrite (FeS_2, reservoir 6). Atmospheric oxygen is consumed in oxidation of reduced minerals (fluxes F_{16} and F_{64}) and sedimentary organic matter (fluxes F_{15} and F_{52}).

Table 1.5 Carbon and Oxygen Contents of the Geochemical Reservoirs, for the Model of the $CO_2 - O_2$ Cycle.[a]

A. Reservoir Masses, Fluxes, and Residence Times

Reservoir (number i or j)	Content M_i (mol)	Flux F_{ij} (10^{12} mol yr^{-1})	Residence time τ (yr)
1. Atmosphere O_2	3.8×10^{19}	$F_{13} = 2496.5$	$\tau_{13} = 1.52 \times 10^4$
		$F_{15} = 2.5$	$\tau_{15} = 1.52 \times 10^7$
		$F_{16} = 1.0$	$\tau_{16} = 3.8 \times 10^7$
2. Atmosphere CO_2	5.5×10^{16}	$F_{23} = F_{21} = 2500$	$\tau_{23} = 22$
3. Oceanic biota C	6×10^{14}	$F_{32} = 2496.5$	$\tau_{32} = 0.24$
		$F_{34} = 3.5$	$\tau_{34} = 170$
4. Ocean water C			
C (as CO_2 and HCO_3^-)	3.234×10^{18}		
O_2 (as SO_4^{2-})	8.4×10^{19}		
Ca	1.4×10^{19}		
		(CO_2) $F_{42} = 1.0$	
		(O_2 equivalent) $F_{46} = 1.0$	
		(C) $F_{45} = 2.5$	
5. Sedimentary organics C	1×10^{21}	$F_{52} = 2.5$	$\tau_{52} = 4 \times 10^8$
6. Reduced substances (FeS$_2$ as O_2 demand equivalent)	4×10^{20}	$F_{64} = F_{16} = 1.0$	$\tau_{64} = 4 \times 10^8$

B. Equations for the Fluxes and Rate Constants

1. $\log F_{34(C)} - \log F_{23(C)} = -k_1 M_{1(O_2)}$ $\qquad$ $k_1 = 7.5102 \times 10^{-20}$ mol^{-1}
2. $F_{16(O_2)} = k_{16} M_{6(O_2 \text{ demand})}$ $\qquad$ $k_{16} = 2.5 \times 10^{-9}$ yr^{-1}
3. $F_{15(O_2)} = k_{15} M_{5(C)}$ $\qquad$ $k_{15} = 2.5 \times 10^{-9}$ yr^{-1}
4. $F_{64(O_2)} = F_{16(O_2)}$
5. $F_{52(O_2)} = F_{15(O_2)}$
6. $F_{46(rd^b)} = k_{46} F_{34(C)} M_{4(O_2 \text{ in } SO_4^{2-})}$ $\qquad$ $k_{46} = 3.40 \times 10^{-21}$ mol^{-1}
7. $F_{45(C)} = F_{34(C)} - F_{46(rd)}$
8. $M_{2(CO_2)} = k_2 M_{4(Ca)} M_{4(HCO_3)}^2$ $\qquad$ $k_2 = 3.78 \times 10^{-40}$ mol^{-2}
9. $M_{4(CO_2)} = k_3 M_{2(CO_2)}$ $\qquad$ $k_3 = 0.254$
10. $F_{21(O_2)} = F_{23(C)}$
11. $F_{13(O_2)} = F_{32(CO_2)}$

[a] From Garrels et al. (1976).
[b] Reduced substances.

The fluxes listed in Table 1.5 describe a balanced CO_2–O_2 cycle. The physical significance of the flux relationships is explained in the next paragraphs.

The little variation in the organic carbon content of old sediments can be interpreted as indicating that any increase in the biomass, at constant oxygen content of the atmosphere, is accompanied by a proportional increase in its oxidation rate. Conversely, in the absence of free O_2 in the atmosphere, nearly all biomass would sink to the ocean floor. The latter condition is $F_{23} \simeq F_{34}$ and it would hold if only much slower fermentative decay of organic material were responsible for the return of CO_2 to the atmosphere. These relationships are represented by equation 1 of Table 1.5B.

The fluxes of oxygen from the atmosphere to the reduced and organic matter reservoirs assume that the oxygen demand is proportional to the reservoir sizes, and that neither pyrite nor organic material from older sediments has been redeposited during the last 2.5×10^9 years in oceanic sediments. These fluxes are represented by equations 2 and 3 of Table 1.5B.

Equations 4 and 5 of Table 1.5B correspond to mass balance relationships for oxidation and reduction fluxes out and into reservoir 6, and oxidation fluxes through reservoir 5.

The flux of reduced material to the pyrite reservoir is taken as proportional to the sulfate concentration in sea water and the flux of organic material (F_{34}) to the ocean floor. Equations 7, 10, 11, and 12 of Table 1.5B are mass balance relationships between the fluxes. Equations 8 and 9 correspond to the conditions of a thermodynamic equilibrium between the dissolved calcium, CO_2 species, and $CaCO_3$, and an equilibrium between the atmospheric and dissolved CO_2.

For the two species involved in the CO_2–O_2 cycle—SO_4^{2-} and Ca^{2+}— calcium sulfate minerals are a significant source and sink in sediments. For example, at present there is little gypsum precipitation on a global scale, but a significant volume of gypsum in old sediments is exposed to erosion, contributing Ca^{2+} and SO_4^{2-} to the ocean. The perturbations of the CO_2–O_2 cycle, considered in the subsequent sections of this chapter, were computed for the cycle model without a $CaSO_4$ reservoir, in a form shown in Figure 1.8. The potential significance of the $CaSO_4$ reservoir to the model would have been indicated if, as a result of computations, large changes emerged in the SO_4^{2-} concentration in the ocean. This, however, was not the case, and in none of the three models of the cycle perturbation did the SO_4^{2-} content of the ocean depart from its present value (Table 1.5) more than 12%. Insofar as the goals of the model aimed at a long time scale cannot be too precise, such small variations in the SO_4^{2-} concentration in ocean water do not justify inclusion of an additional reservoir and additional fluxes.

It should be emphasized that, in principle, the steady-state cycle as shown in Figure 1.8 can "run" without free oxygen in the atmosphere; the contents of reservoir 1 can be zero if all the oxygen produced in photosynthesis were consumed in respiration. The possible existence of such a world, with photosynthesis but no free oxygen in the atmosphere, is at the heart of the question of what caused oxygen accumulation in the atmoshpere in the first place. Hypotheses and discussions of this question are given in, for example, Walker (1974), Garrels et al. (1976), Holland (1978). Conceivably, during the early stages of evolution of photosynthesis there was more than one attempt by an oxygen-producing evolutionary lineage to establish itself in the world, until finally an attempt succeeded. Another line of speculation may suggest that at early stages, all the free oxygen produced was consumed by reduced materials on the surface of the earth, until they became exhausted and oxygen could begin to accumulate.

Three perturbations of the CO_2–O_2 cycle considered below are increased rates of erosion, increased rates of photosynthesis, and cessation of biological productivity (doomsday).

The atmospheric reservoirs of CO_2 and O_2 are coupled with reservoirs in which the residence times are very long (Table 1.5). The residence time of reduced materials in reservoir 6 with respect to oxidation, the residence time of organic matter in sediments (reservoir 5), and the residence time of CO_2 in the ocean and limestones (reservoirs 4 and 7 combined) are all of the order of 10^8 yr. Because of the long residence times, changes in the existing conditions in any one of the reservoirs would, in principle, produce long-term changes in the entire system. The cycle diagram in Figure 1.8 may be considered with a somewhat different arrangement of the fluxes, and qualitative effects of other configurations may be briefly examined. For example, if oceanic CO_2 (reservoir 4) is not at an equilibrium with $CaCO_3$ (reservoir 7) and significant time is required to attain an equilibrium, then the residence time of CO_2 in the ocean would be drastically shorter, only about 3×10^6 yr, with respect to oxidation of organic matter on the ocean floor (M_4/F_{42}). As another example, a reservoir of terrestrial biota may be added to the cycle diagram. Instead of being a closed loop taking CO_2 and returning it to the atmosphere, the terrestrial plant reservoir may be considered with output to the organic matter reservoir 5. This would have effects on the response times of the individual reservoirs if the flux of carbon from the terrestrial reservoir to reservoir 5 were significantly greater than flux F_{45} (net flux from ocean to sediments) during most of Phanerozoic time.

Perturbation: Increased Land Erosion Rate. Deforestation, cultivation of land, and related industrial activities have increased the erosion rate of land. Similar increases might have been effective on a continental scale in .

the geologic past, during the periods of land uplift and mountain building. An increase in the rate of erosion results in a greater exposure of reduced substances and organic matter in sediments to the atmospheric oxygen. A threefold increase in oxygen consumption by reservoirs 5 and 6 (Figure 1.8), assumed for this perturbation scenario, corresponds to trebling of fluxes $F_{15} = F_{52}$ and $F_{16} = F_{64}$, which may be represented by the values of the rate constants k_{15} and k_{16}, three times higher, 7.5×10^{-9} yr^{-1}, than the values listed in Table 1.5. With the new values of k_{15} and k_{16}, the mass balance equations for the CO_2–O_2 cycle can be solved numerically, giving the changes in the reservoir contents of C and O_2 as a function of time. The results of this computation are shown in Figure 1.9.

In response to a greater oxygen consumption by the organic and reduced matter reservoirs, oxygen concentration in the atmosphere declines. After about 3×10^6 yr, the rate of change diminishes sufficiently to consider the system as having come to a new steady state. Because of the lower oxygen concentration in the atmosphere, the flux of organic matter to the ocean floor F_{34} increases. Mathematically, the increase in F_{34} is based on equation 1 of Table 1.5B. The fluxes of CO_2 to the atmosphere from the sediment organic matter and ocean floor (reservoirs 5 and 4, F_{42} and F_{52}) increase, resulting in a higher concentration of CO_2 in the atmosphere.

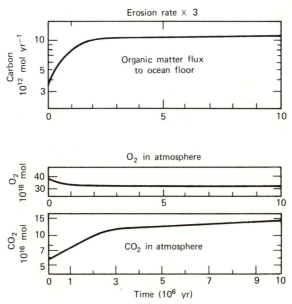

Figure 1.9 Perturbation of the carbon and oxygen cycles. Increased erosion rate on land (data from Garrels et al., 1976).

The flux of organic matter to the ocean floor and the CO_2 content of the atmosphere each increase by a factor of about 3. The O_2 content of the atmosphere declines by about 15%, from 38×10^{18} to 32×10^{18} mol. The changes in the contents of the other reservoirs, not plotted in Figure 1.10, are very small in comparison to the changes in the atmospheric O_2, CO_2, and organic matter flux to the ocean floor. The changes in the reservoirs of pyrite and organic matter are within 0.5% of their masses (which is within rounding-off errors) and the amount of SO_4^{2-} in the ocean increases by 1.5%. The small change in the pyrite reservoir is accounted for mathematically by the condition of the reducing and oxidizing fluxes in the model (F_{46} and F_{64}, Table 1.5). The reducing flux is proportional to the amount of organic matter reaching the ocean floor F_{34} but the oxidizing flux is proportional to the mass of pyrite, such that the higher fluxes nearly balance.

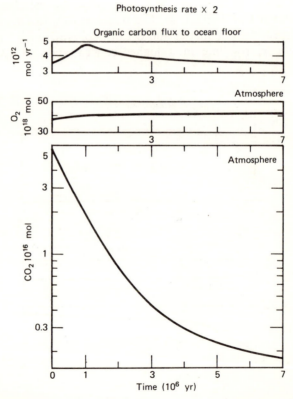

Figure 1.10 Perturbation of the carbon and oxygen cycles. Increased rate of photosynthesis (data from Garrels et al., 1976).

Perturbation: Increased Photosynthesis Rate.   A global increase in the rate of photosynthesis is a perturbation that may be hypothetically achieved by an increased flux of nutrients to the ocean and soils. Less hypothetical is experimental evidence on higher rates of plant growth owing to higher CO_2 concentrations in the air and higher CO_2/O_2 ratios (Bacastow and Keeling, 1973; Skinner, 1976). If a higher proportion of CO_2 relative to O_2 can increase the rates of carbon fixation by plants, then this process would tend to counteract the effects of an accelerated land erosion which, as shown in Figure 1.9, leads to a higher CO_2/O_2 ratio in the atmosphere.

Doubling the rate of photosynthesis (flux F_{23}) produces the following effects. Oxygen concentration in the atmosphere begins to increase because of the higher carbon uptake rate from the CO_2 by the biomass. The O_2 increase amounts to about 10%. At the same time, the mass of the biota increases and the flux of carbon to the ocean floor F_{34} also increases initially, according to equation 1 of Table 1.5B. This increase is counteracted eventually by the rising oxygen content of the atmosphere, such that after about 1×10^6 yr (Figure 1.10) the flux of organics to the ocean floor declines to its initial value. The continual demand on the atmospheric CO_2 depresses its atmospheric level very significantly. The processes eventually lead to a level at which the atmospheric CO_2 may become a limiting nutrient for photosynthesis, about $\frac{1}{30}$ of the present level. Because of the withdrawal of CO_2 from the atmosphere and the ocean, the pH of ocean water would rise to the value of about $pH \simeq 9$; changes in other reservoirs are very small, amounting to a 0.3% increase in the pyrite and organic matter reservoirs, and 1.2% increase in the SO_4^{2-} content of the ocean.

Perturbation: Cessation of Biological Productivity.   This hypothetical doomsday scenario is discussed with reference to the phosphorus cycle in Section 1.4.3. For the CO_2–O_2 cycle, cessation of reproduction in the ocean would end the existence of the plankton within less than 1 yr. Its decay in surface waters would increase the atmospheric CO_2 by less than 0.1% and decrease atmospheric oxygen by about 0.001%. After this, however, oxygen would be continuously consumed by oxidation of reduced minerals (reservoir 6) and sediment organic matter (reservoir 5), without being replenished by photosynthesis, such that the O_2 amount in the atmosphere would decline (Figure 1.11). This calculated rate of decline in O_2 is based on the oxygen consumption rate constants for reservoirs 5 and 6, as given in Table 1.5. If these rates were to become lower owing to the lower oxygen partial pressure in the atmosphere, the process of oxygen depletion would last longer than shown in Figure 1.11. In the absence of organic productivity, the pyrite reservoir would lose material and the SO_4^{2-} concentration in the ocean would rise correspondingly. The sedimentary

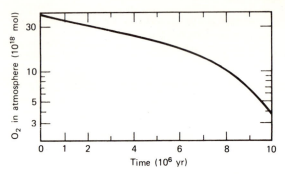

Figure 1.11 Perturbation of the carbon and oxygen cycles. Doomsday or sterilization of all biota. Decrease in atmospheric oxygen (data from Garrels et al., 1976).

organic carbon reservoir would also lose mass because of oxidation. Overall changes in the reservoir masses taking place during the 10×10^6-yr period of the decline in oxygen (Figure 1.11) would be the following: The pyrite and organic carbon reservoirs would diminish by 3%, oceanic sulfate would increase 12%, but the atmospheric CO_2 would increase by a factor of 50.

The long time it would take the atmospheric oxygen to diminish significantly should be noted. If the terrestrial biomass were included in the doomsday model computation, its decay would not speed up the oxygen decline much; the mass of carbon in the terrestrial biota is of an order of 1.7×10^{16} mol C, which is only 0.05% of the 3.8×10^{19} mol O_2 present in the atmosphere.

1.5.2 CO_2—Shorter Time Scale

Burning of Fuels. The geochemical cycle of CO_2 has been affected on a global scale by human activity, which was historically a part of mankind's intellectual, technological, and economic development. Burning of fuels and clearing of forests are two major types of processes in which humans persist and which add CO_2 and oxides of other gases to the atmosphere. Chemical composition of the main natural fuels is given in Table 1.6.

The atmosphere is the first reservoir that receives the products of fuel burning. The major points of interest and concern to people arising from the addition of the products of burning to the atmosphere include: (1) warming of the earth's surface because of a greater absorption of infrared radiation by CO_2 and other gaseous products in the atmosphere (Wang et al., 1976); (2) direct effects of the higher CO_2 content of the atmosphere on the rates of biological productivity on land and in the ocean; (3) damage to life caused by some products of fuel burning (sulfur

**Table 1.6   Major Constituents of Terrestrial Plant Material
and Natural Fuels (wt %)**

Element	Land plant material[a]	Bituminous coal[a]	Crude oil[b]	Asphalt[b]	Natural gas[b]
Carbon	45.0	80.0	82.2–87.1	80–85	65–80
Oxygen	43.0	5.0	0.1–4.5	–	–
Hydrogen	5.5	5.0	11.7–14.7	8.5–11	1–25
Nitrogen	3.1	1.5	0.1–1.5	0–2	1–15
Sulfur	0.1–0.3	1.0	0.1–5.5	2–8	trace–0.2
Phosphorus	0.2–0.3				

[a]Bowen (1966). Dry plant material of angiosperms and gymnosperms.
[b]Levorsen (1967).

oxides, carbon monoxide, heavy metals); and (4) changes in the chemical composition of atmospheric precipitation and surface waters, and the effects these changes may have on the weathering of earth's crust and transport of dissolved materials to the oceans.

Concentrations of the major constituents in wood, petroleum, and related materials, given in Table 1.6, illustrate the dominant role of the oxides of carbon among the products of burning. The amount of fuel burned annually has been increasing since the middle of the nineteenth century at the rate of about 4.3% per year (Figure 1.12). This is about twice as fast as the rate of increase in the human population of the globe. Between the years 1915 and 1945 the lower rates of fuel burning are attributable to two world wars and an economic depression. In late 1970s, 4×10^{14} mol CO_2 yr^{-1} are generated by fuel burning. This value itself is relatively small in comparison with other natural fluxes of carbon in the exogenic cycle, such as uptake of CO_2 by plants on land and in the ocean, and return of CO_2 by oxidation of organic material on the earth's surface (Tables 1.5 and 1.7). The amounts of carbon in the known reserves of fossil fuels, land and marine biosphere, and in adjoining reservoirs, as well as the natural and man-produced fluxes of carbon, are summarized in Table 1.7. The upper limits of the CO_2 production owing to burning of all the fuels or all the forests are the values in moles of C of the fuel reserves and forests given in Table 1.7. These values may be compared with the amounts of CO_2 contained in the atmosphere, the upper 200 m of the ocean, and deep ocean water.

The course of addition of CO_2 from fuel burning to the atmosphere during the second half of the nineteenth and first half of the twentieth centuries has been demonstrated by Suess (1955) in a study of the isotopic ratio $^{14}C/^{12}C$ in tree rings. Radioactive isotope ^{14}C is present in the atmosphere, and it is taken up with the stable isotopes ^{12}C and ^{13}C by growing

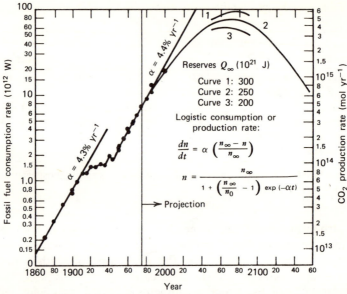

Figure 1.12 Production of CO_2 due to burning of fossil fuels. Consumption of fuels assumed to follow a logistic function (from Zimen et al., 1977). By permission of Prof. K. E. Zimen, Berlin, Germany. (J = joule, W = watt, Appendix D.)

plants. Because of the relatively short radioactive decay half-life of ^{14}C (5730 yr), coal and petroleum are free of ^{14}C. The CO_2 formed by fuel burning contains no ^{14}C and when added to the atmosphere it lowers the $^{14}C/^{12}C$ abundance ratio. This dilution of the $^{14}C/^{12}C$ ratio in the atmosphere owing to addition of CO_2 from fuel burning is known as the Suess effect. The stable isotopes of carbon ^{12}C and ^{13}C occur in somewhat different proportions in natural combustible gases, crude oils, and coals. The abundance ratios $^{13}C/^{12}C$, expressed as $\delta^{13}C$ in per mil and plotted in Figure 1.13, show the differences between the atmospheric CO_2, combustible fuels, and other materials. Gases, coal, and petroleum are lighter in their carbon isotopic composition than the atmospheric carbon dioxide. The isotopically lighter carbon in plants and fossil fuels derived from them, in comparison to the CO_2 of the atmosphere and ocean water, reflects the processes of the biological fractionation of the isotopes. Burning of fossil fuels should produce carbon dioxide of isotopically lighter carbon composition than the atmospheric CO_2 characterized by $\delta^{13}C \simeq -6‰$. The reservoir of carbon contained in fossil fuels is large in comparison to the atmosphere (Table 1.7), and according to some forecasts the amount of CO_2 in the atmosphere may rise four- to fivefold, as shown in Figure 1.14. This process would continue to lower both the $^{14}C/^{12}C$ and

Table 1.7   Fuel Reserves and Geochemical Reservoirs Receiving CO₂ from Burning

Reservoir (number i or j)	Reservoir content (10^{16} mol C)	Fluxes F_{ij} (10^{16} mol C yr^{-1})
1. Fuel reserves[a]	60	$F_{12} \simeq 0.04$ to atmosphere
Consumed by 1970	1.2	in 1970s
2. Atmosphere		
CO₂ preindustrial[a]	5.1	
CO₂ present ($\simeq 10\%$ increase)	5.6	
O₂ free (10^{16} mol)	3800	
3. Land		
Biomass ($=$ plants)[b]	6.9	$F_{\text{atm-biomass}} \simeq 0.3 \pm 0.1$
All forests	6.2	
All the rest	0.7	
Soil humus[c]	25	
4. Ocean		
Biomass (plants)[b,d]	0.03 ± 0.01	$F_{\text{atm-biomass}} \simeq 0.2$
Surface layer[e]	5.1	
Deep ocean[a,f]	$(60\text{–}63) \times 5.1 = 300$	
5. Sediments[g]	500,000	
Organic C	100,000	
CaCO₃	400,000	

[a]Zimen and Altenhein (1973).
[b]Whittaker and Likens (1973).
[c]Bolin (1977).
[d]Bolin (1960).
[e]Oeschger et al. (1975); 58 m contains equivalent of preindustrial atmospheric CO₂ (5.1×10^6 mol).
[f]Ekdahl and Keeling (1973).
[g]Garrels et al. (1976).

$^{13}C/^{12}C$ ratios of the atmospheric carbon dioxide, through dilution with dead and isotopically light carbon.

Distribution of Added CO₂.   Measurements of the CO₂ concentration in the atmosphere, done since 1958 on the Mount Mona Loa on the Island of Hawaii and more recently at the South Pole, show that the CO₂ content of the atmosphere has been increasing in the 1970s at the rate of approximately 0.8 ppm (by volume) per year (Ekdahl and Keeling, 1973). Since preindustrial times, the total increase in the atmospheric CO₂ has been of the order of 10%, from an extrapolated value of 290 ppm about 120 yr ago to the present value of 325 ppm.

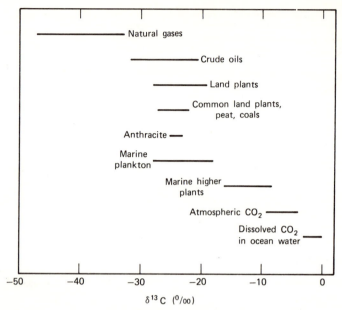

Figure 1.13 Depletion of heavy carbon ^{13}C relative to CO_2 in plants and fuels (from Silverman, 1963; Degens, 1969; Parker, 1971).

The measured increase in the CO_2 content of the atmosphere accounts for only about 50% of the amount burned annually. Where the remaining 50% goes is an open question, although the ocean and terrestrial plants are the first choice for the reservoirs that can accept excess CO_2. Estimates by several investigators of how the CO_2 from fuel burning distributes itself between the atmosphere, ocean, and biota are listed in Table 1.8. The fractions of CO_2 assigned to the ocean and biota vary considerably, depending on the assumptions made concerning the rates of transfer between the different reservoirs and residence times within them.

In a system of three major reservoirs—atmosphere, surface ocean, and land biota—the important fluxes determining the distribution of the CO_2 added to the atmosphere are those between the atmosphere and the ocean, and the atmosphere and land plants.

The flux between the atmosphere and surface ocean has been based on a thermodynamic equilibrium distribution of CO_2 between air and ocean water. The distribution is commonly expressed by the so-called buffer factor (Bolin and Eriksson, 1959; Broecker et al., 1971; Keeling, 1973; Bacastow and Keeling, 1973), defined as

$$\frac{dp}{p_0} = \xi' \frac{d\Sigma C}{\Sigma C_0} \qquad (1.24)$$

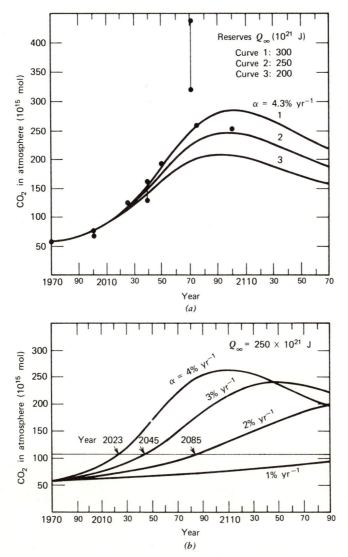

Figure 1.14 Projected increases in the amount of CO_2 in the atmosphere. (*a*) Reserves of different magnitude, in joules, constant annual rate α of consumption of fuel reserves. (*b*) Different consumption rates (from Zimen et al., 1977). By permission of Prof. K. E. Zimen, Berlin, Germany.

Table 1.8   Distribution of CO_2 from Fuel Burning (in Percent of Total CO_2 Produced Per Year Around 1970) in the Atmosphere, Ocean, and Biomass According to Different Authors

Reservoir	Bolin and Bischof (1970)	Broecker et al. (1971)	Machta (1972)	Bacastow and Keeling (1973)	Oeschger et al. (1975)	Niehaus (1976)	Bolin (1977)
Atmosphere	30–40	60 ± 10	55	50	55	50	40 ± 5
Ocean	~50	40 ± 10	30	14–22	33	19 ⎫	
Land plants	15–20		15	29–36	12	31 ⎭	60 ± 5

The buffer factor ξ' relates the increase in the partial pressure of CO_2 in the atmosphere dp to the increase in the total concentration of dissolved inorganic CO_2 species in ocean water $d\Sigma C$. p_0 is the partial pressure of CO_2 in preindustrial times and ΣC_0 is the preindustrial concentration of total inorganic carbon in the surface ocean. The name buffer factor stems from the fact that the distribution of the inorganic carbon species in ocean water [such as HCO_3^-, CO_3^{2-}, and $CO_2(aq)$] is related to the pH-buffering capacity of the carbonate and borate ions in solution.

The values of ξ' used in various models of the carbon dioxide partition between the atmosphere and ocean are in the range

$$8 < \xi' < 12$$

A value of $\xi' = 10$ means that for a 10% increase in the partial pressure of CO_2 in the atmosphere ($dp/p_0 = 0.1$) there is a 1% increase in the total amount of dissolved inorganic carbon in surface ocean water:

$$\frac{d\Sigma C}{\Sigma C_0} = \frac{0.1}{10} = 0.01$$

A slightly different definition of the buffer factor, similar to equation 1.24, is (Keeling, 1973)

$$\frac{dp}{p_0} = \xi \frac{\Sigma C}{\Sigma C_0} \tag{1.25}$$

which relates the increase in the partial pressure of CO_2 to the total amount of dissolved inorganic carbon in ocean water ΣC. Irrespective of which of the preceding two relationships between the atmospheric CO_2 and dissolved CO_2 is used, the problem reduces to the question: What is the concentration of total dissolved carbon ΣC in ocean water as a function of p_{CO_2} in the atmosphere? A direct relationship between p_{CO_2} and ΣC is given in Section 1.5.3.

The amount of CO_2 taken up by plants in response to a higher partial pressure of CO_2 in the atmosphere has been considered as generally increasing. The increment of the flux of CO_2 from the atmosphere to the photosynthetic reservoir can be related to the CO_2 content of the atmosphere by the so-called biota growth factor, analogous to the buffer factor for the distribution of the CO_2 between the atmosphere and ocean. The biota growth factor β appears in the following equation defining an increase ΔF of the photosynthetic flux:

$$\Delta F = F_0 \beta \left(\frac{M}{M_0} - 1 \right) \qquad (\text{mol yr}^{-1}) \qquad (1.26)$$

where ΔF is the increment of the net photosynthetic flux (that is, the CO_2 flux due to photosynthesis less the return flux due to respiration), F_0 is the flux in preindustrial times, M_0 is the mass of CO_2 in the atmosphere in preindustrial times, and M is the new mass. The values of the biota growth factor β, used in different models (Keeling, 1973; Oeschger et al., 1975; Siegenthaler and Oeschger, 1978) varied between 0 and 0.4. No response of the plant world to the higher CO_2 levels in the atmosphere corresponds to $\beta = 0$, and an increase in the net flux to plants by 40% corresponds to $\beta = 0.4$.

Models and Predictions. Models of the CO_2 addition to the atmosphere have essentially been based on four to six reservoir boxes: the atmosphere (subdivided into the upper and lower in some models), biomass (subdivided into short lived and long lived in some models), surface ocean, and deep ocean (compare geochemical cycle diagrams in Figures 1.5 and 1.8). Mathematical variety in the functional relationships describing the fluxes between the reservoirs and transport of CO_2 within them is considerable, ranging from first-order fluxes and well-mixed reservoirs to nonlinear inputs and fluxes, and diffusional dispersal within reservoirs. Models of this type have been developed by the following authors, and some others listed in their bibliographies: Bolin and Bischof (1970), Broecker, Li, and Peng (1971), Bacastow and Keeling (1973), Ekdahl and Keeling (1973), Machta (1972), Niehaus (1976), Zimen and Altenhein (1973), Bolin (1977), Zimen et al. (1977), and Siegenthaler and Oeschger (1978).

Some of the more recent projections of the rising CO_2 levels in the atmosphere, shown in Figure 1.14, are based on (1) the consumption of fuels taking place as an S-shaped or logistic function of time (a nearly exponential rise followed by an asymptotic approach to a steady level), (2) equilibration between CO_2 in the atmosphere and the surface ocean, and (3) uptake of CO_2 from the atmosphere by plants. The CO_2 content of the atmosphere is projected to rise for the next 100 years (Figure 1.14*a*) because the rate of fuel burning has not yet reached a steady value. Subsequently the amount of CO_2 in the atmosphere should go through a peak and decline, the decline attributable to the oceanic and biological sinks. It may be noted that a fivefold increase in the partial pressure of CO_2 in the atmosphere after a century would come to the limit of CO_2 acceptable for photosynthesis of at least some plant species (Niehaus, 1976).

If the rate of increase in fuel consumption and burning had proceeded according to a logistic curve growth law, but at lower rates than the present rate of about 4% per year, then the CO_2 increase in the atmosphere would have been considerably slower, as shown in Figure 1.14b. Doubling of the CO_2 level in the atmosphere would be attained by year 2020 if the fuel consumption rate increased logistically at 4% per year, and by year 2085 if the growth were 2% per year.

The assumption used in the models that terrestrial plants are an important sink for the CO_2 added to the atmosphere needs to be clarified. If about 25% of the CO_2 amount added to the atmosphere annually (4×10^{14} mol yr^{-1}) is taken by the terrestrial biomass, the increment is about 2.5–5% of the photosynthetic flux:

$$\frac{1 \times 10^{14} \text{ mol yr}^{-1}}{(20 \text{ to } 40) \times 10^{14} \text{ mol yr}^{-1}} \times 100 = 2.5 \text{ to } 5\%.$$

In the earlier part of this century, when less fuel was burned, the percent amounts presumably added to the biosphere were lower (Figure 1.12). One percent or so annual increase rate in the terrestrial biosphere could be detectable on a time scale of several decades if it were looked for and if mankind did not interfere with nature. However, extensive deforestation that has taken place in industrial times and replacement of forests by grasslands and newly planted forests make it difficult to detect the increment in the biomass which may be due to the higher CO_2 levels in the atmosphere.

To illustrate the physical magnitude of the 25–50% of the CO_2 added to the atmosphere annually, the following comparisons can be made. About 1×10^{14} mol C is equivalent to 25% of the standing crop of the oceanic biomass. The same amount, made into $CaCO_3$ and deposited over 85% of the ocean surface (3×10^{18} cm^2) corresponds to the $CaCO_3$ sedimentation rate of about 1.5 cm per 1000 yr, which is comparable to the rates in the carbonate sedimentation areas of the oceans. Made into organic matter (CH_2O, of density 1 g cm^{-3}), the amount of 1×10^{14} mol C would cover the continents (1.5×10^{18} cm^2) in a layer 20 μm thick.

1.5.3 Distribution of CO_2 between Atmosphere and Ocean

For a given partial pressure of CO_2 in the atmosphere, total concentration of dissolved inorganic carbon ΣC at equilibrium with the atmospheric CO_2 can be computed from the following relationships (Keeling, 1973).

A quantity known as alkalinity or "titration alkalinity" [Alk] (see Section 4.4.3) is defined for ocean water as

$$[\text{Alk}] = \frac{K_0 K_1}{[H^+]}\left(1 + \frac{2K_2}{[H^+]}\right)p_{CO_2} + \frac{K_B \Sigma B}{[H^+] + K_B} + \frac{K_w}{[H^+]} - [H^+]$$

(1.27)

where [ ] denotes concentrations in mol l^{-1}, ΣB is the total borate concentration $\{\Sigma B = [B(OH)_3] + [B(OH)_4^-] \text{ mol } l^{-1}\}$, and K_i are dissociation constants for CO_2, borate, and water at a given temperature. For average oceanic conditions, the values of [Alk], K_i, and ΣB are listed below. At a known p_{CO_2}, the value of [Alk] can be determined by acid titration of sea water and, subsequently, the hydrogen-ion concentration $[H^+]$ and the pH can be computed from equation 1.27. The relationship between total dissolved inorganic carbon ΣC

$$\Sigma C = [H_2 CO_3] + [HCO_3^-] + [CO_3^{2-}]$$

(1.28)

and partial pressure of CO_2 in the atmosphere p is

$$\Sigma C = \frac{p}{\Phi} \quad (\text{mol } l^{-1})$$

(1.29)

where Φ is a parameter dependent on the equilibrium dissociation constants and $[H^+]$

$$\frac{1}{\Phi} = K_0\left(1 + \frac{K_1}{[H^+]} + \frac{K_1 K_2}{[H^+]^2}\right)$$

(1.30)

After $[H^+]$ has been computed from equation 1.27, the quantity $1/\Phi$ can be computed and, finally, ΣC determined using equation 1.29. For a given p and ΣC, the quotients dp/p_0, defined in equations 1.24 and 1.25 with the aid of the buffer factors, can also be computed.

The equilibrium dissociation constants and other parameters needed in equation 1.27 are the following, for average oceanic conditions (Keeling,

1973):

$$[\text{Alk}] = 2.43 \times 10^{-3} \text{ equivalents } l^{-1}$$

$$\Sigma B = 0.0409 \times 10^{-3} \text{ mol } l^{-1}$$

$$K_0 = 3.347 \times 10^{-2} \qquad K_B = 1.88 \times 10^{-9}$$

$$K_1 = 9.85 \times 10^{-7} \qquad K_w = 6.46 \times 10^{-15}$$

$$K_2 = 7.8 \times 10^{-10}$$

It should be noted that the relationship between ΣC and p is nonlinear. The value of [Alk] is conservative if no carbonate (and borate) minerals dissolve when the CO_2 pressure is raised. The value given above, $[\text{Alk}] = 2.43 \times 10^{-3}$ equivalents l^{-1} can become higher if $CaCO_3$ dissolves in ocean water due to a higher CO_2 partial pressure in the atmosphere.

CHAPTER 2

Fluxes and Transport:
Advection and Dispersal

This chapter and Chapter 3 deal with some fundamentals of transport processes and the mathematical relationships that provide a basis for part of the material in subsequent chapters. The essential mathematical formulations are assembled in Chapters 2 and 3, to allow the other chapters to follow more closely their subject matter whenever geochemical transport is discussed. Advection and dispersal by turbulence are fast processes in comparison to molecular diffusion, and Chapters 2 and 3 follow this distinction.

2.1   FLUX

A measure of the rate of transfer of material from one geochemical reservoir to another and from one physical or chemical state to another is the flux. A general definition of the flux is

$$\text{flux} = \text{proportionality factor} \times \text{driving force} \qquad (2.1)$$

The dimensions of the flux are $ML^{-2}T^{-1}$ or MT^{-1}, where M is a measure of the quantity of material carried by the flux (not necessarily the mass), L is a linear dimension, and T is time. Depending on the nature of the flux—a flux of mass, energy, volume, or particles—the dimensions of the driving force and the factor of proportionality must correspond to the flux dimensions. The driving forces are the mechanisms responsible for the flux, and the mechanisms considered in this chapter are advection, dispersal, and diffusion. Although the distinctions between these three are not sharp, it is convenient to deal with them separately because of the differences in the physical scales and times associated with each type of process.

2.2 ADVECTION

Advection is flow or, more generally, displacement relative to the observer of a parcel of material under the influence of forces. Flow of water and wind, lowering of the land relief because of erosion, atmospheric precipitation, evaporation of water, deposition of sediments, and movements of crustal plates are advective processes. From this short list it is obvious that the major advective flows are driven by forces of global nature, and the magnitude of the flows does not strongly depend on the chemical composition of the reservoirs from and to which material is being transported.

2.2.1 Magnitudes of Advection

Flow of material is associated with velocity, either a velocity of its own or a velocity of the medium through which the material is transported. A range of advective velocities characteristic of different environments near the surface of the earth is shown in Figure 2.1. From the rates of sediment accumulation on the deep ocean floor (measurable in millimeters to centimeters per 1000 years), to the velocities of raindrops, and wind at the earth's surface (measurable in meters per second), the speeds vary by a factor of 10^{13}. The ranges of advective velocities shown in the figure are the more representative values. To indicate an entire range of the velocities associated with any particular process, almost all the lines would extend over the diagram field to the left, to very small values. For such processes as the settling of particles in water and ground water flow, higher values than shown in the figure are encountered for large particles, surface runoff, and certain rivers. The velocities of surface current in bodies of water $(10^{0}-10^{2} \text{ cm sec}^{-1})$ grade continuously into the lower velocities typical of soil and deeper ground waters. At the lower end of the velocity scale, the rates of sedimentation include the slower deposition on the deep ocean floor and the faster accumulation rates in near-shore and lake environments.

Order of magnitude estimates of the residence times in different geochemical reservoirs are deducible from Figure 2.1. For example, the residence time of falling raindrops in the atmosphere (1–3 km) is less than 1 hr; the residence time of particles settling through the ocean (4 km) varies, depending on their size, from years to centuries; the residence time of newly formed ocean floor material, moving away from a ridge at the rate of 2 cm yr^{-1} over a distance of 2000 km, is 100 million years. Additional comparisons between linear dimensions of different geochemical reservoirs and advective velocities associated with them can be made by reference to Table 1.2. It may be noted in Figure 2.1 that the advective

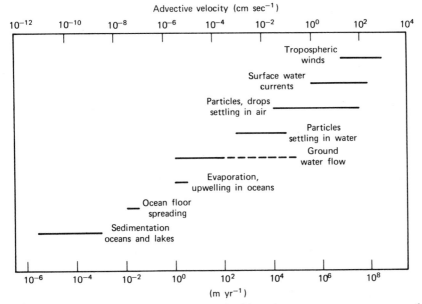

Figure 2.1 Characteristic velocities associated with major transport processes near the earth's surface.

velocities in the horizontal and veritical dimensions in waters—surface currents against upwelling—differ by several orders of magnitude, reflecting basically the circumglobal motions within the fluid reservoirs.

2.2.2 Advective Fluxes

Flow of material of density ρ (g cm^{-3}) with velocity U (cm sec^{-1}) results in the flux F of

$$F = \rho U \quad \left(\text{g cm}^{-2} \text{ sec}^{-1}\right) \tag{2.2}$$

Similarly, if the material flowing with velocity U (cm sec^{-1}) contains another substance at concentration C (g cm^{-3}), then the flux of the latter is

$$F = CU \quad \left(\text{g cm}^{-2} \text{ sec}^{-1}\right) \tag{2.3}$$

The flow velocity U may be regarded as the driving force for the flux. Essentially all fluxes, irrespective of the detailed nature of the driving mechanisms, can be written as products of a term having the dimensions of velocity and a term having the dimensions of concentration or density. This also applies to dispersive and diffusional fluxes treated in other sections of this chapter.

The advective fluxes of greater importance to the geochemical processes in sedimentary environments are the flow of water in rivers, through sediments and crustal rocks (ground and pore water flow), and the flow of solids through the atmosphere and water (sedimentation). Pore water flow is treated in the next section; sediments and settling are discussed in Chapters 5 and 6.

2.2.3 Pore Water Flow

Darcy's Law. Flow of water through a column of porous material, formulated by Henry Darcy in 1856 on the basis of experiments, can be described by a flux relationship in which the volume of water q flowing per unit area of a porous bed is proportional to the hydrostatic pressure difference across the bed (see Figure 2.2),

$$q = \frac{\mathcal{K}(\varphi_1 - \varphi_2)}{L} \qquad (\text{cm}^3\,\text{cm}^{-2}\,\text{sec}^{-1}) \qquad (2.4)$$

where $\mathcal{K}$ is a coefficient of hydraulic conductivity (of dimensions cm sec^{-1}), L is the length of the porous column (cm), and φ is

$$\varphi = z + \frac{P}{\rho g} \qquad (\text{cm}) \qquad (2.5)$$

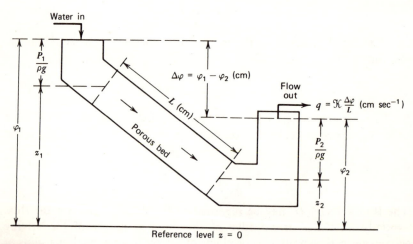

Figure 2.2 Darcy's law of water flow through a porous bed (adopted from Todd, 1964; Bear, 1972).

where P is hydrostatic pressure (g cm^{-1} sec^{-2}), ρ is the density of the fluid (g cm^{-3}), and g is the acceleration due to gravity (cm sec^{-2}). The parameter φ is numerically equal to the elevations of the entry and exit points of the flow system above some reference level $z=0$. Although φ has the dimensions of distance, it can be thought of as a driving force by considering it as energy of water per unit weight of the water column (i.e., dyn cm dyn^{-1}). By a similar reasoning, the flow velocity U in equations 2.2 and 2.3 is a driving force represented by the momentum per unit mass of the following water (g cm sec^{-1} g^{-1}). The above definition of φ is one of the forms used in computations of water flow through porous beds (e.g., Engelhardt, 1960; Bear, 1972). Another term, called the potential φ', is the driving force expressed as energy per unit mass of fluid (Hubbert, 1940),

$$\varphi' = gz + \frac{P}{\rho} \equiv g\varphi \qquad (\text{cm}^2 \text{ sec}^{-2}) \tag{2.6}$$

The inclined flow path L in Figure 2.2 can be set upright and the elevations of the entry and exit points above the column base then become φ_1 and φ_2, respectively. Note that the driving energy $\varphi_1 - \varphi_2$ in Figure 2.2 is simply the difference in elevation of the inflow and outflow planes. In a horizontal aquifer, the elevations of the entry and exit points are equal ($z_1 = z_2$), and the water flow can be driven only if there is a difference in pressure along the horizontal,

$$q = \frac{\mathcal{K}(P_1 - P_2)}{\rho g L} \qquad (\text{cm sec}^{-1})$$

The mean velocity U of the flow through the porous column is

$$U = \frac{q}{\phi} \qquad (\text{cm sec}^{-1}) \tag{2.7}$$

where ϕ is the volume fraction of sediment occupied by flowing water. The fractional quantity ϕ will be referred to as sediment *porosity* or the pore volume fraction of bulk sediment. Existence of small blind pockets on the surface of sediment particles and, in general, the occurrence of very small pores can make the sediment porosity ϕ considerably greater than the volume fraction occupied by flowing water. Such differences between the volume fractions of pore space are not considered in this section.

The coefficient of hydraulic conductivity $\mathcal{K}$ combines the effects of the solid particles on the flow and the effects of the physical characteristics of the fluid. On dimensional grounds, the properties of the flowing fluid can be represented by its density ρ (g cm^{-3}), acceleration due to gravity g (cm sec^{-2}), and the fluid viscosity η (g cm^{-1} sec^{-1}), whereas the effects of the solid particles on the flow can be represented by a parameter k, as in the following relationship:

$$\mathcal{K} = \frac{k\rho g}{\eta} = \frac{kg}{\nu} \qquad (\text{cm sec}^{-1}) \qquad (2.8)$$

where the parameter k, called the permeability, has the dimensions of L^2 and $\nu = \eta/\rho$ ($\text{cm}^2 \text{ sec}^{-1}$) is the kinematic viscosity.

For water near 20°C, the viscosity is $\eta = 0.01$ g cm^{-1} sec^{-1} ($= 1$ centipoise) and the product $\rho g = 980$ g cm^{-2} sec^{-2}. Thus it follows from equation 2.8 that the coefficients of hydraulic conductivity $\mathcal{K}$ and permeability k differ by a factor of about

$$\frac{k}{\mathcal{K}} = 10^{-5}$$

Equation 2.4 can be rewritten as a product of a gradient and a coefficient of proportionality,

$$q = -\mathcal{K}\frac{d\varphi}{dx} \qquad (\text{cm}^3 \text{ cm}^{-2} \text{ sec}^{-1}) \qquad (2.9)$$

where x denotes the distance coordinate along which the gradient of φ is measured (the length of the column L in the configuration of Figure 2.2). Equation 2.9 is a conventional form of a one-dimensional flux, as defined in equation 2.1.

Example. A generalized pattern of water flow in a major artesian aquifer in central Florida is shown in Figure 2.3. The northern tips of the arrows lie in the recharge area, from which the ground water flows south and toward the two coasts of the peninsula. The aquifer is a major source of ground water in central Florida (Back and Hanshaw, 1971). The contour lines on the map show the elevation in meters of the aquifer above sea level. The elevation is the hydraulic head or, effectively, the difference $\Delta\varphi = \varphi_1 - \varphi_2$ shown in Figure 2.2. The gradients amount to a vertical drop of about 10 m over a map distance of about 100 km in the coastal regions. Thus the gradient $-d\varphi/dx$ or $\Delta\varphi/L$ is of the order of

$$\frac{\Delta\varphi}{L} \simeq \frac{10 \text{ m}}{100 \text{ km}} = 10^{-4}$$

From the values of the hydraulic conductivity coefficient $\mathcal{K}$ given in Figure 2.4, a representative value for a "good" aquifer can be taken as

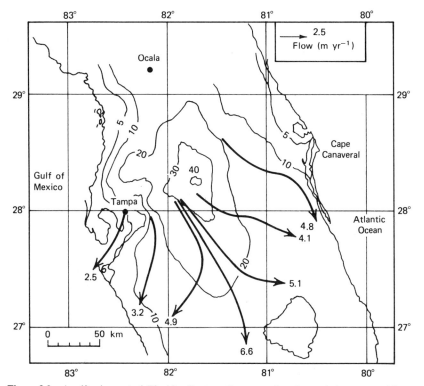

Figure 2.3 Aquifer in central Florida. Contour lines are elevations of the water table (m) above sea level. Arrows indicate flow direction and velocity (adopted from Back and Hanshaw, 1971).

$\mathcal{K} = 10^{-1}$ cm sec^{-1}. The flow rate q, using equation 2.4 or 2.9, is

$$q = (10^{-1} \text{ cm sec}^{-1})(3 \times 10^7 \text{ sec yr}^{-1}) \times 10^{-4} = 300 \text{ cm yr}^{-1}$$

The flow rate $q = 3$ m yr^{-1} is of the same order of magnitude as the flow rates indicated in Figure 2.3. The residence time of water in a 100-km-long aquifer is

$$\frac{10^5 \text{ m}}{3 \text{ m yr}^{-1}} = 30,000 \text{ yr}$$

This estimate of the residence time assumes that there is no recharge of the aquifer from other entry points along the 100-km route, and that there are no significant evaporative losses. Estimates of the coefficients of hydraulic

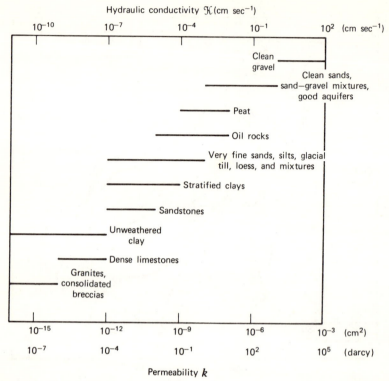

Figure 2.4 The coefficients of hydraulic conductivity and permeability of sediments and rocks (from numerous literature sources).

conductivity $\mathcal{K}$, listed in Figure 2.4, vary by a factor of 100 or more for a given sediment type. This introduces a comparable margin of uncertainty in rough computations of ground water discharge. Cracks and joints in a poorly permeable rock matrix can allow water to flow much faster than expected on the basis of the hydraulic conductivity value determined on a coherent sediment sample.

Permeability k.   Some characteristic values of the coefficient of permeability k of various sediment and rock types are given in Figure 2.4. The difference of ten or more orders of magnitude between the permeabilities of gravel and clay, as well as the factor of 10^2-10^3 variation in k within one sediment type indicate that the permeability is highly sensitive to the packing arrangement of the sediment particles.

The coefficient of permeability k as defined in equation 2.8, is a function of the texture of the porous solid. Permeabilities are measured in the laboratory and in the field using Darcy's law, equation 2.9, and they can also be computed from theoretical models. The models are based on flow through geometrically different arrangements of pores, such as capillaries arranged in bundles and three-dimensional lattices, tortuous channels, and branching channels, to mention some of the possible geometries. The coefficient of permeability k depends on the following characteristics of the sediment: (1) grain size of the sediment, (2) sediment porosity, and (3) the packing arrangement of particles, which itself is some function of the particle shape and coordination. Several explicit relationships for k, theoretically or experimentally derived, that give its dependence on the particle-grain size, sediment porosity, and packing are listed with explanatory notes in Table 2.1. As a whole, the permeability is directly related to the cross-sectional area of the particles (r^2, where r is the equivalent sphere radius of the particle), to some power of the sediment porosity ϕ^n, and to a parameter that represents the packing of particles.

The dependence of permeability k on the sediment porosity is strong, although the equations of different investigators, derived from theoretical models and experimental work on different sediments, do not agree on the nature of the dependence of k on ϕ. The equation of Krumbein and Monk (1942), equation 1 in Table 2.1, makes k dependent on particle size only, whereas other equations (2, 4, and 5) give k as dependent on ϕ, without an explicit dependence on particle size.

In the literature, the permeabilities are often expressed either as power functions of ϕ or of the ratio $\phi/(1-\phi)$. The ratio is also known as the *void ratio*. The dependence of k on both ϕ and $\phi/(1-\phi)$ is implicit in equations 6 and 7 in Table 2.1. The dependence of permeability k on the void ratio $\phi/(1-\phi)$ rather than only on porosity ϕ arises in theoretical derivations of permeability of channel networks of different geometries. One such derivation leads to the Carman-Kozeny equation for permeability, listed as equation 6 in Table 2.1. The ratio $\phi/(1-\phi)$ figures explicitly in equation 4, based on many sediment samples from the Gulf of Mexico, and in the Carman-Kozeny equations, 6 and 7.

The functional relationships for k summarized in Table 2.1 relate to some characteristic particle size r that can be interpreted as the mean particle radius. For a given particle size, the dependence of permeability k on porosity ϕ is shown in Figure 2.5; the plotted values of $k(\phi)$ are the ϕ-dependent terms in equations 2–9 in Table 2.1. It can be noted that equations 6–9, applying to a randomly textured sediment and sediment made of cylindrical particles, show very similar trends in the change of permeability with porosity. Equations 4 and 5, for clayey and aragonitic sediments, respectively, show a trend similar to unconsolidated sands, equation 3, in the range of porosity between 0.2 and 0.7. The strong

Table 2.1 Coefficient of Hydraulic Permeability k (cm^2 or approximately 10^8 darcy) as a Function of Sediment Porosity ϕ (fraction) and Particle Radius r (cm)

Equation	Explanatory notes
1. $k = 2.47 \times 10^{-19} r^2$	Krumbein and Monk (1942), sands, $0.005 < r < 0.1$ cm
2. $k = \text{const} \times \phi^5$	Terzaghi (1925) in Rieke and Chilingarian (1974, p. 148), $0.2 < \phi < 0.8$
3. $k \propto \phi^9 r^2$	Sands, from data of Beard and Weyl (1973); proportionality factor depends on ϕ (Figure 2.5), $0.25 < \phi < 0.4$
4. $k = 10^{-9} \left(\dfrac{\phi}{1-\phi} \right)^7$	Marine clayey sediments (Bryant et al., 1974)
5. $k = 7.25 \times 10^{-11} \left(\dfrac{\phi}{1-\phi} \right)^7$	Aragonitic sediment, $0.2 < \phi < 0.7$, from data of Robertson (1967). Original data also obey $\log k = -17.0 + 13.6\phi$
6. $k = \dfrac{\phi^3}{(1-\phi)^2} \cdot \dfrac{c_K}{S_s^2}$	The Carman-Kozeny equation, where c_K is the Kozeny constant and S_s is the surface area of the pore space per unit volume of solid (Bear, 1972, p. 166). For beds made of spherical and cylindrical particles, see equations 7, 8, and 9 below
7. $k = \dfrac{\phi^3}{(1-\phi)^2} \cdot \dfrac{r^2}{45}$	Spherical particles, $c_K = \frac{1}{5}$ and $S_s = 3/r$ (cm^{-1}) in equation 6
8. $k = \dfrac{-2\ln(1-\phi) + 4(1-\phi) - (1-\phi)^2 - 3}{1-\phi} \cdot \dfrac{r^2}{8}$	
	Circular cylinders, flow parallel to cylinder axis. r is cylinder radius. The Kozeny constant c_K given by Happel and Brenner (1973, p. 393). For straight circular cylinders, $S_s = 2/r$ (cm^{-1}) in equation 6
9. $k = \dfrac{-\ln(1-\phi) - \dfrac{1-(1-\phi)^2}{1+(1-\phi)^2}}{1-\phi} \cdot \dfrac{r^2}{8}$	
	Circular cylinders, flow perpendicular to cylinder axis, parameters as in equations 8 and 6. For a random network of cylinders, Happel and Brenner (1973) recommend a weighted sum of k from equations 8 and 9, in proportions $\frac{2}{3}$ and $\frac{1}{3}$

dependence of k on ϕ, such as $k \simeq \phi^n$ with $n = 9$, emerges from the empirical determinations of k, as represented by equations 3–5 in Table 2.1. The strong variation in k over a relatively narrow range of porosities reflects the dependence of permeability on other textural characteristics of sediments, such as packing and shape of the grains.

Permeability of unconsolidated silicate sands as a function of mean

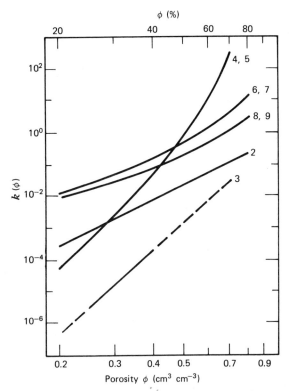

Figure 2.5 Dependence of the permeability k on sediment porosity ϕ, according to the models listed in Table 2.1.

grain size, porosity, and sorting is shown in Figure 2.6. The degree of sorting is given by the sorting coefficient So, determinable from the cumulative frequency distribution curves of sediment samples [So $= (Q_{75}/Q_{25})^{1/2}$, where Q_{75} is the third quartile and Q_{25} is the first quartile of a cumulative distribution]. The lower values of the sorting coefficient So correspond to very well sorted sediments, So $\simeq 1.7$ is moderately well sorted, 2.4 is poorly sorted, and 4.2 is very poorly sorted. Poor sorting represents a large variance of grain sizes in the sediment, whereas a high degree of sorting represents a more narrow size distribution. In the data shown in Figure 2.6, the permeability is highest for the well sorted and large grain-size sediment. It decreases as the degree of sorting deteriorates and the particles become smaller. The overall decrease in porosity is not large, from about 42 to 25%, dependent primarily on the degree of sediment sorting rather than on the particle size. In all cases, the permeability increases with r^2. The relationship between k, r, and ϕ is given in equation 3 in Table 2.1.

Deviations from Darcy Flow.   The linear relationship between the flow rate q and the hydraulic gradient, given in Equation 2.9, does not hold at higher flow rates: an increase in the hydrostatic head produces a nonlinear increase in the flow rate. It has been observed that Darcy's law is obeyed as long as the Reynolds number for the flow does not exceed some value between 1 and 10 (Bear, 1972). The dimensionless Reynolds number Re, representing the quotient of inertial and viscous forces in a flowing fluid, is

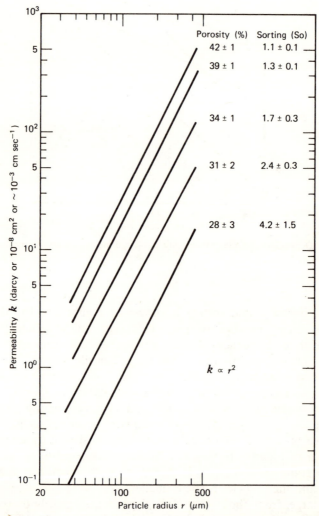

Figure 2.6   Dependence of the permeability on the particle size in sands of different degrees of sorting (data from Beard and Weyl, 1973).

defined for a porous medium made of particles as (see also Section 6.2.2)

$$\mathrm{Re} = \frac{2rU}{v} \tag{2.10}$$

where the parameters are, as defined previously, the particle radius r, water flow velocity U, and the kinematic viscosity of water v. For water ($v \simeq 10^{-2}$ cm^2 sec^{-1}) flowing through a sediment made of particles of radius $r = 100$ μm $= 10^{-2}$ cm, the value of $\mathrm{Re} < 5$ can characterize a flow of velocity $U \lesssim 2$ cm sec^{-1}. This is a relatively fast flow, as may be seen from the range of water flow velocities in sediments given in Figure 2.1.

A number of nonlinear relationships between the flow rate q and the hydrostatic driving head $\Delta\varphi/L$ are given in Table 2.2. As the nonlinearity is expressed in the higher powers of the flow rate q, this means that q is proportional to some fractional power of the hydraulic gradient, $q \propto (\Delta\varphi/L)^{1/n}$, where $n > 1$. Thus increase an or a decrease in the driving pressure gradient by some factor produces an increase or decrease in the flow rate that is smaller than the factor. Accordingly the response of the flow rate to variations in the hydraulic gradient is weaker than in the region of Darcy's law, and in this sense the flow is more stable.

The flow rate equations in Table 2.2 are written in the form

$$q + a_1 q^2 = a_2 G$$

Table 2.2 Some Nonlinear Relationships between Flow Rate q (cm^3 cm^{-2} sec^{-1}) through a Porous Bed and Hydraulic Gradient $G \equiv \Delta\varphi/L > 0$ (See Figure 2.2)[a]

	Flow rate equation[b]	q^2 term is negligible if:
1.	$q + \dfrac{3.5r}{150v(1-\phi)} q^2 = \dfrac{g\phi^3 r^2}{37.5v(1-\phi)^2} G$	$q \ll \dfrac{43v(1-\phi)}{r}$
2.	$q + \dfrac{r}{150v(1-\phi)} q^2 = \dfrac{g\phi^3 r^2}{45v(1-\phi)^2} G$	$q \ll \dfrac{150v(1-\phi)}{r}$
3.	$q + \dfrac{5.36 r^{0.9}}{180 v^{0.9}(1-\phi)^{0.9}} q^{1.9} = \dfrac{g\phi^3 r^2}{45v(1-\phi)^2} G$	$q \ll \dfrac{50v(1-\phi)}{r}$
4.	$q + \dfrac{24r}{1100v} q^2 = \dfrac{gr^2}{275v} G$	$q \ll \dfrac{46v}{r}$
5.	$q + \dfrac{0.55 k^{1/2}}{v} q^2 = \dfrac{gk}{v} G$	$q \ll \dfrac{1.82v}{k^{1/2}}$

[a] Other parameters—r: particle radius (cm); v: kinematic viscosity (cm^2 sec^{-1}); g: gravitational acceleration (cm sec^{-2}); and k: permeability (cm^2).
[b] From Bear (1972).

where $G = \Delta\varphi/L$ is the hydraulic gradient, and a_1 and a_2 are parameters dependent on sediment porosity, water viscosity, and sediment particle size. If the term a_1q^2 can be neglected then the nonlinear equation reduces to the ordinary Darcy-law equation. The higher-power term can be ignored if the following inequalities hold

or

$$a_1q^n \ll q$$

$$q \ll a_1^{1/(1-n)}$$

The latter relationship is given for each flow equation in Table 2.2. For water in sediments ($\nu \simeq 10^{-2}$ cm^2 sec^{-1}, $\phi \simeq 0.5$, and $r \simeq 10^{-2}$–10^{-1} cm) these values correspond to flow rates q of the order of 10^0–10^1 cm sec^{-1}. Thus the nonlinear flow rate equations may apply to shallow near-surface ground waters, whereas the Darcy-law equations may be expected to be obeyed by the deeper and slower flowing ground waters.

2.3   DIFFUSION

Thermally-induced random motions of molecules of one material within another and random motions of small parcels of the medium cause dispersal and migration. Dispersal of molecules, atoms, or ions by forces of intermolecular nature in a gas, fluid, or solid is molecular diffusion. In a medium that is turbulent rather than still, migration of foreign material is referred to as dispersal, or turbulent or eddy diffusion. A diffusional flux can be considered as a product of the concentration gradient and a proportionality factor called the diffusion coefficient. The flux in one dimension is

$$F = -\text{diffusion coefficient} \times \frac{dC}{dz} \qquad (2.11)$$

where the dimensions of the diffusion coefficient are L^2T^{-1} and those of the concentration gradient are ML^{-4}.

As the meaning of the word conveys, diffusion spreads material through a medium. The driving force for the molecular diffusion is the gradient of the chemical potential of the diffusing species (see Section 3.1). The driving force for the eddy diffusion is the formation of eddies and microspic velocity fluctuations within the medium, caused by dissipation of kinetic energy. Molecular diffusion tends to spread the material and decrease the chemical potential gradient, whereas eddy diffusion operates on concentration, decreasing the concentration gradients as an end result.

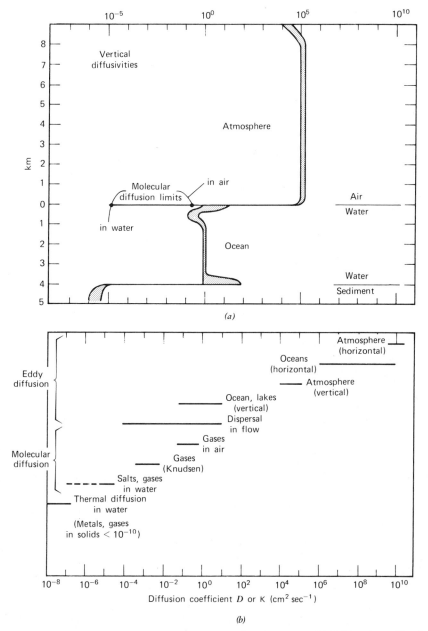

Figure 2.7 Molecular and eddy diffusion in natural environments. (*a*) Schematic profile of the coefficients of vertical eddy diffusion κ_z through the atmosphere, ocean water, and of molecular D in sediment pore water. (*b*) Range of diffusion coefficients in various media.

Relationships of the type given in equation 2.11 apply to dispersal processes of very different magnitudes, from molecular diffusion in crystals to turbulent diffusion in the atmosphere.

The order of magnitude values of the diffusion coefficients characteristic of different environments near the earth's surface—atmosphere, oceans and lakes, pore waters of sediments, and solids—are summarized graphically in Figure 2.7. The great differences in the magnitudes of the diffusion coefficients and, consequently, in the fluxes of materials between different environments may be compared with similar differences in the advective velocities shown in Figure 2.1. The anisotropic nature of water and atmosphere is reflected in the different orders of magnitude of the vertical and horizontal eddy diffusion coefficients: the horizontal diffusivities are much higher. There is a superficial analogy between this relationship and the observations on stream flow (in pipes, channels, and aquifers), which establish that dispersal in the flow direction is stronger than in the direction transversal to the flow.

2.4   RELATIVE EFFECTIVENESS OF DIFFUSION AND ADVECTION

When dissolved material diffuses in flowing water, the flux across a plane perpendicular to the flow direction is a sum of the diffusional F_d and advective F_a fluxes. From equations 2.3 and 2.11, the total flux F is

$$F = F_d + F_a$$

$$= -D\frac{dC}{dz} + UC \qquad \left(\text{g cm}^{-2}\,\text{sec}^{-1}\right) \qquad (2.12)$$

where D $(\text{cm}^2\,\text{sec}^{-1})$ is the diffusion coefficient. A diagram of the diffusional and advective fluxes is shown in Figure 2.8: the concentration difference between the plane at z (concentration C) and the plane at $z + \Delta z$ (concentration $C + \Delta C$) establishes the concentration gradient $\Delta C/\Delta z$. The flow of velocity U is parallel to the z coordinate.

If there is no transport of dissolved material across the plane at any location z, then the flux is zero and equation 2.12 with $F = 0$ becomes

$$\frac{1}{C}\frac{dC}{dz} = \frac{U}{D} \qquad (2.13)$$

Equation 2.13 shows that at zero flux, the concentration gradient dC/dz and the flow velocity U are of the same algebraic sign (concentration C

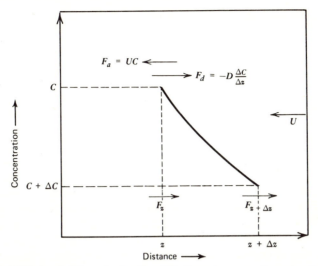

Figure 2.8 Diffusional (F_d) and advective (F_a) fluxes in a one-dimensional system with molecular diffusion and flow.

and diffusion coefficient D are positive quantities, by definition). When the direction of the diffusional flux F_d is from left to right in the coordinate scheme of Figure 2.8, the flux is positive but the concentration gradient is negative: $dC/dz < 0$ gives $F > 0$, as given by the conventional definition of the diffusional flux in equation 2.11. Therefore U must be negative if the concentration gradient is $dC/dz < 0$, which means that the flow direction is from right to left in the coordinate scheme of Figure 2.8. Effectively, the diffusional flux in one direction is balanced by the advective flux in the opposite direction.

The quotient D/U (cm) is referred to in the literature as the scale distance of a system characterized by particular values of D and U.

The quotient $D/\Delta z$ has the dimensions of velocity (cm sec^{-1}) and it may be thought of as the velocity of diffusional transport of material through a layer of thickness Δz.

Flow velocity U transports material in time t a distance L_a,

$$L_a = Ut \quad \text{(cm)} \tag{2.14}$$

A characteristic distance L_d for dispersal of material by diffusion is defined, on dimensional grounds, as

$$L_d = (Dt)^{1/2} \quad \text{(cm)} \tag{2.15}$$

The quotient of the two distances is

$$\frac{L_a}{L_d} = U\left(\frac{t}{D}\right)^{1/2} \tag{2.16}$$

which shows that for constant flow velocity U and diffusion coefficient D, the distance transported by advection can become greater than the distance transported by diffusion after sufficiently long time. This is equivalent to stating that, given enough time, advective transport becomes more important than transport by diffusion. A modifier to this statement is needed: by definition of distance L_a, there is no transported material beyond L_a at time t; in diffusion, however, some material is also present beyond distance L_d at time t (in diffusion from a constant source, L_d is a distance from the source, where concentration is $1/e$ of its value at the source boundary). The comparison of the distances L_a to L_d is a measure of the relative effectiveness of the two processes operating independently.

Examples. The following three examples demonstrate the relative effectiveness of advective and diffusional transport. The time needed for the two characteristic distances to become equal ($L_a/L_d = 1$) is, from equation 2.16, $t = D/U^2$.

For a molecular diffusion coefficient $D = 1 \times 10^{-5}$ cm^2 sec^{-1} and water flow velocity $U = 1$ m yr$^{-1} = 3 \times 10^{-6}$ cm sec^{-1}, the time is $t = 1 \times 10^{-5}/(3 \times 10^{-6})^2 = 10^6$ sec $= 14$ days. The length of time $t = 14$ days may be interpreted as indicating that within 14 days the water flow would transport dissolved material about as far as it can be transported by molecular diffusion alone.

For a molecular diffusion coefficient ($D = 1 \times 10^{-5}$ cm^2 sec^{-1}) and water flow velocity comparable to the rates of sedimentation in the deep ocean ($U = 1$ cm/1000 yr $= 3 \times 10^{-11}$ cm sec^{-1}), the time is $t = 300 \times 10^6$ yr. This indicates that molecular diffusion is an important process in comparison with pore water flow on a time scale of sediment life on the deep ocean floor, about 100×10^6 yr.

For an eddy diffusion coefficient $\kappa = 10^5$ cm^2 sec^{-1} and water flow $U = 10$ cm sec^{-1} representative of surface ocean waters, the time is $t = 10^3$ sec $\simeq 20$ min.

Other combinations of diffusion coefficients and advective velocities can be taken from the order of magnitude values plotted in Figures 2.1 and 2.7.

2.5   DISPERSAL IN FLOW

A dispersal process of magnitude intermediate between molecular and eddy diffusion is associated with water flow through porous beds. Friction of water at the walls of pore channels leads to development of velocity gradients and results in a relatively faster flow near the channel axis. Mixing in the directions parallel and perpendicular to the flow results from small fluctuations in velocity in the presence of velocity gradients. Original studies of dispersal of dissolved materials in a flow through a straight capillary were done by Taylor (1953, 1954). Later investigators have extended the studies of dispersal to pore channel systems of more complicated geometries, where the effects of wall friction and departures of the flow from a straight line jointly contribute to mixing. Extensive coverage of the subject on a monograph level has been given by Scheidegger (1960) and Bear (1972).

2.5.1 Mixing at a Moving Front

A sharp boundary between pore water containing a dissolved chemical species and pore water free of the species cannot exist very long because of molecular diffusion of the solute and diffusion of water in the opposite direction. Thus some blending and smearing of the initially sharp boundary between the solute-containing and solute-free regions will always take place. If there is flow of pore water in the direction perpendicular to the boundary plane, the plane will move with the flow, and molecular diffusion (or a faster dispersal process) will also contribute to the blending of the boundary. The concept of "piston flow" mentioned in Section 1.3.2 describes the flow of solute-containing water that pushes without mixing the solute-free water ahead of it. Such a flow is only an approximation to a real system, because molecular diffusion takes place independently of whether water flows or stands still, and will cause blending of the interface between the two solutions. The faster the flow, the more efficient is the mixing between the solute-free and solute-containing parts of the pore water system. A fast blending of the boundary means faster dispersal and transport of dissolved material forward, relative to the original boundary. Schematically, the blending of an originally sharp boundary between two pore water regions can be treated as a diffusional process, characterized by certain values of a dispersal coefficient (analogous to the molecular diffusion coefficient D).

With reference to Figure 2.9, dispersal of dissolved material is a diffusion-like process, characterized by a dispersal or diffusion coefficient D (cm^2 sec^{-1}), and by advection of water taking place with a velocity U (cm sec^{-1}) in the x direction. This is a case of diffusion and advection that can be described by the following differential equation, based on Fick's second law of diffusion and derived in Section 3.1:

$$\frac{\partial C}{\partial t} = D \frac{\partial^2 C}{\partial x^2} - U \frac{\partial C}{\partial x} \qquad (g\ cm^{-3}\ sec^{-1}) \qquad (2.17)$$

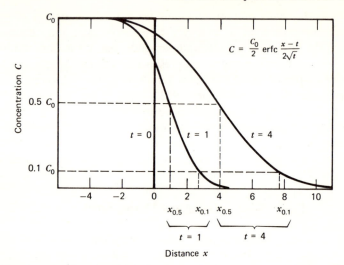

Figure 2.9 Concentration of a dissolved substance in the presence of diffusion and flow in a one-dimensional system. Concentration against distance profiles shown for three consecutive values of time t. Dispersal of concentration shown on the horizontal axis for times $t = 1$ and $t = 4$.

where C is concentration (g cm^{-3}), the diffusion coefficient D and flow velocity U are taken as constant, and x is the distance coordinate. Implicit in equation 2.17 is that the chemical species is conservative: no chemical reaction terms are included in the equation. The initial conditions needed for the solution of equation 2.17 for an infinitely long column, such as shown in Figure 2.9, are

$$\text{at } t = 0: \quad C = C_0 \quad \text{at} \quad x < 0 \tag{2.18}$$

$$C = 0 \quad \text{at} \quad x > 0 \tag{2.19}$$

The solution is (e.g., Dankwerts, 1953; Bear, 1972)

$$C = \frac{C_0}{2} \operatorname{erfc} \frac{x - Ut}{2(Dt)^{1/2}} \tag{2.20}$$

where function erfc denotes the error function complement, which is defined and tabulated in Appendix A. Distance x is measured relative to a fixed point $x = 0$, positive x increasing to the right and negative to the left. The solution given in relationship 2.20 is valid both for positive and negative values of x. Three concentration-against-distance profiles computed for three different times t (Figure 2.9) give an idea of how the concentration profiles change in a system where flow and dispersal take place simultaneously.

At $x=0$, concentration C increases from a mean value of $C=C_0/2$ to $C=C_0$ after a long time.

Owing to the simultaneous operation of the flow and diffusional processes, the initially steplike concentration profile becomes progressively smoother with time, and the distance along the coordinate x between any two concentrations increases with time. This means that points of different concentration move relative to $x=0$ with different velocities, as can be demonstrated in the following examples.

The position $x_{0.5}$ of a point where concentration is one-half of the initial concentration can be obtained from equation 2.20, using $x=x_{0.5}$ and $C=C_0/2$,

$$\operatorname{erfc} \frac{x_{0.5}-Ut}{2(Dt)^{1/2}} = 1$$

From Table A.1 in Appendix A, one finds that erfc $0=1$. Therefore

$$\frac{x_{0.5}-Ut}{2(Dt)^{1/2}} = 0$$

$$x_{0.5}=Ut$$

and   (2.21)

$$\frac{dx_{0.5}}{dt} = U$$

The latter relationship shows that the point $x_{0.5}$, at which concentration is always $C_0/2$, propagates in the x direction with velocity U.

A point $x_{0.1}$ where concentration is one-tenth of the initial concentration $(C=0.1C_0)$ can be determined from

$$\operatorname{erfc} \frac{x_{0.1}-Ut}{2(Dt)^{1/2}} = 0.2$$

From Table A.1 in Appendix A, we have erfc $0.9=0.2$. Therefore

$$x_{0.1}=0.4(Dt)^{1/2}+Ut$$

and   (2.22)

$$\frac{dx_{0.1}}{dt} = 0.2\left(\frac{D}{t}\right)^{1/2}+U$$

Equations 2.21 and 2.22 state that the lower the concentration, the faster it moves away from the reference point $x=0$. In the particular case of the two concentrations $C=0.5C_0$ and $C=0.1C_0$, the velocities of propagation and the positions on the x coordinate of the two points are related as

$$\frac{dx_{0.1}}{dt} > \frac{dx_{0.5}}{dt} \quad (\text{cm sec}^{-1})$$

and

$$x_{0.1} - x_{0.5} = 0.4(Dt)^{1/2} \quad (\text{cm})$$

The latter relationship shows that isoconcentration points disperse and the distance between them increases with time in proportion to $t^{1/2}$.

2.5.2 Dispersal Coefficients

Flow through porous spaces is characterized by dispersal stronger than dispersal due to molecular diffusion in still water, and some experimentally determined values of the dispersal coefficients D_f are shown in Figure 2.10 for different flow velocities and particle sizes. The coefficient D_f denotes dispersal in the flow direction and is called the longitudinal dispersal coefficient, to distinguish it from dispersal in the direction perpendicular or

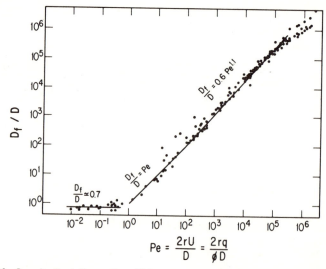

Figure 2.10 Longitudinal dispersal coefficient D_f in flow, as a function of the Peclet number Pe. D_f/D is the ratio of dispersal to molecular diffusion coefficient (from Bear, 1972). Reproduced by permission of American Elsevier Publishing Company, New York.

transverse to the flow. The vertical axis in Figure 2.10 shows the ratio D_f/D, which is the ratio of the dispersal coefficient to the molecular diffusion coefficient of the chemical species dispersed in the flow. The horizontal axis gives the Peclet number Pe, a dimensionless quotient of the flow velocity U, particle diameter $2r$, and molecular diffusion coefficient D,

$$\text{Pe} = \frac{2rU}{D} = \frac{2rq}{\phi D} \qquad (2.23)$$

where the relationship $U = q/\phi$ between the flow rate, water discharge, and porosity is given in equation 2.7.

The Peclet number can be interpreted as flow velocity U (cm sec^{-1}) or discharge q (cm^3 cm^{-2} sec^{-1}) for a given chemical species in a given porous bed; the molecular diffusion coefficient D characterizes the chemical species in water, and the particle radius r and porosity of the bed ϕ characterize the sediment. Some limiting values of Pe for water in sediments can be derived for the following conditions. On the low end, a flow velocity of the order of magnitude of meters per year is $U \simeq 10^{-5}$ cm sec^{-1}; for sand-size particles, the particle radius $r \simeq 10^{-1}$ cm; molecular diffusion coefficient $D \simeq 10^{-5}$ cm^2 sec^{-1}. From these values, the Peclet number is Pe $\simeq 10^{-1}$. For finer-grained sediments and for slower water flow, the values of Pe are lower, whereas for faster flow rates in coarse-grained sediments they are correspondingly higher. For fine-grained sediments ($r < 10^{-2}$ cm) and slowly moving ground or sediment-pore waters, the Peclet numbers are small.

At low Peclet numbers, such as Pe < 1, dispersal is ineffective in comparison to molecular diffusion: the ratio D_f/D is less than 1. The value of the ratio D_f/D falls below unity because the effective diffusion coefficient in a porous medium is smaller than the molecular diffusion coefficient D in a bulk solution (Chapter 3). In the region of small Pe, the effective diffusion coefficient is about $D_f \simeq 0.7D$.

As Pe increases from Pe $\simeq 1$ to Pe $\simeq 10^2$, the dispersal coefficient D_f increases approximately linearly with Pe. In the region of higher Pe values, D_f increases approximately in proportion to Pe$^{1.1}$, slightly faster than in the linear region.

2.6   EDDY DIFFUSION IN WATER

2.6.1   A Short Summary

Turbulence, in the atmosphere and water, is the result of dissipation of kinetic energy leading to random motions of parcels of air or water of varying size. The notion that turbulence is a spectrum of eddies of various sizes exchanging water, and dissolved and suspended materials is behind the concept of turbulent or eddy diffusion, analogous to molecular diffusion. Dispersal of dissolved and suspended materials by turbulence can be treated mathematically in a manner similar to Fick's second law of diffusion, equation 3.7, allowing for anisotropy of eddy diffusion. In the three-dimensional case, when the eddy diffusion coefficients, denoted K_x, K_y, and K_z (cm^2 sec^{-1}), have different values, the relationship describing the rate of change of some conservative component $\partial C/\partial t$ at a point of coordinates (x, y, z) is

$$\frac{\partial C}{\partial t} = \frac{\partial}{\partial x}\left(K_x \frac{\partial C}{\partial x}\right) + \frac{\partial}{\partial y}\left(K_y \frac{\partial C}{\partial y}\right) + \frac{\partial}{\partial z}\left(K_z \frac{\partial C}{\partial z}\right) \qquad (2.24)$$

Eddy diffusion coefficients in water and air are usually determined through introduction of dyes, floats, and by use of man-made and natural radioactive tracers. The dispersal of the dyes or tracers from the source is followed over a period of time. Summaries of how such experiments are carried out and how the observational results are treated, including references to the very extensive literature in this field, may be found in Okubo (1971) and Csanady (1973).

A cloud of dissolved material of some initial size in a still medium would gradually increase in size owing to molecular diffusion. As the mass of material contained within the original homogeneous cloud spreads over a greater area or volume, concentration of material within the initial volume diminishes with time. In the presence of turbulent eddies, those eddies which are either much smaller or much bigger than the cloud dimensions are not effective in breaking it up, whereas only those eddies whose dimensions are comparable to the cloud dimensions contribute effectively to its spreading. As the process of spreading continues, progressively bigger eddies contribute to the breaking up of the cloud, such that the area of the spread increases as some function of time.

The unequal rates of spreading in the horizontal and vertical dimensions in the atmosphere, oceans, and lakes are represented by the different values of the eddy diffusion coefficients for horizontal (K_x and K_y) and vertical (K_z) eddy diffusion. The magnitudes of the horizontal and vertical eddy diffusion coefficients, given in Figure 2.7, show that the coefficients of vertical eddy diffusion K_z are several orders of magnitude smaller than the horizontal eddy diffusion coefficients. A general reason for this is the mechanical stability of a water column with respect to vertical mixing by eddies; the presence of density gradients, when density increases downward because of lower temperatures or higher salt concentrations or both, stabilizes the water column. The effect of density gradients on reducing the vertical eddy diffusivity is particularly pronounced at the thermoclines of lakes and oceans (see Figures 2.7 and 2.12).

2.6.2   Horizontal Eddy Diffusion

A measure of horizontal spreading by eddy diffusion is the size of an area covered by material at different times. Material originally confined to a small area covers progressively larger areas with time. If the rates of eddy diffusional spreading in the two horizontal dimensions x and y are not equal, the material will spread as a more or less elliptical patch. The linear dimensions of spread can be taken as the standard deviations σ_x and σ_y of the concentration of dispersed material. Then a measure of an equivalent circular area is

$$\sigma^2 = 2\sigma_x \sigma_y \quad (\text{cm}^2) \tag{2.25}$$

Okubo (1971) has summarized the results of dye dispersal studies in surface waters of some rivers, the North Sea, and the Atlantic and Pacific Oceans obtaining a relationship between the size of dispersed cloud σ^2 and time t, shown in Figure 2.11a. The straight line drawn through the

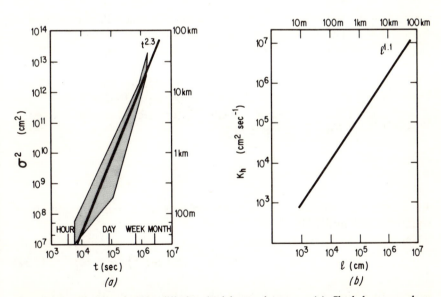

Figure 2.11 Horizontal eddy diffusion in lakes and oceans. (a) Shaded area encloses observed values of dispersal at different times. (b) Coefficient of horizontal eddy diffusion as a function of distance of dispersal (from Pritchard et al., 1971; Okubo, 1971).

measured points is

$$\sigma^2 = 0.015t^{2.3} \quad (cm^2) \tag{2.26}$$

where t is in seconds. The equation shows that the area of the spread increases in proportion to the power 2.3 of time and this indicates that the rate of spreading becomes faster after longer times. The relationships between the area σ^2, mean horizontal eddy diffusion coefficient K_h, and diffusion scale ℓ are

$$K_h = \frac{\sigma^2}{4t} \quad (cm^2\ sec^{-1}) \tag{2.27}$$

$$\ell = 3\sigma \quad (cm) \tag{2.28}$$

Substitution of equation 2.26 in equations 2.27 and 2.28 gives

$$K_h = 0.004t^{1.3} \quad (cm^2\ sec^{-1}) \tag{2.29}$$

$$K_h = 0.45\ell^{1.1} \quad (cm^2\ sec^{-1}) \tag{2.30}$$

where t is the seconds and ℓ in centimeters. Equation 2.30 is plotted in Figure 2.11b. The latter two relationships show that the eddy diffusion coefficient increases with time as well as with distance, and that larger areas are characterized by greater values of horizontal eddy diffusivity. Csanady (1973, p. 102) pointed out that large-scale horizontal diffusion in Lake Ontario also follows the relationships shown in Figure 2.11.

A mean horizontal dispersal velocity U_h (cm sec^{-1}) can be defined as $U_h = K_h / \ell$ and, from equation 2.30,

$$U_h = \frac{K_h}{\ell} = 0.45\ell^{0.1} \quad (cm\ sec^{-1}) \tag{2.31}$$

An alternative definition of dispersal velocity is $U_h = d\ell/dt$ and, from equations 2.26 and 2.28,

$$U_h = \frac{d\ell}{dt} = 0.42t^{0.15} \quad (cm\ sec^{-1}) \tag{2.32}$$

Both definitions of U_h show that horizontal dispersal velocity increases with diffusion scale and time, although the increase is not strong: a hundredfold increase in ℓ or t produces a twofold increase in U_h. For example, after 1 day ($t \simeq 10^5$ sec), the mean dispersal velocity is $U_h \simeq 2.4$ cm sec^{-1}; after 1 year ($t = 3 \times 10^7$ sec), the velocity is $U_h \simeq 5.6$ cm sec^{-1}. The increase in diffusion scale or, in other words, in the mean dispersal velocity with time is not restricted to eddy diffusion. A similar phenomenon also occurs in transport of dissolved materials by flow (see Section 2.5.1).

2.6.3 Vertical Eddy Diffusion

Turbulence generated by wind in surface waters is dissipated by work against buoyancy. Intuitively it is clear that turbulence near the surface of a water column can neither penetrate nor mix the water column if the column is mechanically stable. One of the criteria of vertical stability is a quantity known as the buoyancy frequency or the Brunt-Väisälä stability frequency N, defined as

$$N^2 = \frac{g}{\rho}\frac{\partial\rho}{\partial z} \qquad (\text{sec}^{-2}) \qquad\qquad (2.33)$$

where g is the gravitational acceleration (cm sec^{-2}), ρ is the density of water (g cm^{-3}), and z is the vertical distance coordinate (cm), defined as positive and increasing downward. In the notation of equation 2.33, density increasing with depth gives a positive density gradient $\partial\rho/\partial z > 0$. The terminology of stability frequency is borrowed from a mechanical analog of an oscillating parcel of water that has been displaced from a position of a hydrostatic equilibrium in a density gradient. The buoyant force would tend to push the displaced parcel from the denser surroundings upward, the parcel would go beyond its equilibrium position, come to a halt in the less dense surroundings, and sink downward. The parameter N has the dimensions of frequency (sec^{-1}) and is derivable from the solutions of motion equations (Eckart, 1960; Turner, 1973). The higher the value of N^2 or the stronger the density gradient $\partial\ln\rho/\partial z$ cm^{-1}, the more stable is the water column against vertical mixing. Representative values of N^2 at different depths in an oceanic water column and in a lake are shown in Figure 2.12.

In ocean water, the density is a function of both temperature and salinity, and the same can be said of natural brines in lakes and in the subsurface. In freshwater lakes salt concentrations are low, of the order of $10^{-2} - 10^{-1}$ g l^{-1}, in comparison to 35 g l^{-1} in ocean water. Water density in freshwater lakes is controlled by temperature, such that the density gradient $\partial\rho/\partial z$ in equation 2.33 can be replaced by the temperature

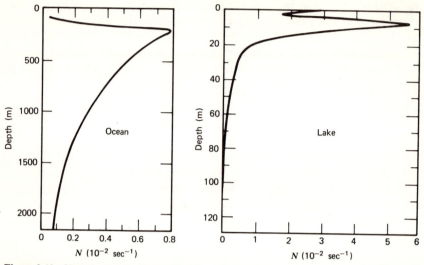

Figure 2.12 Variation of the Brunt-Väisälä stability frequency N (sec^{-1}) with depth in the ocean and in a lake during the period of seasonal stratification (data from Eckart, 1960; Zimmermann, 1961).

gradient $\partial T/\partial z$ and the coefficient of thermal expansion of water α,

$$\alpha = -\frac{1}{\rho}\frac{\partial \rho}{\partial T} \qquad (\text{deg}^{-1}) \qquad (2.34)$$

$$N^2 = -g\alpha\frac{\partial T}{\partial z} \qquad (\text{sec}^{-2}) \qquad (2.35)$$

A negative temperature gradient $\partial T/\partial z < 0$ corresponds to a decrease in temperature with depth. This gives a positive value of the stability frequency N^2, as in the general case of density increasing with depth.

In the presence of horizontal currents, the stability of a water column also depends on the vertical gradient of a horizontal velocity $\partial U_h/\partial z$ sec^{-1} within the water layer, in addition to its dependence on vertical density gradients. The velocity gradient has the dimensions of frequency, and $(\partial U_h/\partial z)^2$ and N^2 can be combined in a dimensionless quotient known as the Richardson gradient number Ri,

$$\text{Ri} = \frac{N^2}{(\partial U_h/\partial z)^2} \qquad (2.36)$$

As the vertical gradients of horizontal velocities are related to dissipation of turbulent energy generated in a water layer by wind stress at the surface, attempts have been made to correlate the Richardson number Ri or other quotients based on it with wind velocities and vertical eddy diffusion coefficients κ_z computed from observational data (Kullenberg, 1971). The difficulty in a general extension of this approach is that while data on vertical distributions of temperature and salinity are easily available (from which N^2 can be computed), data on horizontal velocities and their vertical gradients are considerably more difficult to obtain.

For some water environments, the vertical eddy diffusion coefficients κ_z and stability frequencies N^2, plotted against each other in Figure 2.13, show certain interrelationships. First it may be noted that, in general, the vertical eddy coefficient κ_z is inversely related to the stability frequency N^2. This relationship is to be expected, as the more stable the water column is, the less susceptible it is to turbulent mixing. Second, the vertical eddy diffusion coefficients κ_z in the thermocline region of lakes and

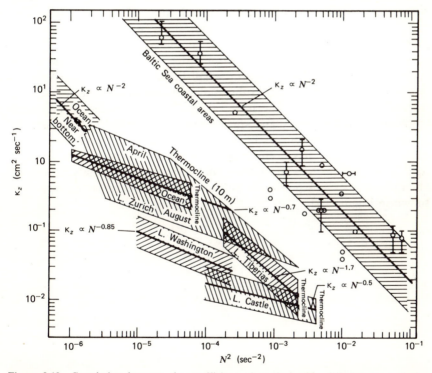

Figure 2.13 Correlation between the coefficient of vertical eddy diffusion κ_z and the Brunt-Väisälä stability frequency N (data from Eckart, 1960; Jassby and Powell, 1975; Kullenberg, 1971; Lerman, 1971; Li, 1973; Munk, 1966; Quay, 1976; Rooth and Oestlund, 1972; Sarmiento et al., 1976).

oceans are lower than in the water layer below the thermocline; conversely, the stability frequencies are higher at the thermocline.

The fact that the κ_z and N^2 values for different lakes and seas fall in different regions of the diagram probably indicates that the vertical eddy diffusivity depends on other factors, in addition to the stability frequency. However, the data for different environments shown in Figure 2.13 have been obtained by different methods over different lengths of time, so that comparisons between them are difficult. The Baltic Sea data are based on dye dispersal experiments; the data for lakes are based on vertical temperature profiles averaged over periods ranging from days to months; the oceanic box represents the values of κ_z reported in the literature for the thermocline and deeper regions (values based on temperature, salinity, and some radioactive tracer profiles), and on a generalized profile of N^2 shown in Figure 2.12. Within short distances of the deep ocean floor, up to about 300 m above the bottom, the vertical eddy diffusion is faster than it is higher up in the water column. In this near-bottom layer, in the deep Atlantic and Pacific, values of the vertical eddy diffusion coefficient κ_z have been reported between 5 and over 100 cm^2 sec^{-1} (Sarmiento et al., 1976). The diffusion coefficients were computed from the vertical concentration profiles of radon (^{222}Rn) in water, diffusing out of the ocean floor sediments and decaying with a half-life of about 4 days. (Applications of the ^{222}Rn technique to the studies of eddy diffusion in lakes are discussed in Chapter 7.) The high values of κ_z are related to the small gradients of water density and small values of the stability frequency N. An empirical correlation of the vertical eddy diffusion coefficients and stability frequencies has been reported in the form of $\kappa_z = 4.6 \times 10^{-6} N^{-2}$ (cm^2 sec^{-1}), where N is in units of cm^{-1}, with an uncertainty margin of about a factor of 3. A portion of this correlation for deep bottom waters is shown in Figure 2.13.

CHAPTER 3

Fluxes and Transport: Molecular Diffusion

The contrast between molecular and eddy diffusion lies in the scale of the phenomena: the distances over which an atomic species can jump within a unit of time in a gas at atmospheric pressure, in water, and in a solid are small in comparison to the physical dimensions of turbulent eddies. The gradients of the chemical potential drive the molecular diffusion, and the thermal kinetic energy of a diffusing molecule is of the order of kT. Molecular diffusion is relatively fast in a gas, but it is slower in water, and very much slower in solids. In a given medium, the higher the temperature, the faster the diffusion. On the geological time scales, molecular diffusion becomes an important mechanism of transport in interstitial solutions and gases in sediments, and in certain reactions between minerals and water. At the temperatures of the earth's surface, diffusion in solids is too slow to be a significant transport mechanism. This chapter discusses molecular diffusion of dissolved ionic and gaseous species in water, diffusion of gases in still air, and diffusion of various chemical species in mineral lattices, zeolites, and glasses. A complement of chemical diffusion is thermal diffusion, discussed with reference to aqueous solutions and temperature gradients in sediments. For these diffusional processes, this chapter gives some basic theory, data, and numerical examples.

3.1 BASIC EQUATIONS

3.1.1 Fick's Experiments

Studies of salt migration across animal tissue membranes separating solutions of different concentrations led Adolf Fick in 1855 to formulation of an empirical relationship between the flux of a dissolved species F, area S

across which the flux takes place, concentration gradient $\Delta C/\Delta z$, and a coefficient of proportionality D, as

$$F = SD\frac{\Delta C}{\Delta z} \qquad (\text{g sec}^{-1}) \qquad\qquad (3.1)$$

where the parameters and their dimensions are as defined in equations 2.11 and 2.12. Relationship 3.1 in its present form or in a form referred to a unit of area,

$$F = -D\frac{dC}{dz} \qquad (\text{g cm}^{-2}\,\text{sec}^{-1}) \qquad\qquad (3.2)$$

is known as Fick's first law of diffusion. Fick (1855) postulated the diffusional flux equation on the basis of an analogy with the equation for the conductance of heat, developed earlier by Fourier, and the conductance of electric current, developed earlier by Ohm.

With reference to the diagram of fluxes in Figure 2.8, Fick's first law of diffusion stipulates that the flux is directly related to the magnitude of the gradient, to cross-sectional area, and to a coefficient D that depends on the nature of the diffusing material and the medium. If the diffusing species is neither produced nor destroyed, then conservation of mass requires that any change in the flux within the layer between z and $z+\Delta z$ must be accounted for by a change in concentration within the layer, that is

$$\frac{\Delta C}{\Delta t} = -\frac{\Delta F}{\Delta z} \qquad (\text{g cm}^{-3}\,\text{sec}^{-1}) \qquad\qquad (3.3)$$

where t is time. Equation 3.3 states that the rate of change in concentration within the layer of thickness Δz is equal to the change in the flux between the points z and $z+\Delta z$. In the differential notation

$$\left(\frac{\partial C}{\partial t}\right)_z = -\left(\frac{\partial F}{\partial z}\right)_t \qquad\qquad (3.4)$$

where the partial derivatives are used because concentration C and flux F are both functions of distance z and time t. (The subscripts z and t, denoting the value of the partial derivative taken at a fixed location z or at a fixed time t, will be dropped from subsequent equations.)

If concentration C remains constant with time then the system is at a steady state, which corresponds to the condition

$$\frac{\partial C}{\partial t} = 0 \qquad\qquad (3.5)$$

At the steady state, the flux F is a constant,

$$F = -D\frac{dC}{dz} = \text{const} \tag{3.6}$$

which means that a concentration gradient dC/dz must be maintained for the flux to be nonzero. Substitution of $F = -D\,dC/dz$ in equation 3.4 gives

$$\frac{\partial C}{\partial t} = \frac{\partial}{\partial z}\left(D\frac{\partial C}{\partial z}\right) \qquad (\text{g cm}^{-3}\ \text{sec}^{-1}) \tag{3.7}$$

Equation 3.7 is known as Fick's second law of diffusion and relates the rate of change in concentration at a fixed point to the second derivative of concentration with respect to distance at that point.

To test equations 3.2 and 3.7, Fick (1855) set up experiments on diffusion of salt in water, in systems of a constant cross section (cylinder) and a variable cross section (inverted conical funnel). In each experiment concentrated salt solution was maintained by an excess of salt crystals at the bottom of the apparatus—at the base of the cylinder or near the apex of the inverted cone. Dissolved salt diffused upward, and a zero concentration near the upper boundary of the apparatus was achieved by having the entire system immersed in a large container of replenishable pure water. Salt concentration in the vertical water column was determined as solution density, and the difference between the densities of the solution and pure water (1.0 g cm^{-3}) was used as a measure of concentration of the diffusing salt.

Assuming the diffusion coefficient D to be independent of concentration, concentration C at the steady state along the axial direction of the cylinder is given by, from equations 3.2 and 3.7,

$$\frac{d^2C}{dz^2} = 0 \tag{3.8}$$

The boundary conditions for diffusion along the axial direction of a cylinder are

at $z = 0$ (cylinder bottom): $C = C_s$

at $z = h$ (cylinder top): $C = 0$

With these boundary conditions, the solution of equation 3.8 is

$$C = C_s\left(1 - \frac{z}{h}\right) \qquad (\text{g cm}^{-3}) \tag{3.9}$$

For the case of salt diffusion in a cone, the cross-section area of the cone is $S = \text{const} \times z^2$, where z is the vertical dimension of the straight cone. The differential equation describing the concentration change along the vertical distance at the steady state is

$$\frac{d}{dz}\left(SD\frac{dC}{dz}\right) = \frac{d}{dz}\left(z^2\frac{dC}{dz}\right) = 0 \qquad (3.10)$$

The boundary conditions are

$$\text{at } z = z_1 \text{ (near cone apex):} \qquad C = C_s \qquad (3.11)$$

$$\text{at } z = h \text{ (at the cone base):} \qquad C = 0 \qquad (3.12)$$

With these boundary conditions, the solution of equation 3.10 is

$$C = C_s\frac{1 - h/z}{1 - h/z_1} \qquad (\text{g cm}^{-3}) \qquad (3.13)$$

The steady-state concentration in a cylinder varies nearly linearly with distance from the bottom to the top, whereas in a cone the concentration decreases nonlinearly along the same direction. The results of Fick's (1955) experiments are shown in Figure 3.1, and they were considered by their author as indicating that the diffusion law is valid.

In the experiment on the diffusion of dissolved salt in the cylinder, the concentration gradient in the lower part of the water column is steeper, as shown in the figure. This suggests faster diffusion, and it was noted by Fick (1855) as indicating that a steady state had not yet been attained fully. In fact, Fick's results were probably better than he suspected: the diffusion coefficient of NaCl in water increases somewhat with an increasing concentration (Table 3.3), so that a curvilinear profile, with the curvature of the magnitude and direction as recorded by Fick, is to be expected at a steady state.

3.1.2 Chemical Reactions, Advection, and Diffusion

Fick's second law of diffusion can be written as follows for the case of diffusion and advection in a one-dimensional system, from equations 2.12 and 3.4:

$$\frac{\partial C}{\partial t} = \frac{\partial}{\partial z}\left(D\frac{\partial C}{\partial z}\right) - \frac{\partial}{\partial z}(UC) \qquad (3.14)$$

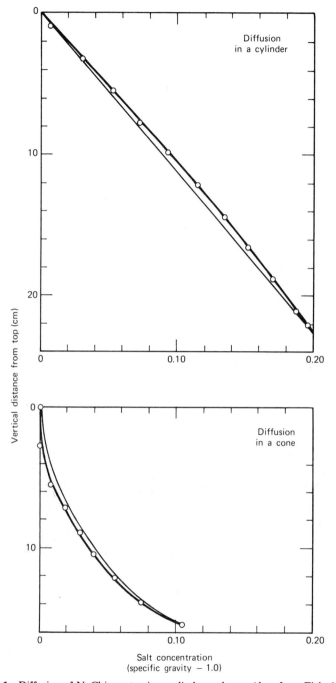

Figure 3.1 Diffusion of NaCl in water, in a cylinder and cone (data from Fick, 1855).

A chemical reaction taking place within a layer element Δz, shown in Figure 2.8, may produce or consume a chemical species, thereby generating a change in concentration $\Delta C/\Delta t$ independently of diffusive or advective fluxes. Chemical reaction rates $\pm \mathcal{R}$ (g cm^{-3} sec^{-1}) can be added as terms to the right-hand side of equation 3.14, giving a more general form of Fick's second law of diffusion,

$$\frac{\partial C}{\partial t} = \frac{\partial}{\partial z}\left(D\frac{\partial C}{\partial z}\right) - \frac{\partial}{\partial z}(UC) \pm \mathcal{R} \qquad (3.15)$$

where $+\mathcal{R}$ denotes production and $-\mathcal{R}$ removal of the chemical species. In general, the chemical reaction rates $\mathcal{R}$ can be position and time dependent. Solutions of the diffusion-advection-reaction equation 3.15 give concentration C as a function of position z and time t. Explicit solutions exist, depending on the boundary conditions and the nature of the terms D, U, and $\mathcal{R}$. If the diffusion coefficient D and advective velocity are constant and independent of z, then equation 3.15 is a differential equation with constant coefficients, which can be solved explicitly for certain boundary conditions and certain forms of the chemical term $\mathcal{R}$. Two examples of solving equation 3.15, for the case of a steady state $(\partial C/\partial t = 0)$ and the case of a transient state $(\partial C/\partial t \neq 0)$, are given in Appendix B. Other solutions, describing migration and chemical reactions in water and sediments, are given in Chapter 8.

A closer comparison between the diffusional and advective fluxes, $-D\,dC/dz$ and CU, can illustrate the concept of "diffusional velocity" as a driving force for molecular diffusion (Jost, 1960; Katchalsky and Curran, 1965; Robinson and Stokes, 1970). The diffusional flux relationship $-D\,dC/dz$ can be written, after multiplication and division by RTC (R is the gas constant, in units of energy mol^{-1} deg^{-1}, and T is temperature in deg K) as

$$-D\frac{dC}{dz} = -\frac{DC}{RT}\frac{d\mu}{dz} \qquad (3.16)$$

where μ is the chemical potential of the diffusing species defined as

$$\mu = \mu_0 + RT\ln a \qquad \text{(energy mol}^{-1}\text{)} \qquad (3.17)$$

$$\mu = \mu_0 + RT\ln\gamma C \qquad (3.18)$$

$$\mu \simeq \mu_0 + RT\ln C \qquad (3.19)$$

In the above relationships a is the thermodynamic activity, γ is the activity coefficient, C is concentration, and μ_0 is the value of the chemical potential in a standard reference state. μ_0 is independent of concentration C and, consequently, of the position on the distance coordinate z, but it is dependent on temperature and pressure. In the definition of the thermodynamic activity—$a = \gamma C$—the activity coefficient γ tends to 1 when $C \rightarrow 0$, from which follows equation 3.19, valid for a dilute solution. The relationships between the activity coefficients referred to different concentration scales (such as mol l^{-1} and mol per 1000 g H_2O) are given by Robinson and Stokes (1970). In dilute solutions, the difference between the concentration scales is not significant, but in saline brines it can amount to 20–30%. The quotient $(D/RT)d\mu/dz$ has the dimensions of velocity, when the coefficient D is of dimensions $L^2 T^{-1}$. This "diffusional velocity" is driven by the gradient of the chemical potential of the diffusing species. The product of the "diffusional velocity" and concentration C is analogous to the product of C and advective velocity U, both giving the flux.

3.2   DIFFUSION COEFFICIENTS IN SOLUTION

3.2.1   Frame of Reference

In a water column where a concentration gradient of a solute exists, the volume fluxes of solute and water must balance if there is no net flow of the entire solution volume: $(vF)_{\text{water}} = (vF)_{\text{solute}}$, where v is the volume fraction and F is the flux (Katchalsky and Curran, 1965). The diffusion coefficient of the solute D describing the flux of solute relative to a fixed frame of reference (an arbitrarily fixed level in sediment, an interface between water and some other phase, or the bottom of an experimental column) is not identical with the diffusion coefficient D' describing the flux relative to the solvent, as the solvent (water) is also subject to diffusional flow. The two coefficients are interrelated through

$$D = D' v_w \qquad (3.20)$$

where v_w is the volume fraction of water in solution. In dilute solutions v_w is essentially unity, such that there is no difference in the reference frame chosen. In sea water $v_w = 0.98$ and in concentrated saline brines the volume fraction of water is lower, $v_w \simeq 0.75$. Even the latter value, however, is probably not large enough to be considered an important correction in thermodynamic calculations of diffusional fluxes of dissolved solids in natural waters.

3.2.2 Tracer and Self-Diffusion Coefficients

Diffusion of one chemical species at low concentration into another is called tracer diffusion. Self-diffusion designates diffusion of a chemical species into a phase of a different composition or into a phase containing only the same species. For example, NaCl in water dissociates into Na^+ and Cl^-, but both diffuse into pure H_2O along the same concentration gradient, as the condition of electrical neutrality must be maintained. Similarly, diffusion of H_2O molecules in water is referred to as self-diffusion, irrespective of the isotopic composition of the diffusing species. An example of a tracer diffusion may be diffusion of tritium-containing (^3H or T) water molecules HTO into H_2O, or any other isotopically labeled species into a medium of similar elemental composition, or diffusion of a tracer into a medium of different composition. In such cases, tracer diffusion and self-diffusion of a tracer are synonymous.

For aqueous ions of electrolytes, the limiting value of the tracer or self-diffusion coefficient D^0 at infinite dilution is

$$D^0 = \frac{RT\lambda^0}{\mathscr{F}^2 |z|} = 8.925 \times 10^{-10} \frac{T\lambda^0}{|z|} \qquad (\text{cm}^2 \text{ sec}^{-1}) \qquad (3.21)$$

where R is the gas constant, $\mathscr{F}$ is the Faraday, z is the valence charge of the ion, T is temperature in degrees Kelvin, and λ^0 is the equivalent limiting conductance of the ion (cm^2 ohm^{-1} equivalent^{-1}). Values of λ^0 for many ions in aqueous solutions have been tabulated by Robinson and Stokes (1959).

For a number of ions occurring in natural waters, the values of D^0 in the temperature range from 0 to 25°C are listed in Table 3.1.

An electrolyte dissociating into ν_1 ions of charge z_1 and ν_2 ions of charge z_2 requires that electroneutrality be maintained: $\nu_1 z_1 + \nu_2 z_2 = 0$. The self-diffusion coefficient D^0 of such an electrolyte at infinite dilution is

$$D^0 = \frac{RT}{\mathscr{F}^2} \frac{\nu_1 + \nu_2}{\nu_1 |z_1|} \frac{\lambda_1^0 \lambda_2^0}{\lambda_1^0 + \lambda_2^0} \qquad (3.22)$$

where λ_1^0 and λ_2^0 are the limiting ionic conductivities of the positive and negative ions. An alternative way of writing equation 3.22 is, by using the electroneutrality condition given above, in terms of ionic charges,

$$D^0 = \frac{RT}{\mathscr{F}^2} \cdot \left(\frac{1}{|z_1|} + \frac{1}{|z_2|} \right) \frac{\lambda_1^0 \lambda_2^0}{\lambda_1^0 + \lambda_2^0} \qquad (3.23)$$

Table 3.1 Tracer Diffusion Coefficients of Ions at
Infinite Dilution in Water[a]

Cation	D^0 (10^{-6} cm^2 sec^{-1})			Anion	D^0 (10^{-6} cm^2 sec^{-1})		
	0°C	18°C	25°C		0°C	18°C	25°C
H^+	56.1	81.7	93.1	OH^-	25.6	44.9	52.7
Li^+	4.72	8.69	10.3	F^-	—	12.1	14.6
Na^+	6.27	11.3	13.3	Cl^-	10.1	17.1	20.3
K^+	9.86	16.7	19.6	Br^-	10.5	17.6	20.1
Rb^+	10.6	17.6	20.6	I^-	10.3	17.2	20.0
Cs^+	10.6	17.7	20.7	IO_3^-	5.05	8.79	10.6
NH_4^+	9.80	16.8	19.8	HS^-	9.75	14.8	17.3
Ag^+	8.50	14.0	16.6	S^{2-}	—	6.95	—
Tl^+	10.6	17.0	20.1	HSO_4^-	—	—	13.3
$Cu(OH)^+$	—	—	8.30	SO_4^{2-}	5.00	8.90	10.7
$Zn(OH)^+$	—	—	8.54	SeO_4^{2-}	4.14	8.45	9.46
Be^{2+}	—	3.64	5.85	NO_2^-	—	15.3	19.1
Mg^{2+}	3.56	5.94	7.05	NO_3^-	9.78	16.1	19.0
Ca^{2+}	3.73	6.73	7.93	HCO_3^-	—	—	11.8
Sr^{2+}	3.72	6.70	7.94	CO_3^{2-}	4.39	7.80	9.55
Ba^{2+}	4.04	7.13	8.48	$H_2PO_4^-$	—	7.15	8.46
Ra^{2+}	4.02	7.45	8.89	HPO_4^{2-}	—	—	7.34
Mn^{2+}	3.05	5.75	6.88	PO_4^{3-}	—	—	6.12
Fe^{2+}	3.41	5.82	7.19	$H_2AsO_4^-$	—	—	9.05
Co^{2+}	3.41	5.72	6.99	$H_2SbO_4^-$	—	—	8.25
Ni^{2+}	3.11	5.81	6.79	CrO_4^{2-}	5.12	9.36	11.2
Cu^{2+}	3.41	5.88	7.33	MoO_4^{2-}	—	—	9.91
Zn^{2+}	3.35	6.13	7.15	WO_4^{2-}	4.27	7.67	9.23
Cd^{2+}	3.41	6.03	7.17				
Pb^{2+}	4.56	7.95	9.45				
UO_2^{2+}	—	—	4.26				
Sc^{3+}	—	—	5.74				
Y^{3+}	2.60	—	5.50				
La^{3+}	2.76	5.14	6.17				
Yb^{3+}	—	—	5.82				
Cr^{3+}	—	3.90	5.94				
Fe^{3+}	—	5.28	6.07				
Al^{3+}	2.36	3.46	5.59				
Th^{4+}	—	1.53	—				

[a]From Li and Gregory (1974). (By permission of Pergamon Press, Ltd. Copyright
1974)

Equations 3.20 and 3.23 are known as the Nernst equations.

With an increasing concentration of the diffusing solute, concentrations are no longer identical to thermodynamic activities, and the activity coefficients and electrophoretic corrections can be introduced into equation 3.22 to obtain the diffusion coefficient D at concentrations higher than the very dilute,

$$D = D^0 f(C) \tag{3.24}$$

where $f(C)$ represents functional relationships whose terms depend on concentration. Robinson and Stokes (1970) discuss a number of mathematical models used for estimation of diffusion coefficients of electrolytes in concentrated solutions. Agreement between computed and observed values is good for $1:1$ strong electrolytes, such as NaCl, up to moderately high concentrations (about 1 molal), but differences of an order of a factor of 1.4 build up at higher concentrations near 4 molal. For electrolytes of the type $CaCl_2$, Li_2SO_4, and Na_2SO_4 the agreement between theory and observations is good only in the dilute range up to 0.005 mol l^{-1}.

In dilute solutions, the tracer diffusion coefficient, defined in equations 3.22 and 3.23, can be used as a reasonable approximation to the diffusion coefficient. For example, the diffusion coefficients of ions in sea water are only 0–8% lower than their values at infinite dilution. For the major ions in sea water, the tracer diffusion coefficients are listed in Table 3.2.

3.2.3   Interdiffusion

When two chemical species are present at different concentrations within a system (e.g., a higher concentration of KCl at one end and a higher concentration of NaCl at the other end of a water column), the condition of electrical neutrality requires that the number of ionic charges diffusing in one direction must be equal to the number of charges diffusing in the opposite direction. This process is called interdiffusion or counterdiffusion, and the diffusion coefficient characterizing it is often denoted D_{12}. The main difference between counterdiffusion and tracer or self-diffusion is that the two species diffusing counter to one another occur at comparable concentrations. The interdiffusion coefficient D_{12} for two ions of the same charge can be defined as (e.g., Jost, 1960, p. 146; Manning, 1968, p. 21, intrinsic diffusion coefficient)

$$D_{12} = \frac{(C_1 + C_2) D_1 D_2}{D_1 C_1 + D_2 C_2} \tag{3.25}$$

where D_1 and D_2 are the self-diffusion coefficients of the two counterdiffusing ions, and C_1 and C_2 are their concentrations. If the concentration of one of the diffusing species is very small, then the interdiffusion coefficient

Table 3.2 **Molecular Diffusion Coefficients of Some Dissolved Species in Bulk Sea Water and Sediment–Sea Water Mixtures Approximating Pore Waters (sediment porosity $\phi \simeq 0.7$)**

Aqueous species	Temperature (°C)	Diffusion coefficient D (10^{-6} cm^2 sec^{-1})	
		Sea water	Pore water
Na^{+} [a]	5	8.0	3.5
	24	13.4	5.8
K^{+} [a]	5	11.4	
	24	17.9	
Ca^{2+} [a]	5	5.0	8.5
	24	7.5	15.5
Cl^{-} [a]	5	11.5	5.9
	24	18.6	10.2
SO$_4^{2-}$ [a]	5	5.8	3.3
	24	9.8	5.3
	7		2.0–2.8[b]
	19		3.7–3.9[b]
SiO$_2$	5		3.3[c]
	25	10.0[d]	

[a] Li and Gregory (1974).
[b] Goldhaber et al. (1977).
[c] Fanning and Pilson (1974).
[d] Wollast and Garrels (1971).

D_{12} tends to the tracer diffusion coefficient D_1 or D_2 of the low-concentration species.

Another way of treating the fluxes of diffusing species when their concentrations are of comparable magnitudes is by means of the Onsager reciprocal relationships and the entropy dissipation function. Fundamentals of this procedure are discussed in textbooks on irreversible thermodynamics (e.g., Fitts, 1962; Katchalsky and Curran, 1965), and its applications to the determination of diffusion coefficients in solutions of two salts have been dealt with by, among others, Fitts (1962), Miller (1967), and Wendt and Shamim (1970).

For ternary systems (solute 1–solute 2–H$_2$O) the essential relationships between the diffusion coefficients are the following. One-dimensional diffusional fluxes of the two dissolved components are

$$F_1' = -D_{11}' \frac{dC_1}{dz} - D_{12}' \frac{dC_2}{dz} \tag{3.26}$$

$$F_2' = -D_{21}' \frac{dC_1}{dz} - D_{22}' \frac{dC_2}{dz} \tag{3.27}$$

where the prime indicates that the fluxes and diffusion coefficients are measured relative to the solvent frame of reference, as explained in Section 3.2.1. The flux of each component depends on the concentration gradients of both components, and the direct (D'_{ii}) and cross (D'_{ij}) diffusion coefficients. The flux equations 3.26 and 3.27 can be converted to a fixed frame of reference by multiplying both sides by the water volume fraction v_w, which gives

$$F_1 = -D_{11}\frac{dC_1}{dz} - D_{12}\frac{dC_2}{dz} \tag{3.26a}$$

$$F_2 = -D_{21}\frac{dC_1}{dz} - D_{22}\frac{dC_2}{dz} \tag{3.27a}$$

The cross coefficient D_{12} in equation 3.26a should not be confused with the interdiffusion coefficient defined in equation 3.25.

In terms of the Onsager phenomenological relationships, the fluxes of the two dissolved species F'_1 and F'_2 can also be written as

$$F'_1 = -L_{11}\frac{d\mu_1}{dz} - L_{12}\frac{d\mu_2}{dz} \tag{3.28}$$

$$F'_2 = -L_{21}\frac{d\mu_1}{dz} - L_{22}\frac{d\mu_2}{dz} \tag{3.29}$$

where the L's are the phenomenological coefficients and μ is the chemical potential defined in equation 3.17. If the chemical potential gradient $d\mu/dz$ is given in units of energy mol^{-1} cm^{-1} and the flux is in units of mol cm^{-2} sec^{-1}, then the units of the coefficients L are mol^2 $energy^{-1}$ cm^{-1} sec^{-1} or mol^2 g^{-1} cm^{-3} sec. As the chemical potential of each species is a function of C_1 and C_2, the potential gradients can be written as

$$\frac{d\mu_1}{dz} = \frac{\partial\mu_1}{\partial C_1}\frac{dC_1}{dz} + \frac{\partial\mu_1}{\partial C_2}\frac{dC_2}{dz} \tag{3.30}$$

$$\frac{d\mu_2}{dz} = \frac{\partial\mu_2}{\partial C_2}\frac{dC_2}{dz} + \frac{\partial\mu_2}{\partial C_1}\frac{dC_1}{dz} \tag{3.31}$$

Using a shorthand notation, $\mu_{ij} \equiv \partial\mu_i/\partial C_j$, and substituting equations 3.30 and 3.31 in 3.28 and 3.29, the diffusional fluxes become

$$F'_1 = -(L_{11}\mu_{11} + L_{12}\mu_{21})\frac{dC_1}{dz} - (L_{11}\mu_{12} + L_{12}\mu_{22})\frac{dC_2}{dz} \tag{3.32}$$

$$F'_2 = -(L_{21}\mu_{11} + L_{22}\mu_{21})\frac{dC_1}{dz} - (L_{21}\mu_{12} + L_{22}\mu_{22})\frac{dC_2}{dz} \tag{3.33}$$

The coefficients of the concentration gradients in the pairs of the flux equations 3.32 and 3.26, and 3.33 and 3.27 define the diffusion coefficients in terms of Onsager's phenomenological coefficients,

$$D'_{11} = L_{11} \mu_{11} + L_{12} \mu_{21} \tag{3.34}$$

$$D'_{12} = L_{11} \mu_{12} + L_{12} \mu_{22} \tag{3.35}$$

$$D'_{21} = L_{21} \mu_{11} + L_{22} \mu_{21} \tag{3.36}$$

$$D'_{22} = L_{21} \mu_{12} + L_{22} \mu_{22} \tag{3.37}$$

The direct and cross diffusion coefficients D_{ii} and D_{ij} must be determined experimentally, and if such data are available the fluxes of each of the two dissolved components can be computed using equations 3.26 and 3.27. In general, the direct and cross diffusion coefficients are concentration and temperature dependent. A more involved procedure is the computation of the phenomenological coefficients L_{ij} when the diffusion coefficients D_{ij} and the dependence of the chemical potentials on concentration μ_{ij} are known. Then the four equations can be solved to give the four unknowns L_{11}, L_{12}, L_{21}, and L_{22}. Onsager's theory requires that the cross coefficients must be equal—$L_{12} = L_{21}$—and this condition is closely met by experimental data.

In Table 3.3 are given the four diffusion coefficients D_{11}, D_{12}, D_{21}, and D_{22} of the solutes in the systems KCl–NaCl–H$_2$O and MgCl$_2$–NaCl–H$_2$O. The coefficients refer to a fixed frame of reference that applies to the fluxes defined in equations 3.26a and 3.27a. In NaCl–KCl solution at the

Table 3.3   Direct D_{ii} and Cross D_{ij} Diffusion Coefficients in Aqueous Solutions Containing Two Solutes at 25°C

Components		Concentration (mol l^{-1})		Ionic Strength	Diffusion coefficients (10^{-5} cm^2 sec^{-1})			
1	2	C_1	C_2	(mol l^{-1})	D_{11}	D_{12}	D_{21}	D_{22}
KCl–NaCl[a]		0.25	0.25	0.50	1.82	0.18	0.00	1.37
		0.50	0.25	0.75	1.84	0.24	0.00	1.36
		0.25	0.50	0.75	1.82	0.12	0.00	1.42
		0.50	0.50	1.00	1.83	0.20	0.00	1.39
MgCl$_2$–NaCl[b]		0.25	0.25	1.00	0.89	0.105	0.35	1.20
		0.15	0.30	0.75	0.84	0.045	0.48	1.36
		0.10	0.46	0.76	0.79	0.014	0.63	1.43

[a] Fitts (1962, p. 96).
[b] Wendt and Shamim (1970).

ionic strength between 0.5 and 1.0 the interaction between the two dissolved components would contribute about 10% to the flux of KCl if the concentration gradients were equal and of the same sign. Conversely, the effect of KCl on the diffusional flux of NaCl is null. In the $MgCl_2$–NaCl solution, each species affects the other. The effect of NaCl on the diffusional flux of $MgCl_2$ is relatively small: D_{12} contributes between 2 and 11% to D_{11} within the range of concentrations given in the table. The effect of $MgCl_2$ on diffusion of NaCl is stronger, as the values of the coefficient D_{21} amount to 30–40% of D_{22}.

3.3   EFFECTS OF ENVIRONMENTAL VARIABLES ON DIFFUSION

3.3.1   Aqueous Solutions: Temperature, Concentration, Pressure

For molecular diffusion of spheres in a viscous fluid, the diffusion coefficient D is related to temperature T, viscosity of the medium η, and radius of the diffusing particle r by the Stokes-Einstein relationship

$$D = \frac{kT}{6\pi\eta r} \qquad (cm^2 \ sec^{-1}) \tag{3.38}$$

where $k = 1.38 \times 10^{-16}$ erg deg^{-1} is the Boltzmann constant and other parameters are in cgs units. According to the Stokes-Einstein equation, the diffusion coefficient is inversely related to the particle size r. For a given particle size, the quotient $D\eta/T$ is a constant,

$$\frac{D\eta}{T} = \frac{7.324 \times 10^{-18}}{r} \qquad (g \ cm \ sec^{-2} \ deg^{-1})$$

$$\frac{D\eta}{T} = \frac{1.181 \times 10^{-17}}{V^{1/3}} \qquad (g \ cm \ sec^{-2} \ deg^{-1}) \tag{3.39}$$

where $V = \frac{4}{3}\pi r^3$ is the volume of the diffusing particle.

The Stokes-Einstein equation does not take into account any chemical interactions between the solutes and solvent that are not reflected in the viscosity of the solution. However, it gives a correct order of magnitude of D for many dissolved species in water ($\eta \simeq 10^{-2}$ poise or g cm^{-1} sec^{-1}, $T = 300°K$, $r \simeq 10^{-8}$cm)

$$D \simeq \frac{7.324 \times 10^{-18} \times 300}{10^{-2} \times 10^{-8}} \simeq 10^{-5} \ cm^2 \ sec^{-1}$$

The relationship $D\eta/T = $ const allows one to extrapolate the values of D from one temperature to another, if the viscosity of the fluid medium is known at the other temperature for the same concentration of the diffusing species. More generally, as viscosity is a function of solute concentration and composition of the mixture, temperature, and pressure, the following uses of equation 3.38 have been made.

At variable temperature but fixed composition and pressure:

$$\frac{D_{T_2}}{D_{T_1}} = \frac{(\eta/T)_1}{(\eta/T)_2} \tag{3.40}$$

from which one of the D values can be estimated if the value at the other temperature is known.

At variable concentration but fixed temperature and pressure:

$$\frac{D_{C_2}}{D_{C_1}} = \frac{\eta_{C_1}}{\eta_{C_2}} \tag{3.41}$$

where subscripts C_1 and C_2 denote different concentrations.

At variable pressure but fixed composition and temperature:

$$\frac{D_{P_2}}{D_{P_1}} = \frac{\eta_{P_1}}{\eta_{P_2}} \tag{3.42}$$

The pressure dependence of diffusion coefficients of electrolytes in water is light: for pure water and relatively dilute salt solutions, such as sea water, the viscosity at atmospheric pressure *decreases* 5–8% as the pressure rises to about 1500 atm for pure water and about 750 atm for sea water (Bett and Cappi, 1965; Horne and Johnson, 1966). At higher pressures the viscosity *rises* with increasing pressure. The viscosity minimum observed in water and dilute salt solutions with increasing pressure is known as the viscosity anomaly of water. The decrease in viscosity of the order of 5–8%, mentioned above, is most pronounced at the lower temperatures, near 2°C in pure water and near −2°C in sea water. At higher temperatures, the viscosity minimum diminishes in magnitude such that near 30°C there is almost no minimum on the viscosity against pressure plot. Also, addition of dissolved electrolytes causes the viscosity minimum to be less pronounced, and for sea water it vanishes within about the same temperature range.

The self-diffusion coefficient of water at 30°C follows relationship 3.38 in the pressure range from 1 to 1000 atm; the value of D decreases 2% (Benedek and Purcell, 1954).

The relationship $D\eta/T=$ const and other relationships based on it apply to diffusion of gases in water (Section 3.4).

For ionic species in water, the relationship $D\eta/T=$ const is not ordinarily obeyed: in Figure 3.2 are shown the viscosity η of NaCl–H_2O solutions, diffusion coefficient D of NaCl, and the product $D\eta$ at 25°C. The product $D\eta$ varies with concentration more strongly (percent wise) than the viscosity of the solution and the diffusion coefficient of NaCl. The mean diffusion coefficients of a number of electrolytes, the cations or anions of which are common constituents of natural waters and brines, are listed in Table 3.4. The dependence of the diffusion coefficients on concentration is not very strong, and the values of D increase 10–20% between dilute and more concentrated solutions. For the univalent 1:1

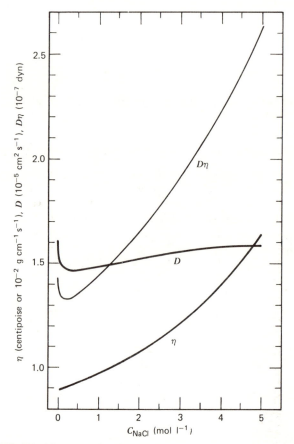

Figure 3.2 Self-diffusion of NaCl and viscosity of aqueous NaCl solutions (data from Robinson and Stokes, 1970; Stokes and Mills, 1965).

electrolytes, such as NaCl or KBr, the values of the diffusion coefficients at different concentrations are between 1.5×10^{-5} and 2.4×10^{-5} cm^2 sec^{-1}. For salts containing cations or anions of higher valence, such as CaCl$_2$ and (NH$_4$)$_2$SO$_4$, the diffusion coefficients are lower, between 0.8×10^{-5} and 1.3×10^{-5} cm^2 sec^{-1}.

The self-diffusion coefficients of ions at infinite dilution D^0 increase by a factor of about 2 between 0 and 25°C,

$$D_{25}^0 = D_0^0 (1 + \alpha t) \qquad (\text{cm}^2 \text{ sec}^{-1}) \tag{3.43}$$

where t is temperature in degrees Centigrade, $\alpha = 0.048 \pm 0.02$ deg^{-1}, and $\alpha = 0.040 \pm 0.02$ deg^{-1} for the cations and anions that diffuse faster than F$^-$, including Cs$^+$, Rb$^+$, NH$_4^+$, K$^+$, Ag$^+$, I$^-$, Br$^-$, Cl$^-$, and NO$_3^-$ (Table 3.1).

Table 3.4 **Mean Diffusion Coefficients D (10^{-5} cm^2 sec^{-1}) of Dissolved Ionic Solids at Different Concentrations in Water at 25°C**[a]

Concentration (mol l^{-1})	NaCl	NaBr	KCl	KBr	NH$_4$NO$_3$	CaCl$_2$	(NH$_4$)$_2$SO$_4$
0	1.610	1.625	1.993	2.016	1.929	1.335	1.530
0.05	1.507	1.53	1.864	1.89	1.788	1.121	0.802
0.1	1.483	1.52	1.844	1.87	1.769	1.110	0.825
0.2	1.475	1.51	1.838	1.87	1.749	1.107	0.867
0.3	1.475	1.52	1.838	1.87	1.739	1.116	0.897
0.5	1.474	1.54	1.850	1.89	1.724	1.140	0.938
0.7	1.475	1.57	1.866	1.92	1.709	1.168	0.972
1.0	1.484	1.60	1.892	1.98	1.690	1.203	1.011
1.5	1.495	1.63	1.943	2.06	1.661	1.263	1.047
2.0	1.516	1.67	1.999	2.13	1.633	1.307	1.069
2.5	—	1.70	2.057	2.20	1.605	1.306	1.088
3.0	1.565	—	2.112	2.28	1.578	1.265	1.106
3.5	—	—	2.160	2.35	—	1.195	1.122
4.0	1.594	—	2.196	2.43	1.524	—	1.135
4.5	—	—	—	—	—	—	—
5.0	1.590	—	—	—	1.472	—	—
6.0	—	—	—	—	1.421	—	—
7.0	—	—	—	—	1.370	—	—
8.0	—	—	—	—	1.320	—	—

[a]From Robinson and Stokes (1970); Landolt-Börnstein (1969).

3.3.2   Pore Space of Sediments

Porosity and Diffusion.   Molecules of material diffusing through a solution-filled pore space encounter in their path an irregular network of branching pore canals with whose walls they collide. Observations on dissolved solids (and gases) show that diffusion through a porous space is slower than through the solution (or gaseous) medium in the absence of the solid framework. Slower diffusional fluxes are the result of lower values of the diffusion coefficients in comparison to those of the bulk solution. The physical characteristics of the porous medium that are responsible for the hindrance of molecular diffusion are the porosity itself and tortuosity of the diffusional path. Porosity ϕ was defined in Section 2.2.3. Tortuosity θ is the ratio of the mean length L_p of the path through the porous space between some two points to the straight line distance L between the same points.

$$\text{Tortuosity:}\quad \theta = L_p/L \geqslant 1 \tag{3.44}$$

$$\text{Porosity:}\quad 0 \leqslant \phi \leqslant 1 \tag{3.45}$$

Tortuosity θ depends evidently on the sediment porosity ϕ, although the nature of the relationship between θ and such factors as the particle-grain size and packing is not well understood.

The diffusion coefficients of dissolved species, can be related to the porosities and tortuosities of sediments by means of a quantity known as the formation factor f (Archie, 1942; Klinkenberg, 1951; Bear, 1972). The formation factor f is the ratio of the electrical resistance of a brine-filled porous medium R_p (ohms) to the electrical resistance R_w (ohms) of the brine occupying the same volume as the bulk porous medium,

$$f = \frac{R_p}{R_w} \tag{3.46}$$

If the solids do not conduct electricity whereas only the solution does, then the resistance of the porous medium is

$$R_p = \frac{R_{sw} L_p}{A_p} \quad \text{(ohms)} \tag{3.47}$$

where R_{sw} is the specific resistance of the solution (ohm cm), L_p is the length of the conducting path within the porous space (cm), and A_p is the fractional cross section available to conductance of current (cm^2). The resistance of a volume of solution, with length L (cm) and cross section A

(cm^2) the same as the external dimensions of the porous material, is

$$R_w = \frac{R_{sw}L}{A} \quad \text{(ohms)} \tag{3.48}$$

Substitution of the relationships from equations 3.47 and 3.48 in equation 3.46 defines the formation factor f in terms of the sediment porosity and tortuosity,

$$f = \frac{L_p/L}{A_p/A} = \frac{\theta}{\phi} \tag{3.49}$$

where the ratio of the water filled to total cross section of sediment is the porosity $\phi = A_p/A$.

Other relationships between the formation factor f, tortuosity θ, and porosity ϕ, based on experimental and theoretical work (the different approaches of the theory involve primarily different assumptions about the pore shape and size configurations) are listed below.

$$f = \frac{\theta^{5/3}}{\phi} \quad \text{(Winsauer et al., 1952)} \tag{3.50}$$

$$f = \frac{\theta}{\phi^2} \quad \text{(Bear, 1972)} \tag{3.51}$$

$$f = \frac{1}{\phi^2} \quad \text{for sandstones (Archie, 1942)} \tag{3.52}$$

$$f = \frac{1}{\phi^{1.3}} \quad \text{for sands (Archie, 1942)} \tag{3.53}$$

$$f = \frac{1}{\phi^n} \quad n = 2.5\text{--}5.4 \text{ for clays (Manheim and Waterman, 1974)} \tag{3.54}$$

Equation 3.51 becomes identical to 3.52 if the lengths of the diffusional path and straight line distance are the same, $L_p = L$.

As long as the flow of electric current through the pore space follows the same pattern of the tortuous channels as in diffusion, diffusion coefficients and formation factors can be correlated—$D \propto 1/f$—and a relationship between diffusion coefficients and porosity can be established. Figure 3.3

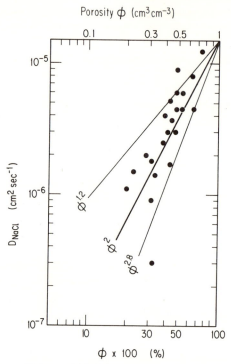

Figure 3.3 Diffusion coefficients of NaCl in water-filled sediments of different porosity ϕ (data from Manheim, 1970; Lerman, 1978). Reproduced with permission from *Annual Review of Earth & Planetary Sciences*, Vol. 6. © 1978 by Annual Reviews Inc.

shows a plot of diffusion coefficients in pore waters of different sediments as a function of the sediment porosity. The value of the diffusion coefficient is approximately directly related to ϕ^2,

$$D = D_0 \phi^2 \qquad (\text{cm}^2 \text{ sec}^{-1}) \tag{3.55}$$

where D_0 is the value of the diffusion coefficient in the bulk solution (that is, when $\phi = 1$). Equation 3.55 may be used to estimate molecular diffusion coefficients in pore waters of sediments, for lack of a better model. See data in Table 3.2.

Correlation of electrical resistances of water-filled sediments with diffusivities of dissolved species is meaningful (1) if the electrical conductivity of the solids is negligible in comparison to the conductivity of the pore solution, (2) if the pore space is occupied only by the conducting solution (in mixtures of gases or oils with brines and in partially filled pore spaces the volume fraction conducting is generally smaller than the sediment porosity ϕ), and (3) if the surface conductance is also insignificant in comparison to the conductance by solution in the pore space. Surface conduction in porous materials represents some fraction of electric current conducted by a layer of solution adjacent to the solid surface (the ionic double layer or adsorbed layer). Electrical conduction by a solution layer of molecular dimensions close to a solid surface and diffusion of ions in such layers do not ordinarily account for a significant fraction of the ionic fluxes through the pore space (van Olphen, 1957, 1963; Lerman, 1977).

Diffusion in Small Pores.   The linear dimensions of ions and simple organic molecules in water are measurable in angstroms (10^{-8}–10^{-7} cm). Linear dimensions of interstices in sediments, contained between sediment particles, may be thought of as being of the same order of magnitude as the particle dimensions: microns (10^{-4}–10^{-3} cm) for fine-grained clayey sediments. Even for such small pores, the dimensions of dissolved species are only about 10^{-4} of the pore diameters. As the pores become smaller, tens to hundreds of angstroms in radius, diffusion of dissolved species is retarded by a combination of geometric and hydrodynamic effects (hindrance of pore walls through an increased drag force). One equation relating diffusion coefficients of dissolved species in small pores D and in a bulk solution D_0 is (Renkin, 1954)

$$D = D_0\left(1 - \frac{r}{r_p}\right)^2\left[1 - 2.10\frac{r}{r_p} + 2.09\left(\frac{r}{r_p}\right)^3 - 0.95\left(\frac{r}{r_p}\right)^5\right] \qquad (3.56)$$

where r is the radius of the dissolved species and r_p is the radius of the pore. For relatively large pores we have $r_p \gg r$, and the pore diffusion coefficient D approaches the values of D_0 for the bulk solution.

Diffusion of organic molecules (including urea, glucose, and ribonuclease) of dimensions r between 2.6 and 22 Å, through cylindrical pores of radius r_p between about 46 and 306 Å, has been described as obeying the following relationship (Beck and Schultz, 1970):

$$D \simeq D_0\left(1 - \frac{r}{r_p}\right)^4 \qquad (3.57)$$

where the notation is as defined following equation 3.56. According to the latter equation and experimental results, the ratio $D/D_0 = 1$ for $r/r_p = 0$, and the ratio D/D_0 decreases to 0.1 for the radius ratio $r/r_p = 0.4$. Thus there is a 90% reduction of the diffusional flux when the diffusing species occupies about 16% (0.4^2) of the pore cross-section area. The diffusion coefficients of organic substances in bulk solution at 25°C range from $D_0 = 1.2 \times 10^{-6}$ (ribonuclease, $r = 21.6$ Å) to 14×10^{-6} cm^2 sec^{-1} (urea, $r = 2.64$ Å).

3.4   DIFFUSION OF GASES IN WATER

Diffusion coefficients of gases in water have been related to the nature of the gas, temperature, and viscosity of the solution by relationships similar to the Stokes-Einstein equation 3.39 for D. The following two relationships have been shown to produce about equally acceptable correlations of the diffusion coefficients of gases in water with other parameters. One is the Wilke and Chang (1955) equation

$$D = 5.06 \times 10^{-9} \frac{T}{\eta V_b^{0.6}} \tag{3.58}$$

and the other is the Othmer and Thakar (1953) equation

$$D = \frac{8.83 \times 10^{-7}}{\eta_w^{1.1} V_b^{0.6}} \tag{3.59}$$

where T is temperature in degrees Kelvin, η is the viscosity of the solution (poise of g cm^{-1} sec^{-1}), η_w is the viscosity of water at a given temperature (same units as for η), and V_b is the molal volume of the gas (cm^3 mol^{-1}). The molal volumes V_b (also called the normal boiling point volumes) for some gases are listed in Table 3.5. The difference between the two relationships for D given in equations 3.58 and 3.59 in a dilute solution near 20°C ($\eta \simeq \eta_w = 0.89 \times 10^{-2}$ poise) amounts to about 5%, or a factor of

$$\frac{5.06 \times 10^{-9} \times 293 / 0.89 \times 10^{-2}}{8.83 \times 10^{-7} / (0.89 \times 10^{-2})^{1.1}} = 1.05$$

Inherently there is a greater appeal to use the Wilke and Chang equation 3.58 because it explicitly introduces a dependence of D on the temperature and viscosity of a solution, and viscosity depends on the chemical nature of the dissolved species. Both the Wilke and Chang, and Othmer and Thakar equations 3.58 and 3.59, as well as equation 3.39 based on the Stokes-Einstein relationship, have been demonstrated to hold for common gases in water, for temperatures between 0 and about 50°C (Himmelblau, 1964; Ferrell and Himmelblau, 1967).

Table 3.5 Molar Volumes of Gases V_b (cm³ mol⁻¹)

Gas	V_b estimated[a,b]	V_b experimental[a]
Air	29.9	—
Ar	—	29.2
Br_2	53.2	—
Cl_2	48.8	45.5
CH_4	43.9	44.8
CO	30.7	—
CO_2	34.0	37.3
H_2	14.3	28.5
H_2O	18.9	—
H_2S	32.9	35.2
He	—	31.9
N_2	31.2	34.7
NH_3	25.8	24.5
N_2O	36.4	36.0
NO	23.6	—
NO_2	—	31.7
O_2	25.6	27.9
SO_2	42.2	43.8

[a] Himmelblau (1964).
[b] Satterfield (1970).

In Table 3.6 are given the values of D for a number of gases in water, computed from the experimentally determined values of V_b (Table 3.5) or, when experimental values were not available, from the estimated V_b. The measured values of D at 25°C, also given in Table 3.6, have an uncertainty margin of about ±10%; for some gases, however, the values of D reported by different sources vary by as much as a factor of 2 to 3. The diffusivities of gases and ionic dissolved solids in water are the same to an order of magnitude, 10^{-6}–10^{-5} cm² sec⁻¹.

In the presence of ionized solutes, such as NaCl, $MgCl_2$, Na_2SO_4, and $MgSO_4$, diffusivity of CO_2 in water is slower, as shown by the data plotted in Figure 3.4. For NaCl solutions, the decrease in D_{CO_2} with an increasing NaCl concentration is roughly predictable from the Stokes-Einstein equation, as the product $D_{CO_2}\eta$ is nearly constant. For $MgSO_4$ and Na_2SO_4 solutions, the product $D_{CO_2}\eta$ increases with an increasing salt concentration; in other words, D_{CO_2} is greater than might have been expected from the Stokes-Einstein or Wilke-Chang equations. This fact suggests that if diffusion coefficients of CO_2 are estimated for saline brines using the Wilke-Chang equation, equation 3.58, the values may be somewhat lower than in reality, although the difference would be within the limits of variation between the different experimental results.

Table 3.6. Diffusion Coefficients D of Gases in Water Computed from Equation 3.58. Experimentally Determined Values at 25°C are in Parentheses. Units of D: 10^{-5} cm^2 sec^{-1}

Gas in H$_2$O	Temperature (°C)			
	5	15	25	35
Air	1.21	1.67	2.21	2.82
Ar	1.23	1.69	2.24 (1.46)[a] (0.78)[b]	2.87
Br$_2$	0.86	1.18	1.56	2.00
Cl$_2$	0.94	1.30	1.72 (1.45)[a]	2.20
CH$_4$	0.95	1.31	1.73	2.22
CO	1.19	1.64	2.17	2.78
CO$_2$	1.06	1.46	1.93 (1.90)[a]	2.47
H$_2$[c]	1.88	2.60	3.44 (5.2 ± 1.8)[a]	4.40
H$_2$S	1.10	1.51	2.00 (1.36)[a]	2.56
He	1.16	1.61	2.12 (5.8)[a] (2.38)[b]	2.72
Kr			(0.80)[b]	
Ne			(2.8)[a] (1.82)[b]	
N$_2$	1.11	1.53	2.02 (2.0)[a]	2.58
NH$_3$	1.36	1.88	2.49	3.18
N$_2$O	1.08	1.49	1.98 (2.57)[a]	2.53
NO	1.39	1.92	2.55	3.26
NO$_2$	1.17	1.61	2.13	2.73
O$_2$	1.26	1.74	2.30 (2.35)[a]	2.94
Rn[d]	0.79	1.05	1.37	
SO$_2$	0.96	1.33	1.76 (1.9)[a]	2.25
Xe			(0.83)[b]	

[a] Himmelblau (1964).
[b] Boerboom and Kleyn (1969).
[c] Computed using $V_b = 14.3$ cm^3 mol^{-1} from Table 3.5.
[d] Broecker and Peng (1974).

3.5   THERMAL DIFFUSION IN SOLUTIONS

A temperature gradient imposed on an initially homogeneous solution causes a diffusional flux of solute that produces a concentration gradient. This phenomenon is known as the Ludwig-Soret or the Soret effect. The flux of solute induced by the heat flux or temperature gradient is added algebraically to the diffusional flux driven by the chemical potential gradient,

$$F = -D\frac{dC}{dz} - D_T C\frac{dT}{dz} \quad \text{(g cm}^{-2}\text{ sec}^{-1}\text{)} \qquad (3.60)$$

$$\underset{\text{Chemical diffusion}}{\phantom{-D\frac{dC}{dz}}} \quad \underset{\text{Thermal diffusion}}{\phantom{-D_T C\frac{dT}{dz}}}$$

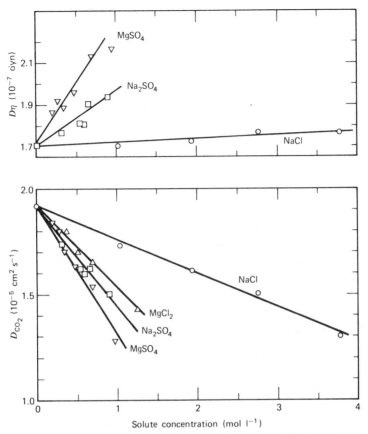

Figure 3.4 Diffusion coefficient of CO_2 and the product of the diffusion coefficient and viscosity in aqueous solutions of ionic salts (data from Himmelblau, 1964; Stokes and Mills, 1965).

where T is temperature and D_T is the coefficient of thermal diffusion. By factoring D out, equation 3.60 becomes

$$F = -D\left(\frac{dC}{dz} + sC\frac{dT}{dz}\right) \tag{3.61}$$

where

$$s = \frac{D_T}{D} \quad (\text{deg}^{-1}) \tag{3.62}$$

and s is called the Soret coefficient, defined as a positive quantity according to equation 3.62. The flux relationship in equation 3.61 shows that in the presence of a thermal gradient the diffusional flux is supplemented by another term, enhancing the flux if the concentration dC/dz and temperature dT/dz gradients are of the same sign, and reducing the flux if the gradients are of opposite sign.

A detailed derivation of relationship 3.60, based on the theorem of entropy production due to flow of heat and of solutes, can be found in books dealing with irreversible thermodynamics, such as the textbook by Katchalsky and Curran (1965) cited previously.

At the steady state, the fluxes of solutes produced by the chemical potential and temperature gradients are balanced. Then $F=0$ and equation 3.61 becomes

$$\frac{1}{C}\frac{dC/dz}{dT/dz} = \frac{1}{C}\frac{dC}{dT} = -s \tag{3.63}$$

The latter relationship shows that at the steady state the temperature and concentration gradients are of opposite sign, and that the diffusing material tends to migrate toward the colder pole or, in other words, the concentration at the colder end is higher.

For the common electrolytes and constituents of natural waters, the values of the Soret coefficient s are between 10^{-3} and 10^{-2} deg^{-1}. For a number of species in water, their s coefficients are given in Table 3.7.

Table 3.7   Soret Coefficients s (10^{-3} deg^{-1}) of Some Dissolved
Species in Water, at Different Concentrations
and Temperatures (deg C)[a]
A.   Sodium Salts

Concentration	NaCl			NaBr		Na$_2$SO$_4$	
(mol l^{-1})	25°	31–34°	37°	25°	31–34°	25°	31°
0.0005						6.70	
0.001	2.31						
0.002						6.66	
0.005						6.50	
0.01	2.05			2.11		6.39	
0.02				1.91			
0.05	1.59						
0.5							5.50
1.0		1.46	2.48		2.06		8.00
1.5		1.70	2.37				
2.0		1.77	2.34		2.14		
3.0		1.88	2.32		2.27		
4.0		2.03	2.23		2.39		
5.0			2.01		2.37		
6.0					2.36		

B. Potassium Salts

Concentration (mol l^{-1})	KCl 15°	KCl 25°	KCl 30°	KCl 35°	KCl 45°	KBr 25°	KBr 31–34°	KBr 37–40°	K$_2$SO$_4$ 25°	K$_2$SO$_4$ 31°
0.0005									6.25	
0.0006		1.76								
0.002									5.95	
0.005		1.53							5.86	
0.01						1.44			5.63	
0.02		1.22								
0.05		0.99					1.12	2.25		
0.1			0.50							
0.2			0.95							
0.4			1.25							
0.50										4.3
0.57								1.88		
0.96							1.33			
0.99										5.95
1.0		0.35		1.09	1.78			2.06		
1.5								2.00		
1.93							1.70			
2.0	0.08	0.76		1.40	1.92			2.10		
2.5								2.26		
3.0	0.45	1.05		1.58	2.03		2.13	2.33		
3.5								2.47		
3.85							2.36			
4.0	0.86	1.33		1.73	2.10					
4.04								2.60		

C. Magnesium and Strontium Salts

Concentration (mol l^{-1})	MgCl$_2$ 25°	MgSO$_4$ 31–33°	SrCl$_2$ 31°	SrCl$_2$ 39°
0.01	0.18			
0.1	0.19			1.32
0.2				1.35
0.4				1.59
0.5		6.87		
0.8				1.75
1.0		6.17	2.06	
1.6				1.98
2.0		4.73	1.50	
3.2				2.09
4.0		2.95	0.99	

[a] Data from various sources compiled in Landolt-Börnstein (1968, pp. 219–237).

In pore waters of sediments, thermal diffusion may contribute somewhat to molecular diffusion, but the contribution is small in most of the cases where molecular diffusional fluxes are pronounced. The following example illustrates this.

Example. Concentration of dissolved salts increases vertically in a pore water column from $10\,g\,l^{-1}$ at depth 0 to $11\,g\,l^{-1}$ at depth 100 m. The mean diffusion coefficient of the dissolved species is $D = 3 \times 10^{-6}\ cm^2\ sec^{-1}$. The Soret coefficient is $s = 2 \times 10^{-3}\ deg^{-1}$. The temperature gradient is twice the value of the mean geothermal gradient in the continental crust, $dT/dz = 6 \times 10^{-4}\ deg\ cm^{-1}$. With the preceding values one can estimate the molecular diffusional flux F_d,

$$F_d = D \frac{\Delta C}{\Delta z} = \frac{3 \times 10^{-6} \times (11 - 10) \times 10^{-3}}{10^4} = 3 \times 10^{-13}\ g\ cm^2\ sec^{-1}$$

Note that because of the thermal gradient, the temperature at the top of the sediment-pore water column T_0 increases by 6 deg at the bottom to temperature T_1,

$$T_1 - T_0 = (6 \times 10^{-4}\ deg\ cm^{-1})\ (10^4\ cm) = 6\ \ deg$$

According to equation 3.43 defining the dependence of diffusion coefficients on temperature, a 6 deg rise in temperature increases the value of D by about 30%,

$$\frac{D(T_1)}{D(T_0)} \simeq 1 + 0.05 \times 6 = 1.3$$

The average value of $D = 3 \times 10^{-6}\ cm^2\ sec^{-1}$ taken in this example is, therefore, the mean of $D = 2.55 \times 10^{-6}$ at the top and $D = 3.45 \times 10^{-6}\ cm^2\ sec^{-1}$ at depth 100 m.

The flux of material due to thermal diffusion F_T is ($\bar{C}$ is the mean concentration within the 100-m-long pore water column),

$$F_T = Ds\bar{C}\frac{\Delta T}{\Delta z} = 3 \times 10^{-6} \times 2 \times 10^{-3} \times \frac{11 + 10}{2} \times 10^{-3} \times 6 \times 10^{-4}$$

$$= 4 \times 10^{-14}\ g\ cm^{-2}\ sec^{-1}$$

The flux by molecular diffusion exceeds the thermal diffusion flux by a factor of about 10, from the preceding computation,

$$\frac{F_d}{F_T} \simeq 10$$

If both concentration and temperature increase in the downward direction, the thermal diffusion flux accounts for about 10% of the total flux upward.

3.6   DIFFUSION IN GASES

The atmosphere is a turbulent medium where dispersal by turbulence exceeds the rates of dispersal by molecular diffusion by many orders of magnitude (Figure 2.7). The environmental domains where molecular diffusion of gases in air may apply are the deeper pore space of soils and thin boundary layers of nonturbulent air at the interfaces between land and atmosphere, and water and atmosphere.

3.6.1   Diffusion of Gases in Air

Fick's first law of diffusion for a gas (component 1) diffusing into a stagnant layer of air or another gas (component 2) is

$$F_1 = -D_{12}\frac{dC_1}{dz} \tag{3.64}$$

where C_1 is concentration of the diffusing gas in units of mol cm^{-3} or g cm^{-3} and D_{12} (cm^2 sec^{-1}) is the interdiffusion coefficient (Section 3.2.3) of the gas in the gas-air mixture. If concentration of the diffusing gas is low and the total gas pressure is not affected by the added flux, the distinction between the frames of reference for the diffusional flux, relative to the other gas or relative to a fixed level, can be ignored (Section 3.2.1). If the gas behaves as an ideal gas, then its concentration C_1 is

$$C_1 = \frac{n_1}{V} = \frac{p_1}{RT} \qquad (\text{mol cm}^{-3}) \tag{3.65}$$

$$C_1 = \frac{M_1 p_1}{RT} \qquad (\text{g cm}^{-3}) \tag{3.66}$$

where p_1 is the partial pressure of the gas and M_1 is its gram-formula weight (g mol^{-1}). Equations 3.64 and 3.66 define the flux in terms of the partial pressure

$$F_1 = -\frac{D_{12}M_1}{RT}\frac{dp_1}{dz} \qquad (\text{g cm}^{-2}\ \text{sec}^{-1}) \tag{3.67}$$

The interdiffusion coefficient of gases in a binary gas mixture D_{12} is given by the following relationship:

$$D_{12} = \frac{B(T)T^{3/2}}{P}\left(\frac{1}{M_1} + \frac{1}{M_2}\right)^{1/2} \qquad (\text{cm}^2\ \text{sec}^{-1}) \tag{3.68}$$

where the parameter $B(T)$ is a quantity dependent on temperature and nature of the gases. The values of $B(T)$ must be calculated from tabulated parameters based on kinetic molecular characteristics of the gases. Such tabulations and computational examples are given, for example, in Satterfield (1970) and Perry (1963). Other parameters in equation 3.68: T is temperature in degrees Kelvin, P is total pressure (atm), M_1 and M_2 are the gram-formula weights (g mol^{-1}) of the gas and air or, in general, of the two interdiffusing gases.

For several gases in air at the total pressure $P = 1$ atm, the computed values of the interdiffusion coefficients in the temperature range between 5 and 35°C are listed in Table 3.8. According to the definition of D_{12}, the heavier the gas (larger M_1), the smaller is the diffusion coefficient. It may also be noted that diffusion coefficients depend on $T^{3/2}$ in gases, whereas in aqueous solutions D is proportional to T in the range of earth-surface temperatures (equation 3.43). The decrease in D_{12} with an increasing pressure has been reported as valid up to approximately $P \simeq 20$ atm (Satterfield, 1970).

In comparison to the diffusion coefficients of ionic solutes and gases in water (10^{-5} cm^2 sec^{-1}), the diffusion coefficients in the air are three to four orders of magnitude higher (10^{-2}–10^{-1} cm^2 sec^{-1}).

3.6.2   Gases in Pore Space

Porosity and Diffusion.   A molecular diffusional flux of a gas through the pore space of a material, written as Fick's first law of diffusion, is

$$F = -D\frac{dC_1}{dz} \qquad (\text{g cm}^{-2}\ \text{sec}^{-1}) \qquad (3.69)$$

where the flux F is units of mass per unit time and per unit of total area (i.e., a unit of area including the pore space and solid framework), D is the diffusion coefficient (cm^2 sec^{-1}), which includes the effects of porosity and tortuosity, and concentration C_1 is in units of mass per unit volume of interstitial space or gas filling the pore space.

The ratio D/D_{12} relates the diffusion coefficient D of one gas in another *within a pore space* to the diffusion coefficient D_{12} of the same gas in the other gas in the *absence of a solid framework*. Thus the ratio D/D_{12} is a function of the porosity and tortuosity of the sediment. The values of D are obtainable from measurements of the flux F under the conditions of a known concentration gradient, as given in equation 3.69. The diffusion coefficient D_{12} in a free gas space can be computed from equation 3.68 or taken from Table 3.8.

Figure 3.5 shows the values of the ratio D/D_{12} for hydrogen-air diffusion in various porous materials, including glass spheres, mica, vermiculite, kaolin, talc, pumex, soil crumbs, sodium chloride, and sand, in the porosity range $0.2 < \phi < 1.0$. The diffusion coefficient in the pore space decreases as a power of porosity,

$$D = D_{12}\phi^n \quad \text{or} \quad D = \phi D_{12}\phi^{n-1} \qquad (3.70)$$

with $n = 1.5\text{--}10.0$. The latter relationship is similar to the dependence of the ionic diffusion coefficients in pore waters of sediments on the sediment porosity, as given in equation 3.55 and Figure 3.3.

An alternative route to equation 3.70 is to write a relationship from the data in Figure 3.5,

$$\frac{D}{D_{12}} = \frac{\phi}{\theta} \qquad (3.71)$$

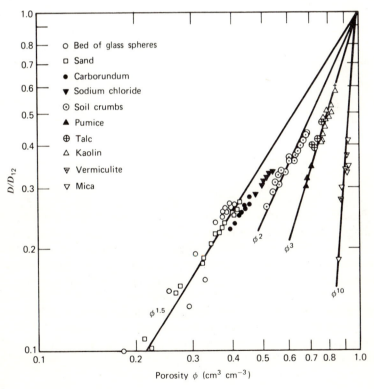

Figure 3.5 Diffusion coefficients of hydrogen in air, in various porous materials (adopted from Currie, 1960; Satterfield, 1970).

where θ is a factor analogous to tortuosity, as defined in equation 3.44. The value of $1/\theta$ is obtained by dividing the D/D_{12} value, read from a curve, by the corresponding value of ϕ. In this procedure, if $1/\theta = \phi$ then the relationship $D = D_{12}\phi^2$ is obtained, indicating a dependence of the diffusion coefficient in the pore space on ϕ^2. The following ranges of tortuosity values θ at atmospheric pressure ($P=1$ atm) are useful. For the data shown in Figure 3.5, within the porosity range $0.2 < \phi < 1.0$, θ varies between about 1.5 and 2.5. Within a porosity range $0.2 < \phi < 0.7$ for a variety of industrial porous materials (catalysts, powders), θ has been reported between 2 and 8.

In summary, the diffusion coefficients of gases in air in unconsolidated porous materials are lowered by a factor of 2 to 10, so that the ratio D/D_{12} becomes $0.5–0.1$, within the range of porosities applicable to dry sediments. The flux equation 3.69 can be written as

$$F = -D_{12}\phi^n \frac{dC_1}{dz} \tag{3.72}$$

or, to emphasize the fact that the flux F refers to a unit of total area and the diffusion coefficient in free gas decreases due to the porosity and tortuosity of sediment,

$$F = -\phi D_{12}\phi^{n-1} \frac{dC_1}{dz} \tag{3.73}$$

or, alternatively,

$$F = -\frac{D_{12}\phi}{\theta} \frac{dC_1}{dz} \tag{3.74}$$

Diffusion in Small Pores.   At atmospheric pressure, diffusional fluxes of gas through submicron-size pores are smaller than the fluxes predictable from equation 3.64, for a given concentration gradient and the diffusion coefficient D_{12}. The slower fluxes are attributable to the phenomenon of Knudsen diffusion: in small pores (or in gases at low pressures) the number of collisions between the gas molecules and the solid walls is large compared to the number of collisions between the molecules. The loss of energy during the molecule-wall collisions results in a smaller value of the diffusion coefficient describing the gas flow through the pore. The diffusional flux is

$$F = -\phi D_K \frac{dC_1}{dz} = -\frac{\phi D_K M_1}{RT} \frac{dp_1}{dz} \quad (\text{g cm}^{-2}\text{ sec}^{-1}) \tag{3.75}$$

where D_K (cm^{-2}sec^{-1}) is the diffusion coefficient for the Knudsen diffusion, F refers to the flux per unit of total area because of the porosity multiplier ϕ, and concentration C_1 is as defined in equation 3.66.

The diffusion coefficient D_K is defined for a cylindrical pore as

$$D_K = 9700 \, r_p \left(\frac{T}{M_1} \right)^{1/2} \qquad (\text{cm}^2 \text{ sec}^{-1}) \qquad (3.76)$$

where the numerical factor is $9700 = \frac{2}{3}(8R/\pi)^{1/2}$ and r_p is the radius (cm) of a cylindrical pore.

Two points may be noted in a comparison of the diffusion coefficients D_K and D_{12}: in the Knudsen diffusion range, the diffusion coefficient D_K

Table 3.8.   Diffusion Coefficients D_{12} (10^{-1} cm^2 sec^{-1})
of Components in Gas-Air Mixtures at Total Pressure
$P = 1$ atm[a]

Gas pair	Temperature (°C)			
	5	15	25	35
Air–Ar	1.72	1.83	1.94	2.06
–Br$_2$	0.88	0.94	1.00	1.07
–Cl$_2$	1.07	1.15	1.22	1.30
–CH$_4$	1.94	2.07	2.20	2.33
–CO	1.79	1.91	2.03	2.16
–CO$_2$	1.37	1.46	1.55	1.64
–F$_2$	1.79	1.91	2.02	2.14
–He	6.33	6.68	7.03	7.43
–Hg	1.14	1.23	1.31	1.40
–H$_2$	6.67	7.08	7.51	7.95
–HCl	1.51	1.61	1.73	1.83
–HCN	1.35	1.44	1.54	1.65
–H$_2$O	1.86	2.00	2.14	2.28
–H$_2$S	1.46	1.55	1.66	1.76
–Kr	1.34	1.43	1.52	1.62
–Ne	2.83	3.01	3.20	3.37
–N$_2$	1.80	1.91	2.04	2.15
–NH$_3$	1.91	2.05	2.19	2.34
–N$_2$O	1.37	1.46	1.55	1.64
–NO	1.81	1.92	2.04	2.17
–O$_2$	1.82	1.93	2.05	2.17
–SO$_2$	1.11	1.19	1.26	1.35

[a]Equation 3.68. Parameter $B(T)$ in the equation computed from tabulations in Perry (1963) and Satterfield (1970).

depends on the pore size r_p and on $T^{1/2}$ (for large pores the flux relationship 3.75 does not hold, as is shown below); in larger pores and in a free mixture of gases, the interdiffusion coefficient D_{12} depends on $T^{3/2}$ and on the total gas pressure P.

Example. An order of magnitude estimate of D_K. For CO_2 ($M_1 =$ 44 g mol^{-1}) at $T = 298°K$, diffusing through pores of radius $r_p = 10$ Å, the Knudsen diffusion coefficient D_K is, from equation 3.76,

$$9700 \times 10 \times 10^{-8} \times \left(\frac{298}{44}\right)^{1/2} = 2.5 \times 10^{-3} \qquad (\text{cm}^2 \text{ sec}^{-1})$$

The diffusion coefficients in the Knudsen diffusion range (10^{-3} cm^2 sec^{-1}) are 100 times greater than the diffusion coefficients for ionic and gaseous species in water (10^{-5}), and about 100 times smaller than the diffusion coefficients D_{12} for gases in a free space.

The pore radius r_p can be expressed through more easily measurable parameters, such as porosity ϕ and the specific surface area of the solid S_s (cm^3 g^{-1}). In a cylindrical pore the volume/surface-area ratio is equal to $r_p/2$. The volume of all the pores per unit volume of bulk sediment is ϕ, and the surface area per unit volume of bulk sediment is

$$S = S_s \rho_b = S_s \left[\rho_s(1-\phi) + \rho_g \phi\right]$$

$$S \simeq S_s \rho_s (1-\phi) \qquad (\text{cm}^2 \text{ cm}^{-3}) \qquad (3.77)$$

where ρ_b is the density of bulk sediment (g cm^{-3}), ρ_s is the density of the solid, and ρ_g is the density of the gas that is much less dense than the solid, or $\rho_g \ll \rho_s$. Thus an equation for the pore radius r_p, sediment surface area S, and pore volume ϕ is

$$\frac{r_p}{2} = \frac{\phi}{S} \qquad \text{or} \qquad r_p = \frac{2\phi}{S_s \rho_s (1-\phi)} \qquad (\text{cm}) \qquad (3.78)$$

Relationship 3.78 can be substituted for r_p in equation 3.76, giving D_K as a function of porosity and solid surface area (or total pore space area)

$$D_K = \frac{1.94 \times 10^4 \phi}{S_s \rho_s (1-\phi)} \left(\frac{T}{M_1}\right)^{1/2} \qquad (3.79)$$

Example. An estimate of pore size, equation 3.78. What is the mean pore radius of a sediment, the porosity of which is $\phi = 0.10$ or 10%, the surface area $S_s = 50$ m^2 g^{-1} (a value fairly typical of fine-grained sediments), and the density of solid particles $\rho_s = 2.5$ g cm^{-3}?

$$r_p = \frac{2 \times 0.1}{50 \times 10^4 \times 2.5 \times (1 - 0.1)} = 1.78 \times 10^{-7} \text{ cm} = 18 \text{ Å}$$

Wide Range of Porosities. Diffusion in the range of pore sizes between the small pores, where Knudsen diffusion applies, and the larger pore size, where ordinary molecular diffusion of gases takes place, is controlled by the following diffusion coefficient D (Pollard and Present, 1948; Evans et al., 1961; Scott and Dullien, 1962; Satterfield, 1970)

$$D = \frac{D_K D_{12} \phi^{n-1}}{D_K + D_{12} \phi^{n-1}} \quad \text{(cm}^2 \text{ sec}^{-1}) \tag{3.80}$$

where D_K is given by equation 3.76 or 3.79, and $D_{12}\phi^{n-1}$ is the gas interdiffusion coefficient, corrected for the effects of porosity, as given in equation 3.70. In the notation of equation 3.71, D_{12} can also be expressed by means of tortuosity, giving an alternative form of equation 3.80,

$$D = \frac{D_K D_{12}/\theta}{D_K + D_{12}/\theta} \quad \text{(cm}^2 \text{ sec}^{-1}) \tag{3.81}$$

If the pore radius r_p is large then D_K becomes large and the diffusion coefficient D in equation 3.80 or 3.81 tends to

$$D \rightarrow D_{12}\phi^{n-1} \quad \text{or} \quad D \rightarrow D_{12}/\theta$$

The latter indicates that if the porosity is large, diffusion takes place in accordance with the flux relationship 3.73. Conversely, if the pores and porosity are small then D_K becomes small and

$$D \rightarrow D_K$$

indicating that the flux is controlled by the Knudsen diffusion mechanism.

For the entire range of porosities (or pore sizes) the diffusional flux can be written as

$$F = -\phi D \frac{dC_1}{dz} = -\frac{\phi D}{RT} \frac{dp_1}{dz} \quad \text{(mol cm}^{-2} \text{ sec}^{-1}) \tag{3.82}$$

where D is as given either in equation 3.80 or 3.81.

Example.   Partial pressures of carbon dioxide in soils can be up to 30 times higher than the atmospheric pressure of CO_2, 1×10^{-2} against 3×10^{-4} atm. What is the diffusional flux of CO_2 through a 100-cm-thick layer, for different porosities of the layer, from 5% ($\phi = 0.05$) up?

Data for the example (see insert in Figure 3.6):

$\Delta z = 100$ cm

$\Delta p = 3 \times 10^{-4} - 5 \times 10^{-3} = -4.7 \times 10^{-3}$ atm, that is, the assumed CO_2 pressure difference between the top and bottom of the layer

Pressure gradient $\Delta p / \Delta z = -4.7 \times 10^{-5}$ atm cm^{-1}

$T = 298 \,^{\circ} K$

$$\frac{1}{RT} \frac{dp_1}{dz} = \frac{-4.7 \times 10^{-5} \text{ atm cm}^{-1}}{(82.06 \text{ cm}^3 \text{ atm deg}^{-1} \text{ mol}^{-1})(298\,^{\circ}K)}$$

$$= -1.922 \times 10^{-9} \text{ mol cm}^{-4}$$

$\rho_s = 2.5$ g cm^{-3}

$S_s = 20$ m^2 g$^{-1} = 2 \times 10^5$ cm^2 g^{-1}

$M_{CO_2} = 44$ g mol^{-1}

For the low porosity range, the diffusion coefficient D_K is, from equation 3.79,

$$D_K = \frac{(1.94 \times 10^4)(298^{1/2})}{(2 \times 10^5)(2.5)(44^{1/2})} \cdot \frac{\phi}{1 - \phi} = \frac{0.101\phi}{1 - \phi} \qquad (\text{cm}^2 \text{ sec}^{-1})$$

For the higher porosity range, the value of D_{12} for the CO_2 – air pair at $298\,^{\circ}K$ is taken from Table 3.8. The effective diffusion coefficient is, from equation 3.73 with $n = 2$,

$$D_{12}\phi = 0.155\phi \qquad (\text{cm}^2 \text{ sec}^{-1})$$

The value of the diffusion coefficient for the entire range of porosities, from equation 3.80, is

$$D = \frac{0.101\phi \times 0.155\phi}{(1 - \phi)[0.101\phi/(1 - \phi) + 0.155\phi]} = \frac{0.01561\phi}{0.256 - 0.155\phi} \qquad (\text{cm}^2 \text{ sec}^{-1})$$

Finally, the flux of CO_2 through air-filled pore space can be computed using equation 3.82 with the explicit values of the diffusion coefficient D and concentration gradient $(dp_1/dz)/RT$ given above.

$$F = \frac{0.01561\phi^2 \times 1 \times (2 \times 10^{-9}) \times (3.16 \times 10^7 \text{ sec yr}^{-1})}{0.256 - 0.155\phi}$$

$$= \frac{3.703 \times 10^{-3}\phi^2}{1 - 0.606\phi} \qquad (\text{mol cm}^{-2} \text{ yr}^{-1})$$

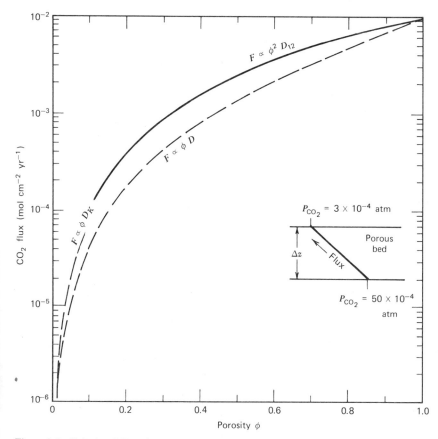

Figure 3.6 Calculated flux of CO_2 through a porous bed as a function of the bed porosity ϕ.

The computed CO_2 flux as a function of the soil bed porosity is shown in Figure 3.6.

For low porosities, the flux shown in Figure 3.6 was computed using equation 3.75 with the values of D_K and concentration gradient derived in this example:

$$F = \frac{(0.101\phi^2) \times (1.922 \times 10^{-9}) \times (3.16 \times 10^7)}{1 - \phi} = \frac{6.134 \times 10^{-3}\phi^2}{1 - \phi}$$

$$= \frac{6.134 \times 10^{-3}\phi^2}{1 - \phi} \quad (\text{mol cm}^{-2}\,\text{yr}^{-1})$$

Similar substitution of $D_{12}\phi$ and concentration gradient relationships in equation 3.73 gives the flux according to the interdiffusion model,

$$F = (0.155\phi^2) \times (1.922 \times 10^{-9}) \times (3.16 \times 10^7)$$

$$= 9.414 \times 10^{-3} \phi^2 \quad (\text{mol cm}^{-2} \text{ yr}^{-1})$$

The computed (hypothetical) fluxes may be used to estimate the residence time of CO_2 in the atmosphere with respect to the flux from soils to air. From Figure 3.6, at 50% porosity ($\phi = 0.5$), the flux is $F \simeq 2 \times 10^{-3}$ mol cm^{-2} yr^{-1}. Multiplied by the entire dry-land area of the globe (1.3×10^{18} cm^2) the flux is 2.6×10^{15} mol yr^{-1}. The mass of CO_2 in the earth's atmosphere is 5.5×10^{16} mol (Table 1.5). The renewal time of CO_2 in the atmosphere by the flux from soils is

$$\frac{5.5 \times 10^{16} \text{ mol}}{2.6 \times 10^{15} \text{ mol yr}^{-1}} = 20 \text{ yr}$$

The estimate of the renewal time can be made longer if the average pressure gradient of CO_2 in soils is smaller and the total area contributing CO_2 is smaller than the values used in this example.

A difference of a factor of 2 in the values of the flux emerges from the different models, as shown by the curves in Figure 3.6, labeled $F \propto \phi^2 D_{12}$ and $F \propto \phi D$. The reason for the lower values of the flux according to the model of equations 3.80 and 3.82 is that the relative weight of diffusion through cylindrical pores in the total flux is too great for loosely packed unconsolidated materials. For the latter, the diffusional flux of gases that follows proportionality relationship $F \propto \phi^2 D_{12}$ shown in Figure 3.6 may be a better approximation to air-filled sediments.

3.7 DIFFUSION IN SOLIDS

Diffusion in minerals at temperatures below 200°C is a poorly explored field because of the slowness of diffusional processes. Theoretical treatment of diffusion in solids, with reference to the different atomic mechanisms responsible for the migration of different chemical species within crystal lattices, can be found in books by Jost (1960), Shewmon (1963), and Manning (1968).

The tracer or intrinsic diffusion coefficients D are usually computed from the measured concentration gradients and fluxes. When the experimentally determined values of $\log D$ are plotted against the reciprocals of temperature $1/T$, a straight line is often obtained for a certain temperature range (compare Figure 3.7). This so-called Arrhenius plot is represented by the relationship between the diffusion coefficient, activation energy for diffusion, and temperature

$$D = D_0 \exp \left(- \frac{E}{RT} \right) \quad (\text{cm}^2 \text{ sec}^{-1}) \quad (3.83)$$

where E is interpreted as the activation energy for diffusion (kcal mol^{-1}), T is temperature in degrees Kelvin, R is the gas constant, and the value of

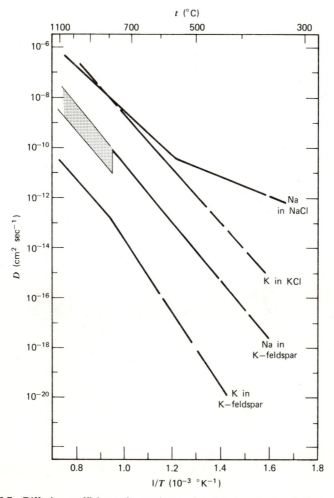

Figure 3.7 Diffusion coefficients of potassium and sodium in potassium feldspar, halite, and sylvite (from data in Table 3.9).

Table 3.9 Diffusion Coefficients in Solids. Equation 3.83

	Temperature range (°C)	D_0 (cm² sec⁻¹)	E (kcal mol⁻¹)	Extrapolated D at 25°C (cm² sec⁻¹)	Data source and notes
In NaCl					
Na	390–550	1.6×10^{-6}	17.7	1.7×10^{-18}	a
	550–1000	3.13	41.4		
SO₄²⁻	540–740	0.065	25.8	7.9×10^{-21}	c
In KCl					
K	600–730	137.	49.6	6.0×10^{-35}	b
Cl	500–730	178.	51.9	1.6×10^{-36}	b
SO₄²⁻	570–740	0.187	28.4	2.8×10^{-22}	c
He	80–250	0.035	9.0	8.8×10^{-9}	c
In SiO₂ (quartz)					
Na	300–570	0.7 ± 0.1	20.2	1.1×10^{-15}	d
	600–800	$(7 \pm 1) \times 10^{-3}$	11.5		∥ c axis
	440–570	200 ± 100	41	1.8×10^{-28}	∥ c axis
	600–790	$(4 \pm 2) \times 10^{-2}$	27		⊥ c axis
Ca	600–800	$(1 \pm 0.8) \times 10^{5}$	68	1.4×10^{-45}	⊥ c axis
O₂	1010–1220	3.7×10^{-9}	55		
In Na₂SiO₃ glass					
Na	100–600	3×10^{-3}	18.5	8.2×10^{-17}	d
		$(0.7-12) \times 10^{-3}$	15–22		D_0 and E concentration dependent (see Fig. 3.8)

Table 3.9 (*Continued*)

	T (°C)	D_0	E	D		Comments
In K_2SiO_3 glass						
K	250–500	4×10^{-4}	17.5	5.9×10^{-17}	d	D_0 and E concentration dependent (see Fig. 3.8)
		$(2\text{–}24) \times 10^{-4}$	15.5–19.5		d	
In SiO_2 glasses						
H_2O	200–1000	6.5×10^{-4}	11.3	3.4×10^{-12}		Different measurement techniques using H_2; H^+ and OH^- in glasses
	300–1000	5.6×10^{-4}	10.4			
	600–1000	1.0×10^{-6}	18.3			
	700–1200	2.7×10^{-7}	17.3			
	800–1500	9.5×10^{-4}	15.8			
wt % $H_2O = 0$	600–1200	3.4×10^{-7}	13	1.0×10^{-16}	e	H_2O in obsidian
$= 0.5$	600–1200	2.7×10^{-6}	13	8.0×10^{-16}		
$= 6.0$	600–1200	3.4×10^{-5}	13	1.0×10^{-14}		
In feldspars						
K	600–800	16.1	68.	1.6×10^{-49}	f	Or_{94} orthoclase
	700–1300	0.02	50		g	Or_{86} adularia
Na	500–800	8.9	53	2.1×10^{-38}	f	Or_{94} orthoclase
	800–1300	0.4	51		g	Or_{86} adularia
Rb	700–800	38.0	73.0		f	Or_{94} orthoclase
^{40}Ar	500–800	0.014	43.1	3.6×10^{-34}		Or_{94} orthoclase
	400–700	2.8×10^{-6}	29.6	5.6×10^{-28}	h	Or_{100} microline
		5.3×10^{-7}	29.6	1.1×10^{-28}		Or_{100} adularia
O_2	350–700	4.5×10^{-8}	25.6	7.7×10^{-27}	i	Or_{98} adularia
	350–800	2.3×10^{-9}	21.3	5.6×10^{-25}		Ab_{97-99} albite
		1.4×10^{-7}	26.2	8.7×10^{-27}		An_{96} anorthite

Table 3.9 (*Continued*)

In mica						
Ar	600–900	0.75	57.9	2.7×10^{-43}	j	phlogopite with 4% annite, $P = 2$ kbars
	225–700	1.7×10^{-5}	23.3	1.4×10^{-22}		phlogopite in vacuum
In zeolites						
H_2O	25			10^{-7}–10^{-8}	a	heulandite, parallel to structure
				10^{-11}		perpendicular to structure
	45			2.1×10^{-8}	k	Ca-heulandite
				1.3×10^{-7}		Ca-chabazite
				2.0×10^{-13}		Na-analcite
				1.5×10^{-12}		Na-analcite
Na	75			10^{-12}–10^{-13}		analcite, chabazite
Cs	25			10^{-24}, 10^{-13}		analcite, chabazite
Ca				4×10^{-16}		chabazite
In silicates, reactive surface layer						
Na, K, Ca, Si	25			5×10^{-21} to 1×10^{-22}	l	in orthoclase, albite, oligoclase, labradorite, bytownite, andesine, anorthite
Mg	25			10^{-13}–10^{-18}	m	in serpentine, fosterite, enstatite

Table 3.9 (*Continued*)

Si				$10^{-15}\text{--}10^{-18}$		
In calcite						
CO_2	690–850	4.5×10^{-4}	58.0	1.4×10^{-46}	[n]	$CaCO_3$ 99.5%, $MgCO_3$ 0.5%, $FeCO_3$ trace. Lattice diffusion

[a] Jost (1960).
[b] *Diffusion and Defect Data* (1971, p. 108).
[c] *Diffusion and Defect Data* (1974, p. 103).
[d] Frischat (1975).
[e] Shaw (1974).
[f] Foland (1974).
[g] Petrović (1974).
[h] Yund and Anderson (1974).
[i] Giletti et al. (1978).
[j] Giletti (1974).
[k] Sherry (1975).
[l] Busenberg and Clemency (1976).
[m] Luce et al. (1972).
[n] Haul and Stein (1955).

the constant coefficient D_0 (cm^2 sec^{-1}) is obtained by extrapolation of a plot of $\ln D$ against $1/T$ to the value of $1/T=0$ or $T\to\infty$. A constant slope of such a plot indicates that the activation energy for diffusion E is a constant independent of temperature and composition within some temperature range. In Figure 3.7 are shown the tracer diffusion coefficients D of sodium and potassium in potassium feldspars (orthoclase, nominal composition $KAlSi_3O_8$), halite (NaCl), and sylvite (KCl). For diffusion in potassium feldspar, the slopes of the lines change near 800°C, and this is reflected in the change of the activation energy value E, which is higher in the lower temperature range.

For a number of chemical species occurring in minerals and some glasses, the values of the parameters D_0 and E of equation 3.83 are listed in Table 3.9.

Breaks in the slope of the plot of $\log D$ against $1/T$, such as the one shown in Figure 3.7, are characteristic of diffusion in solids in general. The breaks are usually interpreted as indicating either a solid phase transformation, or a change in the mechanism of diffusion, or both. A decrease in the value of the activation energy parameter E at a higher temperature is fairly common. An opposite effect of E becoming higher at higher temperatures has been reported for diffusion of sodium in NaCl crystals (Table 3.9). The activation energy parameter E for diffusion in solids at temperatures above ~ 100°C is of the order of magnitude of tens of kilocalories per mole. For diffusion in aqueous solutions below 100°C, E is of the order of a few (< 10) kilocalories per mole. The difference in the magnitudes of E for solids and aqueous solutions means that diffusion in solids is more sensitive to temperature changes. The slower diffusion is related to the higher activation energies needed to cause a displacement of an atom, ion, or a vacancy within a solid lattice.

Experimentally determined values of diffusion coefficients D, for a given diffusing species and a given solid, may vary by a factor of 5 or greater within or between reported experiments. Thus the values of D that are calculated from equation 3.83 using the reported values of the parameters D_0 and E should be regarded as order of magnitude estimates.

An alternative form of equation 3.83 can be useful. The value of the diffusion coefficient D_{T_2} at some temperature T_2 can be estimated from a known value D_{T_1} at temperature T_1 as

$$\ln\frac{D_{T_2}}{D_{T_1}} = \frac{E}{R}\left(\frac{1}{T_1} - \frac{1}{T_2}\right) \tag{3.84}$$

Writing

$$T_2 = nT_1 \quad \text{(deg)}$$

where $n \lesssim 1$ is a factor relating the two temperatures, and using decimal instead of natural logarithms ($\ln = 2.303 \log$, and $R = 1.987 \times 10^{-3}$ kcal deg^{-1} mol^{-1}), equation 3.84 becomes

$$\log \frac{D_{T_2}}{D_{T_1}} = \frac{218.5E}{T_1} \frac{n-1}{n} \tag{3.85}$$

If $n = 1$ then the temperature and diffusion coefficient are unchanged. An increase in temperature by a factor of 2 (i.e., $n = 2$) increases $\log D_{T_2}/D_{T_1}$ from 0 to $0.5E/RT$, and an increase by a factor of 4 ($n = 4$) raises $\log D_{T_2}/D_{T_1}$ to $0.75E/RT$. This is another way of stating that the greater the activation energy parameter E, the stronger is the increase in D with increasing temperature.

In sodium and potassium silicate glasses of variable composition, $(Na_2O)_x (SiO_2)_{1-x}$ and $(K_2O)_x(SiO_2)_{1-x}$, the activation energy E for diffusion of sodium and potassium depends on the chemical composition of the glass: the values of E shown in Figure 3.8 vary with the mole fraction of sodium and potassium, although the scatter of the reported E values for a given composition is large. For example, a variation in E about the value of 18 ± 2.5 kcal mol^{-1} results in a $\pm 15\%$ variation in $\ln D/D_0$; the latter

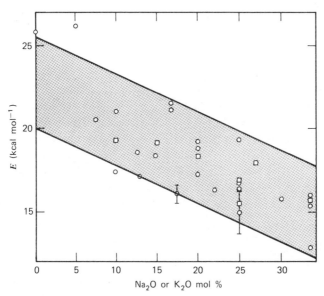

Figure 3.8   Dependence of the activation energy E for diffusion of Na (○) and K (□) on the chemical composition of sodium–potassium glasses (data from Frischat, 1975, pp. 138, 147).

corresponds to a fiftyfold variation in D, between 7×10^{-10} and 3×10^{-8} cm^2 sec^{-1} at 400°C. However, as D_0 and E are both determined from the log D-against-$1/T$ plots, the scatter of D_0 values for sodium and potassium in glasses of different composition is also great. For the Na data plotted in Figure 3.8 (see also Table 3.9), 21 out of 23 values of D_0 fall between 0.7×10^{-3} and 12×10^{-3} cm^2 sec^{-1}. (The extreme high and low values, 0.37 ± 0.19 and 1.6×10^{-4} are associated with the highest and lowest values of E, near 0 and 33.3 mol % Na). The mean of D_0 given in Table 3.9 is $D_0 = 3 \times 10^{-3}$ cm^2 sec^{-1}. The number of data points for K in glass is only seven, and the mean D_0 within a range of a factor of 2 is $D_0 = 4 \times 10^{-4}$ cm^2 sec^{-1}.

Chemical reactions between aqueous solutions and solids depend on the transport of reacting species from the interior of the solid to the surface (slow) and from the solution to the solid surface (relatively fast). From experiments on leaching of Na and K from silicate glasses in water it has been inferred that diffusion through a layer a few microns thick at the solid surface is much faster than in the bulk glass (Frischat, 1975, p. 92). Diffusional fluxes out of the surface layer are characterized by small concentration gradients and diffusion coefficients that are several orders of magnitude higher than in the solid. The diffusion coefficients of alkali ions in leaching experiments are in the range 10^{-14}–10^{-17} cm^2 sec^{-1} at 40°C and 10^{-12}–10^{-16} cm^2 sec^{-1} at 80°C. At the same temperature, the extrapolated value of diffusion coefficient D of sodium in potassium feldspar is much lower, near 10^{-38} cm^2 sec^{-1}. A possible structural reason for a faster diffusion in a surface layer in contact with an aqueous solution is penetration of water and formation of a gel-like structure of the glass, with water filling some fraction of the space within the surface layer. Any structural deformation of a glass surface, such as fracturing or cracking, should presumably promote water penetration. When silicate minerals react with water, the cations and silica are leached at unequal rates from the surface of the solid during the early stages of dissolution (Section 5.5). This phenomenon has been interpreted as the formation of a reactive surface layer through which cations and silica diffuse to solution. The diffusion coefficients near 25°C are of the order of 10^{-13}–10^{-18} cm^2 sec^{-1}, comparable to the diffusion coefficients in hydrated glass. Some values of D for diffusion through the reactive surface layer of silicates are listed in Table 3.9.

Diffusion of water in silicate glass and in volcanic glass (obsidian) is faster than diffusion of metal ions at similar temperatures in ordered silicate structures. The data on diffusion of water in silicate glass given in Table 3.9 show that the diffusion coefficient of H_2O depends on the water concentration in the glass: an increase in the H_2O mass fraction from 0 to 6% raises the diffusion coefficient about a hundredfold.

Diffusion of water in zeolites is much faster than in silicate glasses, as can be anticipated on the basis of the open-channel framework of natural and synthetic zeolites. Zeolites form a mineralogically distinct group of aluminosilicates that are characterized by (1) a well-defined porous space within the crystal lattice, with pores of angstrom dimensions, and (2) exchangeability of water and some of the cations. As removal of water and exchange of some of the cations do not destroy the crystal lattice of a zeolite, zeolites have found wide applications in water filtration, purification, and petroleum processing operations. The presence of well-defined porous channels makes zeolites molecular sieves. The possible water content of zeolites and cross-section dimensions of the pores are listed in Table 3.10.

Near 25°C, the diffusion coefficient D of water in the zeolite heulandite is of the order of $10^{-8}-10^{-7}$ cm^2 sec^{-1}, in the direction parallel to the layered structure. In the perpendicular direction, $D \simeq 10^{-11}$ cm^2 sec^{-1} has been reported for room temperature (Jost, 1960). These values of D for H$_2$O in zeolites may include the effects of water uptake or adsorption on lattice sites, as such processes tend to slow down the diffusional flux of a chemical species through a medium with which it enters in a chemical reaction (see Sections 7.5 and 8.2). From geometric considerations of the sizes of water molecules (radius $r = 1.4$ Å) and pores ($r = 3$ Å), a reduction of a factor of 10 may be expected for diffusion of H$_2$O through zeolite pores, from equation 3.57,

$$\frac{D}{D_0} \simeq \left(1 - \frac{1.4}{3}\right)^4 = 0.08$$

Table 3.10.   Pore Space as Water Volume and Pore Opening Dimensions of Main Zeolite Groups[a]

Zeolite group	cm^3 H$_2$O per 1 cm^3 zeolite	Pore opening dimensions (Å)
Analcite	0.18	–
Mordenite (ptilolite)	0.20–0.33	2.9×5.7–6.7×7.0
Phillipsite	0.34–0.49	2.4×4.8–4.2×4.4
Chabasite	0.30–0.46	2.6×2.6–6.9×6.9
Faujasite	0.49–0.54	3.9×3.9–7.4×7.4
Heulandite	0.33–0.37	2.3×5.0–3.2×7.8
Natrolite	0.21–0.33	2.6×3.9–3.5×3.9

[a]From Weiss (1969).

Thus the self-diffusion coefficient of H_2O at 25°C, $D = 2.5 \times 10^{-5}$ cm^2 sec^{-1} (Robinson and Stokes, 1970; Eisenberg and Kauzmann, 1969), may be reduced to $D \simeq 2 \times 10^{-6}$ cm^2 sec^{-1}. A reduction of D by another factor of 10 may mean either a hindering effect by uptake of H_2O molecules on the zeolite lattice, or diffusion of H_2O units consisting effectively of more than one molecule, each unit having a radius of about $r \simeq 2$ Å. The radius of a water molecule calculated from a water density of 1 g cm^{-3} and molecular weight 18 g mol^{-1}, gives a value close to $r \simeq 2$ Å,

$$r = \left(\frac{18 \times 3}{1 \times 6.023 \times 10^{23} \times 4\pi} \right)^{1/3} \times 10^8 = 1.93 \text{ Å}$$

which is larger than the radius of water molecules $r \simeq 1.4$ Å.

Diffusion of small gaseous molecules, such as nitrogen and simple hydrocarbons in zeolites of 4–5 Å pore size has been reported (Satterfield, 1970) to take place much slower than expected from Knudsen diffusion through porous solids: the diffusion coefficients for gases at temperatures between 20 and 200°C are 10^{-14}–10^{-12} cm^2 sec^{-1}, whereas on the basis of Knudsen diffusion, equation 3.76, values of D between 10^{-4} and 10^{-3} cm^2 sec^{-1} would be obtained. Monovalent and divalent cations, such as Na, Cs, Ca, and Ba, diffuse in the small-pore zeolites (2×4 Å pore dimensions) at the rates comparable to diffusion in the reactive surface layers of silicate minerals. Some values of the diffusion coefficients of water and cations in various zeolites are listed in Table 3.9.

To conclude this section and chapter, a point of historical interest may be mentioned. In 1913, R. E. Liesegang published a book, the title page of which is reproduced in Figure 3.9. Liesegang's earlier work had dealt with the formation of rhythmic and concentric zones of precipitates in gels, from which he drew comparisons between the experiments and natural occurrences of rhythmic banded structures in concretions and agates. The structures, called the Liesegang rings, bear his name. In the 170-page book, Liesegang stresses the role of diffusion in a wide variety of natural processes, ranging from the formation of minerals and ore bodies to the mixing of water layers in the ocean and saline lakes. Many of Liesegang's own experiments on precipitation of banded structures in gels are described in detail. Although a respectable volume of theoretical and experimental information on diffusion existed by 1912, Liesegang's book gives neither the references to the earlier theoretical work on diffusion nor any equations for diffusion whatsoever. The ability of a person to write a book on the role of diffusion in geology without a need to deal even with the simplest Fick's equation can only be admired.

GEOLOGISCHE DIFFUSIONEN

VON

RAPHAEL ED. LIESEGANG

MIT 44 ABBILDUNGEN

DRESDEN UND LEIPZIG
VERLAG VON THEODOR STEINKOPFF
1913

Figure 3.9   Title page of Liesegang's book.

CHAPTER 4

Atmospheric Processes

The approach of this chapter to the geochemical processes in the atmosphere is cast in terms of the reactions involving water. The concept of a well-mixed reservoir (Chapter 1) applies to the atmosphere better than to any other reservoir of the exogenic geochemical cycle. The dispersal and mixing within the atmosphere translate the local inputs from land to the hemispheric and global averages of the return fluxes to the earth's surface. During condensation and nucleation of water vapor, gaseous and solid species are delivered with the atmospheric precipitation to the earth's surface. A summary of the growth and evaporation of rain droplets, and of the uptake of solids and gaseous components by rain provides the basic background for the processes dealt with later in the chapter. Discussion of the more specific cases centers on the chemical composition of atmospheric precipitation, acidity, and alkalinity of rain waters, and the fluxes of a volatile element (mercury) in the atmosphere-land-ocean system.

4.1 WATER IN THE ATMOSPHERE

Rivers, lakes, and ground waters are supplied from the atmospheric water reservoir. The flux of water from the atmosphere to the land surface in the form of atmospheric precipitation and the fluxes of reactive substances associated with it are to a large extent responsible for the chemical weathering of the earth's crust, as well as for the transport of the weathered materials. The total mass of water vapor present in the atmosphere is relatively small: an average figure for the earth is about 2.5 g cm^{-2}, which accounts for 0.25% of the total mass of the air column. In Figure 4.1 is shown the distribution of water vapor in the atmosphere of the Northern Hemisphere. The mass of water vapor is highest in the tropics, decreasing toward high latitudes. The global mean coincides with the water vapor concentration in the 30–40°N belt, where it is by a factor of 2 lower than the highest value near the equator.

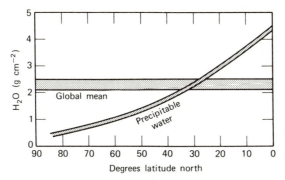

Figure 4.1   Water in the atmosphere. Variation in the amount of water vapor with latitude (from Junge, 1963; Eagleson, 1970) and range of estimates for the global mean (from Nace, 1967; Garrels et al., 1975).

At the mean temperature of the earth, taken as 15°C, the partial pressure of water vapor at equilibrium with pure water is 17 mbar or 0.017 atm. The equilibrium pressure is also called the saturation vapor pressure. Precipitable water vapor is, on the average, 0.0025 atm, and this quantity amounts to 0.25%/0.017 = 15% of the saturation water vapor pressure. The ratio of the water vapor pressure to the saturation vapor pressure at the same temperature is the relative humidity. From the preceding, the mean relative humidity of the earth's atmosphere is only 15%, a value that is low in human and biological terms.

The global average rate of atmospheric precipitation per 1 cm^2 of the earth's surface is between 84 and 88 cm yr^{-1}. The same value applies to the mean rate of evaporation and, perhaps surprisingly, a similar estimate —$\frac{1}{10}$ inch per day—was arrived at near the end of the seventeenth century by Halley (1687). The rate of precipitation divided into the mean water vapor content of the atmosphere gives the residence time of about 10 days The residence time of water vapor can be interpreted as the time between successive rain showers, its shortness being responsible for the local varia- tions in the chemical composition of rain water. Materials from the oceans and continents are raised into the atmosphere and returned to the earth's surface via atmospheric precipitation and as dry fallout. Wind and turbu- lence tend to average the composition of the atmosphere that receives inputs of different nature and magnitude from continental and oceanic sources. Horizontal dispersal in the troposphere is characterized by eddy diffusion coefficients of the order of 10^{10} cm^2 sec^{-1} (Figure 2.7). During a period of 10 days, corresponding to the mean residence time of water in the atmosphere, horizontal eddy turbulence disperses materials over dis- tances up to 1000 km. Dispersal of materials over much larger distances, such as global dispersal of volcanic dust from major eruptions, is accom- plished by high-altitude winds and via the stratosphere. In relation to the sources of gaseous and solid materials at the earth's surface, the shorter dispersal distances are characteristic of the lower atmosphere. Opposed to the averaging trend of dispersal, atmospheric precipitation removes material from the atmosphere before it has been uniformly dispersed, with the result that rain and snow may be chemically imprinted by local sources. The better known examples of this phenomenon are the greater abundance of oceanic salts in rains over coastal areas, and higher concentrations of certain gases near the centers of human industrial activity and in the vicinity of active volcanos.

The ideal gas law is used as an equation of state of water vapor in the atmosphere in either one of the following two forms:

$$p = \frac{nRT}{V} \tag{4.1}$$

$$p = \rho_v R_v T \quad (\text{g cm}^{-1} \text{ sec}^{-2}) \tag{4.2}$$

where p is the partial pressure of water vapor, n is the number of moles of H_2O in volume V, T is temperature in degrees Kelvin, ρ_v is the density of water vapor, and R_v is the gas constant for water vapor, defined as the ratio of the gas constant and the gram-formula weight of water vapor ($R = 8.3147 \times 10^7$ erg deg^{-1} mol^{-1} and $M_w = 18.016$ g mol^{-1}, physical constants are listed in Appendix D)

$$R_v = \frac{R}{M_w} = 0.462 \times 10^7 \quad (\text{erg deg}^{-1} \text{ g}^{-1}) \tag{4.3}$$

Saturation vapor pressure p_o is the vapor pressure at equilibrium with liquid water at a given temperature, corresponding to the equilibrium reaction

$$H_2O(l) = H_2O(g)$$

Values of p_o at different temperatures and 1 bar total pressure are listed in Table 4.1. Relative humidity R.H., referrred to earlier in this section, is the ratio of vapor pressure p to the equilibrium pressure p_o at the same temperature,

$$\text{R.H.} = \frac{100p}{p_o} = \frac{100\rho_v}{\rho_{vo}} \quad (\%) \tag{4.4}$$

where ρ_{vo} is the density of the equilibrium vapor.

Table 4.1 Values of Parameters Pertaining to Condensation and Evaporation of Water[a]

Temperature (°C)	σ (erg cm^{-2})	p_o (mbar or 10^3 erg cm^{-3})	ρ_{∞} (10^{-6} g cm^{-3})	L (10^{10} erg g^{-1})	κ (10^3 erg cm^{-1} sec^{-1} deg^{-1})	D (cm^2 sec^{-1})
−10	77.10	2.863	2.355	2.523	2.36	0.211
− 5	76.40	4.215	3.402			
0	75.62	6.108	4.840	2.499	2.43	0.226
5	74.90	8.719	6.785			
10	74.20	12.272	9.381	2.476	2.50	0.241
15	73.48	17.044	12.803			
20	72.75	23.373	17.257	2.452	2.57	0.257
25	71.96	31.671	22.992			
30	71.15	42.430	30.294	2.428	2.64	0.273
35	70.35	56.236	39.500			
40	69.55	73.777	50.993	2.405	2.70	0.289

[a]Landolt-Börnstein (1956), Smithsonian Meterological Tables (1958), Rogers (1976). σ: surface tension between pure water and air; p_o: water vapor pressure at equilibrium with liquid water; ρ_{∞}: density of water vapor at equilibrium with liquid; L: heat of condensation of water vapor; κ: coefficient of thermal conductivity of air; D: diffusion coefficient of water vapor in air at 1000 mbar or approximately 1 atm total pressure. Vapor density $\rho_{\infty} = 2.165 \times 10^{-7} \ p_o/T$; p_o in erg cm^{-3} and T in degrees Kelvin.

4.2   RAIN AS A TRANSPORT AGENT

4.2.1   Solid Particles and Liquid Droplets in Air

Solid particles and liquid droplets in the atmosphere are collectively known as aerosols. The origins of aerosols are diverse: oceans contribute salt particles forming by evaporation of seawater spray, continents contribute mineral and organic detritus from natural and man-produced sources, and particles are also produced by oxidation of gases. In the photochemically active environment of the higher altitudes, numerous reactions among gases and small particles lead to production of new chemical species and charged ions. The range of sizes of particles in the atmosphere, shown in Figure 4.2, extends from the submicron-size Aitken particles (so named after their discoverer) into the macroscopic range. Irrespective of the size and name, concentration of particles over the oceans is the lowest, and it is the highest in the areas affected by human activity. Order of magnitude values of concentrations are given in Figure 4.2.

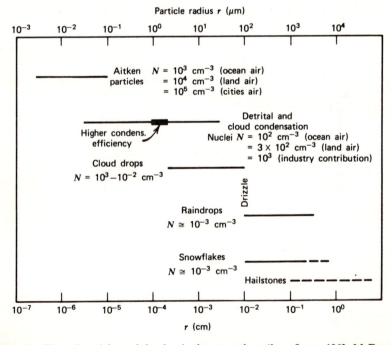

Figure 4.2   Sizes of particles and droplets in the atmosphere (from Junge, 1963; McDonald, 1958; Radke and Hobbs, 1976).

An estimate of the total mass of particulate materials in the atmosphere can be derived from the global production rate of particles in the atmosphere $(2000 \times 10^{12}$ g yr$^{-1})$ and their mean residence time in the atmosphere (about 10 days). The mass of particulates is

$$2 \times 10^{15} \text{ g yr}^{-1} \times 0.03 \text{ yr} = 6 \times 10^{13} \text{ g}$$

Input of particulate materials to the atmosphere includes mineral dust, sea salt spray, products of oxidation of gases, and direct human-made and natural inputs, such as forest fires and volcanic emanations. Estimates of the rate of particle input range from 800×10^{12} to 2750×10^{12} g yr^{-1} (Goldberg, 1971; Schneider and Kellogg, 1973). Residence times of particles in the atmosphere range from about 1 week in the lower troposphere up to 40 days at higher altitudes (Cadle, 1973).

Mean concentration of particulate matter C_a in the atmosphere of volume 4×10^{18} m^3 is

$$C_a = \frac{6 \times 10^{19} \ \mu\text{g}}{4 \times 10^{18} \ \text{m}^3} = 15 \ \mu\text{g m}^{-3}$$

The latter figure can be compared to the mean concentrations of particulates in the air of nonurban areas in the continental United States, 23–83 μg m^{-3} (3–10% of the particulates is organic matter), and to the mean concentration of particulates at urban areas, 111 μg m^{-3}, with a variation factor of 1.8 (Cadle, 1973, 1975).

A fraction of aerosol particles acts as nuclei for condensation of water vapor. These are called cloud condensation nuclei, and their radii are greater than 0.1 μm. Liquid or frozen water droplets forming clouds range in size from $r \simeq 2$–100 μm. By convention, the size $r = 100$ μm $= 0.1$ mm is often taken as a dividing line between cloud and rain droplets. The number concentration of cloud droplets varies, depending on their sizes, from 10^3 cm^{-3} for the smaller droplets to 10^{-2} cm^{-3}, or 10 liter^{-1}, for the biggest droplets. Rain droplets form as a result of condensation and coalescence of the smaller cloud droplets. A typical size range of rain droplets is from $r = 0.1$ mm to about 6 mm. At the upper size limit, water droplets become unstable and break into smaller droplets. A characteristic number concentration for rain is about 10^{-3} cm^{-3}, or 1 droplet per liter of air.

Water condensing in the atmosphere to rain includes (1) strongly soluble solids, mostly of oceanic salt origin, (2) weakly soluble or insoluble solids (dust), and (3) dissolved gases. The mechanisms responsible for the formation of liquid water droplets and incorporation in them of dissolved and solid materials are discussed in the following sections.

4.2.2   Conditions of Growth and Evaporation of Rain Droplets

Droplet Size and Vapor Pressure.   Attachment of water molecules to the surfaces of hygroscopic particles, lowering of temperature, or both, produce nuclei of precipitation. A liquid droplet can stably exist if the water vapor pressure at its surface is equal to the water vapor pressure in its surroundings. If the vapor pressure at the droplet surface is higher than in the surroundings, the droplet evaporates. Conversely, if the vapor pressure at the surface is lower than in the surrounding atmosphere, the droplet grows by condensation. The problem of condensation of water vapor and the formation of liquid and solid nuclei in the atmosphere is encountered in such fields as meterology, atmospheric physics, and atmospheric chemistry. An advanced thermodynamic treatment of the theory of condensation, approached as a phenomenon in a liquid-vapor system, is given in the monograph by Dufour and Defay (1963). Mason's (1971) book on the physics of clouds deals with the formation of rain and ice nuclei from a more general point of view—leaning toward physics and meteorology. Shorter but clear and readable discussions of the problem of vapor condensation and nucleation are found in the books by Byers (1965) and Rogers (1976). The material in this section deals primarily with those features of growth and evaporation of rain droplets that are relevant to the occurrence of dissolved and solid substances in rain water.

Surface tension at the interface between liquid water and air affects the vapor pressure of small droplets. The relationship between the vapor pressure of a spherical droplet and its radius is given by the Kelvin equation,

$$p_o(r) = p_o e^{a/r} \qquad (4.5)$$

where $p_o(r)$ is the water vapor pressure at equilibrium with a droplet of radius r, p_o is the equilibrium pressure at a flat water surface ($r \to \infty$), and a is a parameter defined below. Both $p_o(r)$ and p_o can be taken in any mutually consistent pressure units. Parameter a is a quotient of the surface tension, liquid water density, gas constant, and temperature,

$$a = \frac{2\sigma}{\rho_L R_v T} \qquad \text{(cm)} \qquad (4.6)$$

where σ is surface tension between water and air (in units of dyn cm^{-1} or erg cm^{-2}), ρ_L is the density of liquid water (g cm^{-3}), R_v is the water vapor gas constant defined in equation 4.3, and T is temperature in degrees Kelvin.

At 10°C, the value of a is

$$a = \frac{2 \times 74.2}{1.0 \times (0.462 \times 10^7) \times 283} = 1.1 \times 10^{-7} \, \text{cm} = 1.1 \times 10^{-3} \, \mu\text{m}$$

The values of interfacial water-air tension at several temperatures are given in Table 4.1. For water droplets sufficiently larger than 0.001 μm, equation 4.5 becomes

$$p_o(r) = p_o\left(1 + \frac{a}{r}\right) \tag{4.7}$$

For water droplets of radius larger than $r = 0.1$ μm the term $a/r = 0.01$, such that the vapor pressure at the droplet surface is approximately equal to the vapor pressure at a flat water surface, and the droplet size does not affect its water vapor pressure. As a is a positive quantity, an equilibrium vapor pressure at a droplet surface is always higher than at a flat surface.

The conditions of droplet growth, equilibrium, and evaporation can be formulated as differences between the water vapor pressure in the atmosphere p and the equilibrium vapor pressure $p_o(r)$ at the same temperature as

$$p - p_o(r) > 0 \qquad \text{growth}$$

$$p - p_o(r) = 0 \qquad \text{equilibrium}$$

$$p - p_o(r) < 0 \qquad \text{evaporation}$$

A certain critical value of the droplet radius r_c can satisfy the condition of equilibrium at a given temperature and environmental vapor pressure p. This condition is, from equation 4.5,

$$p - p_o e^{a/r_c} = 0 \tag{4.8}$$

The quotient p/p_o is often called supersaturation S_p, although it can be either greater than unity (supersaturation) or less than unity (undersaturation). The ratio

$$S_p = \frac{p}{p_o} = \frac{\rho_v}{\rho_{vo}} \tag{4.9}$$

is similar to the definition of relative humidity in equation 4.4. From the latter two equations, a critical radius of a droplet at equilibrium is

$$r_c = \frac{a}{\ln S_p} \qquad \text{(cm)} \tag{4.10}$$

or, from the approximate relationship given in equation 4.7,

$$r_c = \frac{a}{S_p - 1} \quad \text{(cm)} \tag{4.11}$$

For a degree of supersaturation of 0.1% at 10°C, $S_p = 1.001$ and $a = 1 \times 10^{-7}$ cm, as computed in an earlier example. Then the critical radius is $r_c = 1 \times 10^{-4}$ cm, or 1 μm.

Dissolved Salts and Vapor Pressure. Substances dissolved in water lower its vapor pressure. For dilute solutions, the vapor pressure of H_2O obeys Raoult's law, which states that the vapor pressure of water is proportional to its mole fraction in solution,

$$p_o(s) = p_o \frac{n_w}{n_w + \nu n_s} \tag{4.12}$$

where $p_o(s)$ is the H_2O vapor pressure at equilibrium with solution, p_o is the vapor pressure of pure water, n_w is the number of moles of water, and n_s is the number of moles of solute dissociating into ν cations and anions. Among strongly soluble salts, the coefficient ν is $\nu = 2$ for NaCl and other 1:1 ionic solids, and $\nu = 3$ for $MgCl_2$ and similar compounds.

In a dilute solution the mole fraction of water is

$$\frac{n_w}{n_w + \nu n_s} \simeq 1 - \frac{\nu n_s}{n_w}$$

and the vapor pressure equation becomes

$$p_o(s) = p_o \left(1 - \frac{\nu n_s}{n_w}\right) \tag{4.13}$$

In the case of water droplets condensing around a soluble particle, it is convenient to relate the masses of the particle and water to spheres of certain size and density. A solid particle can be thought of as a sphere of initial radius r_s, density ρ_s, and gram-formula weight M_s (g mol^{-1}), each mole of it dissociating into ν cations and anions. Then the number of moles of solute νn_s is

$$\nu n_s = \frac{4\pi r_s^3 \nu \rho_s}{3 M_s} \quad \text{(mol)} \tag{4.14}$$

and, similarly, the number of moles of water is

$$n_w = \frac{4\pi r^3 \rho_L}{3 M_w} \quad \text{(mol)} \tag{4.15}$$

where r is the droplet radius, ρ_L is the liquid water density, and M_w is the gram-formula weight of H_2O. Substitution of the relationships for n_w and n_s in equation 4.13 gives

$$p_o(s) = p_o \left(1 - \frac{\nu \rho_s / M_s}{\rho_L / M_w} \cdot \frac{r_s^3}{r^3} \right) \tag{4.16}$$

A simpler form of 4.16 is

$$p_o(s) = p_o \left(1 - \frac{b}{r^3} \right) \tag{4.17}$$

where b is defined as

$$b = \frac{\nu r_s^3 \rho_s / M_s}{\rho_L / M_w} \quad (cm^3) \tag{4.18}$$

Parameter b depends only slightly on temperature, as the solid density at atmospheric temperatures is practically constant, and the density of liquid water varies within 1% of the value $\rho_L = 1.0$ between -10 and $10°C$. The H_2O vapor pressure over a dilute salt solution changes with temperature in essentially the same way as the vapor pressure of pure water p_o (Table 4.1).

An order of magnitude estimate of b can be made as follows: for a particle made of NaCl of initial radius $r_s = 0.5$ μm (5×10^{-5} cm), $\rho_s = 2.17$ g cm^{-3}, $M_s = 58.4$ g mol^{-1}, and $\nu = 2$,

$$b = \frac{2 \times (5 \times 10^{-5})^3 \times 2.17/58.4}{1.0/18.02} = 1.67 \times 10^{-13} \quad (cm^3)$$

An NaCl particle of radius $r_s = 0.5$ μm dissolved in a water droplet of radius $r = 50$ μm gives a concentration of about 50 μmol NaCl liter^{-1}, a value within the range of salt concentrations in rain waters.

As a limiting case of application of equation 4.13, a saturated NaCl solution may be considered. The saturation concentration of NaCl is 6.1 mol per 1000 g H_2O or 6.1 mol NaCl per 55.5 mol H_2O. In the temperature range between 0 and 70°C the solubility of NaCl is nearly constant. Using the value of 6.1 molal in equation 4.13,

$$p(s) = p_o \left(1 - \frac{2 \times 6.1}{55.5} \right) = 0.78 p_o$$

which means that the water vapor pressure over a saturated NaCl solution is 22% lower than over pure water. Stated differently, the relative humidity of the atmosphere at equilibrium with a saturated NaCl solution is 78%. The measured value of $p(s)$ at 25°C, given in Table 4.2, is $0.75 p_o$, close to

Table 4.2   Activity of Water, Solute Concentration, and Density
of Saturated Solutions of Highly Soluble Salts[a]

Solid phase	Solubility[b] (wt %)	Density of solution ρ (g cm^{-3})	Water activity[c] $a_w = \dfrac{p(s)}{p_o}$
CaCl$_2$·6H$_2$O	45.3	1.45	0.31
Ca(NO$_3$)$_2$·4H$_2$O	58.0	1.6	0.50
KBr	40.6	1.38	0.81
KCl	26.5	1.18	0.84
KNO$_3$	27.2	1.19	0.92
K$_2$CO$_3$·2H$_2$O	52.9	1.57	0.43
MgCl$_2$·6H$_2$O	35.4	1.31	0.33
Mg(NO$_3$)$_2$·6H$_2$O	42.1	1.39	0.53
MgSO$_4$·7H$_2$O	26.7	1.30	0.88
NaBr·2H$_2$O	48.6	1.54	0.58
NaCl	26.5	1.20	0.75
NaNO$_3$	47.9	1.39	0.74
Na$_2$CO$_3$·10H$_2$O	22.7		0.87
Na$_2$SO$_4$·10H$_2$O (20°C)	16.0	1.15	0.93
NH$_4$Cl	28.0		0.77
NH$_4$NO$_3$	68.3	1.32	0.62
(NH$_4$)$_2$SO$_4$ (20°C)	42.6	1.25	0.80

[a]At 25°C, other temperatures noted. From Robinson and Stokes (1970), International Critical Tables, Stephen and Stephen (1963), and Gmelin's Handbuch der Anorganischen Chemie, Vol. Mg.
[b]When H$_2$O is part of the stoichiometric formula of the solid phase, weight percent in solution refers only to the anhydrous part of the formula: grams per 100 g solution.
[c]$p(s)$ is H$_2$O vapor pressure of solution, p_o of pure water. To obtain $p(s)$ in atmospheres or millimeters of mercury, multiply values in the table by the vapor pressure of pure water at 25°C, $p_o = 23.753$ mm Hg = 0.031 atm; and at 20°C, $p_o = 17.529$ mm Hg = 0.023 atm.

the preceding value. The equilibrium vapor pressures of some other salt solutions are also listed in Table 4.2 as water activity a_w, which is the ratio of the vapor pressure of the solution to that of pure water, $p(s)/p_o$. The list includes salts that occur in the atmosphere as well as salts precipitating in saline brines.

When a soluble salt particle comes in contact with a small water droplet or when water begins to condense on a salt particle, a state of saturation may exist within the droplet until it has grown to a large enough volume. Such a state of saturation is probably of little significance to dissolution of gases and other chemical reactions taking place in water droplets, as droplets grow at a rate inversely related to their size, and the lifetime of a small droplet is short. The rates of growth of droplets are dealt with in Section 4.2.3. It is instructive to ask what is the size of a water droplet within which a particle of NaCl has dissolved, making the droplet a saturated NaCl solution. A salt particle of radius r_s and density ρ_s has the mass

$$\tfrac{4}{3}\pi r_s^3 \rho_s \quad \text{(g)}$$

The mass of water needed to make a saturated solution by dissolving the entire particle is

$$\tfrac{4}{3}\pi r_s^3 \rho_s \times \frac{1000}{mM_s}$$

where m is the molal concentration of salt in its saturated aqueous solution (moles solute per 1000 g H_2O) and M_s is the gram-formula weight of solute. Another commonly tabulated measure of solubility is G, grams of solute per 100 g solution (G wt %). Then the mass of water needed to make a saturated solution is

$$\tfrac{4}{3}\pi r_s^3 \rho_s \times \frac{100-G}{G}$$

The mass of a saturated solution droplet of radius r and density ρ is the sum of the masses of solute and H_2O,

$$\tfrac{4}{3}\pi r^3 \rho = \tfrac{4}{3}\pi r_s^3 \rho_s \left(1 + \frac{1000}{mM_s}\right)$$

from which the droplet radius is

$$r = r_s \left[\frac{\rho_s}{\rho}\left(1 + \frac{1000}{mM_s}\right)\right]^{1/3} \qquad (4.19)$$

If the solubility is in units of G wt % (Table 4.2) then the ratio $(100-G)/G$ replaces $1000/mM_s$ in equation 4.19.

For a saturated NaCl solution, $\rho = 1.2$ g cm^{-3}, $m = 6.1$ mol per 1000 g H_2O, and other parameters are given in a preceding computational example for b. The ratio of the solution droplet radius to initial radius of the solid particles is

$$\frac{r}{r_s} = \frac{2.17}{1.2}\left(1 + \frac{1000}{6.1 \times 58.4}\right)^{1/3} = 1.9$$

The result $r/r_s = 1.9$ means that the droplet radius of a saturated solution of NaCl is about twice as large as the radius of the NaCl particle that dissolved in the droplet, making it a saturated solution. For much less soluble minerals, complete dissolution is either possible only in larger droplets, or it may never be attained because of the physical limits of droplet size.

Droplet Size and Salt Effects Combined. In equation 4.7 the term p_o denotes an equilibrium water vapor pressure over a flat surface. To allow for the fact that the surface may represent an aqueous salt solution, the equilibrium vapor pressure $p_o(s)$, defined in equation 4.17, can be substituted for p_o in 4.7, giving

$$p_o(r) = p_o\left(1 - \frac{b}{r^3}\right)\left(1 + \frac{a}{r}\right) \qquad (4.20)$$

or, as a further approximation,

$$p_o(r) = p_o\left(1 + \frac{a}{r} - \frac{b}{r^3}\right) \qquad (4.21)$$

Parameter a, defined in equation 4.6, is a function of surface tension σ between an aqueous solution and air. The values of σ listed in Table 4.1 refer to pure water. For dilute salt solution, σ is slightly higher, but in a saturated NaCl solution it is about 14% higher than for pure water at 20°C.

For the values of parameters a and b computed in this section, the droplet size loses its effect on vapor pressure for droplets of $r > 0.1$ μm, and the effect of salt concentration diminishes rapidly for droplets of $r > 1$ μm (the value of b is smaller for smaller salt particles, and therefore the effects of salt concentration may become negligible even at smaller droplet sizes). When the effects of droplet size and salt concentration can be neglected, the equilibrium vapor pressure of a droplet is equal to the vapor pressure of bulk water,

$$p_o(r) \simeq p_o.$$

4.2.3 Rates of Droplet Growth and Evaporation

A model describing growth of water droplets by condensation of water vapor at the droplet surface (or shrinking of droplets through evaporation), attributed to Maxwell, is based on Fick's first law of diffusion, and it states that the rate of increase in the droplet mass dM/dt (g sec^{-1}) is proportional to the droplet surface area $4\pi r^2$ (cm^2) and diffusional flux of water vapor at the droplet surface (Duguid and Stampfer, 1971),

$$\frac{dM}{dt} = 4\pi r^2 D \left(\frac{\partial \rho_v}{\partial R} \right)_{R=r} \qquad (\text{g sec}^{-1}) \qquad (4.22)$$

where D is the diffusion coefficient of water vapor in air ($\text{cm}^2 \text{ sec}^{-1}$), ρ_v is the water vapor concentration or density in air (g cm^{-3}), R is the radial distance coordinate normal to the sphere surface, and the flux $D(\partial \rho_v / \partial R)$ is at the distance of radius r from the sphere center. Diffusion of vapor to or from a sphere surface at a steady state is given by the following equation (for example, Carslaw and Jaeger, 1959):

$$D \left(\frac{\partial^2 \rho_v}{\partial R^2} + \frac{2}{R} \frac{\partial \rho_v}{\partial R} \right) = 0 \qquad (\text{g cm}^{-3} \text{ sec}^{-1}) \qquad (4.23)$$

With the boundary conditions of a constant concentration of water vapor in air

$$\rho_v = \rho_\infty \qquad \text{at} \quad R \to \infty$$

and constant concentration at the droplet surface

$$\rho_v = \rho_{vo} \qquad \text{at} \quad R = r$$

the solution of equation 4.23 is

$$\rho_v = \rho_\infty - \frac{(\rho_\infty - \rho_{vo})r}{R} \qquad (\text{g cm}^{-3}) \qquad (4.24)$$

The vapor concentration gradient at the sphere surface $(\partial \rho_v / \partial R)_{R=r}$ can be evaluated from equation 4.24 and used in 4.22. Further, the mass of a spherical droplet is $M = \frac{4}{3}\pi r^3 \rho_L$, the derivative of which dM/dt can also be substituted in equation 4.22, giving the rate of change in the droplet radius dr/dt as a function of other parameters,

$$\frac{dr}{dt} = \frac{D(\rho_\infty - \rho_{vo})}{r \rho_L} \qquad (\text{cm sec}^{-1}) \qquad (4.25)$$

Equations for the rate of change in mass dM/dt and radius dr/dt of a sphere can in principle also be used for systems other than water droplets in air. Dissolution of mineral particles in water and the rate of decrease in particle size, in terms of the model behind equation 4.25, are discussed in Section 5.5.

Droplets grow—$dr/dt > 0$—when the vapor density in the environment is greater than the equilibrium vapor density at the droplet surface

$(\rho_\infty > \rho_{vo})$, and evaporation takes place when the inequality sign is reversed. An important feature of equation 4.25 is that the droplets grow or evaporate at a rate inversely related to their size r. In other words, the smaller a droplet is, the faster it changes its dimensions when other factors are constant. A corollary of this growth law is that a size distribution of droplets tends with time to become narrower in its size range, as the smaller droplets grow faster than the larger ones.

Using the definition of the supersaturation ratio in equation 4.9, the rate of change in radius dr/dt becomes

$$\frac{dr}{dt} = \frac{D\rho_{vo}(S_p - 1)}{r\rho_L} \tag{4.26}$$

and the time required for a droplet to change its size from r_0 to r is, by integration of 4.26,

$$t = \frac{(r^2 - r_0^2)\rho_L}{2D\rho_{vo}(S_p - 1)} \quad \text{(sec)} \tag{4.27}$$

Under a given supersaturation of undersaturation S_p, the time required to change the droplet size by some factor r/r_0 increases directly with r_0^2.

The preceding discussion of growth $(S_p > 1)$ and evaporation $(S_p < 1)$ of droplets is based only on the process of molecular diffusion of water vapor molecules at the droplet surface. However, an additional process also plays a role: water vapor condensing on the droplet surface releases heat, which raises the temperature of the droplet surface in some measure above the temperature of the surrounding air. Back transfer of heat involves migration of water molecules away from the droplet surface. Combination of the diffusional flux of water molecules and the flux due to heat transfer leads to an equation for the rate of growth or evaporation that is similar, except for one term, to equation 4.26,

$$\frac{dr}{dt} = \frac{D\rho_{vo}(S_p - 1)}{r\rho_L(1 + \chi)} \quad \text{(cm sec}^{-1}) \tag{4.28}$$

The dimensionless parameter χ, derivable from the Clausius-Clapeyron equation (Rogers, 1976), is

$$\chi = \frac{D\rho_{vo}L^2}{\kappa R_v T^2} \tag{4.29}$$

where D, ρ_{vo}, R_v, and T are as defined previously, L is the heat of condensation of water vapor, and κ is the coefficient of thermal conductivity of air. For temperatures between -10 and $40°C$, the values of L, κ, D, and ρ_{vo} are given in Table 4.1. All the values listed are in cgs units and can be used directly in equations 4.26 through 4.30. From equation 4.28, the time required for a droplet to change in size from r_0 to r is

$$t = \frac{(r^2 - r_0^2)(1 + \chi)\rho_L}{2D\rho_{vo}(S_p - 1)} \quad \text{(sec)} \qquad (4.30)$$

Example. What is the time required for water droplet to increase in size from $r_0 = 5$ μm to $r = 50$ μm in air at $10°C$ and water vapor supersaturation of 1% ($S_p = 1.01$)?

An approximate solution that takes into account diffusion of water vapor only, without the heat of condensation effect, can be derived from equation 4.27. Using for the density of liquid water $\rho_L = 1.0$ g cm^{-3}, the values of r, r_0, and S_p as defined in the question, and the remaining parameters from Table 4.1, the growth time is

$$t = \frac{\left[(5 \times 10^{-3})^2 - (5 \times 10^{-4})^2\right] \times 1.0}{2 \times 0.24 \times 9.38 \times 10^{-6}(1.01 - 1)} = 550 \text{ sec, or 9 min}$$

A somewhat more accurate computation, using equation 4.30, will give a value of t greater than the preceding value by a factor of $(1 + \chi)$. From data in Table 4.1 for $10°C$ and equation 4.29, the value of χ is

$$\chi = \frac{0.24 \times 9.38 \times 10^{-6} \times (2.476 \times 10^{10})^2}{2.50 \times 10^3 \times 0.462 \times 10^7 \times (283.2)^2} = 1.49$$

$$1 + \chi = 2.49$$

According to equation 4.30, the time required for the increase in the droplet size is

$$t = 2.49 \times 550 \text{ sec} = 1370 \text{ sec, or 23 min}$$

4.3 UPTAKE OF GASES AND SOLIDS BY RAIN

4.3.1 Solubility of Gases

A chemical thermodynamic equilibrium between species A in a gaseous state and A in an aqueous solution, written as a chemical reaction,

$$A(g) = A(aq)$$

has an equilibrium constant K,

$$K = \frac{a_{A(aq)}}{f_{A(g)}} \tag{4.31}$$

where a_A is the thermodynamic activity of the dissolved species A with respect to some chosen standard reference state and f_A is the fugacity of the gaseous A. At room temperature and total pressure near 1 atm, the fugacity of the gas can be equated to its partial pressure—$f_A = p_A$— and the activity of the aqueous species in a dilute solution can be equated to its concentration—$a_A = [A]$, where brackets denote concentration on a molal scale, moles A per 1000 g H_2O. For a detailed discussion of the concepts of activity and fugacity, and their relationships to concentration and partial pressure units the reader may consult such textbooks as Lewis et al. (1961), Klotz (1964), Stumm and Morgan (1970), and Garrels and Christ (1965).

Replacing, in equation 4.31, activity by concentration and fugacity by partial pressure, the equilibrium constant can be written as a dimensioned quotient

$$K = \frac{[A]}{p_A} \qquad \text{mol (kg } H_2O)^{-1} \text{ atm}^{-1} \quad \text{or} \quad \text{mol } l^{-1} \text{ atm}^{-1} \tag{4.32}$$

For dilute solutions and atmospheric pressures, the interchange of concentration units between moles per 1000 g H_2O and moles per liter, and of pressure units between atmospheres and bars introduces no significant error. At higher concentrations and temperatures, and at pressures appreciably different from atmospheric, concentration in solution is not equal to the thermodynamic activity, and partial pressure of a gas may not be equal to its fugacity. Under such conditions, corrections in the form of activity and fugacity coefficients enter in the definition of the equilibrium constant given in equation 4.32. From the latter definition of K, concentration of a volatile species in solution at equilibrium with its gaseous phase is

$$[A] = K p_A \tag{4.33}$$

where K is effectively a gas solubility coefficient, dependent on temperature and total pressure. Many other coefficients of gas solubility have been, and are being used in the literature. The more commonly encountered are Henry's-law coefficient K_H and Bunsen's solubility coefficient α. For these two and some other solubility coefficients, equations relating them to the equilibrium constant K are summarized in Table 4.3. Note that K and Λ are virtually identical coefficients.

**Table 4.3 Coefficients Expressing Solubility of Gases in Water
and Some of their Conversion Factors** [a, b]

K $\left(\dfrac{mol}{kg\ H_2O\ atm}\right)$	K_H (atm)	α $\left(\dfrac{liters\ gas\ NTP}{liter\ H_2O \cdot atm}\right)$	λ $\left(\dfrac{liters\ gas\ NTP}{kg\ H_2O \cdot atm}\right)$	Λ $\left(\dfrac{mol}{kg\ H_2O \cdot bar}\right)$
$\dfrac{m}{p}$	$\dfrac{p}{X}$	$22.414\rho K$	$21.692K$	$0.9869K$
$\dfrac{m+55.51}{K_H}$		$1.033\rho\lambda$	$\dfrac{0.9678\alpha}{\rho}$	$\dfrac{0.0440\alpha}{\rho}$
$\dfrac{0.0446\alpha}{\rho}$		$22.712\rho\Lambda$	21.979Λ	0.0445λ
0.0461λ				
1.013Λ				

[a] From Landolt-Börnstein (1976).
[b] Definitions: K is an equilibrium constant, equation 4.32; m is gas concentration is solution (moles in 1 kg H_2O), and p is partial pressure (atm). K_H is Henry's-law coefficient; X is mole fraction of dissolved gas. α is Bunsen's solubility coefficient, liters of gas at NTP ($T = 273.15°K$; total $P = 1$ atm) dissolved in 1 liter H_2O; ρ is density of gas-free water (kg l^{-1}). Λ and λ are solubility coefficients in Figure 4.3. Note the similarity of Λ to K.

Solubilities of the major and trace gases occurring in the atmosphere are plotted as a function of temperature in Figure 4.3. The solubility values are given as the coefficients Λ and λ, defined in Table 4.3. The coefficient Λ is essentially the reciprocal of Henry's-law coefficient K_H. Large differences in the solubilities of such gases as SO_2, CO_2, and O_2 over a range of temperatures from 0 to 100°C should be noted. Between 0 and 50°C, solubilities of all the gases shown in Figure 4.3 decrease with an increasing temperature. For most of the gases, the solubility decreases by a factor of about 2 when the temperature increases from 0 to 25°C.

An empirical relationship has been used to describe concentration of a gas in rain water C_r as a function of the gas concentration in air ρ_v, water concentration in air C_w, and a collection efficiency factor σ_w (Junge, 1963, 1975; Hidy, 1973),

$$C_r = \frac{\sigma_w \rho_v}{C_w} \quad (mg\ l^{-1}) \tag{4.34}$$

where ρ_v is in units of $\mu g\ m^{-3}$ and C_w is in units of $g\ m^{-3}$. Water concentrations in rain-forming air $C_w = 0.5$–$6\ g\ m^{-3}$ have been correlated with the effiency factor as $\sigma_w = 0.1C_w\ \mu g\ g^{-1}$. If this relationship were universally true then concentrations of dissolved gases in rain water could have been determined from the equation $C_r = 0.1\rho_v$. The latter means that

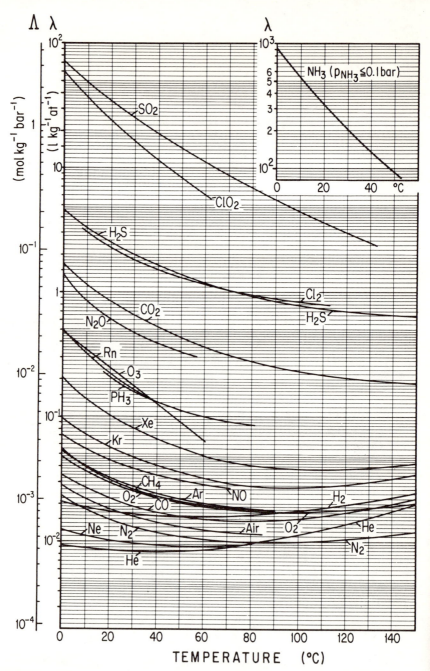

Figure 4.3 Solubilities of gases in water as a function of temperature. Coefficients of solubility Λ and λ are defined in Table 4.3 (from Landolt-Börnstein, 1976; solubility of methane from Perry, 1963). Reproduced by permission of Springer-Verlag, Heidelberg, Germany.

the ratio of concentration of a gas in rain water to its concentration in air is a constant, $C_r/\rho_v = \sigma_w/C_w \simeq 0.1$. In fact, however, the concentration ratio is not always constant; a range of values of C_r/ρ_v between about 2 and 0.15 has been reported for chlorine, and a greater range from about 2.0 to 0.05 for sulfur, the ratio decreasing with an increasing sulfur concentration in air ρ_v (Junge, 1975). An equation of the form

$$C_r = (0.1 \text{ to } 2)\rho_v \qquad (\text{mg } l^{-1}) \qquad (4.35)$$

gives order of magnitude estimates of concentrations of gases scavenged by rain water.

4.3.2 Diffusion of Gases Across Gas-Water Interface

A gas in the atmosphere at a partial pressure p that dissolves in water according to the equilibrium solubility equation 4.33 has a concentration value Kp (mol l^{-1}) at the water side on the air-water interface. This is shown diagrammatically in Figure 4.4a. (Any possible departures from this equilibrium relationship due to the kinetic mechansims of gas adsorption at interfaces are not considered in this model.) In still water, transport of the dissolved gas down from the interface is by molecular diffusion. For a nonreacting gas in a fixed volume of water, a steady-state concentration equal to Kp will ultimately be attained throughout the water volume. The steady state corresponds to a chemical equilibrium between the gas dissolved in water and the gas present in the atmosphere.

The time required for a droplet to equilibrate with an inert gas in the atmosphere can be estimated as follows. In a spherical droplet of a fixed

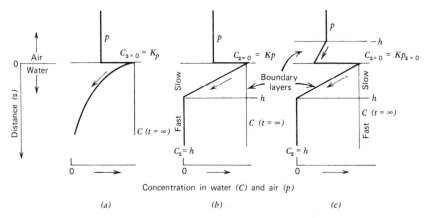

Concentration in water (C) and air (p)

(a) (b) (c)

Figure 4.4 Three schematic models of transport across an air-water interface.

volume, gas diffuses from the surface toward the center, as shown schematically in Figure 4.4a. The gas concentration at the droplet surface $C_{z=0}$ is constant, but its concentration elsewhere within the droplet increases with time. A convenient measure of time required to attain a nearly uniform gas concentration within a still droplet is the time when concentration at the center $C_{z=c}$ has reached 95% of the surface concentration $C_{z=0}$. This is the definition of the time to steady state, discussed in Section 1.3.2. For a droplet initially free of dissolved gas, the steady state will be nearly attained at time t_{ss} when the following condition exists:

$$\frac{C_{z=c}}{C_{z=0}} = 0.95$$

For a droplet initially containing gas at a uniform concentration C_0, the time to steady state t_{ss} corresponds to the condition

$$\frac{C_{z=c} - C_0}{C_{z=0} - C_0} = 0.95 \qquad (4.36)$$

For diffusion into a sphere of fixed dimensions, the time required to fullfill a near-steady-state condition 4.36 is

$$t_{ss} = \frac{0.4 r_0^2}{D} \qquad (\text{sec}) \qquad (4.37)$$

where r_0 is the droplet radius (cm) and D is the diffusion coefficient of the dissolving gas in water (cm^2 sec^{-1}). Equation 4.37 is derivable from the solutions of the diffusion equation in a sphere, given in a graphical form by Carslaw and Jaeger (1959) and Crank (1956). For gases in water, the diffusion coefficients are of the order of $D \simeq 10^{-5}$ cm^2 sec^{-1} (Section 3.4); this value of D and droplet radii from $r_0 = 3$ to 150 μm give the times to steady state t_{ss} for dissolving gases, as listed in Table 4.4. The times required to attain an equilibrium are short, up to about 10 sec for a droplet of $r_0 = 0.15$ mm. A meaningful comparison can be made between the time needed to attain an equilibrium with a dissolved gas, and the length of time a droplet can exist while falling through the atmosphere and evaporating on the way. The life span of falling droplets of different initial sizes and the distances through which they fall before complete evaporation are listed in Table 4.4. The longevity of droplets is greater than the times t_{ss} to equilibration with dissolved gases, from which it can be concluded that gas-water equilibrium can always be attained within the lifetime of a rain droplet. Interference with the processes of gas adsorption and equilibration may be caused by fine particles or impurities adsorbed and attached to the droplet surfaces.

Table 4.4   Time to Equilibration of a Water Droplet with a Nonreacting Gas, Time to Droplet Evaporation, and Distance Fallen Through Air

Droplet initial radius r_0 (μm)	Time to steady state for gas diffusion into droplet[a] t_{ss} (sec)	Life span in an atmosphere of 80% relative humidity[b]	
		Time to complete evaporation (sec)	Distance fallen
3	0.004	0.16	2 μm
10	0.04	1.8	0.2 mm
30	0.12	16	2.1 cm
100	4.0	290	208 m
150	9.0	900	1.05 km

[a]Equation 4.37, with $D = 1 \times 10^{-5}$ cm^2 sec^{-1}.
[b]From Rogers (1976).

Diffusion Across Boundary Layers.   The theories of gas migration across the air-water interface (and some theories of dissolution of solid particles) often assume that transport of material between two phases takes place through a thin layer, the physical characteristics of which differ from the characteristics of the two adjoining phases. The thickness of such postulated boundary layers is small in comparison to the linear dimensions of the bulk phases on either side of it. A model of one boundary layer on the water side of a water-gas interface is shown schematically in Figure 4.4b. Two boundary layers, one on the gas side and the other on the water side of the interface, are shown in Figure 4.4c. Because of the importance of the boundary layer concept in diffusional transport models, and because such models are logical extensions of the diffusion and dissolution case (Figure 4.4a), a brief discussion of the boundary layer models is given in this section.

In the diagram of Figure 4.4b, the air space is well mixed, such that the gas concentration in it is uniform, as shown by the zero gradient of the partial pressure of the gas above the interface. Water below the boundary layer is also well mixed, but within the boundary layer transport takes place by molecular diffusion. In an analogy to a solid crystal in contact with an aqueous solution, a uniform concentration below the boundary layer corresponds to the solid phase and a uniform concentration above the boundary layer represents a well-mixed solution.

For a linear concentration gradient across the boundary layer in Figure 4.4b, $(C_{z=h} - C_{z=0})/h$, the flux downward of the diffusing species is

$$F = \frac{D(C_{z=0} - C_{z=h})}{h} \quad \text{(g cm}^{-2}\text{ sec}^{-1}) \quad (4.38)$$

where D is the diffusion coefficient of the chemical species within the boundary layer. In the case of an aqueous boundary layer, D is usually taken as a molecular diffusion coefficient. Using the gas solubility relationship 4.32, concentration at the top of the boundary layer is

$$C_{z=0} = Kp$$

and the flux is

$$F = \frac{D(Kp - C_{z=h})}{h} \qquad (4.39)$$

Equation 4.39 is used in experimental studies of gas uptake or release by water. The flux F across the water-air interface and concentration $C_{z=h}$ within a well-mixed water column can be measured, and from the known values of the molecular diffusion coefficient D of the gas in water and its solubility K, the thickness of the boundary layer h can be computed. From measured rates of gas transfer across the ocean-atmosphere interface, a boundary layer of mean thickness $h \simeq 50$ μm has been computed for the ocean (Kanwisher, 1963; Broecker and Peng, 1974).

From the configuration of the concentration against distance diagram in Figure 4.4b it is clear that if concentration in bulk water $C_{z=h}$ is greater than concentration at the top of the boundary layer $C_{z=0}$, then the direction of flux is out of the water.

Another possible configuration of two boundary layers, each adjacent to the interface, is shown in Figure 4.4c. At a steady state, the fluxes at the interface must be equal,

$$\left(D \frac{\Delta \rho_v}{\Delta z} \right)_{air} = \left(D \frac{\Delta C}{\Delta z} \right)_{water} \qquad (4.40)$$

As the molecular diffusion coefficients of gases in air are of the order of $D_{air} \simeq 10^{-1}$ cm^2 sec^{-1} but the diffusion coefficients in water are $D_{water} \simeq 10^{-5}$ cm^2 sec^{-1}, it follows from equation 4.40 that the concentration gradient within the upper boundary layer should be much smaller than in the lower boundary layer,

$$\left(\frac{\Delta \rho_v}{\Delta z} \right)_{air} = \frac{D_{water}}{D_{air}} \left(\frac{\Delta C}{\Delta z} \right)_{water}$$

$$\simeq 10^{-4} \left(\frac{\Delta C}{\Delta z} \right)_{water} \qquad (4.41)$$

A model of two boundary layers at the water-air interface has been used in the studies of isotopic fractionation during evaporation of water from the ocean surface (Craig and Gordon, 1965) and in the studies of gas exchange between the atmosphere and lake waters (Liss, 1975).

A solid-liquid analogy of a two-boundary-layer system can be drawn for silicates undergoing partial hydration in water, mentioned in Section 3.7. Ions in a solid, such as a silicate glass, diffuse through a hydrated layer, into a laminar layer of solution adjacent to the solid surface, and further into the bulk solution. The diffusion coefficients of ions within the hydrated layer of the solid are higher than in the bulk solid, yet much lower than the molecular diffusion coefficients in water.

4.3.3 Scavenging of Solid Particles by Rain

Solid particles are taken up by atmospheric precipitation at the time of droplet nucleation and growth in clouds, and during droplet fall from the cloud through the atmosphere. The process of particle inclusion in raindrops within clouds is known as *rainout*, whereas scavenging of particles by falling raindrops and snow flakes is known as *washout*. As liquid droplets can collide with solid particles anywhere in the atmosphere, the washout process can occur everywhere, but rainout is limited to the zones of cloud formation.

The ability of rain to scavenge solid particles can be formulated in the same form as equation 4.34, used for uptake of gases by rain. For solids,

$$C_r = \frac{\sigma_w C_a}{C_w} \quad (\text{mg l}^{-1}) \qquad (4.42)$$

where C_r denotes the particulate matter concentration in rain (mg l^{-1}), C_a is concentration of particles in air (μg m^{-3}), and the collection efficiency factor σ_w and water concentration in air C_w (g m^{-3}) retain their significance, as in the case of gas uptake. An explicit form of equation 4.42,

$$C_r = (0.2 \text{ to } 6) C_a \quad (\text{mg l}^{-1}) \qquad (4.43)$$

has been reported for radioactive products of nuclear bomb explosions and cosmic spallation products in the atmosphere taken up by rain. For very small particles and for areas of high particle concentration in air, the scavenging efficiency of rain and thus the numerical coefficient in equation 4.43 may be lower. The collection efficiency for snowflakes is also generally lower than for rain droplets, $\sigma_w = 0.002$–0.02 (Hidy, 1973; Junge, 1975).

An upper limit of concentration of solid particles in rain can be arrived at in the following two ways. One is from the mean concentration of particulate matter in the atmosphere (Section 4.2.1) $C_a = 15$ μg m^{-3} and equation 4.43,

$$C_r = 3 \quad \text{to} \quad 90 \text{ mg l}^{-1}$$

A second estimate is from the mean rate of input of particulate matter to the atmosphere (2000×10^{12} g yr^{-1}) and annual rate of atmospheric precipitation 4×10^{17} l yr^{-1}),

$$C_r = \frac{2 \times 10^{18} \text{ mg yr}^{-1}}{4 \times 10^{17} \text{ l yr}^{-1}} = 5 \text{ mg l}^{-1}$$

To repeat, the above two estimates of C_r are upper-limit values, as some of the particulate matter in the atmosphere is soluble and some of it deposited at the earth's surface without having been transported by rain.

The removal of particulate matter from the atmosphere falls into three types of process: (1) wet removal, which includes rainout and washout; (2) transport by sedimentation, including settling, impact deposition, and turbulent and molecular diffusion; and (3) growth of particles, caused by condensation of vapors on the particle surfaces, and aggregation by collisions and coagulation. Process (1) involves reduction of the particle size if some of the particles dissolve in rain water, process (2) does not affect the particle size, and process (3) results in an increase in size. The rates of removal by the individual processes are listed in Table 4.5. The removal rates are shown for three particle sizes (radius $r = 0.05$, 0.5, and 5.0 μm) occurring at certain concentrations near the ground and at 2 km altitude near a cloud base. The removal rate by rainout is always relatively high for particles of all sizes. Processes leading to an increase in the particle size are more important for the smaller particles, whereas settling and impact deposition are relatively more important for the larger particles.

In a steady-state atmosphere, the rate of removal of the particulate matter is equal to, or less than the rate of input. (The removal rate is less if some of the particle mass dissolves or evaporates.) If the removal rate of the mass of solids is equal to the input rate, then the removal rate of particles can be estimated using the following values: input rate equal to 2×10^{15} g yr^{-1}, mean particle radius $r = 1$ μm, mean particle mass 10^{-11} g particle^{-1}, and atmosphere volume 4×10^{24} cm^3. The mean removal rate is:

$$\text{removal rate} = \frac{(2 \times 10^{15} \text{ g yr}^{-1})(3 \times 10^{-8} \text{ yr sec}^{-1})}{(10^{-11} \text{ g particle}^{-1})(4 \times 10^{24} \text{ cm}^3)}$$

$$= 10^{-6} \text{ particles cm}^{-3} \text{ sec}^{-1}.$$

Table 4.5 Estimated Removal Rates of Particles of Three Sizes r by Different Processes in the Atmosphere Near Ground Level (Urban Atmosphere Conditions) and Near Cloud Base, 2 km Altitude.[a] N is number of particles cm^{-3}. Removal rates in units of particles cm^{-3} sec^{-1}

Removal process	$r = 0.05$ μm		$r = 0.5$ μm		$r = 5.0$ μm	
	Ground level $N = 10^5$ cm^{-3}	Cloud base (2 km) $N = 10^3$ cm^{-3}	Ground level $N = 10^2$ cm^{-3}	Cloud base (2 km) $N = 1$ cm^{-3}	Ground level $N = 10^{-1}$ cm^{-3}	Cloud base (2 km) $N = 10^{-3}$ cm^{-3}
Wet removal						
Rainout (droplets $r = 10$ μm)	—	10^{-2}	—	10^{-1}	—	10^{-3}
Washout (droplets $r = 1$ mm, $N = 10^{-3}$ cm^{-3})	10^{-3}	10^{-5}	10^{-8}	10^{-10}	10^{-7}	10^{-9}
Sedimentation[b]	$10^{-1} - 10^0$	10^{-4}	10^{-3}	10^{-5}	10^{-6}	10^{-8}
Scavenging and growth[c]	10^4	10^0	10^{-1}	10^{-5}	10^{-4}	10^{-10}

[a]From Hidy (1973).
[b]Includes settling, inertial and diffusional deposition, and convective diffusion.
[c]Includes condensation of vapors on particles, coagulation, and scavenging by particle collisions. The fastest rate for the process in each category is listed in the table.

Although it is difficult to compare too closely the mean rate of removal with the rates of the individual removal processes given in Table 4.5, it can be noted that the mean rate is comparable to the rates of settling and washout near the ground, for particles of radius from 0.5 to 5.0 μm. This also means that the fast uptake of particles in rainout, within the cloud formation zone of the atmosphere, removes only a small fraction of the total mass of particulate matter, whereas most of the mass is removed by processes operating closer to the ground.

4.4   ACIDITY AND CHEMICAL COMPOSITION OF RAIN

4.4.1   General Trends

Concentrations of the major chemical species in atmospheric precipitation from the east of North America and north and east of Europe are listed in Table 4.6. Concentrations of individual chemical species vary considerably within geographically large areas, such as northern Europe and the European part of the USSR. Most of the ionic constituents of atmospheric precipitation are derived from soluble aerosol particles and reactive gases, whose residence times in the atmosphere are short. Consequently, variations in the local input rates to the atmosphere and seasonal variations in the amount of rain are likely to increase the compositional variance of atmospheric precipitation. The shorter the residence time of a gas in the atmosphere, the greater is the statistical variance of its concentration in the atmosphere, measured geographically or as a time series. The inverse relationship between the residence times of gases and the variance of their concentrations in the atmosphere has been demonstrated for H_2O, CO, CO_2, CH_4, O_2, O_3, and some other atmospheric gases (Junge, 1974).

Chemical composition of atmospheric precipitation is occasionally reported as "excess concentration," which is the concentration measured less "contribution from sea salt." Although this method is in principle valid, as it should reveal the sources, other than oceans, of chemical constituents in rain water, in the published literature it is not always clear what and how much is being subtracted from the chemical analysis. As the proportions of the ions in ocean water may change during the processes of spray formation and evaporation, subtraction of some fraction of the mass of ions dissolved in rain water in a relative proportion of their abundance in ocean water may lead to errors.

The SO_4^{2-} concentrations (Tables 4.6 and 4.8) fall into two distinct ranges: the higher range is associated with the areas of industrial activity, but the lower range represents rains in weakly industrialized areas outside of northern Europe. The higher values of SO_4^{2-} are attributable to the

Table 4.6   Chemical Composition of Atmospheric Precipitation

Constituent	Concentration (μmol 1^{-1})			
	Global mean estimate[a]	Northern Europe[b]	Northeast United States[c]	European USSR[d]
Na^+	86	13–85	5	39–96
K^+	8	4–7	2	10–20
Mg^{2+}	11	5–16	2	8–12
Ca^{2+}	2	16–33	4	10–52
NH_4^+	—	6–48	12	22–61
H^+ [e]	2	16–38	74	0.2–79
HCO_3^-	2	21	—	30–79
Cl^-	11	11–98	7	23–107
NO_3^-	—	5–36	12	6–16
SO_4^{2-}	6	10–63	30	38–94
Total				
mean (mg 1^{-1})	3.7	10.2	4.0	16.6

[a]Garrels and Mackenzie (1971).

[b]Stations in France, Belgium, north and southwest Sweden (Granat, 1972, 1976).

[c]Hubbard Brook Experimental Forest, New Hampshire, mean values for years 1963–1974 (Likens, 1976).

[d]Mean values for years 1958–1961, rain and snow, several stations between northern coast and southern plains (Drozdova et al., 1964). pH=4.1 to 6.5 for more than 90% of samples, lower and higher values being rare.

[e]Concentration of H^+ ion taken from reported pH values. For example, pH=5.7 is 2×10^{-6} mol H^+ 1^{-1}.

production of SO_2 gas in the process of fuel burning and its subsequent oxidation to SO_4^{2-} in water.

The hydrogen-ion concentration in rain water is a measure of its acidity. Pure water equilibrated with CO_2 of the atmosphere ($p = 3 \times 10^{-4}$ atm) has a pH of 5.7 at 25°C. The global mean composition of rain water given in Table 4.6 has pH=5.7 or, as listed in the table, H^+-ion concentration of 2 μmol 1^{-1}. The H^+-ion concentrations in rains of industrialized regions are between 10 and 80 μmol 1^{-1}, and the pH values of such rain waters are more acidic, with pH < 5. However, the values of pH > 5.7 are not particularly rare: the upper range of the data for the European part of the USSR extends to pH≃6.5. One of the likely sources of the lower acidity of rain water is uptake and partial dissolution of $CaCO_3$ in the atmospheric dust. Analyses of dew made during the rainless summer months in the Middle East show that concentrations of the calcium and bicarbonate ions are

Table 4.7 Reactive Gases in the Atmosphere

Gas	Mass in atmosphere[a,b] (10^{12} mol)	Production rate for atmosphere (10^{12} mol yr^{-1})	Partial pressure p[g] (atm)	Dissolution reactions	Equilibrium constant (25°C, 1 atm total P) log K
CO_2	56,000		3.1×10^{-4}	$CO_2(g) + H_2O(l) = H_2CO_3(aq)$	-1.43
				$H_2CO_3(aq) = H^+(aq) + HCO_3^-(aq)$	-6.4
HCl	0.22[c]	17.0[d]	1.2×10^{-9}	$HCl(g) = H^+(aq) + Cl^-(aq)$	6.29
NH_3	1.83	8.3[a]	1.0×10^{-8}	$NH_3(g) = NH_3(aq)$	1.76
				$NH_3(aq) + H_2O(l) = NH_4^+(aq) + OH^-(aq)$	-4.74
NO	0.22		1.2×10^{-9}		
NO_2	0.22	1.3[e]	1.2×10^{-9}	$3NO_2(g) + H_2O(g) = NO(g) + 2HNO_3(aq)$	10.7
HNO_3		1.3[a]			
H_2S	0.03	2.75[f]	1.7×10^{-10}	$H_2S(g) = H_2S(aq)$	-1.0
				$H_2S(aq) = H^+(aq) + HS^-(aq)$	-7.0
SO_2	0.02	1.56[a]	1.1×10^{-10}	$SO_2(g) = SO_2(aq)$	0.35 (10°C)
SO_4^{2-}	0.05			$SO_2(g) + \frac{1}{2}O_2(g) + H_2O(l) = H_2SO_4(aq)$	35.8

[a]Garrels et al. (1975).
[b]Svensson and Söderlund (1975).
[c]Junge (1963).
[d]Duce (1969).
[e]Martens et al. (1973).
[f]Seinfeld (1975).
[g]p = (gas mole fraction) × 1 atm = (mass in atmosphere)/$(1.8 \times 10^{20}$ mol = total moles of gases in atmosphere).

higher by a factor of about 2 than their concentrations in the atmospheric precipitation of the rainy seasons (Yaalon and Ganor, 1968). The higher calcium concentrations are likely to be caused by dissolution of $CaCO_3$ dust originating from the limestone terrain, and from the human uses of limestone in construction and industry in the area. The median value of the pH of dew was reported as pH = 7.7, against a generally low background of micromolar concentrations of ions.

The pH of some rivers and lake waters in southern Norway was between 6.5 and 7.8 in 1930s and 1940s. By the first half of 1970s the pH declined down to about 5.6–6.3 (Gjessing et al., 1976). A mean rate of an increase in acidity is about -0.04 pH unit per year, or lowering by 1 pH unit in 25 yr.

4.4.2   Sources of Rain Acidity

Gas-Water Reactions.   In a simplified approach to the sources of acidity of rain water, chemical reactions can be considered for those gases that react with water and form either an excess or deficiency of H^+ ions relative to the neutral condition of $[H^+] \simeq 10^{-7}$ mol l^{-1}. The main natural gases in this category are CO_2, HCl, NH_3, NO_2, H_2S, and SO_2. Dissolution reactions of these gases in water and the equilibrium constants for the reactions are listed in the last two columns of Table 4.7. The strong acids forming as products of dissolution reactions are HCl, HNO_3, and H_2SO_4. Dissolved CO_2, written as a chemical species H_2CO_3, is a weak acid, H_2S is essentially neutral, and NH_3 is a base. The partial pressures of these reactive gases in the atmosphere are low, and the gases fall in the category of trace components of air, despite their pH-determining role in rain water.

The production rates of the reactive gases for the atmosphere, listed in Table 4.7, are estimates of the mean rates of their addition to the atmosphere from all sources, natural and anthropogenic. Production of HCl is attributable to a reaction between NaCl in salt particles or rain droplets and HNO_3 or NO_3^- ion,

$$NaCl + HNO_3 \rightarrow NaNO_3 + HCl$$

Nitric acid or the nitrate ion is produced in a process of oxidation of nitrogen oxides and their dissolution in water. An overall reaction between NO_2 gas and water vapor (Martens et al., 1973) gives HNO_3 in solution as a final product,

$$3NO_2(g) + H_2O(g) = NO(g) + 2HNO_3(aq)$$

This reaction is listed under dissolution reactions in Table 4.7.

The reaction between ammonia and water in Table 4.7 gives the ions NH_4^+ and OH^- as products, without allowing for possible subsequent oxidation to nitrogen oxides. In the same form is listed the overall dissolution reaction for H_2S gas.

For sulfur dioxide, an equilibrium reaction between SO_2, O_2, and H_2O, giving sulfuric acid H_2SO_4, does not take place in atmospheric waters (note the very high value of the equilibrium constant, computed from the standard free energy data). Oxidation of SO_2 gas to SO_3 in the absence of catalysts is a slow process. In the presence of such inorganic catalysts as NaCl, $CuSO_4$, $MnCl_2$, and $MnSO_4$, oxidation of SO_2 to SO_3 takes place rapidly (Seinfeld, 1975), although the nature of catalysts that may be present in atmospheric aerosols is not known. The gas SO_2 is very soluble in water, and its concentration at equilibrium with the gas phase is $[SO_2] = 2.25 p_{SO_2}$ (solubility coefficient $K = 10^{0.35}$, Table 4.7). SO_2 is ultimately oxidized to SO_3 and to sulfate according to an overall reaction

$$SO_2 + \text{oxidizing species} \rightarrow SO_3 + H_2O \rightarrow H_2SO_4(aq)$$

where ozone and nitrogen oxides can serve as oxidizing agents. If H_2S is also oxidized to SO_4^{2-}, it may add as much as a factor of 2 or more to the oxidizable fraction of SO_2, as the atmospheric masses of H_2S and SO_2 are comparable.

pH of Rain Water. Concentrations of individual gases in pure water, under the conditions of equilibrium solubility at 25°C and 1 atm total pressure, can be computed using the values of the equilibrium constant K and gas partial pressure p from Table 4.7 and equation 4.33. Concentrations computed by this method are listed in Table 4.8, column (2).

Another estimate of gas concentrations in rain water, under the idealized conditions of one gas dissolving in water, can be obtained from the ratio of the production rate to the atmospheric precipitation rate, assuming that the gas concentration in the atmosphere is at a steady state. The computed estimates are listed in column (3) of Table 4.8.

The concentration of NH_3 plus NH_4^+, given as 19 μmol 1^{-1}, refers to both species. Literature data reporting NH_3 may in fact refer to NH_4^+ because at pH below 8, typical of rain water, NH_4^+ is the dominant species. For SO_2 and SO_4^{2-}, a concentration of 3.6 μmol 1^{-1}, computed from the atmospheric production rate of SO_2, also refers to both SO_2 and SO_4^{2-}, as if the latter were derived only from oxidation of SO_2. The estimate of 3.6μ mol 1^{-1} is within the range of SO_4^{2-} concentrations in "clean" rains, but it is lower than the values reported for industrialized regions.

Each of the gases listed in Table 4.8, when present in rain water at the concentration computed either from gas-water equilibria or from production rates, is responsible for the lowering or raising of the pH of the

Table 4.8 Concentrations of Reactive Gases in Rain Water and the pH of Rain Water Due to Each Gas

| Gas | Concentration in rain water C_r (μmol l^{-1}) | | | pH of rain water from C_r values in columns (2) and (3) |
	From literature [a] (1)	From gas–H$_2$O equilibria [b] (2)	From production and precipitation scavenging [c] (3)	
CO$_2$	10	11		5.7
HCl			39	4.4
NH$_3$	6–50	0.6	19	8.5–9.1
NH$_4^+$	12–50	3.2		
NO$_2$			3	4.5–5.5
HNO$_3$	5–36	30	3	4.5–5.5
H$_2$S		1.7×10^{-5}	6	7.0
SO$_2$		2×10^{-4}		
SO$_4^{2-}$	10–60 [d] 0.3–4 [e]		3.6	4.1–5.1

[a] Data for northeastern United States, northwestern Europe, and southern Scandinavia. NH$_3$ is likely to be in the NH$_4^+$ form. Some particulate material may be included in the NH$_4^+$ and SO$_4^{2-}$ analytical data (Granat, 1972, 1976; Likens, 1976).

[b] $C_r = 10^6 Kp$, where K and p are given in Table 4.7.

[c] $C_r = $ (production rate)/$(4.4 \times 10^{17} \text{ l yr}^{-1})$, production rate given in Table 4.7.

[d] Rain in industrialized regions.

[e] Rain in unindustrialized regions (Granat, 1976).

153

aqueous solution. The pH values corresponding to the listed gas concentrations are given in the last column of Table 4.8. Each of the acidic gases HCl, NO_2, and SO_2 gives the pH values below the CO_2–H_2O equilibrium pH of 5.7. Atmospheric concentrations or production rates a factor of 2 lower would raise the computed pH values by 0.3 pH units. Ammonia alone raises the pH to 8.5–9. Methods of computing the pH of rain water for each of the dissolved gases and for their mixtures are shown in the following sections. Here, to conclude the discussion of the rain water acidity, an estimate of the pH is given below, based on the empirical relationship between the gas concentrations in air and in rain water, equation 4.35.

The partial pressures of HCl, NO_2, H_2S, and SO_2 in the atmosphere are between 10^{-10} and 10^{-9} atm, as given in Table 4.7. Their gram-formula weights are 36, 44, 34, and 64, or an average of 50 ± 14 g mol^{-1}. Concentrations ρ_v of these gases in the atmosphere can be computed back from the partial pressures, using the ideal gas law equations 4.2 and 4.3,

$$\rho_v = \frac{pM}{RT}$$

$$= \frac{(10^{-10} \text{ to } 10^{-9} \text{ atm})(50 \text{ g mol}^{-1}) \times 10^6 \text{ } \mu\text{g g}^{-1}}{(8.2 \times 10^{-5} \text{ m}^3 \text{ atm deg}^{-1} \text{ mol}^{-1})(298 \text{ deg})}$$

$$= 0.2\text{–}2 \text{ } \mu\text{g m}^{-3}$$

Concentration of gases in rain water, using equation 4.35, is

$$C_r = (0.1 \text{ to } 2)(0.2 \text{ to } 2) = 0.02\text{–}4 \text{ mg l}^{-1}$$

$$C_r = \frac{0.02 \text{ to } 4 \text{ mg l}^{-1}}{50 \times 10^3 \text{ mg mol}^{-1}} = 4 \times 10^{-7}\text{–}8 \times 10^{-5} \text{ mol l}^{-1}$$

Upon complete oxidation, such gases as SO_2 and H_2S dissolved in rain water at concentration C_r mol l^{-1} are electrically balanced by $2C_r$ mol l^{-1} of H^+ ions. The species HCl and HNO_3 at concentration C_r release C_r mol l^{-1} of H^+ ions in solution. For either of the conditions, the pH of water falls in the range of 2 pH units,

$$4.1 < \text{pH} < 6.4$$

When CO_2 alone determines the pH of pure rain water, the pH is about 5.7. Therefore, the range of the pH values for the acidic gases in rain water is

$$4.1 < \text{pH} < 5.7$$

Note that these pH values are essentially the same as those listed in Table 4.8, which were computed by more accurate methods, outlined below.

Computation of Rain Water pH for Individual Gases. The pH of a solution of a gas in water can be computed for gas–H_2O systems at equilibrium. The general method of computation is based on the equations for the electrical charge balance and mass balance of the ionic and undissociated species in solution. For each of the gas–H_2O pairs, the equilibrium constants for the dissolution reactions are listed in Table 4.7. In line with the wide variation in the reported pH and ion-concentration values, an approximation will be introduced: concentrations will be used instead of the thermodynamic activities. Brackets [ ] are used throughout this section to denote concentration in units of moles per liter. The activity of water in solution is taken as unity, because rain water is a very dilute solution (Table 4.6). The solution equilibria between gases and liquid water are treated without considering the fact that the atmosphere may be either undersaturated or supersaturated with respect to the water vapor.

CO_2–H_2O. An overall dissolution and dissociation equilibrium reaction for CO_2, from the two reactions listed in Table 4.7, is

$$\frac{[H^+][HCO_3^-]}{a_{H_2O}p_{CO_2}} = 10^{-7.83}$$

Using $a_{H_2O} = 1$ and $p_{CO_2} = 3.1 \times 10^{-4}$ atm (from Table 4.7),

$$[H^+][HCO_3^-] = 10^{-11.34}$$

The chemical species in solution are H_2O, H_2CO_3, H^+, OH^-, HCO_3^-, and CO_3^{2-}. The charge balance condition is

$$[H^+] = [HCO_3^-] + [OH^-] + 2[CO_3^{2-}]$$

The solution is slightly acidic and, therefore, concentrations of OH^- and CO_3^{2-} can be neglected. This gives

$$[H^+] \simeq [HCO_3^-]$$

and

$$[H^+]^2 \simeq 10^{-11.34}$$

or

$$pH = 5.7$$

HCl–H₂O. Hydrochloric acid is a strongly dissociated acid and the charged species present in solution are H^+, Cl^-, and OH^-. Concentration of OH^- in an acidic solution can be neglected, such that the charge balance condition is

$$[H^+] = [Cl^-]$$

Total concentration of HCl in rain water, given in Table 4.8, is 39×10^{-6} mol l^{-1}. The mass balance condition for total dissolved chloride $\Sigma(Cl)$ is

$$\Sigma(Cl) = [Cl^-] = 39 \times 10^{-6} \qquad (mol\, l^{-1})$$

The latter value can be used in the mass balance condition to obtain $[H^+]$,

$$[H^+] = 39 \times 10^{-6}$$

or

$$pH = 4.4$$

The value of pH=4.4 is listed in Table 4.8, for the case of HCl at equilibrium with rain water.

NH₃–H₂O. In an aqueous ammonia solution, the chemical species are H_2O, NH_3, NH_4^+, H^+, and OH^-. The equilibrium dissolution and dissociation reactions of NH_3 listed in Table 4.7 give

$$\frac{[NH_4^+][OH^-]}{a_{H_2O}p_{NH_3}} = 10^{-2.98}$$

Using $a_{H_2O} = 1$ and $p_{NH_3} = 1 \times 10^{-8}$ atm, and $[OH^-] = 10^{-14}/[H^+]$, we have

$$[NH_4^+] = 10^3[H^+]$$

The charge balance condition is

$$[NH_4^+] + [H^+] = [OH^-]$$

Combining the equilibrium with the charge balance condition,

$$10^3[H^+] + [H^+] = \frac{10^{-14}}{[H^+]}$$

Solution of the latter equation in $[H^+]$ gives

$$[H^+]^2 = 10^{-17}$$

or

$$pH = 8.5$$

Another estimate of the pH can be obtained from the concentration value of NH_4^+ plus NH_3, given in Table 4.8 as 19×10^{-6} mol l^{-1}. The mass balance condition for total dissolved nitrogen $\Sigma(N)$ is

$$\Sigma(N) = [NH_4^+] + [NH_3] = 19 \times 10^{-6} \text{ mol } l^{-1}$$

From the dissociation reaction of $NH_3(aq)$ in water, as listed in Table 4.7,

$$\frac{[NH_4^+][OH^-]}{a_{H_2O}[NH_3]} = 10^{-4.74}$$

or

$$[NH_3] = \frac{10^{-9.28}[NH_4^+]}{[H^+]}$$

Substitution of the latter relationship into the mass balance condition gives $[NH_4^+]$ as a function of $[H^+]$ and $\Sigma(N)$,

$$[NH_4^+] = \frac{1.9 \times 10^{-5}[H^+]}{[H^+] + 5.5 \times 10^{-10}}$$

The latter can be substituted for $[NH_4^+]$ in the charge balance condition, giving an equation in $[H^+]$,

$$[H^+]^3 + 1.9 \times 10^{-5}[H^+]^2 - 10^{-14}[H^+] = 5.5 \times 10^{-24}$$

Solution by trial and error gives

$$[H^+] \simeq 10^{-9.1}$$

or

$$pH = 9.1$$

HNO₃–H₂O. The equilibrium condition for the reaction between $NO_2(g)$ and $H_2O(g)$, listed in Table 4.7, is

$$\frac{[HNO_3]^2 p_{NO}}{p_{NO_2}^2 p_{H_2O}} = 10^{10.7}$$

Using the partial pressures of the nitrogen oxides from Table 4.7, $p_{NO} = p_{NO_2} = 1.2 \times 10^{-9}$ atm, and two values for the partial pressure of water vapor, $p_{H_2O} = 4 \times 10^{-3}$ and 2×10^{-2} atm, the equilibrium concentrations of HNO_3 are

$$[HNO_3] = 1.7 \times 10^{-5} \quad \text{and} \quad 3.8 \times 10^{-5} \quad (\text{mol } l^{-1})$$

The value listed in Table 4.8 is $[HNO_3] = 3 \times 10^{-5}$ mol l^{-1}. Nitric acid is a strongly dissociated acid, such that the mass balance condition for the total dissolved nitrate $\Sigma(NO_3)$ is

$$\Sigma(NO_3) = [NO_3^-] = 1.7 \times 10^{-5} - 3.8 \times 10^{-5} \quad (\text{mol } l^{-1})$$

The charge balance condition, neglecting the OH^- concentration, as in the case of HCl in water, is

$$[H^+] = [NO_3^-] = 1.7 \times 10^{-5} - 3.8 \times 10^{-5}$$

Thus

$$pH = 4.7 - 4.4$$

Another estimate of the HNO_3 concentration in rain water, based on precipitation scavenging, is ten times lower than the value computed above. The lower value of 3×10^{-6} mol l^{-1} gives a higher pH,

$$pH = 5.5$$

For HNO_3, a pH of 4.5–5.5 is listed in Table 4.8.

H₂S–H₂O. The species in an aqueous solution are H_2O, H_2S, H^+, OH^-, HS^-, and S^{2-}. The equilibrium condition for the H_2S dissolution and dissociation reactions listed in Table 4.7 is

$$\frac{[H^+][HS^-]}{a_{H_2O} p_{H_2S}} = 10^{-8}$$

With $a_{H_2O} = 1$ and $p_{H_2S} = 1.7 \times 10^{-10}$ atm,

$$[H^+][HS^-] = 1.7 \times 10^{-18}$$

The charge balance condition is

$$[H^+] = [HS^-] + [OH^-] = [HS^-] + \frac{10^{-14}}{[H^+]}$$

where the concentration of S^{2-} is neglected.

Combination of the mass balance and equilibrium conditions gives

$$[H^+]^2 = 1.7 \times 10^{-18} + 10^{-14} \simeq 10^{-14}$$

Thus

$$pH = 7$$

$SO_4^{2-}-H_2O$. In a dilute solution containing sulfate only, the charge balance condition is

$$[H^+] = 2[SO_4^{2-}]$$

For one estimate of the solution pH, the sulfate concentration $[SO_4^{2-}] = 3.6 \times 10^{-6}$ mol l^{-1}, as listed in Table 4.8, gives

$$[H^+] = 7.2 \times 10^{-6}$$

or

$$pH = 5.1$$

Another estimate of the pH is based on the sulfate concentrations in rains of industrialized regions, $10 \times 10^{-6}-60 \times 10^{-6}$ mol l^{-1}. If a mean value for this range $[SO_4^{2-}] = 35 \times 10^{-6}$ mol l^{-1} represents dissolved sulfate, then

$$pH = 4.1$$

The pH values between 4.1 and 5.1 are given in Table 4.8.

Gas Mixtures and Rain Water pH. The individual cases of gas–water equilibria in the preceding section can be combined into one relationship describing a more comprehensive case of several reactive gases present simultaneously in rain water. The following dissolved components can be considered: H_2CO_3, HCl, HNO_3, H_2SO_4, and NH_3. The condition of electrical charge balance in solution is

$$[H^+] + [NH_4^+] = [HCO_3^-] + [Cl^-] + [NO_3^-] + [HSO_4^-]$$
$$+ 2[SO_4^{2-}] + 2[CO_3^{2-}] + [OH^-]$$

At pH > 3, concentration of the bisulfate ion $[HSO_4^-]$ is negligible in comparison to the sulfate $[SO_4^{2-}]$. At pH $\leqslant$ 7, the carbonate ion concentration $[CO_3^{2-}]$ is negligible in comparison to $[HCO_3^-]$. As the pH values of rain waters usually fall within the range $3 < pH < 7$, the preceding charge balance equation can be simplified to

$$[H^+] + [NH_4^+] = [HCO_3^-] + [Cl^-] + [NO_3^-] + 2[SO_4^{2-}] + [OH^-]$$

$$(4.44)$$

For different given concentrations of Cl^-, NO_3^-, SO_4^{2-}, and NH_4^+, and the concentration of HCO_3^- at equilibrium with atmospheric CO_2, the hydrogen-ion concentration $[H^+]$ can be determined from the charge balance condition, exactly as was done in the individual gas-water systems. Here a question will be asked, what is the concentration of ammonia required to neutralize the acids present in rain water? The neutral point of pH = 7 corresponds to the condition of $[H^+] = [OH^-]$. Thus the charge balance equation becomes

$$[NH_4^+] = [HCO_3^-] + [Cl^-] + [NO_3^-] + 2[SO_4^{2-}]$$

The following concentrations of acid-forming ions can be used from Table 4.8 ($\mu mol\ l^{-1}$): $[HCO_3^-] = 2$, $[Cl^-] = 40$, $[NO_3^-] = 30$, and $[SO_4^{2-}] = 10$. The ammonium ion concentration required to bring the solution pH to 7 is approximately

$$[NH_4^+] = 2 + 40 + 30 + 2 \times 10 = 92\ \mu mol\ l^{-1}.$$

Reported values of $[NH_4^+]$ in rain water (Tables 4.6 and 4.8) are lower, 10 to 50 $\mu mol\ l^{-1}$. Thus the difference between the total concentration of the anions, 90 $\mu mol\ l^{-1}$, and the $[NH_4^+]$ concentration is between 80 and 40 $\mu mol\ l^{-1}$. If this difference is made up by the H^+ ion, then the pH of the solution containing the mixture of gases is

$$pH \simeq 4.1-4.4$$

The latter pH value is within the range of the modern values characteristic of rains in industrialized regions. The preceding computation is simplistic because it does not take into account any of the metal ions that might be present in rain water. The effect of metal cations on the charge balance equation 4.44 is to add mass to the left-hand side of the equation, which may reduce the contribution of $[H^+]$ to the charge balance when cations enter the solution in the form of hydrolyzable compounds.

Summary. Dissolved HCl, HNO_3, and H_2SO_4 in rain water contribute in about equal measure to acidity. Concentrations of each of the acids in rain water, estimated either from gas–water solution equilibria or from uptake by atmospheric precipitation, are comparable to the values reported in the literature for rains of various regions. Ammonia alone can provide a certain measure of neutralization of rain water: about equal amounts of a strong acid HA and total ammonia $\Sigma(N)$ (that is, NH_3 plus NH_4^+) at a concentration level of 10^1 μmol l^{-1} give the solution a pH between 6.5 and 7. As a whole, however, concentrations of acid-forming gases exceed the concentration of ammonia. The model of the acidity of rain water, based on a mixture of dissolved acids and ammonia, generally gives rain water a pH below 5. This raises an interesting question of the global average pH of rain water in preindustrial times: even if concentrations of HCl, HNO_3, H_2SO_4, and NH_4^+ were 10 times lower—of the order of 10^0 rather than 10^1 μmol l^{-1}—the pH of rain water would have been $pH \lesssim 5.5$. This value is close to but somewhat below the pH controlled by the equilibrium CO_2–H_2O at 25°C. Concentrations of acids and ammonia closer to those of modern times would have caused a lower value of the pH of rain water. Insofar as HCl, nitrate, and ammonia are produced by natural processes, their contribution to the acidity of rain is controlled by the magnitude of their natural fluxes to the atmosphere. The acidity of rain water and the pH depend on the relative magnitude of the ammonia flux as compared to the fluxes of the acidic gases. However, the pH model of rain water shows that strong acids might have contributed to the lowering of the pH in rains over certain geographic areas even in preindustrial times.

The sulfate concentrations in rains of industrial and unindustrialized regions suggest that an increase by a factor of 10 might have been caused by industrial activities.

Two factors may cause further increases in the acidity of rain: (1) an increasing rate of input of acid-forming species to the atmosphere; and (2) a decrease in the rate of precipitation. For those reactive gases that are to a large extent removed from the atmosphere by rain, less precipitation would result, at least initially, in higher gas concentrations. On a longer time scale, the effects may be controlled by alternate sinks on land and in water.

4.4.3 Alkalinity of Rain Water

Concentrations of some of the dissolved species in natural waters are not affected by the H^+-ion concentration. In this category belong such ionic species as Na^+ and other metal ions, and such negatively charged ions as Cl^- and SO_4^{2-}. (Concentrations of metal ions are strictly not affected by the solution pH only if such metal-ion complexes as $NaHCO_3^\circ$, $MgHCO_3^+$, and, in general, complexes between metals and pH-dependent ions can be neglected.) Another category of dissolved species is represented by those whose concentrations change as a function of the H^+-ion concentration. Such species as HCO_3^-, CO_3^{2-}, $B(OH)_4^-$, NH_3, HS^-, and residues of phosphoric and organic acids depend on the H^+-ion concentration through their dissociation equilibria.

A quantity called alkalinity [Alk] is defined as the sum total of equivalents of all the chemical species the concentrations of which depend on the H^+ ion, minus the H^+-ion concentration (for example, Stumm and Morgan, 1970). For a solution containing metal ions, carbonate, borate, and ammonia, the alkalinity function [Alk] is

$$[Alk] = [HCO_3^-] + 2[CO_3^{2-}] + [B(OH)_4^-] + [OH^-] + [NH_3] - [H^+]$$

where brackets denote concentrations in units of mol l^{-1}. Because of the charge balance condition, the sum $[Alk]+[H^+]$ is numerically equal to the difference Σ(equivalents of cations) $- \Sigma$(equivalents of anions) of all the species that are *independent* of the H^+-ion concentration. The equation can contain considerably more terms on the right-hand side if phosphates, sulfides, and organic acids occur in significant amounts in solution. In a simpler case, alkalinity of a solution containing only CO_2 and dissolved $CaCO_3$ is

$$[Alk] = [HCO_3^-] + 2[CO_3^{2-}] + [OH^-] - [H^+] \qquad (4.45)$$

or

$$[Alk] = 2[Ca^{2+}]$$

Alkalinity as defined in equation 4.45 is also known as the carbonate alkalinity. For the alkalinity of ocean water, see Section 1.5.3.

Although the alkalinity function can be written as a straightforward charge balance condition for a solution, a practical advantage is that a total concentration of the pH-dependent species can be determined by an acidimetric titration in one operational step.

For rain water containing metal ions, dissolved CO_2, borate, and ammonia, the alkalinity function can be simplified for a solution at pH < 8, typical of most rain waters. At pH below 8, virtually all of the borate ion $B(OH)_4^-$ is present in the form of undissociated boric acid $B(OH)_3$, NH_3 is present in the form of the ammonium ion NH_4^+, and the concentration of the carbonate ion $[CO_3^{2-}]$ is only a small fraction, $< 10^{-3}$, of the bicarbonate ion concentration $[HCO_3^-]$. Thus the alkalinity equation can be simplified to

$$[Alk] = [HCO_3^-] + [OH^-] - [H^+] \qquad (4.46)$$

In pure water containing only CO_2, [Alk]=0. A condition of zero alkalinity can be attained if enough acid is added to solution. Addition of greater quantities of the H^+ ion in the form of a strong dissociated acid makes [Alk]<0. At a given partial pressure of CO_2 in the atmosphere at equilibrium with dissolved CO_2 in water, the concentration terms $[HCO_3^-]$ and $[OH^-]$ in equation 4.46 can be reduced to quotients of the dissociation equilibrium constants and $[H^+]$. The following constants for the dissociation equilibria of H_2O and carbonate species in pure water at 5 and 25°C can be used (Weast, 1974; Keeling, 1973):

$$\frac{[H_2CO_3]}{a_{H_2O}p_{CO_2}} = K_0 \quad 6.38\times10^{-2}\,(5°C); \quad 3.38\times10^{-2}\,(25°C)\ (\text{mol}\,1^{-1}\,\text{atm}^{-1})$$

$$\frac{[H^+][HCO_3^-]}{[H_2CO_3]} = K_1 \quad 3.02\times10^{-7}\,(5°C); \quad 4.47\times10^{-7}\,(25°C)\ (\text{mol}\,1^{-1})$$

$$\frac{[H^+][CO_3^{2-}]}{[HCO_3^-]} = K_2 \quad 2.82\times10^{-11}\,(5°C); \quad 4.68\times10^{-11}\,(25°C)\ (\text{mol}\,1^{-1})$$

$$\frac{[H^+][OH^-]}{a_{H_2O}} = K_w \quad 1.86\times10^{-15}\,(5°C); \quad 1\times10^{-14}\,(25°C)\ (\text{mol}^2\,1^{-2})$$

Using $a_{H_2O}=1$ and the above definitions of equilibrium constants K, equation 4.46 reduces to

$$[\text{Alk}] = \frac{K_0K_1p_{CO_2}}{[H^+]}\left(1 + \frac{K_w}{K_0K_1p_{CO_2}}\right) - [H^+] \qquad (4.47)$$

At the atmospheric pressure of CO_2 ($p_{CO_2}=3\times10^{-4}$ atm) the second term in parentheses is much smaller than 1, and the equations that define [Alk] in terms of $[H^+]$ and numerical constants are:

at 5°C
$$[\text{Alk}] = \frac{5.78\times10^{-12}}{[H^+]} - [H^+] \qquad (4.48)$$

at 25°C
$$[\text{Alk}] = \frac{4.53\times10^{-12}}{[H^+]} - [H^+] \qquad (4.49)$$

In Figure 4.5 are shown the alkalinity and pH values of some rain samples, and the curves computed using equations 4.48 and 4.49. Negative values of [Alk] indicate the presence of excess acid in rain water. At pH values above 5, the computed curves underestimate [Alk]. As the computed

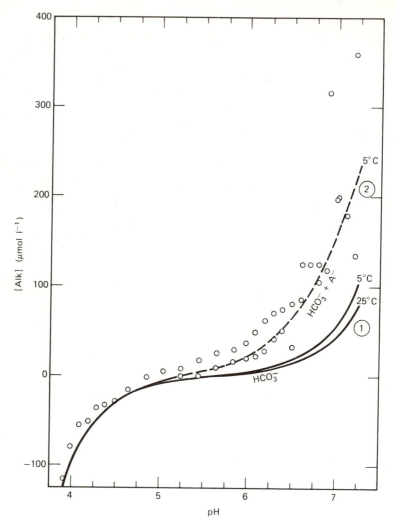

Figure 4.5 Alkalinity and pH of rain waters from northern Europe (from Granat, 1972).

curves are based on a model that takes into account only the carbonate species at pH<7, higher values of alkalinity suggest that additional sources of alkalinity are present in rain water.

One possible source of alkalinity not accounted for by equation 4.46 is dissolution of a solid phase or exchange of H^+ ions for metal ions on a solid during the acid titration of rain water samples. This is effectively an artificial addition of alkalinity in the process of alkalinity determination in the laboratory. At the pH$\simeq$7, the measured [Alk] values are 100 to 200

μmol l^{-1} higher than the values computed from the preceding model, as shown by curve 1 in Figure 4.5. If the additional alkalinity comes from dissolution of a solid phase, of gram-formula weight of about 100 g mol^{-1}, this requires concentration of the solid in rain water of the order of $(2 \times 10^{-4}$ mol l$^{-1})(100$ g mol$^{-1}) = 20$ mg l^{-1}. Concentration of suspended solids in a filtered sample of rain water should be much less than 20 mg l^{-1}. A possible exchange of [H$^+$] for other cations on clay particles in rain water samples, considered as a potential source of higher alkalinity, also requires high concentrations of cation-exchanging materials: a typical value of cation-exchanging clays is about 50 milliequivalents per 100 g of clay, near pH$\simeq$7 (Section 7.5). A clay capable of producing 5×10^{-4} mol g^{-1} of alkalinity should be present at a concentration of 0.4 g liter^{-1} to add 2×10^{-4} mol liter^{-1} of alkalinity to the solution. Again, 400 mg of solids in 1 liter of rain water is an unlikely high concentration.

Estimates of the amount of alkalinity that can be added to rain water by dissolution or exchange reactions with solids, given in the preceding paragraph, indicate that concentrations of the order of milligrams per liter of soluble carbonate and borate minerals can increase significantly the alkalinity of rain water in the areas where sources of such minerals exist. Exchange between the H$^+$ ions and metal ions on clay particles, however, is probably of a more limited significance to rain water composition.

An additional source of alkalinity in rain water may be the presence of organic acids or bases. In a simplest case, an anion A^- of an organic acid in rain water would change the alkalinity equation 4.46 to

$$[Alk] = [A^-] + [HCO_3^-] + [OH^-] - [H^+] \qquad (4.50)$$

The computation that follows is aimed at the magnitude of $[A^-]$ that could have been responsible for the higher alkalinity values plotted in Figure 4.5. A compound HA will be assumed to dissociate according to the reaction

$$HA = H^+ + A^-$$

with an equilibrium dissociation constant

$$K_A = \frac{[H^+][A^-]}{[HA]}$$

The mass balance condition for HA is

$$\Sigma(A) = [HA] + [A^-]$$

$$= [A^-]\left(1 + \frac{[H^+]}{K_A}\right) \qquad (\text{mol l}^{-1})$$

We take the total concentration of HA as $\Sigma(A) = 2 \times 10^{-4}$ mol l^{-1} and the dissociation constant $K_A = 1 \times 10^{-7}$. Then

$$[A^-] = \frac{2 \times 10^{-11}}{1 \times 10^{-7} + [H^+]} \quad (\text{mol l}^{-1})$$

and the alkalinity equation 4.50 becomes

$$[\text{Alk}] = \frac{5.78 \times 10^{-12}}{[H^+]} + \frac{2 \times 10^{-11}}{1 \times 10^{-7} + [H^+]} - [H^+] \quad (4.51)$$

Compare with equation 4.49. The [Alk] against pH curve in Figure 4.5, labeled HCO$_3^-$ + A^-, was computed using equation 4.51. The agreement between the reported and computed values is better than for the model based on HCO$_3^-$ alone. A similar result can in principle be obtained for several chemical species, characterized by the values of the dissociation constants pK from 6 to 8, and occurring at different concentrations in rain water. Small departures from the computed curves at pH < 5 may be due to the presence of dissolved organic acids with the pK values below 5.

The total concentration of a substance that may contribute to the alkalinity of rain water samples plotted in Figure 4.5 is about $\Sigma(A) \simeq$ 200 μmol l^{-1}. A measure of concentration of organic substances dissolved in rain water of industrialized regions are concentrations of carbon in the range between 100 and 1000 μmol l^{-1} (measured as the chemical oxygen demand, or the amount of oxidizable organic matter in solution). The organic substances with the pK between 6 and 8 can, at concentrations comparable to those of carbon, add substantially to the alkalinity of rain waters.

4.5 ATMOSPHERE-OCEAN-LAND MERCURY EXCHANGE

Volatility of mercury accounts for the presence of mercury vapor in the atmosphere. Support of the atmospheric mercury reservoir by oceanic and continental sources, and the fluxes of mercury between the atmosphere and ocean, and atmosphere and land, are the topics dealt with in this section. The discussion of the mercury fluxes between the atmosphere and each of the two other reservoirs focuses on the mechanisms responsible for the fluxes. The answers and solutions given below cannot by any means be considered final: the geochemical cycle of mercury, its abundance in the various geospheres, and the chemical speciation of mercury in waters, soils, and air are known only tentatively. New thermodynamic data, new information on the existence of the mercury species (such as organic and inorganic complexes), and information on the volatility of mercury from minerals and waters may alter the picture of the mercury fluxes more or less drastically. But as the information on mercury is at present in a more complete stage than for other volatile metals, the transport to and from the atmosphere will be discussed with reference to mercury as an example.

4.5.1   Background: The Geochemical Cycle of Mercury

A summary picture of the fluxes of mercury in the three-reservoir system —atmosphere, land, and ocean— is shown in Figure 4.6. The fluxes of mercury (total mercury in units of 10^8 g yr^{-1}) from land to the ocean via rivers, from ocean to sediments, and from sediments back to the exposed land are shown in the figure as estimates for preindustrial times (prehuman fluxes) and for the present. The higher estimates for the present time are the results of industrial activity, burning of fuels, and higher rates of land erosion. In particular, the much higher mercury flux from the continental crust to land (120×10^8 g yr^{-1}) in comparison to the figure for the prehuman time (15×10^8 g yr^{-1}) is attributable to mining and other industrial activities.

The total mass of mercury in the atmosphere, shown in Figure 4.6, is given as $50 \times 10^8 \pm 10 \times 10^8$ g. This figure is based on the mean Hg

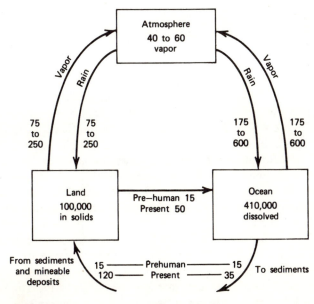

Figure 4.6 The geochemical cycle of mercury and its fluxes between land, atmosphere, ocean, and sediments (from Garrels et al., 1975). Reservoir contents in units of 10^8g Hg, fluxes in units of 10^8g Hg yr^{-1}.

concentration of 1.0–1.5 ng m^{-3} in the atmosphere, and the atmosphere volume of 4×10^{18} m^3. The total flux of mercury from the atmosphere to land and ocean is bracketed by two estimates (Weiss et al., 1971): the lower estimate of 250×10^8 g yr^{-1} and the higher estimate of 850×10^8 g yr^{-1}. In the figure, each of these fluxes is apportioned between land and ocean in the ratio of 0.3/0.7, corresponding to the land/ocean surface area ratio. Apportionment of the total flux gives: atmosphere to land flux, 75×10^8–250×10^8 g yr^{-1}; atmosphere to ocean flux, 175×10^8–600×10^8 g yr^{-1}.

The lower figure of the annual flux is based on the mercury concentration in nineteenth century Greenland ice, 0.06 ± 0.02 ppb and the rate of atmospheric precipitation of 4.2×10^{17} l yr^{-1}: $(0.06 \times 10^{-9}$ g cm$^{-3})(4.2 \times 10^{20}$ cm^3 yr$^{-1}) = 250 \times 10^8$ g yr^{-1}.

The higher figure of the annual flux is based on the same rate of atmospheric precipitation and mean concentration of mercury in rain water as 0.2 ppb (Fleischer, 1970, p. 58; Siegel et al., 1973): $(0.2 \times 10^{-9}$ g cm$^{-3})(4.2 \times 10^{20}$ cm^3 yr$^{-1}) = 850 \times 10^8$ g yr^{-1}.

The third estimate of the mercury flux out of the atmosphere, not shown in Figure 4.6, is the highest: 1500×10^8 g yr^{-1} (Weiss et al., 1971). This figure is based on an assumption that mercury vapor in the atmosphere has the same residence time as water vapor. This means that an "average global rain" removes all the mercury from the atmosphere. With an average time of 10 days between rains, there are 37 "average rains" per year, and the mercury removal rate is: $(40 \times 10^8$ g)(37 yr$^{-1}) = 1500 \times 10^8$ g yr^{-1}.

If the mercury content of the atmosphere is, or was in preindustrial times, at a steady state, then the flux out of the atmosphere (250×10^8–850×10^8 g yr^{-1}) must be maintained by input of an equivalent amount of mercury. The mercury fluxes from land and ocean to the atmosphere are shown in Figure 4.6 as equal in magnitude to the fluxes in the opposite direction. This choice is somewhat arbitrary, although the total flux from the atmosphere to the earth's surface must be balanced by input from either land or ocean or both. That the ocean is likely one of the sources of mercury in the atmosphere is suggested by the balance of mercury transported to the ocean by rivers and removed from the ocean to sediments: if there were no mercury flux out of the ocean to the atmosphere, the mercury fluxes from ocean to sediments (15×10^8–35×10^8 g yr^{-1}) would have to be much higher, of an order of 175×10^8–600×10^8 g yr^{-1}, to remove the amount added from the atmosphere (Garrels et al., 1975).

An estimate of man-produced mercury emissions to the atmosphere is about 100×10^8 g yr^{-1}. This flux is absorbed within the range of estimates from 75×10^8 to 250×10^8 g yr^{-1} for the flux of mercury between the atmosphere and land.

4.5.2   Chemical Species of Hg and Their Occurrence

Mercury in the Atmosphere.   Mercury in the atmosphere presumably occurs in the form of a gaseous species Hg(g). Essentially all the mercury is present in the form of vapor, as the fraction contained in aerosol particles of marine and continental origin is very small (percent to tenths of one percent level reported). In Figure 4.7 are summarized concentrations of mercury in the atmosphere. Over the ocean and several kilometers above

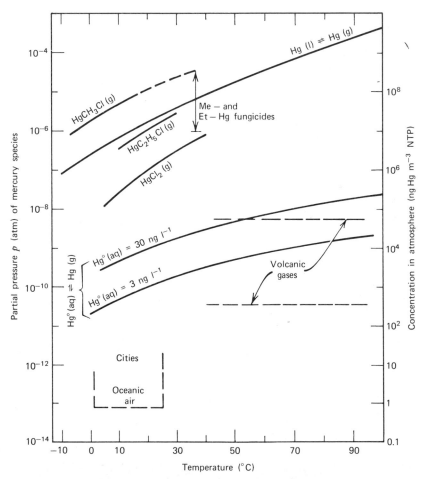

Figure 4.7   Partial pressure of gaseous mercury at equilibrium with various mercury phases and occurrences of mercury in air (equilibria from Charnley and Skinner, 1951; Glew and Hames, 1971, and sources cited in the text). $p_{Hg} = 1.117 \times 10^{-13} \times$ (concentration in ng m^{-3} at NTP), where p is in atmospheres.

ground in the atmosphere, mercury concentration is between 1 and 1.5 ng m^{-3} (conversion from ng m^{-3} to pressure units in atm is given in the caption of Figure 4.7). This concentration level is generally considered as background concentration of mercury in air. The air over cities and over land may have more mercury, of the order of 10^1 ng m^{-3}. Much higher values, up to 10^4 ng m^{-3}, are associated with the air over volcanically active areas and regions rich in mercury minerals.

Concentrations of mercury in the atmosphere can be compared with the vapor pressures of metallic Hg, $HgCl_2$, and several organic mercury compounds, shown in Figure 4.7. All the solid-vapor equilibria are characterized by mercury vapor pressures much higher than the atmospheric concentrations, such that the existing conditions favor volatilization of organic and some inorganic mercury compounds on land. Data summarized by McCarthy et al. (1970) indicate that strong mercury concentration gradients can exist in the pore space of soils and in air near the ground: air pore space of soils can contain 0–100 ng Hg m^{-3}; air near the ground, 30–300 m altitude, can contain 2–9 ng m^{-3}, decreasing to lower levels with an increasing altitude. Some of the higher concentrations of mercury vapor in the soil pore space and above ground may be the result of bacterial degradation of organic Hg compounds used as fungicides in cultivated soils.

Mercury in Ocean Water. Mean concentration of total dissolved mercury in the ocean is often quoted as 30 ng Hg l^{-1}, based on 1930s and 1950s data (Fleischer, 1970). A factor of 3 or 4 variation above or below the mean has been reported, with the higher values associated with near-shore waters. More recent analyses of open sea waters, done in 1970s, give concentrations of total mercury between 5 and 10 ng l^{-1} (Matsunaga et al., 1975; Baker, 1977; Ólafsson, 1978).

The relative abundance of the different complexes of mercury with inorganic ligands in ocean water is shown graphically in Figure 4.8 as log[complex concentration]/[Hg^{2+}-ion concentration]. At the pH$\simeq$8 of ocean water, the three complexes $HgCl_4^{2-}$, $HgCl_3^{-}$, and $HgCl_2^{\circ}$ account for practically all of the dissolved mercury. Significant departures from this distribution may conceivably be caused by the occurrence of complexes of mercury with organic ligands in water. Up to 50% of the total dissolved mercury in ocean water, in an area over the continental shelf of eastern United States, has been reported as associated with some unidentified organic materials, probably derived from the continental runoff (Fitzgerald and Hunt, 1974).

Another chemical species that has to be considered for ocean water is neutral mercury in solution Hg°(aq), that is a potential source of Hg(g) in

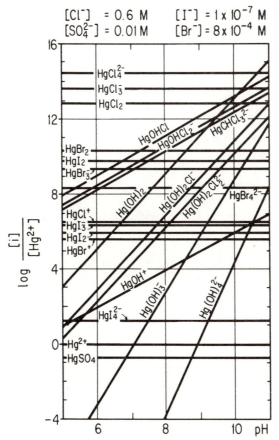

Figure 4.8 Chemical speciation of mercury in a model ocean water. Molal concentration of each species [i] relative to that of the mercuric ion [Hg^{2+}] as a function of pH. Concentrations of main complexing ligands are shown on top of the figure (from Stumm and Brauner, 1975). Reproduced with permission from Chemical Oceanography, 2nd ed., Vol. 1, J. P. Riley and G. Skirrow, Eds. Copyright by Academic Press Inc. (London) Ltd.

the atmosphere. The oxidation-reduction equilibrium

$$Hg^{\circ}(aq) = Hg^{2+}(aq) + 2e^{-} \tag{4.52}$$

takes place fast in pure water (Wollast et al., 1975), and the relative concentrations of the two species or their ratio [Hg^{2+}]/[Hg°] depend on the oxidation potential (E_h) or redox potential (pe) of the solution.

For computation of the equilibrium constants of the reactions involving Hg-species, such as reaction (4.52), the values of the standard free energies of formation of some of the gaseous, aqueous, and solid species are listed in Table 4.9. The working definitions of the oxidation (E_h) and redox (pe) potentials are summarized in Table 4.10.

Under oxidizing conditions in ocean water (aerobic and anaerobic) at pH of between 7 and 9, the oxidation potential is

$$0.4 < E_h < 0.75 \text{ volt}$$

or, equivalently, the redox potential is

$$6.8 < \text{pe} < 12.7$$

For the extreme values of E_h and pe, the equilibrium concentration ratios $[Hg^{2+}]/[Hg^\circ]$, from reaction 4.52 and Table 4.10, are

$$\frac{[Hg^{2+}]}{[Hg^\circ]} \simeq 10^{+3} \quad \text{at} \quad E_h = 0.75 \text{ volt} \quad \text{or} \quad \text{pe} = 12.7$$

$$\frac{[Hg^{2+}]}{[Hg^\circ]} \simeq 10^{-8} \quad \text{at} \quad E_h = 0.40 \text{ volt} \quad \text{or} \quad \text{pe} = 6.8$$

Even if the species Hg^{2+} is much less abundant in ocean water than the neutral species Hg°, by a factor as large as 10^{-8}, this does not affect the dominant role of the mercury-chloride complexes, shown in Figure 4.8; the concentration ratio $[HgCl_4^{2-}]/[Hg^{2+}]$ is about 2.5×10^{14}, which makes Hg° much less abundant than the chloride complexes

$$\frac{[Hg^\circ]}{[HgCl_4^{2-}]} = 10^{-17} \text{ to } 10^{-6}$$

Methylated mercury complexes are produced by bacteria in anoxic sediments. Dimethyl mercury $(CH_3)_2Hg$ and possibly methyl mercury chloride CH_3HgCl have been reported from estuarine sediments containing several percent organic carbon. However, concentrations of methyl-mercury species in the sediment account only for a small fraction of the total mercury, between 0 and 0.07% (Andren and Harriss, 1973). Under oxidizing conditions, methyl-mercury compounds break down into methane and mercury by bacterially mediated processes (Wollast et al., 1975). Decomposition of methyl mercury in oxidized waters works against the occurrence of significant fractions of dissolved mercury in the form of methylated complexes.

Table 4.9 Gibbs Standard Free Energies of Formation ΔG_f° of Some Mercury and Related Species at 25°C and 1 atm Total Pressure[a]

Species	ΔG_f° (kcal mol^{-1})
Hg(l) metal	0
Hg(g)	7.59
Hg°(aq)	9.4
Hg^{2+}(aq)	39.38
Hg(OH)$_2$(aq)	-65.70
HgCl$_4^{2-}$(aq)	-107.7
OH$^-$(aq)	-37.595
Cl$^-$(aq)	-31.35
H$_2$O(l)	-56.69
H$_2$O(g)	-54.636

[a]From Wagman et al. (1969).

Table 4.10 Oxidation-Reduction Equilibrium of Mercury in Aqueous Solution at 25°C [a, b]

Reaction	$Hg°(aq) = Hg^{2+}(aq) + 2e^-$
Standard free energy change	$\Delta G_r^\circ = 39.38 - 9.4 = 29.98$ kcal mol^{-1}
Standard oxidation potential	$E_o = \dfrac{\Delta G_r^\circ}{n\mathscr{F}} = \dfrac{29.98}{2 \times 23.06} = 0.650$ volt
	where $n = 2$ is the number of electrons involved in the reaction and $\mathscr{F} = 23.06$ kcal volt^{-1} equivalent^{-1}
Oxidation potential	$E_h = E_o + \dfrac{2.303\,RT}{n\mathscr{F}} \log Q$
	where $Q = a_{Hg^{2+}} / a_{Hg°}$ is ratio of activities and $RT = 0.592$ kcal mol^{-1} at 25°C

Using $n = 2$, $E_o = 0.650$, and concentrations in mol l^{-1} instead of activities a_i

$$E_h = 0.650 + 0.030 \log \frac{[Hg^{2+}]}{[Hg°]}$$

Redox potential [b]	$pe = pe° + \dfrac{1}{n} \log Q$
	where Q is defined as above and $pe° = \dfrac{\Delta G_r^\circ}{2.303\,nRT} = 11.0$
Thus	$pe = 11.0 + 0.5 \log \dfrac{[Hg^{2+}]}{[Hg°]}$

[a]Garrels and Christ (1965).
[b]Stumm and Morgan (1970).

4.5.3 Volatile Mercury Species in Ocean Water

The aqueous mercury species in ocean water that are likely to have vapor pressures of their own are potentially methyl mercury compounds, such as $(CH_3)_2Hg$ and CH_3HgCl, and the inorganic species $Hg°$ and $HgCl_2°$. Whether methyl mercury compounds form under oxidizing conditions and occur in ocean water is not clear. Thus little can be said at present about their global role as sources of volatile mercury in surface waters of the ocean, with the exception of the restricted areas of near-shore and estuarine environments that are affected by human activity. With regard to the inorganic aqueous species $Hg°$ and $HgCl_2°$ (and other complexes shown in Figure 4.8), these complexes cannot, as a rule, be identified by direct analytical techniques. Their identification is implicit in the complexation and dissociation reactions, and in their equilibrium constants that are computed from the results of titration and electrometric measurements in solutions. Thus the computed concentrations of $Hg°$, $HgCl_2°$, and other mercury species in ocean or fresh water depend basically on the chemical speciation models that have been used to derive the values of the equilibrium constants. Despite the uncertainties of such an approach, it is instructive to examine the vapor pressures of mercury in the atmosphere that may exist if the aqueous species $Hg°$ and $HgCl_2°$ are present in ocean water.

Equilibrium $Hg°(aq) = Hg(g)$.   The equilibrium relationship for a reaction between mercury vapor and neutral mercury in solution

$$Hg°(aq) = Hg(g) \qquad (4.53)$$

is given by Henry's-law constant $K_H = p/X$, where p is the partial pressure of $Hg(g)$ in air (atm) and X is the mole fraction of $Hg°(aq)$ in solution. The temperature dependence of the ratio p/X is (Glew and Hames, 1971)

$$\log \frac{p}{X} = 135.636 - \frac{8031.1}{T} - 42.8848\,T \qquad (4.54)$$

where T is temperature in degrees Kelvin. The equation is valid between 0 and 120°C.

For two concentrations of $Hg°(aq)$—30 and 3 ng l^{-1} (mole fractions $X = 2.7 \times 10^{-12}$ and 2.7×10^{-13})—equilibrium vapor pressures, computed from equation 4.54, are shown as a function of temperature in Figure 4.7. If all the mercury in ocean water (30 ng l^{-1}) were present in the form of $Hg°(aq)$, then the equilibrium vapor pressure at 10°C would be about 3×10^{-10} atm $Hg(g)$. Average vapor pressure of mercury in the atmosphere is about 3000 times lower, 10^{-13} atm. The atmospheric partial pressure of mercury $p \simeq 10^{-13}$ atm would be at equilibrium, according to equation 4.54, with a concentration of $Hg°$ in water

$$X \simeq 1 \times 10^{-15} \qquad \text{or, equivalently, } 0.011 \text{ ng } l^{-1}$$

This concentration would account for 0.03–0.3% of total dissolved mercury in ocean water. But in Section 4.5.2 it is shown that $Hg°$ accounts only for a much smaller fraction $\lesssim 10^{-6}$ of dissolved mercury. On the basis of this computation there is less $Hg°$ in ocean water than can be maintained by an equilibrium with the atmospheric Hg vapor, according to reaction 4.53. Therefore, a condition exists for the flux of Hg(g) from the atmosphere to the ocean, and this flux is estimated in Section 4.5.4.

Equilibrium $HgCl_2°(aq) = HgCl_2(g)$. Existence of the gaseous species $HgCl_2(g)$ is suggested by the vapor pressure of solid mercury chloride $HgCl_2$. The vapor pressure of $HgCl_2$ is described by the relationship (Landolt-Börnstein, 1960)

$$\log p = -\frac{4541}{T} + 10.40 - 0.65 \log T - 1.13 \times 10^{-3} T \qquad (4.55)$$

where p is the partial pressure of $HgCl_2$ in atm, and T is in degrees Kelvin. If the pressure-temperature relationship describes an equilibrium between solid $HgCl_2$ and its vapor, then the reaction can be written as

$$HgCl_2(s) = HgCl_2(g) \qquad (4.56)$$

and its equilibrium constant at 25°C is

$$K_0 = p_{HgCl_2} = 1.66 \times 10^{-7} \qquad \text{(atm)}$$

From reaction 4.56 and two auxiliary reactions written below, an equilibrium constant can be derived for the aqueous $HgCl_2°$ and gaseous $HgCl_2$ exchange:

$HgCl_2(s) = HgCl_2(g)$	$K_0 = 10^{-6.78}$
$Hg^{2+}(aq) + 2Cl^-(aq) = HgCl_2(s)$	$K_1 = 10^{+15.46}$
$HgCl_2°(aq) = Hg^{2+}(aq) + 2Cl^-(aq)$	$K_2 = 10^{-13.24}$

Sum: $HgCl_2°(aq) = HgCl_2(g)$ $\qquad K = K_0 K_1 K_2 = 10^{-4.56} \qquad (4.57)$

The equilibrium constant K_1 was computed from the standard free energy of formation data given in Table 4.9, and K_2 is taken from the data in Figure 4.8. Thus the vapor pressure of $HgCl_2(g)$ at equilibrium with $HgCl_2^\circ(aq)$ is, from equation 4.57,

$$p_{HgCl_2} \simeq 10^{-5}[HgCl_2^\circ] \tag{4.58}$$

Concentration of the aqueous complex $[HgCl_2^\circ]$ in ocean water may be estimated from the relative concentrations of dissolved mercury species, shown in Figure 4.8, three of which account for practically all of dissolved mercury (inorganically bound) in sea water

$$\Sigma(Hg) = [HgCl_4^{2-}] + [HgCl_3^-] + [HgCl_2^\circ]$$

$$= [HgCl_2^\circ]\left(\frac{[HgCl_4^{2-}]}{[HgCl_2^\circ]} + \frac{[HgCl_3^-]}{[HgCl_2^\circ]} + 1\right) \tag{4.59}$$

From Figure 4.8, the concentration ratios are $[HgCl_4^{2-}]/[HgCl_2^\circ] = 10^{1.6} = 40$ and $[HgCl_3^-]/[HgCl_2^\circ] = 10^{0.8} = 6$. Substitution of these values in equation 4.59 gives

$$\Sigma(Hg) = [HgCl_2^\circ](40 + 6 + 1)$$

and

$$[HgCl_2^\circ] = 0.02\Sigma(Hg) \tag{4.60}$$

which means that the neutral species $HgCl_2^\circ$ accounts for 2% of total dissolved mercury in ocean water.
Taking $\Sigma(Hg) = 30$ ng Hg $l^{-1} = 1.5 \times 10^{-10}$ mol l^{-1}, we have, from equations 4.60 and 4.58,

$$[HgCl_2^\circ] = 3 \times 10^{-12} \text{ mol } l^{-1} \quad \text{or} \quad 0.6 \text{ ng Hg } l^{-1}$$

and

$$p_{HgCl_2} \simeq 3 \times 10^{-17} \text{ atm}$$

The latter value of the partial pressure accounts for only $\frac{1}{3000}$ or 0.03% of the mercury vapor pressure in the atmosphere (Figure 4.7). If the gas $HgCl_2(g)$ dissociated in the atmosphere into $Hg(g)$ and $Cl_2(g)$, then the partial pressure of $HgCl_2$ would be virtually 0, and a condition would be set for the flux of $HgCl_2^\circ$ out of water to the atmosphere.

4.5.4   Fluxes of Mercury Between Atmosphere and Ocean

It was shown in Section 4.5.3 that the partial pressure of mercury vapor in the atmosphere and concentration of $Hg°$ in ocean water are such that there may be transfer of $Hg(g)$ from the atmosphere to the ocean. Schematically this is shown in Figure 4.9a. The concentration of $Hg°(aq)$ at the upper surface of the water boundary layer is at equilibrium with the atmospheric $Hg(g)$. The equilibrium concentration in water is about 0.01 ng Hg l^{-1}. Below the boundary layer, the $Hg°$ concentration is near zero, and a linear concentration gradient represents diffusional transport of Hg downward. The magnitude of this diffusional flux can be estimated using equation 4.38 and the following data: boundary layer thickness $h \simeq 50$ μm, 90% of the ocean surface area $S = 3.2 \times 10^{18}$ cm^2, concentration difference across the boundary layer $\Delta C \simeq 0.01$ ng Hg l^{-1}, the diffusion coefficient of dissolved mercury in ocean water $D \simeq 10^{-5}$ cm^2 sec$^{-1} \simeq 300$ cm^2 yr^{-1}. The total flux is

$$F = SD\frac{\Delta C}{\Delta h} = \frac{(3.2 \times 10^{18}) \times 300 \times (0.01 \times 10^{-12})}{5 \times 10^{-3}}$$

$$= 20 \times 10^8 \quad \text{g yr}^{-1}$$

The figure of 20×10^8 g yr^{-1} for the gaseous mercury flux from the atmosphere to the ocean is a small fraction, 3–10%, of the mercury flux that enters the ocean with the atmospheric precipitation.

A schematic flux diagram for a volatile mercury species diffusing out of ocean water is shown in Figure 4.9b. For $HgCl_2°$, its computed concentration in ocean water was given as 0.6 ng Hg l^{-1} at the end of Section 4.5.3.

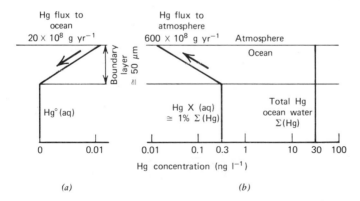

(a)   (b)

Figure 4.9   Schematic models of the mercury flux across the atmosphere-ocean interface.

Taking this value as the concentration difference across the boundary layer $\Delta C \simeq 0.6 \times 10^{-12}$ g cm^{-3} and the value of other parameters as in the preceding flux computation, the flux out of ocean water is

$$F = SD\frac{\Delta C}{\Delta h} = \frac{(3.2 \times 10^{18}) \times 300 \times (0.6 \times 10^{-12})}{5 \times 10^{-3}}$$

$$= 1200 \times 10^8 \quad \text{g yr}^{-1}$$

The estimate of 1200×10^8 g yr^{-1} for the flux ocean-atmosphere is by a factor of two higher than the estimate of 600×10^8 g yr^{-1} quoted in Figure 4.6.

A third estimate of the mercury flux out of the ocean can be based on the observed mercury vapor concentrations over the ocean surface (Eberling et al., 1976). Mercury vapor concentration several meters above the ocean surface over the North Atlantic is near 4 ng m^{-3}, with the total range between 1 and 11 ng m^{-3}. These are mercury concentrations in the gaseous phase. As the background values of mercury concentrations in the atmosphere are between 1 and 1.5 ng m^{-3}, a concentration gradient of mercury between the ocean surface and higher atmosphere is the concentration difference $\Delta C \simeq 4 - 1 = 3$ ng m^{-3} spread over a vertical distance of $h = 1$ to 5 km. Taking the vertical eddy diffusion coefficient in the atmosphere (Figure 2.6) as $D \simeq 10^5$ cm^2 sec$^{-1} \simeq 3 \times 10^{12}$ cm^2 yr^{-1}, the flux from 90% of the ocean surface is

$$F = SD\frac{\Delta C}{\Delta h} = \frac{(3.2 \times 10^{18}) \times (3 \times 10^{12}) \times (3 \times 10^{-15})}{(1 \text{ to } 5) \times 10^5}$$

$$= 600 \times 10^8 \text{ to } 3000 \times 10^8 \quad \text{g yr}^{-1}$$

The four estimates of the mercury flux out of the ocean to the atmosphere—200×10^8, 600×10^8 (Fig. 4.6), 1200×10^8, and 600×10^8–3000×10^8—are spread over one order of magnitude, which is not surprising. A limit on the flux of mercury out of the ocean is approximately set by the amount entering the ocean with atmospheric precipitation, as the fluxes of mercury via river inflow and sedimentation are relatively small, according to the data shown in Figure 4.6.

The problem can be turned around to ask, what fraction of mercury in ocean water should be in a volatile state that can maintain a flux to the atmosphere of the order of 600×10^8 g yr^{-1}?

Concentration difference across the boundary layer ΔC, as shown in Figure 4.9b, is an estimate of concentration of a volatile species:

$$\Delta C = \frac{Fh}{SD} = \frac{600 \times 10^8 \times 5 \times 10^{-3}}{3 \times 10^{18} \times 300} = 0.3 \times 10^{-12} \quad \text{g cm}^{-3}$$

$$= 0.3 \quad \text{ng Hg l}^{-1}$$

Thus a volatile mercury species concentration needed to maintain the flux of 600×10^8 g yr^{-1} is about $0.3/30 = 1\%$ of the total dissolved mercury in ocean water.

4.5.5 Land to Atmosphere Mercury Flux

Mercury concentration in the air over land surface, cited in Section 4.5.2, is between 2 and 9 ng m^{-3}. Following the computational examples for the mercury fluxes from the ocean surface given in the preceding section, the flux from land can be estimated using the following data: dry land surface area $S = 1.3 \times 10^{18}$ cm^2, vertical diffusion coefficient in the atmosphere $D \simeq 10^5$ cm^2 sec$^{-1} \simeq 3 \times 10^{12}$ cm^2 yr^{-1}, concentration difference between ground air and higher altitudes $\Delta C \simeq 5.5 - 1.5$ ng m$^{-3} = 4 \times 10^{-15}$ g cm^{-3}, and the length of the vertical concentration gradient $h = 1$ to 5 km. The flux is

$$F = SD \frac{\Delta C}{\Delta h} = \frac{(1.3 \times 10^{18}) \times (3 \times 10^{12}) \times (4 \times 10^{-15})}{(1 \text{ to } 5) \times 10^5}$$

$$= 300 \times 10^8 \text{ to } 1500 \times 10^8 \quad \text{g yr}^{-1}$$

Summary of Mercury Fluxes. Concentrations of mercury vapor in the air over land and ocean correspond to the fluxes of 300×10^8 g yr^{-1} from land to the atmosphere and 600×10^8 g yr^{-1} from the ocean to the atmosphere. The sum of the two fluxes, 900×10^8 g yr^{-1}, is essentially equal to the mercury flux of 850×10^8 g yr^{-1}, estimated as leaving the atmosphere with the atmospheric precipitation. An uncertainty factor of 5 is likely to be associated with the flux figures.

To maintain a mercury flux out of the ocean to the atmosphere, 1–10% of dissolved mercury in ocean water should be in a chemically volatile form. Although the nature of the volatile species is not known, an inorganic complex $HgCl_2°$ in ocean water has a computed concentration and vapor pressure that come close to the values needed to maintain a flux of about 600×10^8 g yr^{-1}.

CHAPTER 5

Physical and Chemical Weathering

The conditions of the local chemical or physical disequilibrium are ultimately responsible for the weathering and erosion of the earth's crust. The physical and chemical processes, and the interactions among them that control the production and transport of weathered materials are the subject of this chapter. Some background information, preliminary to the discussion of the physical weathering, is given in Section 5.2. The mechanisms of the production and alteration of particles, such as fragmentation and rounding, are discussed with reference to the particle-size spectra of the earth's regolith. The chemical weathering is treated from the point of view of the kinetics of dissolution of minerals. Some models of dissolution and experimental data involving the major rock-forming minerals and constituents of natural waters are found in Section 5.5.

5.1 BEGINNINGS OF WEATHERING

Breakdown and dissolution of solids are the two main mechanisms responsible for the weathering of the continental surface. Physical weathering leads to disaggregation and fragmentation of rocks, making possible the transport of the broken material by running water and wind. Chemical weathering shows itself in dissolution of rock minerals, and in the formation of new mineral phases through chemical and biochemical reactions. The dividing lines between the physical and chemical weathering are not sharp, as each type of the weathering process interacts with the other. Fragmentation and comminution of mineral grains promote dissolution, and the action of water and organic acids on consolidated rocks promotes loosening of the grain-to-grain contacts. Conversely, precipitation of cements in the spaces between grains hinders physical disaggregation.

Igneous rocks and consolidated sediments are aggregates made of different solid phases. A first step needed to overcome the cohesive forces holding an aggregate together is the formation of surfaces of disjunction along which further separation can take place. Cracks of microscopic and macroscopic dimensions are ubiquitously present in solids. Cracks can form at the time when crystals solidify from a melt, they can be inherited from a more loosely packed structure of unconsolidated sediments, or they can form due to mechanical stresses in the crust undergoing deformation. The subsequent growth of cracks leading to disaggregation is a combined result of mechanical and chemical processes. When a crack forms, a new surface is being added to the solid. For this to happen, the conditions must be such that energy can be released in the process and transferred elsewhere within the solid. These two basic conditions of the crack formation and propagation were postulated by A. A. Griffith in the 1920s, in his early studies of the strength of materials (Stookey and Maurer, 1967; Gordon, 1971). The concepts of the Griffith crack (a two-dimensional crack with an elliptical tip) and Griffith crack half-length, used in the theoretical analysis of brittle solids, go back to Griffith's original studies. Cracks form in solids under tension, and ideally they grow in the direction perpendicular to the tensile stress (Figure 5.1). As the crack grows, its direction can be changed by the interfaces, planes of junction or of different strength occurring within the solid. A simplest mechanism for stopping a crack is the occurrence of an interface, the cohesive forces at which are generally weaker than the overall strength of the solid. The crack tip approaching a weaker interface tends to pull the solid apart because of the stress concentration at the advancing crack tip. At an interface of weaker junction, a crack can either stop or change its direction of growth.

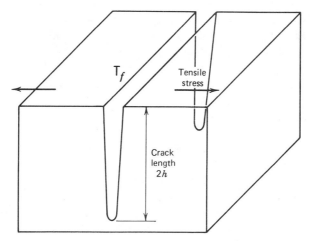

Figure 5.1 Development of a crack in a solid under tensile stress. τ_f is tensile strength.

A measure of brittleness of a solid is its tensile strength T_f expressed in units of stress or pressure (dyn cm^{-2}) by the relationship

$$T_f = \left(\frac{2EW}{\pi h}\right)^{1/2} \qquad (\text{dyn cm}^{-2}) \qquad (5.1)$$

where E is Young's modulus of the solid (dyn cm^{-2}), W is its specific surface energy (erg cm^{-2}), and h is the half-length of the Griffith crack (cm). For many rocks, the tensile strength measured in hand-size specimens is of the order of magnitude $T_f \simeq 10^8$ dyn cm$^{-2} = 100$ bars. Typical values of Young's modulus and specific surface energy are $E \simeq 10^{12}$ dyn cm^{-2} and $W \simeq 10^3$ ergs cm^{-2}. These values correspond to the half-length of the crack of about

$$h = \frac{2 \times 10^{12} \times 10^3}{3.14 \times (10^8)^2} = 0.06 \text{ cm} \quad \text{or} \quad 600 \ \mu\text{m}$$

Thus, to an order of magnitude, millimeter-long cracks can be expected to develop in a rock at the limit of its tensile strength. One of the equations relating the tensile strength of a rock to the half-length of cracks (Oberbeck and Aoyagi, 1972) is of the form

$$\log T_f = 7.402 - \tfrac{1}{2}\log h$$

where T_f and h are in cgs units, as above. A general observation on the tensile strength of solids is that cracks present in the solid reduce its tensile strength at fracture by several orders of magnitude. Crack of the order of $\frac{1}{100}$ of the body dimensions can reduce its tensile strength a hundred- to a thousandfold to $T_f \simeq 10^5 - 10^6$ dyn cm^{-2}.

In addition to the tensile stresses of tectonic origin, temperature variations at the earth's surface can induce stresses comparable to the tensile strength of rocks. Linear contraction of a solid by cooling can be compared to contraction due to volume change at higher pressure, approximately,

$$-\alpha \Delta T = \tfrac{1}{3}\beta \Delta P \qquad (5.2)$$

where α is the linear coefficient of thermal expansion (deg^{-1}), ΔT is the temperature change (deg), β is the coefficient of volume compressibility (bar^{-1}), and ΔP is the pressure change (bar). For common rocks, such as granite, the order of magnitude values of the expansion and compressibility coefficients are $\alpha \simeq 10^{-5}$ deg^{-1} and $\beta \simeq 10^{-6}$ bar^{-1}; for a temperature drop of $\Delta T = -10$ deg, an increase in pressure that would produce an equivalent dilatation is, from equation 5.2,

$$\Delta P = \frac{3 \times 10^{-5} \times 10}{10^{-6}} = 300 \text{ bars} \quad \text{or} \quad 3 \times 10^8 \text{ dyn cm}^{-2}$$

This ΔP value is comparable to the tensile strength of rocks, which shows that significant stresses can be induced by surface cooling of a magnitude comparable to diurnal temperature variations. The crack length of about 0.1 cm, derived in the preceding example, is shorter than the depth of penetration of diurnal cooling and heating: in a 6-hr period ($t = 2 \times 10^4$ sec), in a rock of thermal conductivity $\kappa = 10^{-2}$ cm^2 sec^{-1}, a characteristic penetration depth is $(\kappa t)^{1/2} = 10$ cm.

Small temperature variations contribute virtually nothing to the formation of cracks because the stresses induced in the rock can be easily dissipated along grain contacts and in softer fill of the existing cracks. Looked at differently, the amount of contraction and expansion due to temperature changes in an elastic rock is the product $\alpha \Delta T \simeq 10^{-4}$. This translates into a separation 1 μm wide between two crystals, each 1 cm in length.

Extreme diurnal temperature variations in deserts, and diurnal and seasonal variations in those climates where temperatures periodically drop below freezing at night contribute to the fracturing and fragmentation of rocks, although their role is probably not exclusive. The microflora and other forms of life that can thrive in the more sheltered cracks where water condenses and dew remains, further accelerate the process of breakup and loosening of rock grains.

5.2 PARTICLE-SIZE DISTRIBUTIONS: FUNDAMENTALS

5.2.1 Number, Volume, and Mass Distributions

The sorting of sediments into size fractions classifiable either by linear dimensions or volume or mass is a well-established technique on which much of the analysis of sedimentary textures, sediment transport, and physical diagenesis is based (Krumbein and Pettijohn, 1938). A general equation for a particle-number distribution or a particle-number spectrum relates the number of particles dN of size between r and $r + dr$ to the particle size r,

$$dN = f(r) \, dr \qquad \text{(particles)} \qquad (5.3)$$

where r is, in general, some characteristic linear dimension, and $f(r)$ is a mathematical relationship describing the distribution. Because separation of sediment samples into size fractions is commonly done by settling techniques, there is a tendency to use the settling particle radius (see Section 6.2) as a characteristic particle dimensions r. A similar procedure of approximating a particle to a sphere is commonly followed in optical measurements, where particles of irregular shapes are measured under some magnification and a mean value of a sphere radius is assigned to each particle. The particle radius is used more often than other measures of size in this chapter and throughout the book. Therefore, the terms *particle-size distribution* and *particle-number distribution* will be used interchangeably, always implying a spectrum of the type given in equation 5.3. For other functions of the particle radius, such as the volume or mass, the terminology introduced below will be reserved.

A particle-volume distribution is

$$dV = v(r)\,dN$$

$$= v(r)f(r)\,dr \quad (\text{cm}^3) \tag{5.4}$$

where dV is the volume of dN particles, of size between r and $r+dr$, and $v(r)$ is the volume of an individual particle of size r.

By analogy with the volume-number distributions, a particle-mass spectrum dM is

$$dM = \rho(r)\,dV$$

$$= \rho(r)v(r)f(r)\,dr \quad (\text{g}) \tag{5.5}$$

where $\rho(r)$ is the particle density (g cm^{-3}) which may be particle-size dependent.

The differential form of the equations for dN, dV, and dM gives the spectrum of each parameter as a function of particle size r. Each of the spectra can be converted to a frequency distribution by dividing it by the total number, volume, or mass of particles in the sample. For the particle-number spectrum, the total number of particles N_T is

$$N_T = \int_{r_2}^{r_1} f(r)\,dr \quad (\text{particles}) \tag{5.6}$$

where r_1 and r_2 are the lower and upper limits, respectively, of the particle sizes in the sample. For certain types of functional relationships $f(r)$ it may be mathematically convenient to use zero and infinity as integration limits ($r_1 = 0$ and $r_2 = \infty$). The physical significance of this choice of the integration limits is that the particles in a sample are counted from some very small to some very large size. A frequency size distribution for the particle number is, from equations 5.3 and 5.6,

$$dN = \frac{1}{N_T} f(r)\, dr \qquad \text{(fraction)} \qquad (5.7)$$

and similar relationships can be written for the volume and mass frequency distributions.

A cumulative size distribution of particles of size *greater* than R, in a sample bracketed by the lower r_1 and upper r_2 size limits (that is, $r_1 \leqslant R \leqslant r_2$) is

$$N_{>R} = \int_R^{r_2} f(r)\, dr \qquad \text{(particles)} \qquad (5.8)$$

and a cumulative size distribution of particles of size *smaller* than R is

$$N_{<R} = \int_{r_1}^R f(r)\, dr \qquad \text{(particles)} \qquad (5.9)$$

In both types of the cumulative size distribution, the particle size r varies from the lower to the upper limit. The ratios $N_{>R}/N_T$ and $N_{<R}/N_T$ give the cumulative frequency distributions varying from 0 to 1.

A symmetrical frequency distribution, such as the Gaussian or normal distribution, possesses a characteristic feature of equal probability of deviation toward values higher or lower than the mean. The larger the deviation, the smaller is its probability of occurrence, but the sign of the deviation does not affect the probability. This feature results in the coincidence of the distribution mean and its peak or mode. In an asymmetrical distribution the mean does not coincide with the mode, and this feature can be interpreted as the prevalence of certain processes that preferentially cause some part of the population, either above or below the mean, to have higher probabilities of occurrence. Particle-size distributions of natural materials are, as a rule, pronouncedly asymmetrical. However, certain distributions that have an asymmetric frequency peak can often be mathematically transformed into the Gaussian distribution, for the purpose of more convenient handling or graphical representation. Such transformations do not contradict the fact that the basic parameters, such as the particle number or volume, are asymmetrically distributed with respect to the linear sizes of particles.

5.2.2   Histograms and Continuous Distributions

A bar histogram is a convenient means of graphically displaying a particle-size analysis, where the bar height represents the number, volume, or mass of particles of some size range, and the bar width represents the size range (difference between particle radii). Volume, weight, and number fractions (frequencies), taken as numerical fractions or percentages of the whole sample, are commonly used for the vertical scale of histograms. The height of the bar in a histogram is usually sensitive to the size interval $(r_{i+1} - r_i)$ chosen: a smaller or a larger size interval may contain a different number of particles or a different weight fraction of the sample. To allow for the particle-size variation of several orders of magnitude, from microns to millimeters and higher, a logarithmic size scale, the so-called phi scale (Krumbein, 1934) is used sometimes. The unit of the scale is 1ϕ and, by convention, the definition of ϕ is

$$\phi = -\log_2(\text{diameter in mm})$$
$$= -1 - \log_2 r \tag{5.10}$$

where r is the particle radius in millimeters. For particles of $r = 0.5$ mm, $\phi = 0$. The ϕ values are negative for $r > 0.5$ mm and they are positive for $r < 0.5$ mm. When the particle-size intervals are taken such that the upper size limit r_{i+1} is greater by a factor of 2 than the lower size limit r_i, then each interval is exactly one ϕ unit wide. The difference between $\frac{1}{64}$ and $\frac{1}{128}$ mm is 1ϕ, as is the difference between 64 and 32 mm. The radii of particles are drawn on the linear and phi scales in Figure 5.5, the coordinates of which give the conversion from one scale to another.

An easy way to convert a histogram into a continuous spectrum is the following. The value N_i represented by the bar height can be divided by the size interval for that bar $\Delta r_i = r_{i+1} - r_i$. The result has the units of

$$\frac{N_i}{\Delta r_i} \qquad (\text{units of } N)\,(\text{units of } r)^{-1} \tag{5.11}$$

The mean value of the size interval can be taken as an arithmetic or geometric mean. The geometric mean size $\bar{r}_i$ is

$$\bar{r}_i = (r_i r_{i+1})^{1/2} \qquad (\text{cm}) \tag{5.12}$$

and the geometric mean is preferred for sediments having a broad range of sizes. The results from equations 5.11 and 5.12 can be plotted one against another as a series of discrete points, to which a continuous function can be fitted, giving the equation for dN/dr as a function of r or, generally,

$$\frac{dN}{dr} = f(r)$$

which is identical in form with equations 5.3 through 5.5, describing the particle number, volume, and mass spectra.

A particle-size distribution often encountered in sediments and atmospheric particulates, known as the power-law or Pareto distribution $dN/dr = Ar^{-b}$ (see Section 5.2.3), follows from a special type of volume distribution histogram (Figure 5.2). This type of histogram represents the volume distributions in which the volume of particles V_i of each geometric size interval either increases or remains constant or decreases with an increasing particle size r_i. Such histograms can be represented by the relationship

$$V_i = \text{const} \times r_i^{\alpha} \qquad (\text{cm}^3) \tag{5.13}$$

where α is a constant that can be either positive, zero, or negative. For geometrically increasing size intervals, the ratio of the upper to lower interval limit r_{i+1}/r_i is constant. If the ratio is denoted $r_{i+1}/r_i = n$, then the geometric mean size of the interval is $\bar{r}_i = r_i n^{1/2}$ (on the phi scale, the factor n is taken as $n=2$). The volume V_i is made of N_i particles, all of mean size $\bar{r}_i$. Therefore,

$$V_i = k_v \bar{r}_i^3 N_i$$

$$= k_v n^{3/2} r_i^3 N_i \tag{5.14}$$

where k_v is a shape factor relating the linear size dimension r to the volume V (for a sphere, $k_v = 4\pi/3$).

On the linear scale, the size interval is $\Delta r_i = r_{i+1} - r_i = r_i(n-1)$. The number of particles per unit of size $N_i/\Delta r_i$ is, from equations 5.13 and 5.14,

$$\frac{dN}{dr} \simeq \frac{N_i}{\Delta r_i} = A r_i^{\alpha - 4} \qquad (\text{particles cm}^{-1}) \tag{5.15}$$

where A is a constant combining the constant coefficients k_v and n from the two equations.

A particular case of a flat volume-distribution histogram is $V_i = \text{constant}$ (that is, $\alpha = 0$), and this corresponds to a particle-number distribution with the slope of -4 (that is, $dN/dr = Ar^{-4}$), as shown in Figure 5.2. Size distributions in natural sediments, the number spectra of which have the slope of approximately -4 will be discussed in Section 5.3.

Equation 5.15 in logarithmic form becomes

$$\log \frac{dN}{dr} = \log A + (\alpha - 4) \log r$$

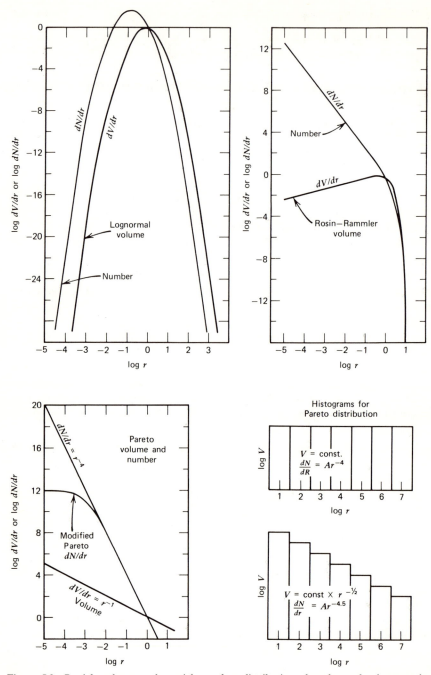

Figure 5.2 Particle-volume and particle-number distributions based on the lognormal, Rosin-Rammler, and Pareto probability density functions.

A plot of $\log dN/dr$ against $\log r$ is a straight line of the slope $(\alpha - 4)$, and constant A is equal to the value of dN/dr at $r = 1$. Relatively small departures from the value of $V_i = \text{const}$ are reflected in some scatter of points about a straight line in the plot of $\log dN/dr$ against $\log r$. The double log scale can absorb much of the scatter, although it should be emphasized that the relationship $dN/dr = Ar^{-b}$ is more suited to represent the general trends of a particle-size distribution, whereas the fine structure of a spectrum is more clearly resolved on a linear scale or in a histogram plot. Deviations from a straight line $\log - \log$ plot, or the presence of two or more straight-line segments, each having a different value of the slope b, can be interpreted as due to such processes as dissolution, fragmentation, or mixing of particles from different sources.

5.2.3   The Lognormal, Weibull, Pareto, and Related Distributions

Three frequency distributions have been extensively used to describe the particle-number or mass spectra of pronouncedly asymmetric shape—the lognormal distribution, the Weibull, or historically more correct, Rosin-Rammler distribution, and the Pareto or power-law distribution. Each of the three distributions and forms derived from the basic distribution types have been applied to particle-size spectra of soils, water transported sediments, dust, and biogenic sediments. The lognormal and the Weibull or Rosin-Rammler distributions are unimodal asymmetric distributions. The Pareto or power-law distribution is a size spectrum decreasing continuously with an increasing particle size. In this section are given some of the main mathematical and physical characteristics of the three size distributions. A summary of their mathematical relationships and other statistical parameters is given in Table 5.1.

A common practice in sedimentology is to express size distributions as weight percentages of the different size fractions in a sediment sample. In work with atmospheric particulates, one commonly deals with the number percent or absolute numbers of particles in the size fractions. The differences between the two practices are mostly related to the techniques used in grain-size separation: the abundance of the coarser materials at the earth's surface makes weighing of sediment samples easy, whereas the collection techniques of dust and other atmospheric particles are aimed at the finer-size materials occurring at lower concentrations in the atmosphere. For particles of constant density, the mass and volume size fractions differ one from another only by the constant factor ρ, the density, as given in equation 5.5: $dM = \rho \, dV$. To transform a particle-volume dV to a particle-number distribution dN, as given in equation 5.4, the particle volume must be known. For spherical particles, the particle volume is

**Table 5.1 The Mean, Median, and Mode
of the Frequency Distributions (Probability Density Functions $f(x)\,dx$
Discussed in the Text**

Function	$f(x)\,dx$ text equation no.	Range of x and constants	Mean x	Median x	Mode x
Gaussian	(5.17)	$-\infty < x < \infty$	μ	μ	μ
Lognormal	(5.16)	$0 < x < \infty$, $\mu > 0,\ \sigma^2 > 0$	$e^{\mu+\sigma^2/2}$	e^{μ}	$e^{\mu-\sigma^2}$
Weibull	(5.25)	$\gamma \leqslant x < \infty$, $m>0,\ c>0$	$\Gamma\!\left(1+\dfrac{1}{m}\right)$	$\left(\dfrac{\ln 2}{c}\right)^{1/m}+\gamma$	$\dfrac{1}{c}\left(1-\dfrac{1}{m}\right)^{1/m}+\gamma$ (for $c>1$) γ (for $0<c<1$)
Rosin-Rammler	(5.28)	$0 \leqslant x < \infty$, $m>0,\ c>0$	$\Gamma\!\left(1+\dfrac{1}{m}\right)$	$\left(\dfrac{\ln 2}{c}\right)^{1/m}$	$\dfrac{1}{c}\left(1-\dfrac{1}{m}\right)^{1/m}$ (for $c>1$) 0 (for $0<c<1$)
Pareto or power law	(5.32)	$c \leqslant x < \infty$, $b>0,\ c>0$	$\dfrac{bc}{b-1}$	$2^{1/c}c$	c

$v(r) = 4\pi r^3/3$, and even this simple relationship can make a volume distribution markedly different in shape from the particle-number distribution. The equations for both dV and dN arising from the different frequency distributions are given in the remainder of this section.

Lognormal Distribution. The probability density function

$$f(x)\,dx = \frac{1}{x\sigma\sqrt{2\pi}}\exp\left(-\frac{(\ln x - \mu)^2}{2\sigma^2}\right)dx \qquad (5.16)$$

is known as the lognormal distribution, where μ and σ are constants. If a variable X is introduced instead of $\ln x$ (that is, $X \equiv \ln x$ and $dX \equiv dx/x$), then the lognormal distribution in x transforms into a normal (Gaussian) distribution in X,

$$f(X)\,dX = \frac{1}{\sigma\sqrt{2\pi}}\exp\left(-\frac{(X - \mu)^2}{2\sigma^2}\right)dX \qquad (5.17)$$

where the constants μ and σ have the same significance as in the lognormal distribution, and they are the mean and the standard deviation from the mean of the Gaussian distribution $f(X)\,dX$. By definition of the probability density function, the integral between the limits of the function's existence is unity, or

$$\int_0^\infty f(x)\,dx = 1 \qquad \text{and} \qquad \int_{-\infty}^{+\infty} f(X)\,dX = 1$$

The cumulative normal distribution that gives the probability of occurrence of all the items of size smaller than x, $F_{<x}$, is

$$F_{<x} = \int_{-\infty}^{x} f(y)\,dy = \frac{1}{\sqrt{\pi}} \int_{y=-\infty}^{y=x} \exp\left(-\frac{(y-\mu)^2}{2\sigma^2}\right)d\left(\frac{y-\mu}{2^{1/2}\sigma}\right)$$

$$F_{<x} = \frac{1}{2} + \frac{1}{2}\operatorname{erf}\left(\frac{x-\mu}{2^{1/2}\sigma}\right) \qquad (5.18)$$

where y is an integration variable and erf is the error function (Appendix A). Tabulations of both the cumulative probability $F_{<x}$ and the error function are available in many handbooks of mathematical tables. By writing $\operatorname{erf}[(x-\mu)/\sigma\sqrt{2}\,]$ as a polynomial in $(x-\mu)/\sigma\sqrt{2}$, explicit equations can be written for the cumulative frequency $F_{<x}$ (Appendix A).

Using the same method of integration as in equation 5.18, the cumulative lognormal probability of all the items smaller than x, $F_{<x}$, is

$$F_{<x} = \int_{y=0}^{y=x} f(y)\,dy = \frac{1}{2} + \frac{1}{2}\mathrm{erf}\left(\frac{\ln x - \mu}{2^{1/2}\sigma}\right) \qquad (5.19)$$

The only difference between equations 5.18 and 5.19 is that in the latter $\ln x$ replaces x. For the lognormal distribution, the limits of x are $0 < x < \infty$, whereas for the normal distribution they are $-\infty < x < +\infty$.

The volume of sediment particles, distributed lognormally as a function of the particle radius r, can be described by the following equation:

$$dV = \frac{A}{r}\exp\left(-\frac{(\ln r - \mu)^2}{2\sigma^2}\right)dr \qquad (\mathrm{cm}^3) \qquad (5.20)$$

where A is a dimensional constant which includes the product $\sigma\sqrt{2\pi}$ of the lognormal function (if r is taken in cm and dV in cm^3, then A is in cm^3). Instead of r, a dimensionless size parameter r/r_1 can be used, where r_1 denotes the lower size limit of the distribution ($r \geqslant r_1$). In this case, as r approaches the lower size limit r_1, the quantity dV/dr tends to a finite value of $dV/dr = A\exp(-\mu^2/2\sigma^2)$.

In the case of $\mu = 0$ and $\sigma^2 = \frac{1}{2}$, the volume distribution simplifies to

$$dV = \frac{A}{r}\exp\left[-(\ln r)^2\right]dr \qquad (5.21)$$

In a double log form, the preceding equation is

$$\ln\frac{dV}{dr} = \ln A - \ln r - (\ln r)^2 \qquad (5.22)$$

The latter produces a nonlinear plot of $\log dV/dr$ against $\log r$, an illustration of which is given in Figure 5.2.

The volume dV of dN spherical particles, of radius between r and $r + dr$, is

$$dV = \tfrac{4}{3}\pi r^3\,dN \qquad (5.23)$$

which in combination with equation 5.20 gives the particle-number distribution,

$$dN = \frac{A}{r^4}\exp\left(-\frac{(\ln r - \mu)^2}{2\sigma^2}\right)dr \qquad (\mathrm{particles}) \qquad (5.24)$$

where A is a constant not identical to A in 5.22.

A plot of $\log dN/dr$ against $\log r$ is nonlinear, as in the case of the particle-volume distribution dV/dr. Note that neither dV in equation 5.20 nor dN in 5.24 are probability density functions, but they are distribution functions or spectra of the particle volume and particle number, respectively, written as functions of the particle radius.

The mean, median, and mode of the lognormal distribution are listed in Table 5.1.

Literature references on applications of the lognormal distribution to sediments, occurrences of trace elements in rocks, and many other occurrences in nature have been summarized by Aitchison and Brown (1963) and Johnson and Kotz (1970). A more recent summary of the applications to sediments is the paper by Dapples (1975).

Weibull and Rosin-Rammler Distributions. The probability density function known as the Weibull distribution is

$$f(x)\,dx = cm(x-\gamma)^{m-1}\exp\{-c(x-\gamma)^m\}\,dx \qquad (5.25)$$

where c, m, and γ are positive constants, and $x>\gamma$. If $m=1$, the Weibull distribution becomes an exponential distribution,

$$f(x)\,dx = c\exp\{-c(x-\gamma)\}\,dx \qquad (5.26)$$

with the values of $f(x)$ decreasing exponentially with an increasing x.

The cumulative distribution of $f(x)$ for all sizes smaller than x is, by integration of equation 5.25,

$$F_{<x} = \int_{y=\gamma}^{y=x} f(y)\,dy = 1 - \exp\{-c(x-\gamma)^m\} \qquad (5.27)$$

The Swedish physicist W. Weibull has derived the probability distribution function with three constant parameters, as given in equation 5.25, from his studies of a chain under load: the chances of a chain breaking due to the failure of an individual link increase with an increasing load or, alternatively, the chances of breaking increase with the length of time that the load has been applied. Weibull recognized the wide applicability of the distribution function to a variety of processes involving fragmentation and comminution of materials. His first contributions were published in the Scandinavian literature in 1939 and subsequently, in 1951, in an English-language periodical. Earlier, however, equation 5.25 with only two constant parameters c and m ($\gamma=0$) was used by Rosin and Rammler (1933), both working in Berlin, to describe the weight distribution of fragments of coal produced by grinding. Following Rosin and Rammler, the distribution function has been used to describe the size distributions of weathered sediments and suspended materials in the ocean (Krumbein and Tisdel, 1940; Dapples, 1975; Carder et al., 1971). The statistical properties of the Weibull distribution function were established in late 1920's, before Rosin and Rammler, and Weibull published their results (Harris, 1971). The Weibull distribution is a special case of a broader class of the so-called generalized gamma distributions, and it has found fairly wide application in studies of sediment particles, comminution of natural and human-made materials, and the formation of crystals in solutions.

In equation 5.25 for the Weibull distribution, the constant γ is essentially the lower limit of x, because $x \geqslant \gamma$ by definition. When x is much larger than γ, the latter affects the distribution function $f(x)$ very little. At the lower values of x, the difference $x - \gamma$ is small, and the shape of the curve $f(x)$ drawn as a function of x can be affected by the choice of γ near the lower end values. With $\gamma = 0$, the function is known as the Rosin-Rammler distribution,

$$f(x)\,dx = cmx^{m-1}\exp\{-cx^m\}\,dx \qquad (5.28)$$

The particle-volume distribution dV, obeying the Rosin-Rammler distribution function, is

$$dV = Ar^{m-1}\exp\{-cr^m\}\,dr \qquad (\text{cm}^3) \qquad (5.29)$$

where r is the particle radius or some characteristic linear dimension and A is a dimensional constant (cm^{3-m}, if r is in cm). If the particle radius is taken as a nondimensional quantity r/r_1, where r_1 is the lower size limit or unity, then A has the dimensions of volume.

The particle-number distribution in a sediment, the particle volumes of which obey the Rosin-Rammler function, is

$$dN = Ar^{m-4}\exp\{-cr^m\}\,dr \qquad (\text{particles}) \qquad (5.30)$$

where the relationship between dN and dV is given in equation 5.23 and A is a dimensional constant, not identical to A in 5.29.

The Weibull distribution has a single mode only if $m > 1$. For $m < 1$, the function $f(x)$ declines steadily from its maximum value at $x = \gamma$ with an increasing x. An example of the Rosin-Rammler or Weibull distribution plot is shown in Figure 5.2. For graphical displays of the cumulative Weibull or Rosin-Rammler distributions, special graph paper is available commercially.

The equations for the mean, median, and mode of the Rosin-Rammler distribution are summarized in Table 5.1.

Depending on the values of the parameters c and m in the Weibull distribution, its ascending or descending limb can be similar to the lognormal function with certain values of μ and σ. This means that parts of certain particle-size distributions can be satisfactorily fitted by both the Weibull and lognormal functions over some range of r. As a point of curiosity, the Weibull distribution with the value of $m \simeq 3.3 - 3.6$ is very similar in shape to the Gaussian distribution curve.

The logarithmic form of equation 5.29 for dV,

$$\ln \frac{dV}{dr} = \ln A + (m-1)\ln r - cr^m \qquad (5.31)$$

shows that a plot of $\log dV/dr$ against $\log r$ is pronouncedly nonlinear, and this also applies to a plot of $\log dN/dR$ against $\log r$

Power-Law or Pareto Distribution. The probability density function

$$f(x)\,dx = bc^b x^{-(b+1)}\,dx \qquad (5.32)$$

where b and c are positive constants, and $x > c$, is known as the Pareto distribution. The distribution function is attributable to the Italian-Swiss economist V. Pareto, whose original work dealt with the distribution of incomes in a population (Johnson and Kotz, 1970).

The cumulative distribution for all values smaller than x is

$$F_{<x} = \int_c^x f(y)\,dy = 1 - c^b x^{-b} \qquad (5.33)$$

The Pareto function for the particle-number distribution dN can be written as

$$dN = Ar^{-b}\,dr \qquad \text{(particles)} \qquad (5.34)$$

where r is a linear dimension of particles and A is a dimensional constant (particle cm^{b-1} if r is in cm).

The cumulative number of particles $N_{>r}$ of radius greater than r, by integration of equation 5.34, is

$$N_{>r} = \int_{r=r}^{r=\infty} dN = \frac{A}{b-1} r^{-(b-1)} \qquad \text{(particles)} \qquad (5.35)$$

which is valid for $b > 1$.

As the particle radius r approaches zero, the quantity dN/dr increases indefinitely. Thus for all practical purposes, a lower size limit of the distribution must be known or stipulated. With the lower size limit r_1, the cumulative number of particles N, of size greater than r_1 and smaller than r $(r \geqslant r_1)$ is, by integration of 5.34,

$$N_{<r} = \int_{r=r_1}^{r=r} dN = \frac{A}{b-1}\left(\frac{1}{r_1^{b-1}} - \frac{1}{r^{b-1}}\right) \tag{5.36}$$

The particle-volume distribution dV is

$$dV = Ar^{3-b} dr \qquad (cm^3) \tag{5.37}$$

where constant A is not identical to A in the equation for dN.

If the particle size r varies from a lower limit $r = r_1$ to $r = \infty$, then the total volume and total mass of the distribution are

$$V = \int_{r=r_1}^{r=\infty} dV \qquad \text{and} \qquad M = \rho \int_{r=r_1}^{r=\infty} dV$$

These are finite only if $b > 4$. For $b \leqslant 4$, the integral of equation 5.37 increases indefinitely with an increasing r.

The logarithmic forms of the particle-number and particle-volume equations are

$$\log\frac{dN}{dr} = \log A - b\log r \tag{5.38}$$

$$\log\frac{dV}{dr} = \log A - (b-3)\log r \tag{5.39}$$

Both equations plot as straight lines in log-log coordinates (Figure 5.2), with a slope of $-b$ for the particle-number spectrum and a slope of $-(b-3)$ for the particle-volume spectrum.

The mean and the median of the power-law distribution are listed in Table 5.1.

Two modified forms of the Pareto distribution deserve mentioning. One is the form of

$$dN = A(r+a)^{-b} dr \tag{5.40}$$

where a is a positive constant. When r is much smaller than a $(r \ll a)$, then dN/dr approaches a constant value Aa^{-b}, essentially independent of the particle size r.

The other form of the power-law distribution is

$$dN = Ar^{-(b-1)} d\ln r \tag{5.41}$$

which can be plotted in log-log coordinates as $dN/d\ln r$ against r. Such plots have been introduced by Junge (1963) in studies of particle-size distributions of atmospheric aerosols and they are used along with the form dN/dr in the literature dealing with atmospheric particles.

The number of different things in nature, the occurrences of which can be approximated by the power-law relationship $Y(x) = Ax^{-b}$, is surprisingly large. Atmospheric and stratospheric aerosol particles of the micron and submicron size obey the power law, at least over some range of the x (particle diameter) values (Junge, 1963; Miranda and Fenn, 1974); fine grained sediments and suspended matter in the ocean, in the submillimeter particle-size range (Lal and Lerman, 1975; Lerman et al., 1977); number concentration of asteroids in the asteroid belt, as a function of the asteroid diameter, in the range between 10 and 1000 km (Chapman, 1974); the frequency of oil field occurrences, within a certain geographic area, as a function of the oil field size, a case cited by Johnson and Kotz (1970, p. 242); the frequency of the crater occurrences on the Moon and on Mars, as a function of the crater diameter (Neukum and Wise, 1976); and the number of natural and man-made lakes in Switzerland as a function of the lake surface area.

5.2.4   Fragmentation

Models.   The theories behind the fragmentation process generally aim at predicting the particle-size distributions of fragmented solids formed under different conditions of applied stress. A common denominator of the theories of fragmentation is the mechanism of random fracturing imposed on the solid in one blow or in a series of repeated stress applications over some period of time (Austin and Klimpel, 1968; Bennett, 1936; Gaudin and Meloy, 1962a, b; Harris, 1968; Klimpel and Austin, 1965; Meloy and Gumtz, 1968; Rosin and Rammler, 1933). On a cosmic scale, a mathematical model of fragmentation has been developed to explain the number of stars, per unit range of mass, as a function of the star mass (Auluck and Kothari, 1968). The size distribution curve obtained from the star model was similar to the curves describing the size distributions of other fractured materials. An early attempt, around 1900, to construct a mathematical model of a fragmented solid has been attributed to the Reverend W. A. Whitworth of London (Whitworth, 1934; Manning, 1952): if a body of mass M is broken at random into n fragments, and the fragments are arranged in order of an increasing mass, then the individual fragment masses are, from the smallest up to the nth particle,

$$\frac{M}{n}\frac{1}{n}, \frac{M}{n}\left(\frac{1}{n}+\frac{1}{n-1}\right),\ldots,\frac{M}{n}\left(\frac{1}{n}+\frac{1}{n-1}+\cdots+\frac{1}{n-r+1}+\cdots+1\right)$$

A particle-size distribution equation based on three parameters has been used to describe the fragment-mass distributions produced during comminution in ball and rod mills (Harris, 1968). The equation, written for the fragment-volume distribution dV, is

$$dV = cmpr^{m-1}(1 - cr^m)^{p-1} dr \qquad (5.42)$$

where c, m, and p are positive constants and $cr > 1$. The cumulative volume $V_{<r}$ of particles of radius less than r is

$$V_{<r} = 1 - (1 - cr^m)^p \qquad (5.43)$$

Equation 5.42 is a probability density function, and when the particle radius r increases indefinitely, the cumulative volume $V_{<\infty}$ tends to unity. The equation is not only of practical but also of conceptual interest because it can be related to the Rosin-Rammler and Pareto distribution functions. Disregarding the mathematical conditions of the probability density function, equation 5.42 can be written as a product of the r-containing terms and a dimensional constant A,

$$dV = Ar^{m-1}(1 - cr^m)^{p-1} dr \qquad (cm^3) \qquad (5.44)$$

If c is small and p large, then the right-hand side of equation 5.44 becomes an exponential function,

$$dV = Ar^{m-1} \exp\{ - c(p-1)r^m \} dr \qquad (5.45)$$

which is identical to the Rosin-Rammler distribution 5.29.

If $p = 1$, then the particle-volume distribution becomes identical to the Pareto or power-law distribution, as given in equation 5.34,

$$dV = Ar^{m-1} dr \qquad (5.46)$$

During a mechanical fragmentation or comminution process, such as grinding taking place in a ball mill, the bigger particles have a higher probability of being struck and fragmented, and continuous grinding produces, therefore, more and more smaller particles. A particle-size distribution of the type shown in equation 5.44 is also a result of the following simple fragmentation model (Lal and Lerman, 1975).

In a parent population of particles, particles undergo fragmentation. The bigger the particle, the more fragments it produces, and the number of fragments is a power function of the parent particle radius:

$$\text{number of fragments per parent particle of radius } r = \left(\frac{r}{a}\right)^\alpha \qquad (5.47)$$

where a and α are positive constants. The physical significance of a is that it is the smallest particle size of the parent population that undergoes, or can undergo, fragmentation. Thus $a < r$. If only some fraction of the parent particle population has been fragmented, what remains is a mixture of fragments and the residual of the parent population. (By visual inspection, even though a tedious process, it is often possible to tell the fragments from unfragmented particles in a sediment.) If the parent particle population has a particle-number distribution of the power-law type, and the parent particles and fragments are regarded as spheres, then the following number-distribution equations apply to the parent, fragmented, and residual materials:

Parent:

$$\frac{dN}{dr} = Ar^{-b} \tag{5.48}$$

Fragments:

$$\frac{dN}{dr} = \frac{3A}{3-\alpha} a^{\alpha(b-4)/(3-\alpha)} r^{(4\alpha-3b)/(3-\alpha)} \left(1 - f a^{\alpha c/(3-\alpha)} r^{-3c/(3-\alpha)}\right)$$

$$\tag{5.49}$$

Residual:

$$\frac{dN}{dr} = fAr^{-(b+c)} \tag{5.50}$$

In the above equations, A is a dimensional constant (the same for all three equations), α is the fragmentation multiplicity defined in equation 5.47, f is the fraction of the original population remaining, and b and c are positive constants appearing in the power exponents.

The fraction $1 - f$ of the parent particle population represents fragments formed according to the mechanism described in equation 5.47. The residual population, represented by fraction f, when plotted as $\log dN/dr$ against $\log r$, has a steeper negative slope $-(b+c)$ than the slope $-b$ of the parent population plot. A similar log-log plot of equation 5.49 for fragments is nonlinear, owing to the difference term in large parentheses. If the fragments and residual particles are counted together, the sum of the two distributions may also deviate more or less pronouncedly from a straight-line log-log plot.

The particle-volume distribution given in equation 5.44 becomes the particle-number distribution dN (with $p = 2$) of the form

$$\frac{dN}{dr} = Ar^{m-4}(1 - cr^m)$$

which is similar to the fragmentation-distribution equation 5.49. If fragmentation is complete, then the residual fraction is $f=0$. In this case, the number distribution of fragments becomes a simple Pareto distribution, similar in shape to the parent particles distribution,

$$\frac{dN}{dr} = A_1 r^{(4\alpha - 3b)/(3 - \alpha)} \tag{5.51}$$

where A_1 is a constant defined as

$$A_1 = \frac{3Aa^{\alpha(b-4)/(3-\alpha)}}{|3-\alpha|} \tag{5.52}$$

and A, a, α, and b retain their meaning. A point should be made about the value of α: for the quantity dN/dr to be decreasing with an increasing particle radius r, the power exponent $(4\alpha - 3b)/(3 - \alpha)$ must be less than 0. This can be satisfied by either one of the following two conditions:

(i) $\alpha < 3b/4$ and $\alpha < 3$

(ii) $\alpha > 3b/4$ and $\alpha > 3$

In a particular case of $b=4$, the power exponent is always $(4\alpha - 3b)/(3 - \alpha) = 4$ and is independent of the fragmentation multiplicity α. In this case, the parent and fragmented populations plot as two parallel lines in the coordinates $\log dN/dr$ against $\log r$.

The conditions of $\alpha < 3$ and $\alpha > 3$ correspond to two somewhat different fragmentation processes. When a parent particle of radius r breaks into $(r/a)^\alpha$ fragments as defined in equation 5.47, each fragment of radius r_f, then the conservation of mass condition requires that the volumes be equal,

$$\tfrac{4}{3}\pi r^3 = \tfrac{4}{3}\pi r_f^3 \left(\frac{r}{a}\right)^\alpha$$

From the preceding equality the relationships between the parent particle radius r and the fragment radius r_f are

$$r_f = r\left(\frac{a}{r}\right)^{\alpha/3} \tag{5.53}$$

$$\frac{dr_f}{dr} = \frac{3-\alpha}{3}\left(\frac{a}{r}\right)^{\alpha/3} \tag{5.54}$$

Since a is the lower size limit of the fragmenting particles, the ratio a/r is a fraction. With $a/r<1$ and the power exponent $\alpha<3$, a particle of large r produces numerous and relatively large fragments of radius r_f. A parent particle of a smaller r produces less numerous and smaller fragments r_f than in the preceding case. This direct relationship between the parent and fragment particle sizes can also be seen in the ratio dr_f/dr: for $\alpha<3$, the derivative is positive, and this means that the fragment size increases with an increasing parent particle size.

In the case of $\alpha>3$ the relationships between the parent particles and fragments are reversed. A particle of a greater r produces numerous but relatively small fragments r_f; a smaller parent particle r produces less numerous but somewhat larger fragments r_f than in the preceding case; the derivative dr_f/dr is negative, meaning that smaller fragments are produced by larger parent particles.

The size distribution of the fragmented population, equation 5.51, plotted as a straight line in log-log coordinates, can have a slope either more or less steep than the slope $-b$ of the parent population, depending on the values of α and b. The condition for a steeper slope is

$$\alpha > \frac{6b}{4+b} \tag{5.55}$$

A numerical example, illustrating the relationships between the size distributions of a parent and fragmented particle populations follows.

Example. A particle-number distribution is $dN/dr = Ar^{-4.2}$ cm^{-3} μm^{-1} (particles per cm^3 of space per μm of particle radius; constant A not given a number value). In the course of weathering, the size distribution becomes $dN/dr = A_1 r^{-4.6}$ cm^{-3} μm^{-1}. Assuming that disaggregation is complete, the value of the fragmentation multiplicity α can be determined from the slope values:

$$-b = -4.2$$

$$\frac{4\alpha - 3b}{3 - \alpha} = -4.6$$

from which it follows that

$$\alpha = 2$$

Thus disaggregation takes place, according to the model, in direct relation to the particle surface area $(r/a)^2$. Next, the value of a, the smallest fragmentable size, can be determined by using equation 5.52. The constants A and A_1 can be read off the graphs of $\log dN/dr$ against $\log r$ for the parent particles and fragments: at $r=1$, we have $\log dN/dr = A$ or A_1. The parameter a, using $\alpha = 2$ and $b = 4.2$, is

$$a = \left(\frac{A_1 |3 - \alpha|}{3A} \right)^{(3-\alpha)/\alpha(b-4)}$$

$$= 0.064 \left(\frac{A_1}{A} \right)^{2.5} \qquad (\mu m)$$

Thus the value of a depends on the ratio of the intercepts A_1/A of the two straight lines. The ratio can be either greater or less than 1, but for the ratio A_1/A between 3 and 10, the value of a is between 1 and 20 μm. The particle radius of about 5 μm is an observational lower limit for many weathered and fragmented terrestrial materials, as well as for the fine particles in the lunar soils (Krumbein and Tisdel, 1940; Leopold et al., 1964; Butler et al., 1973; Butler and King, 1974). Although the reported "smallest" particle sizes in weathered materials are not necessarily a proof that they represent fragments of the smallest size, a comparison of the observed and order-of-magnitude computed lower limits is noteworthy.

In the preceding computational example, a relatively small change in the size distribution slope from -4.2 to -4.6 was produced by the fragmentation multiplicity value of $\alpha = 2$. A greater change in slope, such as from $-b = 3.8$ to $(4\alpha - 3b)/(3 - \alpha) = -5$, requires a larger value of α, in this case $\alpha = 3.6$.

Disaggregation of Soils and Sediments. The size distributions of a parent and fragmented particle populations discussed in the preceding example are plotted in Figure 5.3a. The particle-number distributions are, for the original material,

$$\frac{dN}{dr} = Ar^{-4.2} \qquad \text{with} \quad A = 1 \; \mu m^{3.2}$$

and for the completely fragmented material

$$\frac{dN}{dr} = A_1 r^{-4.6} \qquad \text{with} \quad A_1 = 3 \; \mu m^{3.6}$$

The minimum fragmentable size in this example is $a = 1$ μm. The cumulative particle-mass distribution $M_{>r}$, for particles of radius greater than r is

$$M_{>r} = \frac{r^{4-b} - r_2^{4-b}}{r_1^{4-b} - r_2^{4-b}} \qquad \text{(fraction)}$$

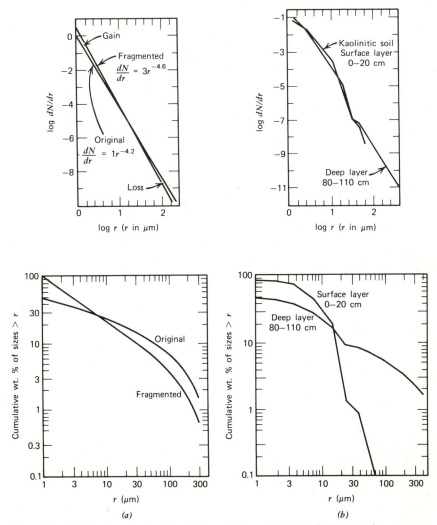

Figure 5.3 Particle-size distributions of fragmented materials. (*a*) Particle number spectrum and cumulative distribution from a fragmentation model. (*b*) Same for a kaolinitic soil from the savannah of northern Surinam (from data of Bakker, 1954).

The equation for $M_{>r}$ follows from the definitions given in equations 5.5 through 5.9. The two cumulative curves in the lower part of Figure 5.3*a* are represented by the preceding equation with the values of $b=4.2$ for the original particle population, $b=4.6$ for the fragmented particles, the lower particle-size limit $r_1=1$ μm, and the upper size limit $r_2=425\mu$m. The curve labeled *original* shows that the fraction of material coarser than 1 μm is

$M_{>1} = 50\%$; an implication of this is that the material finer than 1 μm accounts for about 50% of the sample. In the figure, the curve is placed at the point $M_{>1} = 50\%$ as its origin, instead of $M_{>1} = 100\%$. This serves the purpose of a visual comparison to the particle-mass distributions in a soil, plotted in Figure 5.3b. The deeper soil horizon, between 80 and 110 cm below the surface, contains about 50% by weight of material that is finer than $r = 1$ μm. The finest fraction may contain colloidal material, in parts washed from the surface layer with percolating waters. The crossing over of the mass distribution curves for the two soil layers, the surface layer 0–20 cm and deep layer 80–110 cm, shows the following.

First, there is virtually no colloidal material smaller than 1 μm left in the surface layer of the soil. In the relatively dry climate of the savannah, where the soil of Figure 5.3b originated, the fine colloidal fraction can dissolve and in part be washed away during the rainy seasons. Second, there is very little of the coarse material, larger than 20 μm, in the surface layer. A likely explanation for this is disaggregation of the soil lumps, taking place progressively from the deeper to the surface layer. The coarser material is abundant in the deep soil, and its disaggregation can produce a greater amount of the particles in the size range 1–10 μm, which are more abundant near the soil surface. The decrease in abundance of the coarser particles and the corresponding increase in the fine fraction are easier to observe in the cumulative mass-distribution curves $M_{>r}$ than in the particle-number distributions dN/dr. In the latter (Figure 5.3a), the small differences between the slopes of the plotted lines and departures from linearity obscure some of the changes taking place during weathering. In tropical soils developing on such diverse substrates as beds of volcanic ash and siltstone, the progression of weathering from the source to highly leached material generally coincides with an increase in the abundance of particles of radius $r < 10$ μm, at the expense of the coarser material. A peculiar feature associated with residual soils on volcanic ash is the strong increase in the abundance of the colloidal size fraction ($r < 0.25$ μm) in the most advanced stages of surface weathering (Mohr et al., 1972). The fraction of the colloidal material, increasing from about 10 to 50% of the soil, is probably attributable to chemical and bacterial processes rather than to simple mechanical disaggregation.

In older sediments, the fraction of the finer material decreases with an increasing age or depth in sediment. In sediments of the Oligocene-Miocene age from the Gulf Coast of the United States, Hower et al. (1976) reported that the weight fraction of the material finer than 1 μm decreases with depth, and the decrease is very pronounced for the finest size fraction $r < 0.05$ μm. The data demonstrating this are listed in Table 5.2. Fresh sediments deposited in water contain abundant clay material. Its "disappearance" with time from the sediment is a result of greater compaction

Table 5.2. Cumulative Weight Percent $M_{<r}$ (%) of Particle
of Radius r Smaller than the Radius Value Shown (μm), at Four
Depths in the Gulf Coast Sediments. Percentages of Illite Layers
in the Illite-Smectite Clays are for the Size Fraction $r < 0.05$ μm[a]

Depth below surface	Illite layers in Clay	Size fraction r (μm)				
(m)	(%)	<0.05	<0.25	<1	<5	<largest
1850	16	21	63	71	84	100
3400	64	19	52	64	78	100
4600	80	13	39	58	83	100
5500	78	9	28	52	76	100

[a] Hower et al. (1976).

under the increasing load of the overburden. Some of the decrease in the
abundance of the fine-particle fraction can also be attributed to precipitation of silica cement, forming from SiO_2 released by the mixed layer
illite-smectite clays. With increasing depth, the percentage of the illite
layers (SiO_2 poor) increases in the illite-smectite clays, and silica released
in the process can cement the smaller particle aggregates. An increase in
the relative abundance of illite within the finest-particle-size fraction
($r < 0.05$ μm) is shown in Table 5.2. Comparable increases with depth take
place in clays occurring in the size fractions up to $r \simeq 1$ μm.

Chemical Changes. The first chemical reaction taking place during
fragmentation is the breaking of chemical bonds and formation of new
surfaces. Next, depending on the nature of the solid and surrounding
medium, additional reactions can take place at the newly formed surfaces
of within the solid. Some reactions take place fast, such as, for example, a
visually observable change in color associated with the change in the
oxidation state of Fe(II) that occurs during grinding of the ferrous phosphate vivianite in a humid atmosphere. The more pronounced effects of
grinding and fragmentation include disordering of the crystal lattice,
polymorphic phase transitions in some minerals, and strong increases in
the ion-exchange capacities of micas and clays. The literature on the
mechanical and chemical effects of comminution of minerals is very
extensive. With regard to some of the minerals that have been studied
under laboratory conditions, the following literature references will be
mentioned: for the changes taking place in micas and clay minerals
subjected to comminution, Jackson and Truog (1939), MacKenzie and
Milne (1953 *a, b*), and Takahashi (1959); for the calcite-aragonite polymorphic transition taking place under conditions of grinding, Burns and
Bredig (1956), Jamieson and Goldsmith (1960), Northwood and Lewis

(1970), Schrader and Hoffmann (1969), and Criado and Trillo (1975); for vaterite to calcite transition, Northwood and Lewis (1968); for quartz to amorphous silica transition under grinding, Schrader et al. (1969), Clelland and Ritchie (1952), Moore and Rose (1973), Dempster and Ritchie (1953), and Lin et al. (1975).

A short list of polymorph mineral pairs is given in Table 5.3. The table also lists the following information: the Gibbs standard free energies of formation ($\Delta G°$) and molar volumes (V) at 25°C and 1 atm total pressure of the mineral phases; the pressure P_{eq} that would be required to maintain a thermodynamic equilibrium at 25°C between the two phases or, in other words, P_{eq} is the pressure for the equilibrium polymorphic transition between the two minerals. How the equilibrium pressure P_{eq} is computed is shown in an example given below.

Among the mineral pairs listed in Table 5.3, the transitions are from the less dense to the denser phase for calcite→aragonite, vaterite→calcite,

Table 5.3 Equilibrium Pressures P_{eq} at 25°C
for Some Polymorphic Phase Transitions

Mineral polymorphs		$\Delta G°$ (kcal mol^{-1})[a]	V (cm^3 mol^{-1})[a]	P_{eq} (kbar)[b]
CaCO$_3$	calcite	−269.91	36.93	3.5[c]
	aragonite	−269.68	34.15	
ZnS	wurtzite	−45.76	23.85	—
	sphalerite	−48.62	23.83	
PbO	massicot	−44.93	23.15	10.5
	litharge	−45.12	23.91	
TiO$_2$	rutile	−212.56	18.82	1.7
	anatase	−212.63	20.52	
SiO$_2$	α-quartz	−204.65	22.69	—
	amorphous	−203.30	27.27	
SiO$_2$	α-quartz	−204.65	22.69	26.1
	coesite	−203.37	20.64	
C	graphite	0.0	5.298	15.4[d]
	diamond	+0.693	3.417	

[a]Gibbs standard free energies of formation $\Delta G°$ and molar volumes V at 298°K, 1 atm, from Robie and Waldbaum (1968).
[b]Computed from equation 5.61.
[c]Experimentally determined pressures are $P_{eq} \simeq 3$ kbar (MacDonald, 1956) and 4 kbar (Jamieson, 1953).
[d]Extrapolation of high temperature data (Kennedy and Kennedy, 1976) to 25°C gives $P_{eq} \simeq 16.5$ kbar.

lithage→massicot, graphite→diamond, anatase→rutile, and quartz→ coesite. For the pair quartz→amorphous silica the transition is from the denser to the less dense phase. Although instantaneous pressures that develop when rock fragments of mass between 100 g and 1 kg strike a very small area in such habitats as free fall, landslide, or a pebbly stream bed can be very high, pressure alone is not always sufficient to induce a phase transition. Impacts of meteorites on the earth's surface generate pressures and temperatures that can be sufficiently high to induce some polymorphic transitions, such as the quartz-to-coesite conversion in target sands.

In Figure 5.4 are shown the changes that take place in quartz grains due to comminution in a labortory ball mill. Grains of initial diameter of about 1 mm and density $\rho_s = 2.65$ g cm^{-3} are subjected to grinding in the atmosphere of air or ammonia. The density of the fragments ρ_s decreases with a decreasing particle size, to the value of 2.37 g cm^{-3} for the smallest submicron-size particles. From the density and particle-size data, the computed thickness of the amorphous (disordered) silica layer on the surface of the quartz grains shows that the larger the grain, the thicker is the disordered layer. The larger particles are more susceptible to the surface phase transformation than the smaller fragments because of the higher chance of being struck in the ball mill. The density curve for particles ground in water shows that the change in density is virtually negligible in comparison to the dry grinding experiment. The reason for no observable change in density is likely to be dissolution of the amorphous silica layer: solubility of amorphous silica is much higher than the solubility of quartz (see Figure 8.7). A conclusion that can be drawn from the

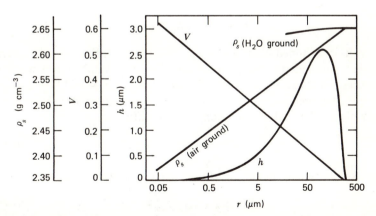

Figure 5.4 Grinding of quartz grains of initial radius $r = 0.5$ mm. Bulk density of grains ground in air and in water (ρ_s), thickness of the amorphous suface layer (h), and volume fraction of different grain sizes (V) (from Lin et al., 1975). By permission of Prof. I. J. Lin, Haifa, Israel.

dry and wet grinding experiments (Figure 5.4) is that in the almost ubiquitous presence of water, the amorphous silica layer—if it forms under natural transport conditions—can be a source of silica dissolving from quartz grains, in excess of the amounts that can be explained by the solubility of quartz.

Example. The equilibrium transition pressures P_{eq} at 25°C for the mineral polymorph pairs listed in Table 5.3 can be evaluated as follows. The Gibbs free energy function is $G(T, P)$, where T is temperature and P is pressure. The total differential of G is

$$dG = \left(\frac{\partial G}{\partial T}\right)_P dT + \left(\frac{\partial G}{\partial P}\right)_T dP$$

$$= -S\,dT + V\,dP \tag{5.56}$$

where S is entropy and V is volume. At constant temperature, the term $S\,dT$ vanishes. The function G can be taken as the Gibbs free energy of formation $\Delta G°$ of a mineral phase at the standard temperature of 25°C (standard reference state: elements at 25°C and 1 atm total pressure). Denoting the standard free energy of formation of phase 1 as $\Delta G_1°$, equation 5.56 for a constant temperature becomes

$$d\Delta G_1° = V_1\,dP$$

which can be integrated between the values of $P = 1$ and $P = P$. The result is

$$\Delta G_{1,P}° = \Delta G_{1,P=1}° + V_1(P - 1) \tag{5.57}$$

where it is assumed that the molar volume of the mineral V_1 remains constant and independent of pressure. This is a valid approximation for pressures not exceeding about 10^5 bar (strictly, 1 atm $= 1.013$ bar but in this example the pressure units can be used interchangeably). For pressures much higher than 1 bar, equation 5.57 simplifies to

$$\Delta G_{1,P}° = \Delta G_{1,P=1}° + V_1 P \tag{5.58}$$

and a similar equation can be written for polymorphic phase 2,

$$\Delta G_{2,P}° = \Delta G_{2,P=1}° + V_2 P \tag{5.59}$$

At equilibrium, the standard free energies of formation of the two phases are equal: $\Delta G_{1,P}° = \Delta G_{2,P}°$. Therefore, the equilibrium pressure P_{eq} can be evaluated from the two preceding equations as

$$P_{eq} = \frac{\Delta G^{\circ}_{2,P=1} - \Delta G^{\circ}_{1,P=1}}{V_1 - V_2} \quad \text{(bar)} \qquad (5.60)$$

Molar volumes V are usually given in units of $cm^3 \ mol^{-1}$, and the free energies of formation ΔG° in units of $kcal \ mol^{-1}$. The conversion from kcal to the energy unit of $kbar \ cm^3$ is $1 \ kcal = 41.84 \ kbar \ cm^3$ (Appendix D). The equation for P_{eq} at a given temperature is therefore

$$P_{eq} = \frac{\Delta G^{\circ}_{2,P=1} - \Delta G^{\circ}_{1,P=1} \quad (kcal \ mol^{-1})}{V_1 - V_2 \quad (cm^3 \ mol^{-1})} \times 41.84 \ kbar \ cm^3 \ kcal^{-1} \quad \text{(kbar)}$$

$$(5.61)$$

For equilibrium reaction at 25°C and 1 atm,

$$\underset{1}{\text{calcite}} \rightleftharpoons \underset{2}{\text{aragonite}}$$

the standard free energies of formation are (Robie and Waldbaum, 1968, p. 20)

$$\Delta G^{\circ}_1 = -269.91 \pm 0.33 \ kcal \ mol^{-1}$$

$$\Delta G^{\circ}_2 = -269.68 \pm 0.35 \ kcal \ mol^{-1}$$

Note that the plus or minus uncertainty on the ΔG° values is greater than the difference $\Delta G^{\circ}_2 - \Delta G^{\circ}_1 = 0.23 \ kcal \ mol^{-1}$, that is the standard free energy of the reaction. The molar volumes of calcite and aragonite, from the same source, are

$$V_1 = 36.934 \pm 0.015 \ cm^3 \ mol^{-1}$$

$$V_2 = 34.15 \pm 0.05 \ cm^3 \ mol^{-1}$$

Substitution of the ΔG° and V values in equation 5.61 gives the equilibrium transition pressure

$$P_{eq} = \frac{(-269.68 + 269.91) \times 41.84}{36.93 - 34.15} = 3.46 \ kbar$$

The value of 3.5 kbar is shown in Table 5.3. The computed value falls between the experimentally determined transition pressures of 3 and 4 kbar.

5.3 PARTICLE-SIZE SPECTRA OF THE WEATHERED CRUST

Solid products of crustal weathering are transported away from the source sites by water, air, and under more specialized conditions, by landslides and moving ice sheets. The weathered material remaining in place or redeposited elsewhere on land forms the regolith—the partly weathered mantlerock of the earth's continental surface. Soils are part of the regolith and are among the more important products of the weathering cycle. Other major sedimentary components of the regolith include alluvium, wind-blown dust, and loess. The mineralogical, chemical, and physical changes produced by weathering show in the mineral composition and particle-size distributions of the regolith materials. This part of the chapter deals with some of the changes taking place in rocks and soils in Section 5.3.1 and with the wind transported materials of mineral origin in Section 5.3.2.

5.3.1 The Regolith

The weathering of rocks progresses from the surface downward, such that the oldest weathered material occurs near the top of an undisturbed soil profile. The breaking of the coarser materials and their disappearance from the surface soils are the result of longer exposure to atmospheric weathering. An example of this process is in the particle-size distributions of a soil profile drawn in Figure 5.3b.

A similar relationship between the grain size and length of exposure to weathering has been reported for the regolith on the Moon (King, 1977). Particles of diameter less than 1 mm on the lunar surface originate from the impacts of meteoroids. The breakup of missiles and comminution of the rocks on the lunar surface taking place during impact contribute to the regolith formation. Thus the weathering of the lunar surface involves comminution by fracturing and melting of rock fragments caused by meteoroid impacts. The mean particle size of the regolith samples decreases with an increasing particle age: particles of mean diameter 82 μm have an age of 3.3×10^9 yr, increasing gradually to 3.8×10^9 yr for particles of 50 μm mean diameter. The gradual decrease in size is attributed to the longer time that was available for comminution by impact on the lunar surface.

For the terrestrial regolith, particle-number spectra and weight-percent frequency histograms for weathered granite, and some residual soils from igneous and metamorphic rocks are shown in Figure 5.5. The particle-number plot of $\log dN/dr$ against $\log r$ for the weathered granite is pronouncedly nonlinear: the smaller particle sizes fall on a straight line with a slope of $b \simeq 2.7$, whereas the coarser sizes fall on the line with the slope of

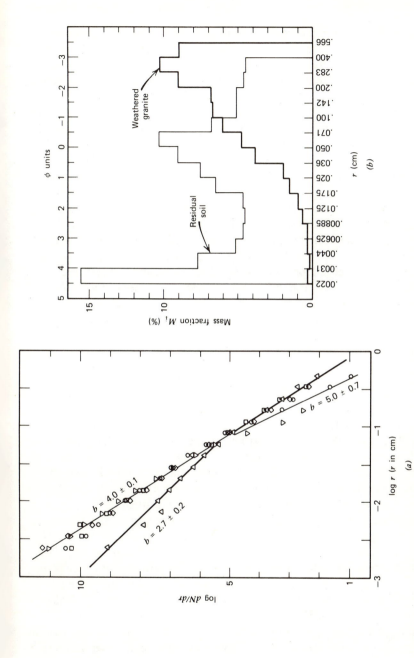

Figure 5.5 Particle-size distributions in soils and weathered rocks (from unpublished data of E. C. Dapples; courtesy of Prof E. C. Dapples, Northwestern University). (a) Particle-number spectra of size. b Is the logarithmic slope of the straight lines drawn through the data points. △ Weathered granite; ◇ residual soil on granite; □ soil from leucogranite; ○ soil from weathered syenite; ▽ residual soil from mica schist. (b) Particle-mass histogram of a weathered granite and residual soil. Note shift to the abundance of smaller particles in the weathered material.

211

$b \simeq 4$. If a smooth curve were drawn through the plotted dN/dr points for the weathered granite, instead of the two straight-line segments shown in the figure, the curvature would be reminiscent of the size distributions originating by fragmentation of the bigger particles, as discussed in Section 5.2.4. An alternative interpretation: the less steep slope of the size distribution in the smaller particle-size range is due to selective removal of the smaller particles by running water. Some departures from the straight-line plot can also be seen for the smaller particle sizes in two of the four residual soils plotted in the figure. The larger-size fractions for the soils from the mica schist and weathered syenite fall on a line of a steeper slope, $b \simeq 5$. Relative to the smaller particles, this indicates a deficiency of the coarser material. A more advanced stage of disaggregation of mechanically weaker rocks may be behind this phenomenon.

The weight-percent histograms of the weathered granite and of a residual soil developing on granite (Figure 5.5b) show a decrease in the fraction of the coarser material from the less (granite) to the more weathered (soil) material. The relatively large differences between the sizes of the individual bars in the histogram are very much absorbed in the log-log plot of the particle-number distribution, where the differences show as relatively small departures from the straight line. A feature of the log-log diagrams of this type is mentioned in Section 5.2.2.

5.3.2   Dust and Ash

The finer-size fraction of the weathered regolith, particles of radius under 10 μm, consists of framework silicate minerals, quartz, clays, and organic debris. The mineralogical composition of the dust in a given geographic area depends on the minerals occurring in the soil and weathered rock sources. For example, quartz is abundant in the dust raised from sandy terrains, and calcite accounts for a significant proportion of the dust blown off limestones and calcareous soils. Records of old dustfalls preserved in snow and ice fields show that the mineral composition of the dust is, as expected, imprinted by those minerals that occur as particles of small size (Table 5.4). Distances over which dust is transported can be very large. Global dispersal of volcanic ash from the more powerful volcanic eruptions, injecting material into the stratosphere, is but one example of long-distance transport occurring sporadically. The eruption of Krakatoa in the East Indies, in 1883, injected sufficient amounts of fine particles into the stratosphere and upper troposphere to cause a measurable decrease in the intensity of solar radiation monitored by European observatories. A less powerful eruption of Laki in Iceland, in 1783, lasted several months and brought haze over Europe and parts of Asia and Africa. In a more recent eruption of Hekla in Iceland, in 1947, transport of ash was recorded over the North Sea, Scandinavia, and Finland (Schwarzbach, 1963). Lamb (1970) provides a comprehensive summary of the recent and historical

Table 5.4   Mineral Components of Global Dust [a]

Wind system	Wt % of mineral fraction						
	Quartz	Feld-spar	Amphi-bole	Mica	Chlo-rite	Kaolin-ite	Smec-tite
Northern Hemisphere							
Polar easterlies	24	20	5	32	11	8	—
Westerlies	22	6	3	56	10	3	—
Trades	54	—	10	17	2	8	7
Southern Hemisphere							
Westerlies	29	19	—	41	7	4	—

[a] Windom (1969).

cases of dispersal of volcanic dust through the atmosphere, and a summary of some of the better studied geological occurrences of volcanic ash can be found in a paper by Eaton (1964).

High-altitude winds transport dust intercontinentally, and clastic materials from the Sahara desert carried by winds across the Atlantic are the external sources of the soils on the islands of Bermuda and Barbados (Jackson et al., 1973; Delany et al., 1967; Prospero and Carlson, 1972). A result of a long-distance dust transport observable by a naked eye is deposition of the reddish colored dust on the snow fields in the Alps, where it is transported by winds from the Sahara. The phenomenon is observable fairly often, when the wind direction and duration are favorable.

Quartz grains of radius <5 μm in the soils of the Hawaiian Islands and in the pelagic sediments of the Pacific Ocean have also been attributed to wind transport from North America (Rex and Goldberg, 1958; Clayton et al., 1972; Churchman et al., 1976). A technique for the identification of the sources of quartz grains is based on the isotopic composition of oxygen in SiO_2. In the Hawaiian soils and Pacific sediments, the ratio $^{18}O/^{16}O$ measures consistently between $\delta^{18}O = +16\%o$ and $+19\%o$ (relative to the SMOW—Standard Mean Ocean Water—calibration standard). These values apply to quartz grains of radius between 0.5 and 5 μm, which are the sizes transportable by wind. The $^{18}O/^{16}O$ ratio values overlap those of the quartz in shales of different geologic ages ($\delta^{18}O$ between $+17$ and $+24\%o$) and they also overlap the upper tail of the ratios in quartz occurring in schists ($+7$ to $+20\%o$). Other potential SiO_2-bearing sources have $\delta^{18}O$ ratio values outside the 16–19%o range: most of the schists, gneisses, and intrusive and extrusive rocks are near 15%o; cherts are between $+20$ and $+35\%o$; and amorphous silica of biogenic origin is between $+37$ and $+45\%o$. Comparisons of the $^{18}O/^{16}O$ ratios in different crustal rocks have led to a conclusion that quartz grains of size

$0.5 < r < 5$ μm, occurring in the Paleozoic and Mesozoic shales in North America, were derived from mixed sources of igneous and metamorphic rocks, and cherts. Quartz occurring in the Hawaiian soils and in the pelagic sediments of the Pacific is similar in its isotopic composition to the quartz in the older shales, from which it was likely to have been derived by erosion and wind transport.

Mineral dust and organic debris in the atmosphere form a suspension of solid particles in a gas. Among the nongaseous components of the atmosphere, all collectively known as aerosols (Chapter 4), the water-soluble and water-insoluble fractions represent to a first approximation two different sources: salts of oceanic origin and mineral dust from the continental surface. A generalized particle-size distribution of an atmospheric aerosol is shown in Figure 5.6. For the smaller particles of radius $r < 0.5$ μm, the particle-number concentration increases more or less pronouncedly with increasing particle size. For particles of radius $r > 0.5$ μm, the number concentration is nearly a straight line on a log-log plot: the particle number decreases as some power of an increasing particle radius, according to equation 5.34 or 5.41. The change in the slope of the curve for the smaller particle sizes indicates either a deficiency or a faster removal mechanism for the finer material. The size distribution of atmospheric aerosols is often viewed as a source and sink balance problem: input of particles to the atmosphere from crustal, oceanic, and internal atmospheric sources is balanced by various removal processes (for more details on removal, see Sections 4.3.3 and 6.3.5). For the smaller particles, coagulation and attachment to the bigger particles may be a more important removal mechanism than settling. (In this case, removal means reduction in number, or disappearance of, particles of some size or sizes, but not necessarily an immediate removal from the atmosphere). Coagulation of small particles is a removal process for particles of certain sizes; fragmentation can also be a removal mechanism for some larger particles. The coarser materials are removed by settling from the atmosphere more efficiently than by coagulation. It has been suggested from dimensional considerations (Friedlander, 1960), that the power exponent of the size-distribution equation $dN/dr = Ar^{-b}$ (particles cm^{-4}) should be $b = 2.5$ for the size range where particles grow by coagulation. The slope b of a log-log plot becomes steeper, $b = 4.75$, for the coarser particles that are removed by settling. The overall range of b between 2.5 and 5.0 agrees with the observed size distributions of atmospheric dust (Figure 5.7). In a detailed analysis, some changes in the slope of the size distribution and the plus or minus errors on b may reflect different input and removal mechanisms. It may be recalled that the curvature of a plot of $\log dN/dr$ against $\log r$ can also be attributed to fragmentation, as well as to the properties of other

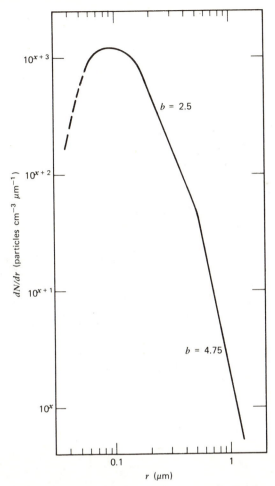

Figure 5.6 Generalized particle-number spectrum of atmospheric aerosols (from Friedlander, 1960).

size distributions, such as the lognormal and Rosin-Rammler (compare Figure 5.2).

Soils contain a large proportion of lumps with radii greater than 10 μm. A beginning of the process of dust transport is the breaking of the soil lumps by particle abrasion and lifting of the particles in the size range between 1 and 10 μm, when the winds attain velocities of a few meters per second. Clay particles ($r \simeq 1$ μm) have been described as clinging to the bigger particles in soils, so that greater wind velocities are needed to

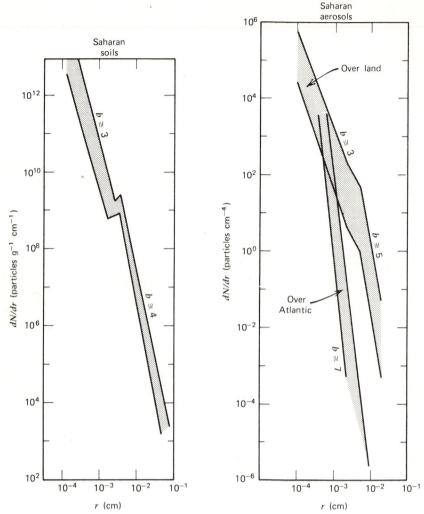

Figure 5.7 Particle-size distributions of aerosols and soils (adopted from Schütz and Jaenicke, 1974).

separate them from the soil lumps. Because of this, the proportion of clay-size particles in the dust is often smaller than can be projected by extrapolation from the larger to the smaller particle sizes. This accounts for a less steep slope of the particle-size distribution for the smaller particle sizes, such as in distributions plotted in Figure 5.7. The so-called giant particles ($r > 30$ μm) also occur in the atmosphere, mostly during dust storms. The residence times of the giant particles are short and, as a whole, the number of the coarser particles in the dust decreases drastically because of settling as the material is being carried farther away from the source. The size distributions of particles in the soils and dust in the Libyan desert, shown in Figure 5.7, are similar to each other in shape, within the size range between 1 and 100 μm. The Saharan dust carried out over the Atlantic Ocean shows a steeper plot of $\log dN/dr$ against $\log r$, indicating a loss of the coarser particles. Particles in the soils and dust near the source have size distributions that can be characterized by two curves, each of the type $dN/dr = Ar^{-b}$ and differing in the value of the power exponent b (Figure 5.7). For the smaller particles of radius $r < 20$ μm, the power exponents are $b \simeq 3$ for soils and aerosols. For the coarser particles of radius $r > 20$ μm, the logarithmic slope is bracketed by the values $4 < b < 5$.

5.4   TRANSPORT OF THE PRODUCTS OF WEATHERING

5.4.1   Dissolved and Suspended Loads in Rivers

Estimates of the global averages of the dissolved, suspended, and wind-blown loads carried from the continents center around the following numbers (dissolved and suspended loads from Gregor, 1977; atmospheric transport from Goldberg, 1971, and Schneider and Kellogg, 1973):

Dissolved load in rivers
From evaporites	0.55×10^9	
From carbonates	1.33×10^9	
From silicates	0.62×10^9	
Total dissolved	2.50×10^9 tons yr^{-1}	2.5×10^9 tons yr^{-1}

Suspended load in rivers
From sands and shales	4.8×10^9	
From crystalline rocks	1.7×10^9	
Total suspended	6.5×10^9 tons yr^{-1}	6.5×10^9 tons yr^{-1}
Total dissolved + suspended		9.0×10^9 tons yr^{-1}

Wind-blown   load $(0.6-1.6) \times 10^9$ tons yr^{-1}

The preceding estimates of the loads carried by rivers correspond to the mean concentrations of about 80 mg l^{-1} of dissolved solids and 200 mg l^{-1} of suspended material (river flow volume 3.2×10^{16} l yr^{-1}). The mass ratio of the loads dissolved : suspended is about 1 : 2.6. A fraction of the dissolved load consists of salts recycled through the atmosphere from the ocean, salts added to the rivers by human activities, and bicarbonate formed by dissolution of CO_2 from the atmosphere. Gregor's (1977) estimate of the dissolved-to-suspended load ratio of $1:2.6$ refers to the materials weathered from the crust, after the contributions due to recycling and human activities have been corrected for in the estimate of the dissolved load. Higher estimates of the input rates of the total dissolved load (4×10^9 tons yr^{-1}) and suspended load (18×10^9 tons yr^{-1}) have been made in the 1960s (Holeman, 1968; Gregor, 1970; Garrels and Mackenzie, 1971).

Of the total mass of crustal rocks that is being weathered, the mass fraction that goes into solution annually is $1/(1+2.6)=0.28$ or about 30%. The remaining 70% of the weathered products are carried as solids. Dissolution results in some decrease of the grain size, and a limiting estimate can be made of how much the mean grain size is affected by the global average rate of dissolution. If all the minerals originally present in the rock were dissolving to the same extent, the mean grain size of the residual solids would be $0.7^{1/3}=0.9$ of the original particle diameter. A 10% decrease in the linear dimensions is not a great change. Greater or smaller changes in linear dimensions would correspond to higher or lower, respectively, ratios of the dissolved to suspended loads. A ratio of $1:8$ corresponds to the residual particle size of $(1-\frac{1}{9})^{1/3}=0.96$ of the original, or to a 4% decrease in size. At an opposite extreme, the dissolved-to-suspended load ratio of $1:1$ corresponds to the residual particle size of $(1-\frac{1}{2})^{1/3}=0.8$ of the original, or to a 20% decrease in diameter.

The rates of chemical weathering of different minerals making the crustal rocks are not uniform. Goldich (1938) observed that in the process of soil formation the common silicate minerals occur in the following sequence, from the least to the most resistent:

$$olivine \rightarrow augite \rightarrow hornblende \rightarrow biotite \rightarrow$$
$$potassium\ feldspar \rightarrow muscovite \rightarrow quartz$$

Among the feldspars, the sequence from the least to the more stable is:

$$calcic\ plagioclase \rightarrow alkalic\ plagioclase \rightarrow potassium\ feldspar$$

The Goldich weathering series is essentially the order of crystallization of silicate minerals from a melt, known as the Bowen reaction series. Olivines, pyroxenes, amphiboles, micas, feldspar, and quartz generally form a sequence of phases that crystallize with a decreasing temperature. The order of increasing stability in weathering coincides with the temperatures of crystallization that are closer to the earth surface conditions. The stability of minerals may be related to how closely the weathering environment compares to the conditions under which the mineral was formed. An overall effect of the variable degree of resistance to weathering is that some minerals disappear from weathered sediments, whereas others are altered little or not at all. Well-weathered sediments contain virtually no high-temperature magnesium and iron silicates, no calcium feldspars, and no detrital calcite or aragonite. (The presence of any of these minerals in soils, river beds, or lake sediments usually indicates proximity to the sediment source and a geologically young age of the sediment.) At the more stable end of the Goldich weathering series, sodium and potassium feldspars, micas, quartz, and a number of abrasion or dissolution resistant oxides and silicates occur as common components of sedimentary rocks.

Strong acids occurring in surface waters and rain should enhance the rates of chemical weathering of crustal rocks. The higher concentrations of the hydrogen ions in rain, caused by the hydrochloric, sulfuric, and nitric acids, are discussed in Chapter 4. The higher content of strong acids can be neutralized by cations leached out of surface rocks by rain water. A balance between strong acids in rain and cations released to river water is

$$C_{H^+} V_{rain} = C_{cat} V_{rivers} \qquad (5.62)$$

where C_{H^+} is the hydrogen-ion concentration due to the presence of strong acids in rain water (that is, a strong acid HL dissociating into H^+ and L^-), V_{rain} is the annual volume of atmospheric precipitation, C_{cat} is the equivalent concentration of cations in river waters, and V_{rivers} is the annual volume of river flow. To determine concentrations of the cations leached by acidic rains of pH $= 4.5$, the following representative values can be used: $C_{H^+} = 3.2 \times 10^{-5}$ equivalents l^{-1}, $V_{rain} = 1.2 \times 10^{17}$ l yr^{-1}, and $V_{rivers} = 0.32 \times 10^{17}$ l yr^{-1}. Thus the cation concentration C_{cat} is, from equation 5.62,

$$C_{cat} = \frac{3.2 \times 10^{-5} \times 1.2 \times 10^{17}}{0.32 \times 10^{17}} = 1.2 \times 10^{-4} \quad \text{equivalents } l^{-1}$$

A global mean concentration of cations in rivers has been given as 1.42×10^{-3} equivalents l^{-1}, of which the amount 0.12×10^{-3} equivalents l^{-1} is attributable to the salts recycled through the atmosphere from the ocean (Livingstone, 1963, Garrels and Mackenzie, 1971). The difference $(1.42 - 0.12) \times 10^{-3} = 1.3 \times 10^{-3}$ equivalents l^{-1} is the cation concentration due to dissolution of crustal materials. The preceding estimate of the cation concentration needed to neutralize the strong acid content of rains is only about $\frac{1}{10}$ of the concentration in river water: $1.2 \times 10^{-4} / 1.3 \times 10^{-3} = 0.1$. The conclusion to be drawn from this computation is that acidic rains contribute about 10% to the dissolved load of rivers. A global average pH value of rain water lower or higher than the pH $\simeq 4.5$, used in the computation as a measure of the strong acid concentration, would raise or depress the estimate of the contribution of rain to weathering. The fact that the pH values of streams and lakes have been declining in some industrialized areas during the last 40 years (Chapter 4) suggests that the rain water acidity is not completely neutralized by leaching and dissolution of rocks.

The dissolved and suspended loads in rivers generally depend on the river discharge, as shown in Figure 5.8. A relationship between the con-

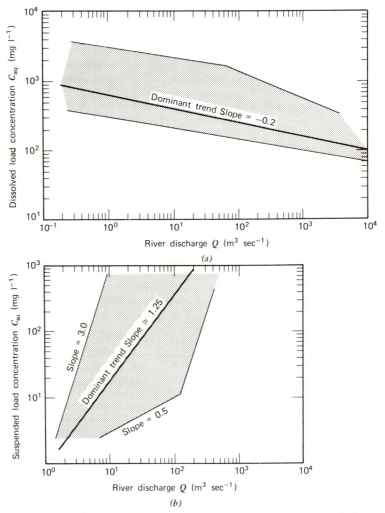

Figure 5.8 Dissolved and suspended loads of rivers as a function of water discharge. (*a*) Dissolved load in rivers of the western and southwestern United States (data from Leopold et al., 1964). (*b*) Suspended load in rivers of Switzerland (data from Peters-Kümmerly, 1973).

centrations of dissolved and suspended materials C and the flow rate Q is

$$C = AQ^m \qquad (5.63)$$

For the dissolved solids concentration C_{aq}, the equation is

$$\log C_{aq} = \log A + m \log Q \qquad m \simeq -0.2 \qquad (5.64)$$

where the flow rate Q is in cubic meters per second and concentration C_{aq} is in milligrams per liter. Concentration decreases slightly with an increasing discharge, an indication that dissolution does not keep pace with the greater volume of flow.

For the suspended solids concentration C_{su} the equation is

$$\log C_{su} = \log A + m \log Q \qquad m \simeq 1.25 \qquad (5.65)$$

The range of m for the bigger rivers is $0.5 < m < 3$. The values of the constant A vary highly from one river to another. The shaded envelope about the dominant trend includes the more important data in Figure 5.8.

From the suspended matter concentrations and the areas of the drainage basins of the individual rivers, the rates of denudation of the crystalline and sedimentary rock terrains in Switzerland have been estimated as follows (Peters-Kümmerly, 1973): 0.05 mm yr^{-1} for crystalline rocks and 0.25 mm yr^{-1} for sedimentary rock terrains, with the ratio of the two denudation rates 5:1. From the global averages of the dissolved and suspended loads in rivers, listed at the beginning of this section, the ratio of the denudation rates of the sedimentary to crystalline rocks is about 3:1, but the rates are about one order of magnitude lower: 0.006 mm yr^{-1} for the crystalline terrains and 0.02 mm yr^{-1} for the sedimentary rock terrains.

5.4.2 Rounding of Particles—an Archimedean Approach

The problem of how mineral particles become rounded during transport by water and wind has held the attention of sedimentologists for a long time because of the relationships between the physical characteristics of grains and their transport histories (Krumbein and Pettijohn, 1938; Kuenen, 1959; Kuenen and Perdok, 1962; Pettijohn, Potter and Siever, 1973). The shape of sand grains also determines to some extent the porosity and permeability of ground water aquifers and oil-bearing rocks, such that the subject of grain rounding or, more generally, change of shape during transport and deposition holds considerable interest in ground-water and petroleum exploration. (Porosities and permeabilities of sediments are dealt with in Chapter 2.) Detailed work on such physical characteristics of mineral grains as their size, sphericity, and roundness has shown that no universal quantitative relationships exist between the physical appearance of sediment particles and their transport histories. However, a decrease in grain size and improvement of the grain rounding (that is, disappearance of sharp edges) in sands generally correlate with prolonged reworking by water. Transport over long distances is not necessarily a condition of better roundness, although a decrease in linear dimensions of grains can sometimes be accounted for by transport. For example, the mean particle radius of the sand in the Mississippi River decreases from about $r = 0.3$ mm at Cairo, Illinois, to 0.1 mm at the distance 1,600 km downstream, the decrease being fairly gradual (Leopold et al., 1964, p. 193).

Rounding of the sharp edges and wearing off the grains to an ellipsoidal or spherical shape is commonly regarded as a result of abrasion achieved by rolling and collisions with smaller particles. A question of interest is, how much mass is lost by a crystal with well-developed faces when it is reduced at the limit to a spheroid or a sphere? Although the shape of the original grain is not known, an estimate of mass removed by rounding is a measure that can be supported by the crystal habits of certain minerals occurring in rocks. Whether the corners and sharp edges are rounded off by dissolution or mechanical abrasion is of secondary significance in this context.

A question of how much volume is removed when a polyhedron is reduced to an inscribed spheroid is reminiscent of a classic problem of geometry, the solution of which was Archimedes' pride. Archimedes' discovery was that the sphere volume can be conservatively divided into the volumes of a cylinder and a cone: $\frac{4}{3}\pi r^3 = \pi r^3 + \frac{1}{3}\pi r^3$, where the base radius of the cylinder and the cone is r, and the height of each body is also r, equal to the sphere radius. A cube can be reduced by rounding to a sphere, the diameter of which is equal to the cube edge; such a sphere touches each of the six faces of the cube. Similarly, a square prism can first be reduced to a circular cylinder of the same height (the cylindrical surface touches the four faces of the prism) and subsequently to an ellipsoid whose surface is tangent to the four faces and two bases of the prism. A number of simple polyhedral forms that can be rounded to a straight circular cylinder, cone, ellipsoid, or sphere are listed in Table 5.5. The last column of the table gives the percentage of the mass or volume loss due to rounding, when rounding is carried out as explained, making an inscribed rounded particle out of a polyhedron. The smallest change shown is for the hexagonal prism rounded to a cylinder, 9% volume loss. An even smaller change would be associated with the rounding of a regular prism having more than six faces. Large losses of volume are associated with the

Table 5.5 Volume Loss of Particles Due to Rounding.
Polyhedra Rounded to Inscribed Spheres, Ellipsoids of Revolution,
Straight Circular Cylinders, or Cones

Particle shape		$R =$	Volume loss $(1-R) \times 100$
Original	Rounded	$\dfrac{\text{volume rounded}}{\text{volume original}}$	(%)
Cube	Sphere	$\dfrac{\pi}{6}$	48
Rhombohedron	Sphere	$\dfrac{0.77\pi}{6}$	60
Tetragonal prism	Cylinder	$\dfrac{\pi}{4}$	21
	Ellipsoid	$\dfrac{\pi}{6}$	48
Hexagonal prism	Cylinder	$\dfrac{\pi}{2\sqrt{3}}$	9
	Ellipsoid	$\dfrac{\pi}{3\sqrt{3}}$	40
Trigonal prism	Cylinder	$\dfrac{\pi}{2\sqrt{3}}$	40
	Ellipsoid	$\dfrac{2\pi}{9\sqrt{3}}$	60
Tetrahedron	Cone	$\dfrac{\pi}{3\sqrt{3}}$	40
	Sphere	$\dfrac{\pi}{6\sqrt{3}}$	70
Double tetrahedron (trigonal bipyramid)	Sphere	$\dfrac{16\pi}{81\sqrt{3}}$	64
Octahedron	Sphere	$\dfrac{\pi}{6\sqrt{3}}$	70

rounding of pyramidal crystals to spheres: 60–70% of volume is lost when an octahedron (a square bipyramid) or a tetrahedron is ground to a sphere. Rounding of prisms and cubes to ellipsoids and spheres also involves substantial losses of mass, between 40 and 60%.

In the weathering cycle, some rounding of mineral grains is probably produced by dissolution, as the sharp broken edges and corners have higher surface energies than the flat crystal faces. However, extensive evidence from laboratory experiments shows that when highly polished spheres made of mineral crystals are placed in corrosive solutions, new crystal faces develop in the process of dissolution, such that the final form is a euhedral crystal (Heimann, 1975). Dissolution controlled by diffusion away from the surface can lead to rounding. Etching and crystal faces are the result of dissolution controlled by chemical reactions specific to certain sites.

A well-rounded grain is likely to be a product of mechanical abrasion. Incomplete rounding, involving some reduction of the corners and edges of crystalline grains, can easily result in a 20–30% loss of mass. In weathering, an average dissolution of 25–30% of the solids, a number given in Section 5.4.1, may be aided by mechanical abrasion and comminution, also producing smaller grain fragments.

5.4.3   Distribution of Minerals in a Soil Profile

In an idealized sequence of the weathering cycle, the progression of chemical weathering of silicate minerals takes place by release of silica, cations, and alumina in different proportions, and the formation of cationic clays (smectites, vermiculite, chlorite, illite), kaolinite [$Al_2Si_2O_5$ $(OH)_4$], and gibbsite [$Al(OH)_3$]. A distribution of mineral phases in a soil forming on a magnesium- and iron-rich silicate rock (peridotite) is given in Table 5.6. The minerals and thickness of the regolith are characteristic of the intensive weathering conditions, warm climate, and abundant rainfall, but without strong denudation of the soil. The bedrock, a serpentinized peridotite, contains serpentine [idealized composition $Mg_3Si_2O_5(OH)_4$] and augite as the major components, and minor amounts of magnetite (Fe_3O_4), spinel ($MgAl_2O_4$), chromite ($FeCr_2O_4$), and rutile (TiO_2) combined under the *miscellaneous* column in Table 5.6. Serpentine and augite do not survive the early stages of weathering, as they are absent from the soil profile. Free silica (SiO_2) occurs in the lower part of the weathered section, where it forms at the expense of the bedrock. The two cationic clays, chlorite and smectite, are more abundant in the lower part of the soil profile, whereas kaolinite (containing no exchangeable cations) increases in abundance toward the upper part of the soil profile. The weathering sequence from smectite through kaolinite to gibbsite is suggested by the relative abundances of the three minerals in the upper part of the profile. The iron mineral goethite (α-FeOOH) is very abundant in the soil, to the extent that the upper 5 m can be described as an iron crust. Some of the silica in the uppermost soil layer, where kaolinite is slightly more abundant, and some of the iron within the entire profile might have been supplied by the decaying vegetation on the surface, through mobilization in organic soluble substances.

**Table 5.6   Mineralogical Composition of a Soil Profile
Under Conditions of Tropical Weathering, Kalimantan Island, Indonesia[a]**

Soil horizon	Thickness (m)	Serpentine	Augite	Silica	Chlorite	Smectite	Kaolinite	Gibbsite	Goethite	Miscellaneous	$\frac{SiO_2}{Al_2O_3}$ (mol/mol)
						(wt %)					
A_1	0–1					1	15	8	74	2	1.2
B_1	1–2					1	6	18	71	4	0.5
B_2	2–5					3	2	17	74	4	0.3
B_3	5–6.5				18	4	2	4	68	4	1.3
C_1	6.5–7			6	16	17			57	4	3.8
C_2	7–7.5	13		22	16	7			40	3	10.3
Bedrock	7.5+	44–77	40–0	6–0	7–0				4–10	12–1	26.8

[a]Mohr et al. (1972).

As the metal cations and silica are leached to a greater extent than aluminum from solids in the process of weathering, an index ratio

$$(\text{cations} + SiO_2)/(Al_2O_3)$$

changes accordingly, decreasing from the bottom layer of a slightly weathered bedrock toward the upper part of the weathered regolith. Below, in parentheses following each mineral, is indicated the molar ratio (cations + Si)/(Al) that corresponds to an ideal stoichiometric composition. An overall decrease in the index ratio from the high values in the bedrock minerals to nil in gibbsite illustrates this point:

serpentine (∞) | chlorite (8) |
           $\rightarrow$ | | $\rightarrow$ kaolinite (1) $\rightarrow$ gibbsite (0)
augite (>3) | smectite (1.7) |

The rates of leaching of silica and cations from individual minerals and the rates of mineral dissolution are discussed in more detail in Section 5.5.

5.5   SOLUBILITIES OF MINERALS

5.5.1   Literature Sources and Thermodynamic Computations

Solubilities of minerals and man-made inorganic phases in water depend to a variable extent on temperature and the presence of other chemical species in solution. Data on solubilities in water at different temperatures are available in extensive tabulations. An emphasis in such tabulations is generally on the solubilities of strongly soluble minerals at atmospheric pressure, and on solutions of simple composition. Time honored sources of solubility data and other properties of aqueous solutions are the *International Critical Tables* (1926–1930) and the *Landolt-Börnstein* tables (1960, 1962, 1976). Additional extensive sources of solubility data are *Seydell's Handbook* (Linke, 1958) and a translated edition of a Russian handbook of solubilities, published in the original between 1961 and 1963 (Stephen and Stephen, 1963–1964). Solubilities of minerals in concentrated salt brines at different temperatures have been compiled for many three-component systems in a Russian-language handbook (Zdanovskii et al., 1953). The book by Braitsch, published in German in 1962 and in the English translation in 1971, summarizes many of the older data on the chemical composition of saline brines and solubilities of evaporitic minerals.

Computation of the solubilities of minerals in aqueous solutions of variable composition is treated in the books by Garrels and Christ (1965) and Stumm and Morgan (1970). Concentrations of chemical species in a solution at a thermodynamic equilibrium with a solid phase can in many cases be computed when certain thermodynamic information is available (the Gibbs free energies of formation of the solid and aqueous species, and relationships between the thermodynamic activities and concentrations for the components of the solution and the solid phase). For many minerals and aqueous species such information is available. For the standard conditions of 25°C and 1 atm total pressure, the foremost source is the publication of the U.S. National Bureau of Standards (Wagman et al., 1968–1971). For minerals and aqueous species of geochemical interest, many thermodynamic data have been compiled by Garrels and Christ (1965) and Berner (1971), with examples of how these data can be used to estimate the thermodynamic parameters at higher temperatures and pressures. A large volume of data on solubilities and ion association in solutions of inorganic and organic compounds has been compiled by Sillén and Martell (1964, 1971). The information is listed in terms of equilibrium constants for solubility and association-dissociation reactions, at specified temperatures and ionic strengths of the medium.

The Gibbs free energies of formation of many minerals at temperatures increasing from 25°C and at 1 atm total pressure have been computed by Robie and Waldbaum (1968) and Robie et al. (1977). Additional sources for minerals are the publications by Helgeson et al. (1978), Mel'nik (1972), and Naumov et al. (1971); for solids, liquid, and gaseous pure phases, see the *JANAF Thermochemical Tables* (1971).

Thermodynamic properties of electrolytes in aqueous solutions have been critically summarized for different ranges of temperatures and pressures in a series of papers by Helgeson and Kirkham (1974–1976). Estimation of the thermodynamic properties of some minerals and dissolved species in concentrated saline brines, of the types occurring in nature, has been dealt with in the papers by Lerman (1970) and Wood (1976). The basic thermodynamic properties of dissolved ionic species, and other derived physical and chemical characteristics of dilute solutions and sea water (such as density, viscosity, and an equation of state) are the subject of a series of papers by Millero and co-workers (Millero, 1971; 1975; Chen and Millero, 1977).

Computer programs are available with the aid of which concentrations of the chemical species in multicomponent solutions can be computed. One such program—WATEQ—performs computations for an equilibrium system containing up to several solid phases, using an extended form of the Debye-Hückel equation for the activity coefficients in solution that is valid up to moderately high concentrations (Truesdell and Jones, 1974; Plummer et al., 1976).

Equilibria in multicomponent gas mixtures have been treated as models of primordial and extraterrestrial atmospheres (Dayhoff et al. 1967). The results are extensive lists of chemical species that can exist at different temperatures in mixtures of seven biologically important elements.

5.5.2   Kinetic Models of Solubility: General

A broad class of problems related to water and sediment environments is connected with the understanding of the kinetics of mineral dissolution, chemical exchange, and precipitation. Dissolution of minerals and man-made solids contributes dissolved materials to surface and ground waters; ground waters reacting with aquifer rocks undergo changes in chemical composition; dissolution of limestones and development of the karst topography is a large-scale phenomenon based on the reactions between water and calcium carbonate; precipitation and dissolution of calcium carbonate and silica in the ocean affects such diverse phenomena as the distribution of trace metals and nutrients, and the carbon dioxide balance of the world. The list of globally or locally important environmental problems where the kinetics of dissolution and precipitation enter as an important factor can be made very long. Studies of the kinetics of reactions aim at such goals as the determination of the rates of transfer of materials between minerals and solutions, and related processes. For example, questions of the time it takes a given mineral to come to an equilibrium with a solution, or the length of time involved in a given dissolution or precipitation process—these are the questions to which kinetics should be able to provide the answers. Through the end of this chapter, dissolution of minerals is discussed with reference to some experimental and theoretical information that deals primarily with dissolution in dilute aqueous solutions resembling fresh waters. Additional material on mineral-solution reactions is given in Chapters 6 through 8.

Four schematic models of dissolution in Figure 5.9 show initial and advanced reaction stages of solids.

In Figure 5.9a, the particle shown is made of several parts joined together. Dissolution affects the joints faster than the bulk material, with the result that the particle disintegrates into several smaller fragments. Such a process has been observed in dissolution of foraminiferal tests made of calcite, where the individual chambers are joined by calcitic sutures (Bé et al., 1975). In water undersaturated with respect to calcite, the joining sutures can dissolve faster than the chamber material. The particle can be characterized by its volume V (cm^3), surface area S (cm^2) and a characteristic linear dimension, radius r (cm). When a particle breaks into a number of fragments because of dissolution, there is some decrease in the total volume V and a decrease in the radii or volumes of the individual fragments, but there is an increase in the total external surface area exposed to solution.

In Figure 5.9b, dissolution of a particle takes place by pitting and etching of the surface, without affecting at first the overall particle dimensions. Such a mode of dissolution has been described for feldspar and amphibole grains in soils (Correns, 1961) and in laboratory experiments

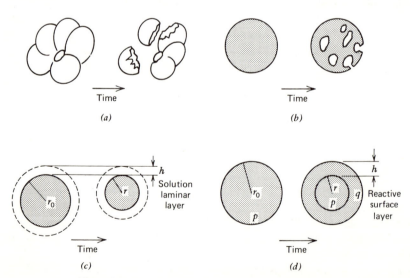

Figure 5.9   Four schemes of dissolution of solid particles.

where feldspars were etched with acids (Berner and Holdren, 1977). Skeletons of certain species of diatoms made of silica are perforated with a regular pattern of submicron-diameter pores. In the course of dissolution, the pore diameters increase in size, similarly to the pattern shown schematically in the figure (Schrader, 1972a). At the early stages of dissolution, the particle radius r remains essentially constant, the surface area may either increase (by pitting of smooth surfaces, such as in feldspar crystals) or decrease (by enlargement of perforation on the shell surface), and the loss of mass results in a lower bulk density of the same macroscopic volume (see Plate I, p. 270).

In Figure 5.9c, dissolution takes place at the particle surface at a uniform rate. In the mathematically simple case of a solid sphere, the volume, surface area, and radius decrease, whereas the shape of the particle remains spherical. A model of dissolution of a sphere is convenient to deal with, and it can also be used for the dissolution scheme in Figure 5.9b, provided the mass loss rather than changes in the linear dimensions are considered.

In Figure 5.9d, a new phase forms at the surface of the particle. The new phase q grows inward at the expense of the original phase p and it can be, for example, a leached residue of p or a new phase formed by reaction of p with an external solution or gas. This model is often used to deal with reactions between pelletized solids and solutions or gases, when a reactive layer or a coat of oxide forms on the original material. The formation of a layer of amorphous silica on the surface of quartz grains, achieved by comminution (Figure 5.4), is a mechanical-chemical reaction that falls in the scheme of Figure 5.9d, although no dissolution is involved in dry grinding.

5.5.3 Kinetic Models of Solubility: Equations

The rate of dissolution, measured as a change of concentration in solution C as a function of time t, is

$$\frac{dC}{dt} = k_m(C_s - C)^m \qquad (C \leqslant C_s) \qquad (5.66)$$

where k_m is the reaction rate parameter of dimensions $(\text{g cm}^{-3})^{1-m} \text{ sec}^{-1}$, when the rate of change dC/dt is in units of $\text{g cm}^{-3} \text{ sec}^{-1}$, and C_s (g cm^{-3}) is an equilibrium or steady-state concentration value to which the solution tends with time.

Power exponent m is called the order of the reaction. For dissolution, the term on the right-hand side of the equation is positive, concentration C

increases with time, and $dC/dt > 0$. For precipitation, concentration decreases with time and $dC/dt < 0$. The latter requires that the term on the right-hand side of the equation be written as $-k_m(C - C_s)^m$, with the condition of $C > C_s$.

The dissolution model of equation 5.66 applies to a closed system where no additional sources or sinks of the dissolved species exist. It effectively states that the rate of dissolution dC/dt is a power function of the distance from saturation, or of the difference between the steady concentration C_s and time-dependent concentration C. Since the early work of Noyes and Whitney (1897), who demonstrated that the dissolution process is a first-order reaction ($m = 1$), equation 5.66 with different values of m has been used often to describe the rates of dissolution or precipitation reactions. In more involved cases, the reaction rate parameter k_m and the reaction order m can be functions of concentration or time, such that the reaction mechanism changes as the dissolution progresses from a state of strong undersaturation to near saturation. One example of such a change in the reaction order and rate is the dissolution of calcite in water at 25°C and 1 atm partial pressure of CO_2 (Plummer and Wigley, 1976): pure water saturated with 1 atm CO_2 at 25°C has a pH of approximately 3.9; calcite introduced into solution begins to dissolve and the pH rises while an equilibrium with 1 atm CO_2 is maintained. The rate of increase in the calcium concentration follows the second-order reaction, with $m = 2$ in equation 5.66. The second-order rate is maintained until the solution pH has risen to about 5.94. From pH $\simeq 5.94$ and to saturation with calcite at pH $\simeq 6$, the reaction order becomes higher, with $m \simeq 4$. In such dissolution rate experiments, the solution pH is measured with a fast-response pH electrode, and calcium concentrations are computed from the pH values and the carbonate dissociation equilibria at the known CO_2 pressure. A plot of $\log dC/dt$ against $\log(C_s - C)$ gives a straight line with a slope of m. The change in the order of the reaction rate shows as a break in the slope of the plot.

A concept of a boundary or laminar layer postulates that a thin layer of solution exists at the surface of the solid, and transport of dissolved material through such a layer is by molecular diffusion from the solid surface to the bulk solution (Figure 5.9c). For a first-order dissolution reaction

$$\frac{dC}{dt} = k(C_s - C) \qquad (\text{g cm}^{-3} \text{ sec}^{-1}) \qquad (5.67)$$

the reaction rate parameter k (sec^{-1}) can be related, on dimensional grounds, to the thickness of the laminar layer h (cm), the diffusion coefficient of the dissolved species D ($\text{cm}^2 \text{ sec}^{-1}$), the reactive solid surface area S (cm^2), and the volume of the solution V (cm^3) by the following equation:

$$k = \frac{D}{h} \times \frac{S}{V} \quad (\text{sec}^{-1})$$

$$= \frac{k_s S}{V} \tag{5.68}$$

where $k_s \equiv D/h$ is a transport or reaction rate parameter of the dimensions of velocity (cm sec^{-1}).

If C_s is interpreted as concentration at the solid surface in contact with the base of the laminar layer and C as concentration at the laminar layer top, the same as in the bulk solution, then the quotient

$$\frac{D(C_s - C)}{h} \quad (\text{g cm}^{-2} \text{ sec}^{-1}) \tag{5.69}$$

is the flux of dissolved material per unit area of the solid surface. The amount dissolving from the entire surface S of the solid into solution of volume V is

$$\frac{dC}{dt} = \frac{DS(C_s - C)}{hV} \tag{5.70}$$

which also follows from substitution of equation 5.68 into 5.67.

The laminar layer model assumes a linear concentration gradient $(C_s - C)/h$ (g cm^{-4}) within the layer and a diffusion coefficient D independent of concentration. In dilute solutions, the diffusion coefficients of dissolved species change little with concentration (Chapter 3). In a solution of constant volume V, a constant reaction rate parameter k requires that the ratio S/h must remain constant and independent of concentration. This can be true in the following cases: if the surface area of the solid decreases appreciably in the course of dissolution, so does the thickness of the laminar layer; alternatively, if the thickness of the laminar layer is independent of the particle size (it may depend on the physical environment within the solution, such as the rate of agitation), then the surface area of the solid S must be constant. In experiments with minerals of low solubilities, of the order of 10^1–10^2 mg l^{-1}, excess of the solid introduced into the solution volume assures that the total surface area S will remain practically constant. Plummer (1972) reported from literature data and from dissolution experiments on calcite in water the range of the laminar layer thicknesses h between 10 and 50 μm, with a mean value of $h \simeq 30 \, \mu$m. As a comparison between the magnitudes of thickness, the electrical potential double layer at the surface of clay particles (computed from certain mathematical models of distribution of the potential) is measurable in tens to hundreds of angstroms, or 10^{-3}–10^{-2} μm (Bolt and Bruggenwert, 1976, pp. 51–53).

For dissolution of calcite in water, in Figure 5.10 are plotted the second-order reaction rate parameters k_2 (mol^{-1} l sec^{-1}) for calcium and hydrogen ions against the specific surface area S/V (cm^2 l^{-1}) of calcite. For calcium, the rate parameter k_2 refers to the rate of addition of calcium ions to solution, whereas for the hydrogen ion, k_2 indicates removal from solution. In the log-log coordinates of Figure 5.10, the slope of the line

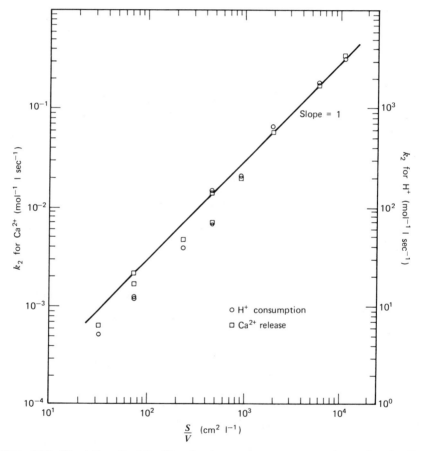

Figure 5.10 Dissolution of calcite. Second-order reaction rate constants k_2 as a function of the surface area of the solid (data from Plummer, 1972, and personal communication, 1975).

drawn through the data points is unity, indicating that the reaction rate is a linear function of the solid surface area. The rate-constant/surface-area ratios are constant for the calcium and hydrogen-ion reactions: for calcium the ratio is $k_2V/S = (2.47 \pm 0.01) \times 10^{-5}$ mol^{-1} l^2 cm^{-2} sec^{-1}, and for the hydrogen-ion plot the ratio is $k_2V/S = 0.23 \pm 0.09$ mol^{-1} l^2 cm^{-2} sec^{-1}.

The rate of dissolution and the rate at which the solution tends to a steady-state value increases with an increasing specific surface area S/V of the solid (see equation 5.70). This phenomenon was shown in a series of experiments on the rates of dissolution of different polymorphic forms of silica (quartz, tridymite, cristoballite, stishovite, and amorphous silica) at various solid/solution ratios (Stöber, 1967). Stated differently, dissolution is faster and steady state can be attained sooner in suspensions of a higher solid/solution ratio.

For a first-order dissolution reaction with a constant reaction rate parameter k, integration of equation 5.67 with the boundary conditions

$$\text{at} \quad t = 0: \qquad C = C_0$$

$$\text{at} \quad t \to \infty: \qquad C = C_s$$

gives concentration in solution C as a function of time:

$$C = C_s + (C_0 - C_s)e^{-kt} \qquad (\text{g cm}^{-3}) \tag{5.71}$$

For a reaction of order higher than 1 ($m > 1$), integration of equation 5.66 with the same boundary conditions as for the first-order dissolution gives

$$C = C_s - \left(\frac{1}{1/(C_s - C_0)^{m-1} + (m-1)k_m t} \right)^{1/(m-1)} \qquad (\text{g cm}^{-3}) \tag{5.72}$$

A dissolution model based on diffusion of the dissolving species through a boundary layer at the solid surface applies to the situations shown schematically in Figures 5.9c and 5.9d. In the situations of Figures 5.9a and 5.9b, however, dissolution takes place preferentially at certain macroscopic or microscopic sites. A controlling factor here is a chemical reaction occurring at certain locations on the solid surface, such that one can speak of a reaction-controlled dissolution. In the case of diffusion through a boundary layer, dissolution is diffusion controlled. Analogous distinctions are made for the process of crystal growth that can be controlled by diffusion, nucleation, or dislocation.

5.5.4 Particle Size Change in Dissolution

First-Order Reaction. This section considers how the size of a spherical particle changes during dissolution, when it takes place according to the first-order reaction as given in equation 5.67. The mass of N particles, all of the same density ρ_s and the same initial radius r_0, is $\frac{4}{3}\pi\rho_s N r_0^3$ grams. When some amount of the solid dissolves, the number of particles N does not decrease but the particle radius becomes smaller. The remaining mass of solid, made of N particles of radius r, is $\frac{4}{3}\pi\rho_s N r^3$ grams. The mass lost from the solid and transferred to solution of volume V gives concentration in solution C

$$C = C_0 + \frac{4\pi\rho_s N}{3V}(r_0^3 - r^3) \qquad (\text{g cm}^{-3}) \qquad (5.73)$$

where C_0 is the initial concentration in solution ($C_0 \geqslant 0$). The equation explicitly relates concentration in solution to the solid particle size, for spherical particles dissolving uniformly at the surface. At an equilibrium or steady state, concentration reaches a value C_s and the particle radius becomes r_s. From the preceding equation, the particle size at saturation r_s is

$$r_s = \left(r_0^3 - \frac{3V(C_s - C_0)}{4\pi\rho_s N}\right)^{1/3} \qquad (\text{cm}) \qquad (5.74)$$

If the final concentration is higher than the initial ($C_s > C_0$), the particles dissolve and $r_s < r_0$. If precipitation takes place from a supersaturated solution ($C_s < C_0$), the particles obviously grow and $r_s > r_0$.

To obtain an explicit relationship between the particle radius, time, and other solution parameters, equation 5.73 can be combined with the first-order dissolution equation 5.71, giving

$$r^3 = r_0^3 - \frac{3V(C_s - C_0)(1 - e^{-kt})}{4\pi\rho_s N} \qquad (\text{cm}^3) \qquad (5.75)$$

It can be verified by inspection that at time $t = 0$, the radius is r_0 and at time $t \to \infty$, the radius tends to the steady-state value r_s.

In a closed system, the rate of dissolution can be measured either as the rate of change of concentration in solution dC/dt or as the rate of decrease in particle radius $-dr/dt$. The latter is obtained by differentiation of equation 5.75 as

$$\frac{dr}{dt} = -\frac{kV(C_s - C_0)e^{-kt}}{4\pi r^2 \rho_s N} \qquad (\text{cm sec}^{-1}) \qquad (5.76)$$

The solid surface area $S=4\pi r^2 N$ decreases with time. In terms of the laminar layer model of dissolution, discussed in Section 5.5.3, the dissolution rate parameter k can remain constant only if the surface-area/laminar-layer-thickness ratio is also constant. This means that the layer thickness $h(t)=D/k_s$ should decrease with time, as the particles become smaller, provided k remains constant. Substitution from equation 5.68 for the coefficient k in 5.76 gives

$$\frac{dr}{dt}=\frac{D(C_s-C_0)e^{-kt}}{\rho_s h(t)} \tag{5.77}$$

which defines theoretically the laminar layer thickness $h(t)=D/k_s$ by the measurable change in particle size dr/dt and time t.

Constant Rate of Dissolution. The proportions of the dissolved and suspended materials in rivers, discussed in Section 5.4.1, were related to the changes in the mean volume and linear dimensions of solid particles undergoing dissolution. The main premise of the computation was that crustal materials in the weathering process behave as individual particles dissolving at a uniform rate over their entire surface. This and other dissolution schemes are also discussed in Section 5.5.2. In this section is given a simple dissolution model, based on a constant rate of mass loss per unit area of the solid, and the results of the model are compared with the rates of chemical weathering on a global scale.

Dissolution taking place by removal of small quantities of solid from a large surface does not, in the first approximation, affect the surface area. A small volume dV removed by dissolution is equal to the product of the surface area S and an increment of the linear dimension dr perpendicular to the surface: $dV=S\,dr$. The mass dM removed by this process is $\rho_s\,dV$ where ρ_s is the solid density, or

$$dM=\rho_s S\,dr \quad\text{(g)} \tag{5.78}$$

Denoting the rate of mass loss per unit area of the solid surface ε' (g cm^{-2} sec^{-1}), another equation can be written for the mass lost during a time interval dt,

$$dM=-\varepsilon' S\,dt \quad\text{(g)} \tag{5.79}$$

Division of equation 5.78 by 5.79 gives the rate of decrease in the linear dimension of the solid dr/dt

$$\frac{dr}{dt}=-\frac{\varepsilon'}{\rho_s}\equiv-\varepsilon \quad\text{(cm sec}^{-1}\text{)} \tag{5.80}$$

When dissolution is diffusion controlled, as given by equation 5.69, then the rate of mass loss by the solid is equal to the flux of the dissolved material into solution, which is

$$\frac{dM}{dt} = -\frac{DS(C_s - C)}{h} \qquad (\text{g sec}^{-1}) \qquad (5.81)$$

From equations 5.78 through 5.81 the linear rate of dissolution ε can be written in terms of other parameters as

$$\varepsilon = \frac{D(C_s - C)}{\rho_s h} \qquad (\text{cm sec}^{-1}) \qquad (5.82)$$

If dissolution takes place into an undersaturated solution of constant concentration C, then the difference $C_s - C$ is constant, and linear rate of dissolution can also be constant.

Order of magnitude estimates of ε can be made from the numerical values of the other parameters in equation 5.82, characteristic of natural water systems. The molecular diffusion coefficient for different chemical species in dilute solutions is $D \simeq 1 \times 10^{-5}$ cm^2 sec$^{-1} \simeq 300$ cm^2 yr^{-1}. The density of solids $\rho_s \simeq 3$ g cm^{-3} and the thickness of the laminar layer $h \simeq 3 \times 10^{-3}$ cm. If the solution is close to an equilibrium with the solid, the difference in concentration $C_s - C$ can be very small and the rate of dissolution ε can also be a small number. Far from an equilibrium, the concentration difference is approximately $C_s - C \simeq C_s$. As an example, the mean linear rate of dissolution ε will be computed for silica, on the basis of the solubilities C_s of some silicate minerals and dissolved SiO_2 concentrations C in surface and subsurface waters. Because of the contact of ground waters with the aquifer rocks and because of their slow rates of flow, concentrations of dissolved species in ground waters are likely to be closer to the solubility values than the concentrations in streams and surface runoff. Mean concentrations of dissolved silica in rivers and subsurface waters flowing through different rock types are listed in Table 5.7. Concentrations in rivers are lower than in subsurface waters that are in contact with silica-rich rocks and sediments. Waters in volcanic areas contain often several hundred mg SiO_2 l^{-1}, and for temperatures less than 100°C these waters are supersaturated with respect to amorphous silica (see the SiO_2 solubility diagram in Figure 8.7). Solubilities of silicate minerals, such as quartz, kaolinite, feldspars, micas, and amorphous silica, in solutions near the neutral pH and at temperatures of the earth's surface, are represented by values in the range between about 6 and 140 mg SiO_2 l^{-1}. For the solubilities C_s of the order of several tens of mg SiO_2 l^{-1}, and concentrations in water C of the same order of magnitude (Table 5.7), the concentration difference $C_s - C$ can vary from nil up to a few or even several tens of

Table 5.7 Concentrations of Dissolved Silica (SiO_2) in Surface and Subsurface Waters

Water type	Dissolved SiO_2 (mg kg^{-1})
Surface	
Rivers, world average[a]	13
Laurentian Great Lakes and smaller lakes in the drainage basin[a]	4
Basaltic soils (tephra)[b]:	
Well-weathered (no glass)	3
Slightly-weathered (glass present)	18
Subsurface[c]:	
Granitic rocks	37
Mafic rocks	41
Metamorphic rocks (slate, schist, gneiss, greenstone)	23
Sandstones (quartzitic and impure)	23
Unconsolidated sand and gravel	31
Clays and shales	28
Limestones and dolomites	13
Volcanic areas[d]	(260)

[a]Livingstone (1963).
[b]Hay and Jones (1972).
[c]White et al. (1963).
[d]Concentration range large, from < 100 to > 600 ppm SiO_2.

mg l^{-1}, or between 10^{-6} and 10^{-5} g SiO_2 cm^{-3}. Using the value $C_s - C \leqslant 3 \times 10^{-6}$ g cm^{-3} together with the values of D, ρ_s, and h given previously, the mean rate of dissolution of silica from crustal rocks is, from equation 5.82,

$$\varepsilon \leqslant \frac{3 \times 10^2 \times 3 \times 10^{-6}}{3 \times 3 \times 10^{-3}} = 0.1 \text{ cm yr}^{-1}$$

Linear dissolution rates one order of magnitude higher or lower than 0.1 cm yr^{-1} can result from some variation in the parameters used. The value of $\varepsilon = 10^{-1}$ cm yr^{-1} is about 100 times higher than the mean global rate of chemical weathering of 10^{-3} cm yr^{-1} (Section 5.4.1). The lower rate of weathering, based on the dissolved load of rivers and river discharge, may reflect the fact that rains do not wet the earth's surface continuously, but fall on the average during only $\frac{1}{10}$ of the year every year. However, the effect of wetting and drying of the surface may be offset by somewhat higher solubilities of the dry surfaces: concentration of the H^+ ions in the adsorbed water layer at the surface of solid can be higher than in the bulk solution, such that in the process of drying the surface may undergo changes that result in a higher solubility at the beginning of a new immersion episode. An alternative reason behind the lower rate of chemical denudation may reside in the fact that only some fraction of the total surface of the solids exposed to waters reacts and dissolves. Poisoning of surfaces by adsorption of metals and other ionic species, and the formation of hydrous iron-oxide coatings can reduce the reactive surface area and the rates of dissolution, when these are estimated as averages.

Dissolution of a Particle-Size Spectrum.   The rate of decrease in linear dimensions of a dissolving solid, given in equation 5.80 as $-dr/dt = \varepsilon$ means for a spherical particle that its radius decreases linearly with time,

$$r = r_0 - \varepsilon t \quad \text{(cm)} \tag{5.83}$$

where r_0 is the initial particle radius. The time to complete dissolution is

$$t = \frac{r_0}{\varepsilon} \quad \text{(sec)} \tag{5.84}$$

and the equation shows that smaller particles dissolve faster. At the rate of radial dissolution $\varepsilon = 1 \times 10^{-3}$ cm yr^{-1}, particles of radius $r_0 = 10$ μm disappear within a year, and the radii of the bigger particles are reduced by the same amount.

A particle-size distribution at time $t = 0$ is given by a general relationship of the type $dN = f(r_0)dr_0$ (number of particles). At any fixed time $t > 0$, the size distribution is, using equation 5.83, $dN = f(r + \varepsilon t)dr$. For a Pareto particle-size spectrum discussed in Section 5.2.3, the equation

$$dN = A(r + \varepsilon t)^{-b} dr \quad \text{(particles)} \tag{5.85}$$

describes the change in dN/dr as a function of time. At time $t = 0$, the equation is identical with the form $dN = Ar^{-b}dr$. Figure 5.11 shows schematically how the shape of a size distribution curve changes with time, when dissolution takes place at the particle surface and at a constant rate. Because the smaller particles dissolve completely before the bigger particles, the values of dN/dr decrease stronger for the smaller particle sizes. The size distribution curve is flat in the region where $r \ll \varepsilon t$, and the near horizontal section of the curve extends further to the right with increasing time.

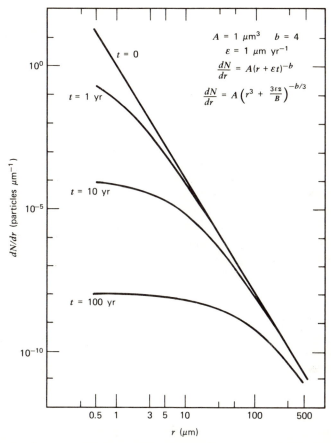

$A = 1\ \mu m^3 \qquad b = 4$

$\varepsilon = 1\ \mu m\ yr^{-1}$

$\dfrac{dN}{dr} = A(r + \varepsilon t)^{-b}$

$\dfrac{dN}{dr} = A\left(r^3 + \dfrac{3\varepsilon z}{B}\right)^{-b/3}$

Figure 5.11 Change in the shape of particle-number distribution undergoing dissolution at the rate ε. The equation in z is discussed in Section 6.2.4.

5.5.5 Reactive Surface Layer

Dissolution of Magnesium Silicates.   Leaching of some of the components of the bulk solid phase, chemical reaction, or exchange with solution are the main mechanisms behind the dissolution scheme shown in Figure 5.9d, where a layer of composition different from the bulk solid forms on the surface. In dissolution experiments, the presence of a surface layer, of composition different from the bulk solid, is usually revealed by the proportions of the dissolved components in solution that differ from their proportions in the bulk. If chemical species go into solution in proportions that are different from those of the bulk solid phase, dissolution is called incongruent. The composition of a surface layer on the solid can be computed from the concentrations in solution and the mass balance between the solid and dissolved components. A general scheme of an incongruent dissolution reaction of a solid of composition $A_m B_n$ can be written as

$$A_m B_n(s) \rightarrow xA(aq) + yB(aq) + A_{m-x}B_{n-y}(s)$$

The composition $A_m B_n$ and mass of the original solid, and the masses of aqueous A and B in solution make it possible to evaluate the mass of the new phase $A_{m-x}B_{n-y}$ by difference.

Dissolution of three magnesium silicate minerals at room temperature—serpentine [idealized composition $Mg_3Si_2O_5(OH)_4$], forsterite (an olivine of composition $Mg_{0.91}Fe_{0.01}SiO_4$), and enstatite (a pyroxene of composition $Mg_{0.925}Fe_{0.075}SiO_3$)—has been described as taking place in three distinct steps (Luce et al., 1972).

In the first dissolution step, the H^+ ions in solution exchange with the Mg^{2+} ions on the solid surface. Mg^{2+} and SiO_2 are released to solution in proportions different from their proportions in the bulk solid phase (more Mg^{2+} than SiO_2 is leached). This stage is short, lasting only a few minutes.

In the next stage, which lasts up to a few days, Mg^{2+} and SiO_2 continue to be released in solution in proportions different from the stoichiometry of the bulk mineral. Migration of Mg^{2+} and SiO_2 takes place through an altered layer at the solid surface.

In the third stage, dissolution of the surface of the solid takes place. As the surface layer contains more Si than Mg relative to their stoichiometric proportions, silica concentration in solution rises faster than magnesium concentration, with a general tendency toward congruent dissolution.

In general, when mineral particles from soils are exposed to water, ions are released initially in nonstoichiometric proportions, but with increasing dissolution time, the released ions approach their stoichiometric ratios in the bulk solid (Correns and von Engelhardt, 1939; McClelland, 1950).

Experiments on dissolution of minerals in closed systems often show that the release of dissolved species into solution during the first several hours to days of dissolution follows the so-called parabolic rate law

$$\frac{dC}{dt} = k_p t^{-1/2} \qquad (g\ cm^{-3}\ sec^{-1}) \qquad (5.86)$$

where k_p is a reaction rate parameter of dimensions $g\ cm^{-3}\ sec^{-1/2}$. By integration, concentration in solution C increases with the square root of time,

$$C = C_0 + 2k_p t^{1/2} \quad (\text{g cm}^{-3}) \tag{5.87}$$

The term C_0 has the significance of concentration developing within the short time of the first dissolution stage, when ion exchange takes place. A plot of C against $t^{1/2}$ produces a straight line with the intercept C_0 at $t=0$. Concentration cannot increase indefinitely according to the parabolic rate law, and other dissolution mechanisms must become effective at the later stage of dissolution when concentration C tends to a steady value as time increases. For dissolution of silicates according to the parabolic rate law at higher temperatures, the dissolution rate parameters have been computed by Helgeson (1971). For three magnesium silicates at room temperature, the rate parameters are summarized in Table 5.8. Because the rates of dissolution depend on the solid/solution mass ratio, or on the surface area of the solid within a given solution volume, the rate constants k_p are often reported in units of mass per unit area of solid. A rate constant k_{ps} can be

Table 5.8 Initial Exchange Quantities ($C_0 V / S$ mol cm^{-2}) and Parabolic Dissolution Rate Parameters (k_{ps}) for Three Magnesium Silicate Minerals. Initial Concentration C_0, Solution Volume V, Solid Surface Area S, and Rate Constant k_{ps} are Defined in Equations 5.87 and 5.88. Experimental Conditions: 1.0 g solid, 1 liter solution, 25°C[a]

Mineral, density, specific surface area	Initial pH	$C_0 V / S$ (10^{-9} mol cm^{-2})		$2k_{ps}$ (10^{-11} mol cm^{-2} sec$^{-1/2}$)	
		Mg	Si	Mg	Si
Enstatite	3.2	4.9	7.3	7.8	2.4
$\rho_s = 3.29$ g cm^{-3}	5.0	2.3	3.0	2.6	2.6
$S_s = 576$ cm^2 g^{-1}	7.0	0.09	0.2	1.6	1.4
	9.6	1.0	0.57	1.7	2.4
Forsterite	3.2	20.0	5.0	180.0	20.0
$\rho_s = 3.13$ g cm^{-3}	5.0	9.0	2.0	61.0	3.6
$S_s = 445$ cm^2 g^{-1}	7.0	2.0	0.7	120.0	2.9
	9.6	4.0	0.6	67.0	0.3
Serpentine	3.2	20.0	0.0	27.0	6.3
$\rho_s = 2.62$ g cm^{-3}	5.0	5.8	0.0	2.2	6.8
$S_s = 914$ cm^2 g^{-1}	7.0	3.4	0.2	1.9	6.8
	9.6	1.0	0.2	1.4	0.7

[a]Luce et al. (1972).

defined as

$$k_{ps} = k_p \times \frac{V}{S} \qquad (\text{g cm}^{-2} \sec^{-1/2}) \qquad (5.88)$$

where V is the volume of solution (cm^3) and S is the solid surface area (cm^2). The rate constant k_{ps} is independent of the solid/solution volume ratio, but it depends, like k_p and the rate of dissolution dC/dt, on such external conditions as the temperature and the H^+-ion activity in solution.

Diffusion in the Reactive Layer. In the model of the reactive surface layer, the difference in composition between the bulk mineral surface at the lower boundary of the layer (the p-q boundary in Figure 5.9d) and the outer boundary at the layer-solution interface requires that dissolving species migrate across the surface layer. A measure of concentration change of a migrating species across the surface layer is the difference between its concentration in the fresh bulk phase ρ_1 and its concentration at the outer surface of the reactive layer ρ_2. If transport across the surface layer takes place by diffusion, the following two mass balance equations express the rate of concentration change in solution dC/dt:

$$\frac{dC}{dt} = \frac{(\rho_1 - \rho_2)}{V} \frac{dh}{dt} \qquad (5.89)$$

$$\frac{dC}{dt} = \frac{D(\rho_1 - \rho_2)S}{hV} \qquad (5.90)$$

where ρ_1 and ρ_2 are concentrations of the migrating component in units of mass per cubic centimeter of solid, S is the reactive solid surface area (cm^2), V is the volume of solution (cm^3), D is the diffusion coefficient within the reactive layer ($\text{cm}^2 \sec^{-1}$), and h is the layer thickness (cm). As dissolution progresses the thickness of the surface layer h increases, although as long as it is small in comparison to the overall diameters of the solid particles, the reactive surface area S remains virtually constant. Equation 5.89 is a mass conservation condition for the material leached out of the surface layer, and is independent of the process by which the material migrates to the solution. Equation 5.90 assumes a diffusional flux and a linear concentration gradient across the surface layer. From the two equations, the rate of growth of the surface layer dh/dt is

$$\frac{dh}{dt} = \frac{D}{h} \qquad (\text{cm} \sec^{-1}) \qquad (5.91)$$

from which it follows, by integration, that the thickness of the layer increases as the square root of time,

$$h = (2Dt)^{1/2} \quad \text{(cm)} \tag{5.92}$$

The parabolic or square-root growth law as given above also applies to the growth of oxide films on metals, where reaction is controlled by the rate of oxygen diffusion through the layer. Differentiating the preceding equation with respect to t, we obtain the rate of growth of the layer as a function of time,

$$\frac{dh}{dt} = \left(\frac{D}{2}\right)^{1/2} t^{-1/2} \quad \text{(cm sec}^{-1}) \tag{5.93}$$

The latter result can be substituted for dh/dt in equation 5.89, giving a parabolic rate law equation for dissolution rate in a different form,

$$\frac{dC}{dt} = \frac{D^{1/2}S(\rho_1 - \rho_2)}{2^{1/2}V} t^{-1/2} \tag{5.94}$$

The coefficient of $t^{-1/2}$ in the preceding equation can be equated to the rate parameter k_p in equation 5.86 to derive an explicit relationship for the diffusion coefficient within the surface layer,

$$D = \frac{2(k_p V/S)^2}{(\rho_1 - \rho_2)^2} \quad \text{(cm}^2 \text{ sec}^{-1}) \tag{5.95}$$

or, in terms of the rate parameter k_{ps},

$$D = \frac{2k_{ps}^2}{(\rho_1 - \rho_2)^2} \quad \text{(cm}^2 \text{ sec}^{-1}) \tag{5.96}$$

To compute the value of the diffusion coefficient D, the reaction rate parameter k_{ps} and the composition difference across the surface layer $\rho_1 - \rho_2$ must be known. The lower limit of D corresponds to the case of $\rho_1 - \rho_2 \simeq \rho_1$, that is, when concentration of the migrating species at the solid-solution interface is zero. Under this assumption, Luce et al. (1972) reported the diffusion coefficients of magnesium and silicon in the surface layer of dissolving magnesium silicates as $D_{Mg} = 10^{-13} - 10^{-18}$ cm^2 sec^{-1} and $D_{Si} = 10^{-15} - 10^{-18}$ cm^2 sec^{-1} (see also Table 3.9).

The inferred thickness of the reactive surface layer that can develop within a day ($t \simeq 10^5$ sec), when the diffusion coefficient for migration through the layer is about $D = 10^{-15}$ cm^2 sec^{-1}, is of the order of $h = (2 \times 10^{-15} \times 10^{-5})^{1/2} \simeq 0.1$ μm.

5.5.6 Dissolution of Feldspars

An early study of Correns and von Engelhardt (1939; also Correns, 1961) on dissolution of potassium feldspar (KAlSi$_3$O$_8$) in aqueous solutions at different pH values has shown that the release of alkali ions takes place initially faster than the release of silica. This can be interpreted as evidence for the formation of a surface reactive layer of composition different from the bulk solid phase, similar to the case in Section 5.5.5 of magnesian silicates. With continued dissolution, the proportions of the dissolving constituents tend to their stoichiometric proportions in the bulk solid. The ion exchange between the K$^+$ ions on the feldspar surface and H$^+$ ions in solution is the first stage of a reaction between feldspar and water (Garrels and Howard, 1959). In a more recent study of the rates of dissolution of feldspars in water, at 25°C and 1 atm pressure of CO$_2$, Busenberg and Clemency (1976) showed that the reaction mechanisms change as dissolution progresses. Potassium feldspar (KAlSi$_3$O$_8$) in the forms of microcline and adularia, and six members of the NaAlSi$_3$O$_8$–CaAl$_2$Si$_2$O$_8$ solid solution series (albite, oligoclase, andesine, labradorite, bytownite, and anorthite) exhibit a pattern of dissolution behavior that can be divided into four stages: the first stage of fast surface exchange, the second stage of concentration increase as some power function of time, the third stage of concentration rise according to the parabolic dissolution rate, and the fourth stage of a linear increase in concentration with time. The dependence on time of concentrations of cations and silica in solution, and the approximate duration of each of the four dissolution stages are as follows:

First stage: Surface solution, H$^+$-cation exchange (<3 min)
Second stage: Concentration rises as a low-power function of time; the stage lasts up to 50 hr,

$$C = k_e t^n, \qquad 0.03 < n < 0.22 \tag{5.97}$$

Third stage: Increase in concentration obeying the parabolic rate law; duration up to 20 days following the second stage,

$$C = C_0 + 2k_p t^{1/2} \tag{5.98}$$

The latter equation is identical to equation 5.87.
Fourth stage: Concentration increases linearly with time; duration up to 30 days following the third stage,

$$C = C_0 + k_0 t \tag{5.99}$$

The rate constants for the individual minerals in the second through fourth dissolution stages are listed in Table 5.9. During the first two stages, more cations than silica are released in solution, resulting in a cation-deficient layer at the feldspar surface. Within the albite-anorthite series, the rate constants for the release of sodium and calcium increase with an increasing mole fraction of each cation in the solid. The faster release of cations continues during the third, parabolic rate dissolution, stage. When dissolution is continuous, the value C_0 at the beginning of each dissolution stage is the concentration value attained at the end of the preceding stage. A computational example of silica release during dissolution of oligoclase is given below.

Example. Compute SiO_2 concentration in solution, in a system closed to water flow, in contact with oligoclase. Specific surface area of the solid $S = 1$ m^2 g^{-1}, solid concentration is 50 g l^{-1}. After the stage of the fast surface exchange, the solubility of silica is given by equations 5.97 through 5.99. For the dissolution process described by equation 5.97, the dissolution rate constant k_e is, from the value of $k_{es} = 7.6 \times 10^{-11}$ mol cm^{-2} sec^{-n} given in Table 5.7,

$$k_e = (7.6 \times 10^{-11} \text{ mol cm}^{-2} \text{ sec}^{-n})(10^4 \text{ cm}^2 \text{ g}^{-1})(50 \text{ g l}^{-1})$$

$$= 3.8 \times 10^{-5} \text{ mol l}^{-1} \text{ sec}^{-n}$$

Allowing this dissolution stage to last 30 hr ($t = 1.08 \times 10^5$ sec), the concentration of SiO_2 at the end of 30 hr is, by equation 5.97 with the power exponent value $n = 0.153$,

$$C = 3.8 \times 10^{-5} \times (1.08 \times 10^5)^{0.153} = 2.23 \times 10^{-4} \text{ mol l}^{-1} = 13.4 \text{ mg l}^{-1}$$

The next stage, parabolic rate dissolution, is characterized by the rate constant value of $k_{ps} = 5.83 \times 10^{-13}$ mol cm^{-2} sec$^{-1/2}$. As in the preceding computation, the rate constant k_p is

$$k_p = 5.83 \ 10^{-13} \times 10^4 \times 50 = 2.92 \times 10^{-7} \quad \text{mol l}^{-1} \text{ sec}^{-1/2}$$

Allowing the parabolic dissolution stage to last 20 days ($t = 17.28 \ 10^5$ sec), we use equation 5.98 with the value of initial concentration $C = 2.23 \times 10^{-4}$ mol l^{-1}, attained at the end of the preceding stage. Thus

$$C = 2.23 \times 10^{-4} + 2.92 \times 10^{-7} \times (17.28 \times 10^5)^{1/2}$$

$$= 6.07 \times 10^{-4} \quad \text{mol l}^{-1}$$

$$= 36.4 \quad \text{mg l}^{-1}$$

Table 5.9 Dissolution Rate Parameters of Feldspars in Water, at 25°C, 1 atm CO_2 Pressure.[a] Rate Constants in Equations 5.97 through 5.99 are $k_i = k_{is}S/V$, where k_{is} is k_{es}, k_{ps} or k_{0s} Given in the Table

Mineral, specific surface area S_s, composition[b]	Power law, equation 5.97 k_{es} (10^{-10} mol cm^{-2} sec^{-n})[c]				Parabolic, equation 5.98 $2k_{ps}$ (10^{-13} mol cm^{-2} sec$^{-1/2}$)				Linear, equation 5.99 k_{0s} (10^{-16} mol cm^{-2} sec^{-1})			
	Na	Ca	K	Si	Na	Ca	K	Si	Na	Ca	K	Si
Albite (0.83 m² g^{-1}) Ab$_{0.989}$An$_{0.001}$Or$_{0.010}$	11.8 (0.059)			0.34 (0.223)	2.98			9.25	25.4			4.41
Oligoclase (1.01 m² g^{-1}) Ab$_{0.696}$An$_{0.240}$Or$_{0.064}$	6.18 (0.059)	3.90 (0.053)		0.76 (0.153)	2.81	1.10		5.83	1.67	0.89		2.60
Andesine (1.49 m² g^{-1}) Ab$_{0.495}$An$_{0.427}$Or$_{0.078}$	2.41 (0.030)	—		0.30 (0.199)	1.14	—		4.05	0.86	—		1.40
Labradorite (1.04 m² g^{-1}) Ab$_{0.445}$An$_{0.530}$Or$_{0.025}$	3.87 (0.039)	3.58 (0.057)		0.72 (0.157)	1.02	1.29		3.27	0.72	1.09		1.26
Bytownite (1.14 m² g^{-1}) Ab$_{0.222}$An$_{0.774}$Or$_{0.004}$	2.31 (0.061)	4.41 (0.064)		0.84 (0.158)	1.34	3.38		3.47	0.54	1.29		1.07
Anorthite (1.84 m² g^{-1}) Ab$_{0.058}$An$_{0.940}$Or$_{0.002}$	1.43 (0.027)	3.63 (0.093)		0.96 (0.173)	0.34	3.36		4.01	0.32	2.18		1.32
Orthoclase (1.52 m² g^{-1}) Ab$_{0.203}$An$_{0.002}$Or$_{0.795}$	1.49 (0.087)		5.43 (0.066)	0.30 (0.134)	1.49		4.19	3.78	0.66		2.70	1.68
Microcline (1.07 m² g^{-1}) Ab$_{0.248}$An$_{0.000}$Or$_{0.752}$	1.85 (0.067)		5.25 (0.075)	0.26 (0.164)	1.16		2.55	3.34	0.55		1.27	1.52

[a] Busenberg and Clemency (1976).
[b] Ab=NaAlSi$_3$O$_8$; An=CaAl$_2$Si$_2$O$_8$; Or=KAlSi$_3$O$_8$.
[c] The value of exponent n is given in parentheses after the value of k_{es}.

In the next, linear stage, the dissolution rate constant in Table 5.7 is $k_{0s} = 2.60 \times 10^{-16}$ mol cm^{-2} sec^{-1}, and the rate constant k_0 is

$$k_0 = 2.60 \times 10^{-16} \times 10^4 \times 50 = 1.30 \times 10^{-10} \quad \text{mol l}^{-1}\text{ sec}^{-1}$$

Allowing the linear-rate dissolution to continue for 30 days (25.92×10^5 sec), we use equation 5.99 with the initial concentration $C = 6.10 \times 10^{-4}$ mol l^{-1}, from the end of the preceding stage. The new concentration value is

$$C = 6.07 \times 10^{-4} + 1.30 \times 10^{-10} \times 25.92 \ 10^5$$

$$= 9.44 \times 10^{-4} \quad \text{mol l}^{-1}$$

$$= 56.6 \quad \text{mg l}^{-1}$$

The main cation in oligoclase is sodium. For the oligoclase of composition given in Table 5.7, the atomic ratio Na/Si is approximately $1/4 = 0.25$. The differences between the dissolution rates of the individual components of feldspars are responsible for the concentration ratios in solution that differ from those of the bulk solid. For Na, using the computational procedure shown above, with the data from Table 5.7, the Na concentrations at the end of each dissolution stage are, consecutively, 6.12×10^{-4}, 7.97×10^{-4}, and 10.13×10^{-4} mol l^{-1}. The Na/Si ratios in solution are 2.7 (after 30 hr), 1.3 (after 21 days), and 1.07 (after 51 days). The higher ratios show that sodium is removed from the solid preferentially to silicon, and although the Na/Si ratio decreases with time, even after almost two months of dissolution it is higher than in the bulk solid, reflecting the faster removal of Na from the solid surface.

The next question, how thick the leached surface layer is, can be answered using the results of the preceding computation. Initially, 50 g of oligoclase was in contact with 1 liter of water, and fifty days later, about 100 mg of dissolved solids are in solution (the amounts of 57 mg SiO$_2$ and 23 mg sodium were computed previously in this example, such that the mass of 100 g of total solids is an orientational figure). Thus the fraction of the original solid that dissolved is 0.1 g/50 g $= 0.002$, or 0.2%. A mass balance between the material in solution (M_{aq}), in the original solid (M_0), and in the residual solid (M_{res}) is

$$M_{aq} = M_0 - M_{res}$$

For spherical particles dissolving at the surface, the ratio of the remaining to the original mass is equal to the ratio of the cubes of the radii:

$M_{res}/M_0 = (r/r_0)^3$. Thus

$$\left(\frac{r}{r_0}\right)^3 = 0.998$$

The fractional decrease in the radius of the fresh feldspar grains is

$$1 - \frac{r_0}{r} = 0.0007 \quad \text{or} \quad 0.07\%$$

For particles of radius $r_0 = 10\ \mu m$, the decrease translates into the thickness of the leached layer of 70 Å.

5.5.7 Dissolution, Supersaturation, and Precipitation

Rise and Decline of Concentration in Solution. The dissolution and leaching behavior of minerals and soils often show that concentrations of dissolved species rise for some period of time during the early stages of the experiment, and subsequently decline to some steady value over a comparable or longer period. In experiments on solubility of three different forms of silica—quartz, silica gel, and silica glass—Morey et al. (1962, 1964) have reported that the solids, tumbled or rotated in water in sealed polyethylene containers at 25°C, continued to dissolve during the period of between one and two years. Very high degrees of supersaturation with respect to quartz were observed in two experiments with this mineral. In one of the experiments, concentration of SiO_2 in solution eventually declined to about 6 mg l^{-1}, a value typical of the solubility of quartz at room temperature. In the case of silica glass, the solubility of which may be comparable to that of amorphous SiO_2 (about 120 mg l^{-1} at 25°C), concentration in solution rose to about 320 mg l^{-1} during the first 600 days, and subsequently declined to about 200 mg l^{-1} during the next 1000 days. With silica gel, concentration in solution rose initially to about 180 mg l^{-1} and declined to the saturation value of 120 mg l^{-1} in 500 days.

Soil suspensions in water, containing between 2:1 and 1:1 by weight of soil to water, also show that the SiO_2 concentration in solution rises during the first several days and subsequently declines (McKeague and Cline, 1963; Kennedy, 1971). An example of a rising and declining concentration of dissolved silica in a soil-water suspension is shown in Figure 5.12.

In a solid plus water system that has neither inflow nor outflow, a rise in concentration can only mean dissolution or desorption from the solids, and a decline in concentration is due to precipitation or adsorption. The fact that concentration in solution goes through a maximum with time implies not only a change in the mechanism of the dissolution reaction, but also

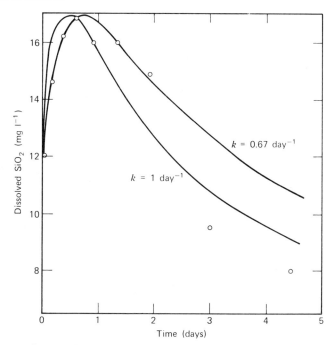

Figure 5.12   Release of silica to solution in aqueous suspensions of soils (data from Kennedy, 1971).

the presence of a source and sink. To identify the sink or detect a new solid phase that may be forming at the expense of the delining concentration in solution is generally difficult, for reasons of the mass relationships between the original and newly forming material. Clay minerals either release or adsorb silica (and cations) in solution, depending on the silica concentration, the pH value of the solution, and the extent of comminution to which the clay was subjected before it was brought in contact with solution (Siever and Woodford, 1973). The last three factors determining whether a mineral will dissolve or adsorb in a given solution reflect the conditions discussed in a different context in the preceding sections of this chapter. These are the degree of undersaturation or supersaturation with respect to a solid phase, exchangeability of the cations in the solid with the H^+ ions in solution, migration of the Si–OH groups from the solid into the solution, and mechanical-chemical alteration of the surface layer of the solid.

Mixed Kinetics of Concentration Peaks.   The mathematical models of the parabolic and zeroth order dissolution rates describe concentrations in solution as increasing with time, according to the $t^{1/2}$ or t law given in equations 5.87 and 5.99. Over long periods of time, such reaction mechanisms would lead to the physically impossible conditions of either total dissolution or an infintely high concentration or, in precipitation, zero concentration. When saturation is attained by a rising concentration from below, or by a declining concentration from above, the parabolic and linear mechanisms must either terminate abruptly (which means that they do not describe adequately an approach to saturation) or they must be replaced by other types of chemical reactions. The models of dissolution or precipitation that are based on the difference between the concentration at saturation and concentration in solution varying with time $(C_s - C)$, as given in equation 5.66, describe smooth concentration changes over the entire range of dissolution or precipitation, from the start at time $t = 0$ to the attainment of saturation C_s as time increases indefinitely $(t \to \infty)$. Solutions of equation 5.66, with the reaction rate parameter k_m constant, give concentration C as a function of time t that can either rise or decline, approaching the steady-state value C_s, but they do not describe those dissolution processes where concentration goes through a maximum. If a graph of concentration C as a function of time t shows a maximum, then the sign of the derivative dC/dt changes from the positive to the negative, whereas the sign of dC/dt in equation 5.66 is either one or the other during the entire time of the reaction.

A simple kinetic model that describes a change in concentration going through a maximum as a function of time is an equation that is the sum of the parabolic and first-order reaction equations 5.86 and 5.67,

$$\frac{dC}{dt} = k_p t^{-1/2} + k(C_s - C) \qquad (5.100)$$

The sum of the two rate terms implies that the two processes operate simultaneously. Another interpretation of the equation is that the first term $k_p t^{-1/2}$ describes dissolution of a solid phase, whereas the second term $k(C_s - C)$ describes precipitation of another phase. At the steady state, the terms dC/dt and $t^{-1/2}$ cancel because

$$\frac{dC}{dt} = 0 \qquad \text{and} \qquad t \to \infty$$

and the steady-state concentration is

$$C = C_s \qquad \text{at} \qquad t \to \infty$$

If dissolution and precipitation are reactions of the same order, written as either one of the two following equations:

$$\frac{dC}{dt} = k_{p,1}t^{-1/2} - k_{p,2}t^{-1/2}$$

$$\frac{dC}{dt} = k_1(C_{s,1} - C) - k_2(C - C_{s,2}) \qquad (C_{s,2} < C < C_{s,1})$$

then in each case, concentration C changes as a function of time t according to either the parabolic or first-order rate law, respectively, because the equations are mathematically identical to 5.86 and 5.67. A system where dissolution and precipitation take place according to the first-order rate law will be dealt with in more detail in Chapter 8.

In equation 5.100, the parameters k_p, k, and C_s are constants, and the equation can therefore be integrated using the standard method applicable to first-order linear differential equations. Integration between the limits $t = 0$ and $t = t$ gives

$$C = e^{-kt}\int_0^t \left(kC_s + k_p t^{-1/2}\right)e^{kt}\,dt + C_0 e^{-kt} \qquad (5.101)$$

where C_0 is concentration at time $t = 0$. Integration of the product $kC_s e^{kt}\,dt$ and rearrangement of the terms on the right-hand side of the equation give

$$C = C_s + (C_0 - C_s)e^{-kt} + k_p e^{-kt}\int_0^t t^{-1/2} e^{kt}\,dt \qquad (5.102)$$

The sum of the first two terms is the solution of the equation for the first-order chemical reaction, equation 5.71. The last term with the integral can be rewritten as

$$\frac{2k_p}{k^{1/2}} e^{-kt}\int_0^{(kt)^{1/2}} e^{kt}\,d(kt)^{1/2} \qquad (5.103)$$

The part of 5.103

$$e^{-kt}\int_0^{(kt)^{1/2}} e^{kt}\,d(kt)^{1/2}$$

is Dawson's integral. The function known as Dawson's integral $\text{Di}(x)$ is defined as

$$\text{Di}(x) = e^{-x^2}\int_0^x e^{y^2}\,dy \qquad (5.104)$$

where y is an integration variable and x is a positive number. Values of Dawson's integral can either be taken from existing tables or computed from series approximations (Miller and Gordon, 1931; Lerman, Mackenzie, and Bricker, 1975). Using the notation of Dawson's integral $Di(x)$, equation 5.102 takes the form

$$C = C_s + (C_0 - C_s)e^{-kt} + \frac{2k_p}{k^{1/2}} Di\left[(kt)^{1/2}\right] \qquad (5.105)$$

The ratio $2k_p/k^{1/2}$ has the dimensions of concentration. The two curves drawn through the measured SiO_2 concentrations in soil-water suspensions, plotted in Figure 5.12, were computed using equation 5.105 with the following values of the constants:

for one curve,

$$C_0 = 12 \text{ mg l}^{-1}, \quad C_s = 5 \text{ mg l}^{-1}, \quad k = 1 \text{ day}^{-1} = 1.16 \times 10^{-5} \text{ sec}^{-1},$$

for the other curve,

$$k = 0.67 \text{ day}^{-1} = 0.77 \times 10^{-5} \text{sec}^{-1}, \quad k_p/k^{1/2} = 7.5 \text{ mg l}^{-1};$$

The resulting equation for either curve is

$$C = 5 + 7e^{-kt} + 15 Di\left[(kt)^{1/2}\right]$$

For different values of the dimensionless product kt (between 0 and 16), the values of Dawson's integral are given in Table 5.10.

Equation 5.105 can be viewed as a kinetic model describing the contributions of two simultaneous processes: a first-order chemical reaction and a parabolic rate dissolution, represented by the Dawson integral term containing the rate constant k_p. The relative weight of the individual processes is shown diagrammatically in Figure 5.13 as a plot of concentration C against time t. The first-order dissolution terms plot as a curve approaching an equilibrium concentration C_s. The term containing Dawson's integral $Di(x)$ and representing the contribution due to the parabolic rate law plots as a curve that goes through a maximum and declines slowly as time increases. How much the parabolic law dissolution contributes to the overall process depends on the quotient of the rate constants $2k_p/k^{1/2}$: a value of the quotient of about 1 produces slight supersaturation, and concentration goes through a broad maximum at time

Table 5.10   Dawson's Integral $Di[(kt)^{1/2}]$ in Equation 5.105 for Selected Values of kt and $(kt)^{1/2}$

kt	$(kt)^{1/2}$	$Di[(kt)^{1/2}]$
0.0	0.0	0.0
0.01	0.1	0.0993
0.04	0.2	0.1947
0.09	0.3	0.2826
0.25	0.5	0.4244
0.49	0.7	0.5105
0.81	0.9	0.5407[a]
1.00	1.0	0.5381
1.21	1.1	0.5262
1.44	1.2	0.5073
1.69	1.3	0.4834
1.96	1.4	0.4565
2.25	1.5	0.4282
2.56	1.6	0.3999
2.89	1.7	0.3726
4.00	2.0	0.3013
6.25	2.5	0.2228
9.00	3.0	0.1783
12.25	3.5	0.1496
16.00	4.0	0.1293

[a]Maximum of $Di(x)$: $Di(0.9241)=0.5410$ (Gautschi, 1964).

corresponding to $kt \simeq 3$. A larger value of the quotient, such as $2k_p/k^{1/2} \simeq 10$, produces a pronounced concentration peak and a sixfold level of supersaturation, followed by a slow decline of concentration with increasing time, approaching the steady-state concentration C_s from above.

Combination of the parabolic and first-order reactions can lead to a slight supersaturation of the order of only of a few percent which, however, sustains itself for a relatively long time, measured as a number of reaction half-lives $1/k$. The occurrence of concentration peaks during dissolution is usually associated with suspensions of relatively high solid-to-liquid ratio. In the more dilute suspensions, the peak phenomenon is not observed or not observed clearly. For example, no concentration maxima of dissolved silica were observed in soil leaching experiments of McKeague and Cline (1963), cited earlier in this section, when the suspensions contained soil in the low ratios between 1:20 and 1:100. Experiments on dissolution of potassium feldspar in buffered solutions at different pH values, done by

Wollast (1967), show that concentration of silica rises in a manner resembling either curve 2 or $1+2$ in Figure 5.13. The duration of the dissolution experiments was about 300 hr. In suspensions of clays and zeolites in sea water, containing solids to liquid in the ratio of $1:200$, no concentration peaks were observed during the period of observation that lasted 8.5 yr. One exception was the zeolite mineral clinoptilolite: during the first 50 days of the experiment, concentration of silica in sea water rose from 0.02 to about 50 mg l^{-1}, and after 300 days it began to decline, reaching the level of about 35 mg l^{-1} after 8.5 yr (Mackenzie and Garrels, 1965; Mackenzie et al., 1967; Lerman, Mackenzie, and Bricker, 1975).

An explanation of the appearance of concentration peaks lies in the dependence of the rate constants k_p and k on the specific surface area of the solid. Each of the rate constants, as defined in equations 5.68 and 5.88, is the product of a constant that refers to a unit of area of the solid (k_{ps} and k_s) and the specific surface area of the solid S/V (cm^2 cm^{-3}): $k_p = k_{ps} S/V$ (mol cm^{-3} sec$^{-1/2}$) and $k = k_s S/V$ (sec^{-1}). The quotient of

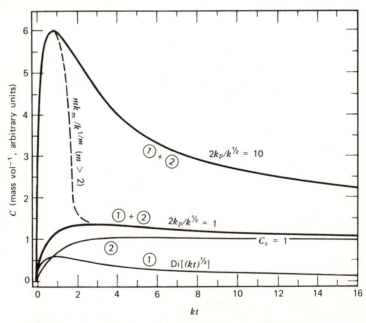

Figure 5.13 Computed concentration against time curves from the dissolution model based on mixed-order kinetics. Curve 1: Dawson's integral. Curve 2: first-order dissolution reaction. Other curves: sums of curve 1 premultiplied by the factor shown and curve 2.

the rate constants that appears in equation 5.105 is

$$\frac{2k_p}{k^{1/2}} = \frac{2k_{ps}}{k_s^{1/2}}\left(\frac{S}{V}\right)^{1/2} \quad (\text{mol cm}^{-3}) \qquad (5.106)$$

As the rate parameter k_{ps} and k_s are independent of the surface area of the solid S, equation 5.106 shows that the quotient $k_p/k^{1/2}$ increases with an increasing specific surface area. For a given volume of solution and a given mass of solid, the specific surface area S/V can be increased by comminution, making more particles of smaller radius. Alternatively, the specific surface area can be increased by having more mass within a given volume of solution, which corresponds to a higher concentration of the suspension.

Extension of the Mixed Kinetics Equation. The computed curves for concentration as a function of time in Figures 5.12 and 5.13 decline toward a steady-state concentration C_s fairly slowly, because of the slow decrease in the value of the Dawson integral function $\text{Di}[(kt)^{1/2}]$ with increasing time t (curve 1 in Figure 5.13). A faster approach to the steady state and a steeper decline of concentration with time can be effected by a model that differs slightly from equation 5.100: if the reaction described by the parabolic rate term $k_p t^{-1/2}$ (mol cm^{-3} sec^{-1}) proceeds instead at a rate directly related to a higher negative power of time t, then the rate term can be written as

$$k_m t^{-(1-1/m)} \quad (\text{mol cm}^{-3} \text{ sec}^{-1}) \qquad (5.107)$$

where

$$m \geqslant 2 \quad \text{or} \quad \frac{1}{2} \leqslant \left(1 - \frac{1}{m}\right) < 1$$

The combined rate equation then becomes

$$\frac{dC}{dt} = k_m t^{-(1-1/m)} + k(C_s - C) \qquad (5.108)$$

which differs from 5.100 only in the power exponent of t. Solution of 5.108 by the same procedure as outlined in equations 5.101 through 5.105 gives the final result in the following form:

$$C = C_s + (C_0 - C_s)e^{-kt} + \frac{mk_m}{k^{1/m}}e^{-kt}\int_0^{(kt)^{1/m}} e^{kt}d(kt)^{1/m} \qquad (5.109)$$

The product of e^{-kt} and the integral is a more general form of Dawson's integral defined in equation 5.104,

$$\mathrm{Di}(x) = e^{-x^m} \int_0^x e^{y^m} dy \qquad (5.110)$$

where y is an integration variable. The case of $m=2$ defines Dawson's integral that was dealt with in the preceding section and shown graphically as curve 1 in Figure 5.13. The property of Dawson's integral with $m>2$ is that the maximum is sharper and the curve descends more steeply to the right with the increasing values of x. Graphical tabulation of equation 5.110, in the form of curves for $\mathrm{Di}(x)$ as a function of x, for $m=2, 3, 4, 5$, and 6, is given by Gautschi (1964, p. 297).

An early transient stage of feldspar dissolution that was given in equation 5.97 proceeds according to the following reaction rate:

$$\frac{dC}{dt} = nk_e t^{n-1} \qquad (\mathrm{mol\ cm}^{-3}\ \mathrm{sec}^{-1})$$

The values of the power exponent n for this dissolution stage, listed in Table 5.9, are between 0.03 and 0.09 for Na, Ca, and K, and between 0.13 and 0.22 for Si. Thus the reaction rate is related to the following fractional negative powers of t: $t^{-0.93 \pm 0.04}$ and $t^{-0.83 \pm 0.04}$. All these are more negative powers than $t^{-0.5}$. A combination of such more negative powers of time dependence with the first-order reaction would produce a supersaturation peak and a subsequent decline of concentration to the steady-state level C_s at a faster rate, as shown schematically in Figure 5.13. However, the first-order and fractional time-power reactions have so far not been demonstrated as taking place simultaneously. Somewhat to the contrary, the first-order dissolution reaction, equation 5.71, may alone be sufficient to describe a long term dissolution process tending to a steady-state or saturation level. In dilute suspensions, the effects of the fractional order kinetics that are observable during the early stages of dissolution may be insignificant in comparison to the longer dissolution times and a slower approach to the steady state (Lerman, Mackenzie, and Bricker, 1975).

Dissolution rate constants k and k_s for a number of silicate minerals are listed in Table 8.2.

CHAPTER 6

Oceans and Lakes: Waters, Solids, and Solutes

The emphasis of this chapter is on the production and removal of particulate materials in the water column of lakes and oceans. Biogenic materials—organic matter and skeletal parts of planktonic organisms made of calcium carbonate and silica—are the main reactive solids that form in water and are in part returned to solution during their settling and after deposition on the bottom. The behavior of suspended materials is treated first from the point of view of the physical processes affecting it: settling, scavenging, and coagulation. Background information is given in the sections on the settling of particles of different shapes and on the settling velocities of biogenic particles. Experimental and observational data on the dissolution of carbonates and silicates are germaine to all models of regeneration of solids in ocean water. The goals of such models are to evaluate the rates of production or regeneration of different materials in the water column, and to estimate their roles in maintaining the chemical composition of waters and sediments.

6.1 THE CYCLE OF WATER

The oceans are the main storage reservoir of water, containing 97% of all water within the surface zone of the earth. The pool of continental waters is continuously supplied from the ocean through evaporation and precipitation. The cycle of evaporation and precipitation is a natural water desalination plant, on the performance of which depends all life on land. In human terms, the natural desalination process is inefficient and wasteful, with numerous losses on the delivery route of fresh water and uneven distribution of the product on the land surface. The natural storage facilities for fresh water on land are the continental ice sheets, pore spaces of sediments containing ground water, and the bigger depressions in the crust that contain lakes. Outside the ocean, ice is the largest reservoir of water on earth, although it contains only 2% of the water volume of the oceans (Table 6.1). The next largest reservoir is ground water, estimated to exist down to a depth of 4000 m. Subsurface waters include a variety of water types, from potable fresh waters to highly concentrated brines associated with salt beds and evaporites. The flow of ground water to the ocean is an estimated 5% of the river flow, and the estimated flow rate of 1600 km^3 yr^{-1} translates into an average residence time of 5000 yr for ground waters. This figure is heavily weighted by the deeper ground waters and by waters in subsurface basins that have internal drainage only. The shallower ground waters of coastal areas can have residence times of an order of months or years, and the figure of 5000 yr is only a mean estimate

Table 6.1 Major Global Reservoirs of Water, Interreservoir Fluxes, and Residence Times

Reservoirs and fluxes	Volume (10³ km³)	Residence time (yr)	Reservoir / Flux
Reservoir [a]			
A. Oceans	1,370,000	4×10^3	A/e
		4×10^4	A/a
		$(1-10) \times 10^7$	A/g
B. Ice caps and glaciers	29,200	1×10^4	B/e
C. Ground water (to depth of 4000 m)	8,350	5×10^3	C/f
D. Freshwater lakes	125	4	D/e
E. Saline lakes and inland seas	104	–	
F. Soil moisture and vadose water	67	1	F/b
		40	F/f
G. Atmosphere	13	0.03	G/(c+d)
H. Rivers (mean instantaneous volume)	1.25	—	

Flux [a]	(10³ km³ yr⁻¹)
a. Evaporation from world oceans	350
b. Evaporation from land	70
c. Precipitation on world ocean	320
d. Precipitation on land	100
e. Inflow to ocean from rivers and ice caps	32
f. Ground water flow to oceans	1.6
g. Flow of ocean water through hydrothermal zones of ocean ridges [b]	100–900

[a]Nace (1967), Menard (1974).
[b]Wolery and Sleep (1976).

of what is a very broad range of residence times with respect to outflow. In deep ocean sediments off the eastern coast of Africa, the salinities of pore waters were reported to decline with depth and fall below the salinity of the overlying ocean water (Gieskes, 1974). This feature was interpreted as an indication of deep flow of continental waters through the continental slope of the ocean floor. A theoretical model of recirculation of ocean waters through the spreading zones of the oceanic ridges gives estimates of the water renewal times between 100 and 900 million years.

The entry in Table 6.1 for saline lakes and inland seas is weighted by such large lakes as the Caspian Sea, with a salinity of about 1 g kg^{-1}. The highly concentrated lakes, where salt and other evaporitic minerals precipitate, account for only a small fraction of the total volume of saline waters. High concentrations of dissolved solids in many of the saline lakes are the result of evaporation of the surface and shallow ground waters that drain into a basin, where further evaporation continues to concentrate the brine. In the areas where evaporation during at least some seasons of the year is strong, preevaporation of waters that flow from the topographically higher areas is the first important step in the formation of continental land brines (Jones et al., 1977; Eugster and Hardie, 1978). Periods of draught cause the more or less saline lakes to shrink in volume, leaving behind a dry surface (playa) covered with evaporation residues of salts and containing concentrated brines in the pore space below the surface of the playa. Relatively small amounts of rain and surface runoff are needed to dissolve and wash into the lake much of the evaporitic mineral residue. In a lake without outflow, concentrations of dissolved solids can fluctuate with lake volume, but a general trend is an increase in concentration with time, because dissolved material is continuously brought in with inflow.

Most of the freshwater lakes are drained by rivers, and the mean residence time for water in lakes is 4 yr, as given in Table 6.1. This is also a measure of the mean residence time of water flowing on land. A lake open to inflow and outflow is essentially an inflated body of a river that is characterized by a longer residence time of water than the sections of the stream above and below the inflated stretch. At the upper end of the water residence time scale are such lakes as Lake Superior, with a residence time of about 200 yr, and Lake Tahoe, with a water residence time of about 700 yr.

The construction of dams on rivers and creation of large water reservoirs and artificial lakes is one type of human activity that affects the flow of water from land to the oceans. The total volume of man-made reservoirs is less than 5000 km^3, or less than 4% of the total volume of freshwater lakes. The potential uses of water in agriculture, in power generation as a coolant, and in industry can divert the river flow from the oceans and add large volumes of water to dry areas. Aside from the technological and economic questions of feasibility of such enterprises, three major problems are associated with the schemes of river diversion and use of their waters on land: first, a general ecological perturbation that would be caused by cutting off the flow of nutrients from land to the oceanic coastal waters; second, the potentially heavy pollution that may result from an extensive use of water by many industrial and agricultural consumers located in series along the route of the flowing water; and third, possible changes in climate that may develop in coastal areas if the mixing of river and ocean waters of somewhat different temperatures are interrupted.

6.2 SEDIMENTATION

6.2.1 Rates of Sedimentation in Oceans and Lakes

On the global scale, the bulk of sediments deposited in water are materials derived from land. The greater mass of the solid products of crustal weathering in comparison to the mass of dissolved materials carried by rivers, and the transport of solids by wind (Chapter 5) account for the large fraction of the terrigenous materials in oceanic and lake sediments. The materials entering the lakes and oceans in solution are the sources for the biogenic and authigenic components of sediments, forming through the biological production and chemical reactions. The relative proportions of the terrigenous and biogenic materials in sediments can vary widely. Strong biological productivity results in deposition of nearly pure calcareous and siliceous oozes on the ocean floor and in some lakes. Nearly pure terrigenous sediments accumulate on the sections of the ocean floor than lie on the path of supply of the major river and wind systems. The rates of deposition of the main types of sediments in the oceans are summarized in Table 6.2. The oceans of the Northern Hemisphere, and the Atlantic Ocean in particular, receive greater amounts of terrigenous materials because of the greater area taken up by continents. The global averages for the rates of sedimentation of the different sediment types, listed in the table, apply to the Quaternary. The rates of sedimentation, determined from the data of the Deep Sea Drilling Project (Davies et al., 1977) have been reported as increasing during the last 40 million years, from about 2–4 mm per 1000 yr to the recent values that are one order of magnitude higher. In near-shore and hemipelagic environments, the rates of sedimentation can be many times faster than in the pelagic ocean. For example, in the Santa Barbara Basin, off the coast of Southern California, sediments accumulate in the basin at depth 600 m, at the rate of about

**Table 6.2   Rates of Sedimentation
of Terrigenous and Biogenic Sediments in the Oceans**

Sediment type	mm per 1000 yr
Terrigenous [a]	
Wind transported (clays, framework silicates)	0.3–6.0
Glacial Antarctic and Arctic River input	0.5–23
(St. Lawrence, South American, and African Rivers)	
from continents to flanks of ridges	1–8
Biogenic [b]	
Calcareous ooze	3–60
Siliceous ooze	2–17

[a] Griffin et al. (1968).
[b] Berger (1974); Turekian (1965); Lisitzin (1974).

4 mm yr^{-1}. This rate is 100–1000 times faster than the mean rates for the pelagic ocean (Koide et al., 1972). Maps showing the geographic distribution of sediment thicknesses and rates of sedimentation in the oceans were published by Lisitzin (1974). The four most abundant clay minerals in oceanic sediments are kaolinite, illite, chlorite, and smectite, or montmorillonite. Maps of their occurrences and abundance on the floor of the world ocean were produced by Griffin et al. (1968). Kaolinite, illite, and chlorite in oceanic sediments have been attributed to transport by wind and rivers from the continents. Smectite is the product of diagenetic alteration of volcanic ash in sea water.

In lakes, the rates of sedimentation fall typically between 10^{-1} and 10^1 mm yr^{-1} (Krishnaswami and Lal, 1978; Bradbury, 1975; Kemp et al., 1976). The rates in the millimeters per year range are more common. The rates of sedimentation expressed in units of mass flux fall in the range 10^1–10^2 mg cm^{-2} yr^{-1}.

Man-made water reservoirs and lakes receive inflow from drainage areas that are usually large in comparison to the reservoir surface areas. Sedimentation in man-made reservoirs is sufficiently fast to create a problem of the reservoir silting and decrease in its water storage capacity. The rates of decrease in the reservoir capacity range from 0.15 to 3.5% yr^{-1}, averaged over periods of observation of 20 yr or less. The reservoirs of smaller volume show higher rates of sediment accumulation than the larger reservoirs. A rate of decrease in volume 1% yr^{-1}, in man-made lakes of mean depth 5–50 m, is equivalent to a mean rate of sedimentation between 50 and 500 mm yr^{-1}. The rates of several mm yr^{-1} to a few tens of mm yr^{-1} apply to the larger reservoirs which lose their water storage capacity due to silting at a slower rate. The rates of sediment loading in man-made reservoirs can be referred to a unit of area of the land drainage basin, giving a mean rate of land erosion. Such rates vary typically between 0.1 and 1 mm yr^{-1}, and these show that the rate of land erosion in areas affected by man is, as expected, much faster than the estimates of the global mean erosion rates cited in Chapter 5. In a series of reservoirs, those upstream can under certain circumstances act as sediment traps reducing the suspended load that flows into the downstream reservoirs (Glymph, 1973; Dendy et al., 1973; Cyberski, 1973).

6.2.2 Settling of Particles of Different Shapes

The Stokes Law.   A particle in a still fluid experiences a force due to gravity and a force due to buoyancy. The difference between the two is the net gravitational force F_G, which depends on the mass of the particle and the mass of fluid displaced by it, as in the equation

$$F_G = g\rho_s v_s - g\rho v_s = g v_s(\rho_s - \rho) \qquad \text{(g cm sec}^{-1}) \qquad (6.1)$$

where g is the acceleration due to the force of gravity, v_s is the volume of the particle, ρ_s is its density, and ρ is the density of the fluid. If the density of the particle is greater than the density of the fluid, the particle obviously sinks. As the particle sinks through the fluid, the flow of the fluid around it creates a resisting or drag force F_D. When the net gravitational and resisting forces become equal, the particle has reached a terminal velocity that is steady. The condition of the equality of the forces $F_G = F_D$ determines the settling velocity of a particle in a given fluid and under given environmental conditions. The drag force F_D acting on a particle generally depends on the viscosity of the fluid η, the settling velocity U_s, the volume of the particle v_s, and its shape, expressible through a parameter α to be defined below. The drag force equation

$$F_D = f(\eta, U_s, v_s, \alpha) \qquad \text{(g cm sec}^{-1}) \qquad (6.2)$$

can be written explicitly for many different shapes, and the settling velocity U_s can be determined from the condition of the equality of forces $F_G = F_D$. Explicit equations for F_D and for U_s for particles of different shapes are given in Appendix C.

The Stokes law of settling, stated in a general form, is that the settling velocity of a particle is directly related to the second power of its characteristic linear dimension,

$$U_s = \alpha B r^2 \qquad \text{(cm sec}^{-1}) \qquad (6.3)$$

where r is a linear dimension, B is a parameter that depends on the nature of the fluid and particle material but not on the particle size, and α is a shape factor. The parameter B is

$$B = \frac{2g(\rho_s - \rho)}{9\eta} \qquad (\text{cm}^{-1}\,\text{sec}^{-1}) \qquad\qquad (6.4)$$

Order of magnitude values of B for water and air at 20°C are as follows. For water, the density is $\rho = 1.0\ \text{g cm}^{-3}$ and viscosity $\eta \simeq 0.01\ \text{g cm}^{-1}\,\text{sec}^{-1}$ (or poise), such that B is

$$B = 2.18 \times 10^4 (\rho_s - 1.0) \qquad (\text{cm}^{-1}\,\text{sec}^{-1})$$

For air, the density is about 1000 times lower than the density of water ($\rho = 1.2 \times 10^{-3}\ \text{g cm}^{-3}$) and the viscosity is about 100 times lower ($\eta = 1.8 \times 10^{-4}\ \text{g cm}^{-1}\,\text{sec}^{-1}$). The density of many solids is much higher than the density of air, and the parameter B is

$$B = 1.19 \times 10^6 (\rho_s - 0.0012)$$

$$\simeq 1.19 \times 10^6 \rho_s \qquad (\text{cm}^{-1}\,\text{sec}^{-1})$$

The shape factor α for the sphere is $\alpha = 1$, and the linear dimension r is the sphere radius. The settling velocity of the sphere is

$$U_s = Br^2 \qquad\qquad (6.5)$$

Equations 6.5 and 6.4 combined are the original equation 127 of Stokes (1851), referring to a slow flow past a sphere. The domains of the Stokes settling for quartz spheres in air and in sea water are shown in Figure 6.1, in a graph of the settling velocity U_s as a function of the particle radius r. The settling of the bigger and heavier particles in air and water, and the settling of very small particles in air deviates more or less appreciably from the Stokes law. The settling of the bigger particles is governed by the impact law, where the settling velocity becomes directly related to the power of the sphere radius $r^{1/2}$. Very small particles in a gas settle somewhat faster than predicted by the Stokes law, because of the slippage of the particles between the gas molecules. The equation for the settling velocity takes this phenomenon into account through the Cunningham correction factor. The impact law of settling and the Cunningham correction to the Stokes law are given in Appendix C.

The Stokes law applies to laminar flow conditions. Transition from the laminar to turbulent flow is characterized by the value of the Reynolds

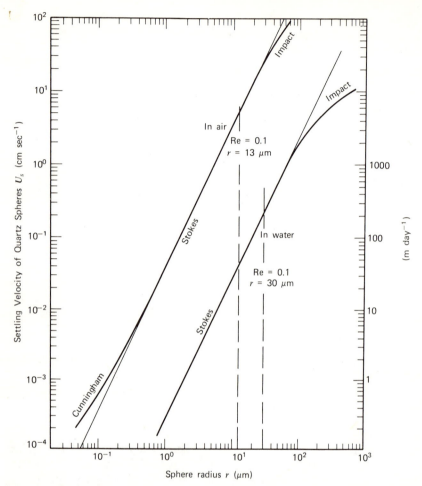

Figure 6.1 Settling velocities of quartz spheres in air and in sea water. Quartz density $\rho_s = 2.65$ g cm^{-3}. Air: $\rho = 1.2 \times 10^{-3}$ g cm^{-3}, $\eta = 1.83 \times 10^{-4}$ g cm^{-1} sec^{-1}. Water: $\rho = 1.03$ g cm^{-3}, $\eta = 1.5 \times 10^{-2}$ g cm^{-1} sec^{-1}.

number in the vicinity of Re$\simeq$0.05 (Chapter 2). In turbulent flows, the resisting force F_D on a settling particle is different from the Stokes drag, and the settling velocity equations have a form different from equation 6.3. For mildly turbulent flows and conditions comparable to flow in natural waters, a summary of equations for the settling velocity of a sphere can be found in Krumbein and Pettijohn (1938). The Reynolds number can be defined in terms of the particle diameter $2r$ and its Stokes settling velocity

U_s as

$$\mathrm{Re} = \frac{2rU_s}{\nu} = \frac{2Br^3}{\nu} \tag{6.6}$$

where $\nu = \eta/\rho$ is the kinematic viscosity of the fluid (cm^2 sec^{-1}). At values of the Reynolds number below 0.05, the agreement between measured settling velocities and the Stokes equation is routinely reported in the literature as being within 1–2%. For the value of the Reynolds number $\mathrm{Re} = 0.1$, taken as an upper limit of the Stokes settling velocity, the radii of the particles (quartz spheres) are $r = 30$ μm in water and $r = 13$ μm in air, as shown in Figure 6.1.

Spheres and Nonspheres. The shape factor α was introduced in the equation for the Stokes settling velocity, to account for the fact that the drag force on a settling particle depends on its shape. The explicit equations for the shape factor of particles of different geometric forms are listed in Table 6.3, and the equations for the settling velocities are derived and given in Appendix C. It is convenient to treat the settling velocities of particles according to equation 6.3 in terms of the radius r of a sphere, the volume of which is equal to the volume of the particle. Such a sphere is called the *equivalent sphere*. The radius r of the equivalent sphere is related to the volume of the particle v_s by the equation

$$r = \left(\frac{3v_s}{4\pi} \right)^{1/3} \quad \text{(cm)} \tag{6.7}$$

The list of particles in Table 6.3 includes a variety of smooth surface and angular forms: sphere, hollow sphere, hemispherical cap, ellipsoids, needle, disc, tetrahedron, cylinder, half ring, and circular ring. The settling velocities of these particles can be compared to the settling velocity of a sphere in two ways: (1) the volume of the particle or the volume of its solid part (for example, the wall of a hollow sphere) can be equated to the sphere volume, as shown in equation 6.7 for the radius of the equivalent sphere; and (2), the longest linear dimension of the particle can be made equal to the diameter of the solid sphere. For the two cases, the settling velocities are shown in Figure 6.2. The effect of the particle shape, when particles are of equal volume, is demonstrated by the graph in Figure 6.2a. The settling velocities of nonspheres vary within a factor of 2 of the settling velocity of the sphere. The spread of the settling velocities is considerably greater when particles of equal length rather than volume are compared, as in Figure 6.2b. This is to be expected, because the mass of a nonspherical

Table 6.3   Shape Factor α for Particles of Different Shape in the Stokes Settling Velocity Equation [a, b]

Particle shape	α	Remarks
Sphere	1	
Hollow sphere	$1.44(\Delta a/a)^{1/3}$	a sphere radius, Δa wall thickness, $\Delta a/a \ll 1$
Hemispherical cap	$1.23(\Delta a/a)^{1/3}$	a cap radius, Δa wall thickness, $\Delta a/a \ll 1$. Settling direction, equatorial plane down
Polyhedra		
Cube-octahedron	0.964	Sphericity $R_s = 0.906$
Octahedron	0.939	$R_s = 0.846$
Cube	0.921	$R_s = 0.806$
Tetrahedron	0.853	$R_s = 0.670$
Ellipsoids		Ellipsoids of revolution, length $= 2c$, equatorial diameter $= 2a$
	($\|c$) $5p^{1/3}/(4+p)$	Prolate, $p = c/a > 1$
	($\perp c$) $2.5p^{1/3}/(1.5+p)$	Oblate, $p = c/a < 1$
Needle	($\|c$) $1.5p^{-2/3}(\ln 2p - 0.5)$	Prolate ellipsoid of revolution, large
	($\perp c$) $1.5p^{-2/3}(\ln 2p + 0.5)$	$p = c/a \gg 1$
Disc	($\|c$) $0.375p^{1/3}f_1(p)$	Oblate ellipsoid of revolution, small
	($\perp c$) $0.375p^{1/3}f_2(p)$	$p = c/a \ll 1$
Cylinder	($\|h$) $1.72p^{-2/3}(\ln 2p - 0.72)$	Approximations for a straight circular cylinder, height h, base radius a.
	($\perp h$) $0.86p^{-2/3}(\ln 2p + 0.5)$	$2p = h/a$
Half ring		
(horizontal)	$0.75p^{-2/3}(\ln 2p + 0.56)$	Half-circular ring bent from a long
(vertical)	$p^{-2/3}(\ln 2p - 0.68)$	prolate ellipsoid, $p = c/a \gg 1$
Ring		
(horizontal)	$0.75p^{-2/3}(\ln 2p + 0.75)$	Circular ring bent from a long prolate ellipsoid, $p = c/a \gg 1$
(vertical)	$p^{-2/3}(\ln 2p - 2.09)$	

[a] $U_s = \alpha B r^2$, where r (cm) is the sphere radius, and parameter B (cm^{-1} yr^{-1}) is defined in equation 6.4.

[b] Shape factor derived in equations given in Appendix C.

particle, such as a needle, of length equal to the diameter of a solid sphere, is much smaller than the mass of the sphere. Similarly, a thin disc, the diameter of which is equal to the diameter of a sphere, is much lighter than the sphere. Obviously, not every needle or disc settles slower than any sphere: for example, a needle 10 μm long and 1 μm thick settles in a vertical orientation faster than a spherical particle of diameter 4 μm.

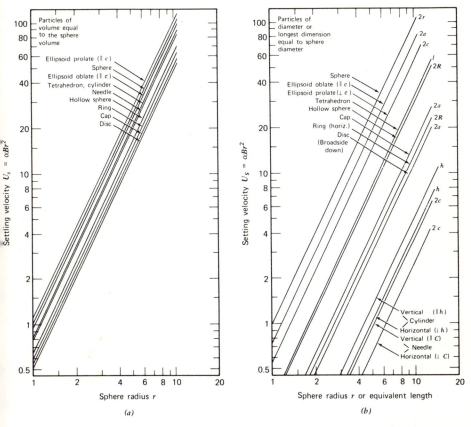

Figure 6.2 Settling velocities of the sphere and of particles of other shapes. (*a*) Particles of volume equal to the sphere volume. Length/width ratios of elongated or flattened particles, and wall thickness of hollow sphere and cap are given in Appendix C. (*b*) Particle of diameter or longest dimension of each particle equal to the diameter of the sphere $2r$. U_s is in units of centimeters per second when r is in micrometers and $B = 1 \times 10^4$ cm^{-1} sec^{-1}.

6.2.3   Settling of Biogenic Particles

Some experimentally determined settling velocities of planktonic organisms and their skeletal parts in sea water are summarized in Figure 6.3 for a range of sizes from microns to millimeters. The shadowed band includes the settling velocities of live and dead cells of flagellates and diatoms, and aggregates of cells. In the log-log coordinates of the graph in Figure 6.3, some of the settling velocities of the cells of different species

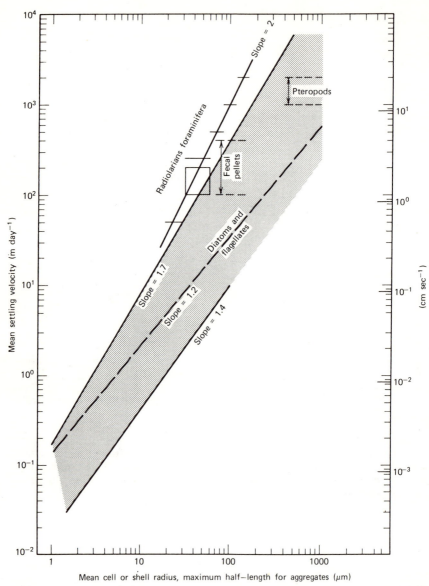

Figure 6.3 Mean settling velocities of biogenic particles: cells with skeletal parts, palmelloid aggregates, and shells (from data in Smayda, 1966; Berger, 1976).

increase with increasing cell size approximately as a power of $r^{1.2}$. Most of the settling velocities can be bracketed between $r^{1.2}$ and $r^{1.7}$. The skeletons of radiolarians and foraminifera fall on a line that corresponds to the settling velocity increase in a direct relation to r^2, although the nature of the data shown in the figure is such that variations within a factor of 2 have been reported for the settling velocities within the individual size fractions. One possible reason for the departure from the Stokes law of r^2 may be the dependence of the particle density on its size. A thin-walled sphere is one type of a particle whose bulk density is radius dependent. Aggregates of organic and mineral material are another type of particles of nonuniform density.

A broader problem of the settling behavior of biogenic particles has to do with the shapes that are very different from those of the simple geometric solids. Some of the variety of shapes characteristic of the oceanic particles can be appreciated at a glance on Plate I. Skeletons of foraminifera and coccoliths made of $CaCO_3$, skeletons of radiolarians and diatoms made of SiO_2, and fecal pellets made by planktonic crustaceans, all differ to a greater or lesser extent from the spherical shape. The protoplasm of the cell extends beyond the skeleton, such as in the specimen of the foraminifera genus *Orbulina* shown in the Plate, making the bulk density of the cell lower and the cell more susceptible to small-scale turbulence in water. Even in the absense of turbulence, a flat disc or a needle that begins to settle in an orientation forming an angle with the vertical (that is, the long axis of the needle or the plane of the disc are neither exactly parallel nor perpendicular to the vertical) will have a certain horizontal component of velocity and will therefore settle along a line at some angle to the vertical (Happel and Brenner, 1973). In the presence of slight turbulence, the falling-leaves motion of settling snowflakes, tree leaves, and similarly shaped objects is well familiar. An analogy may be drawn to the settling velocities of snowflakes in air. The settling velocities of snowflakes increase as a fractional power of the flake radius, $U_s \propto r^{0.3}$ (strictly, r is the radius of the melted snowflake; Langleben, 1954; Mason, 1971). For large flakes resembling thin plates of large diameter, the settling velocity becomes essentially constant and independent of size. Plates tend to settle in a horizontal orientation, face down, and the drag force is directly related to the area of the plate, when the thickness is small and about the same. The increase in the plate diameter is balanced by a higher drag force, which accounts for the constant settling velocity (Byers, 1965). The settling laws, ranging from the r^0 dependence, through $r^{0.3}$ for smaller snowflakes and $r^{0.5}$ for impact settling, to the r^2 dependence of the Stokes law, cover several important conditions that apply to particles in air

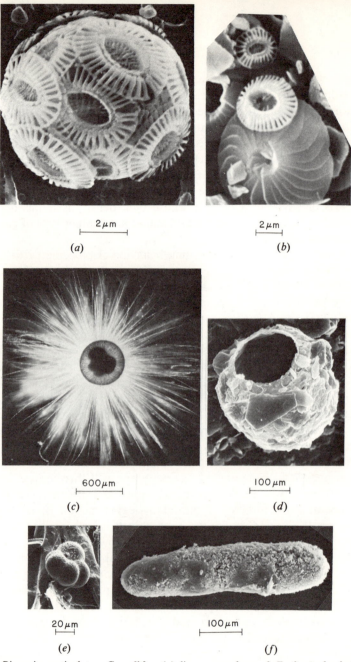

Plate I. Biogenic particulates. **Coccoliths:** (*a*) live coccosphere of *Emiliania huxleyi*; (*b*) individual coccolith platelets from Atlantic Ocean sediments. **Foraminifera:** (*c*) live specimen of *Orbulina universa*, Gulf Stream off Bimini Island; (*d*) Tintinnid (a flagellate) building a shell from suspended particles, Sourth China Sea; (*e*) specimen of a globigerinoid shell. (*f*) Fecal pellet made by a copepod. **Diatoms,** skeletons: (*g*) through (*k*). **Radiolarians,** skeletons: (*l*) through (*o*); the specimen in (*m*) is next to a globigerinoid shell, the portion of which in the lower part of the photograph can be compared to (*d*). Credits: photographs (*a*), (*b*), (*d*), (*e*), (*f*), (*n*), and (*o*) courtesy of Dr. S. Honjo, Woods Hole Oceanographic Institution; photograph (*c*) courtesy of Dr. A. H. W. Bé, Lamont-Doherty Geological Observatory of Columbia University; photographs (*g*) through (*m*) courtesy of Dr. D. Lal, Physical Research Laboratory, Ahmedabad, India.

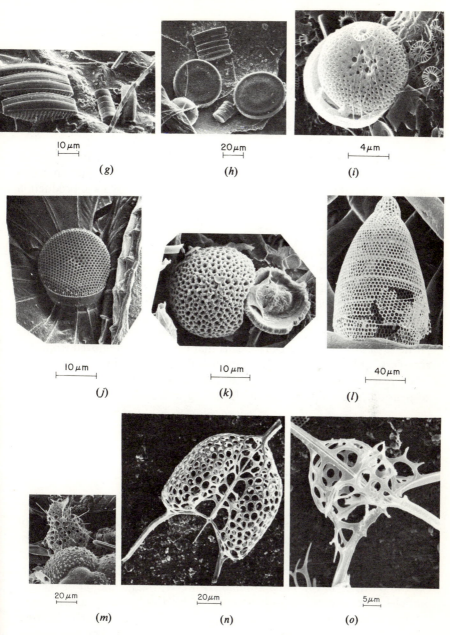

$10\,\mu m$

(g)

$20\,\mu m$

(h)

$4\,\mu m$

(i)

$10\,\mu m$

(j)

$10\,\mu m$

(k)

$40\,\mu m$

(l)

$20\,\mu m$

(m)

$20\,\mu m$

(n)

$5\,\mu m$

(o)

Plate I (*continued*)

271

and water. For biogenic particles, the range of settling velocities between r^1 and r^2 can be bracketed by the theoretical settling velocities of a hollow sphere with a thin wall (Appendix C) and a solid sphere.

6.2.4 Sinks for Settling Particle Assemblages

Sedimentation on the floor of the oceans and lakes is a major sink that removes suspended materials from water. In a steady-state situation, when the rate of input of particles at the water surface either by local production or inflow from outside is balanced by the rate of sedimentation on the bottom, the sedimentation is the only sink, provided nothing affects the particle mass and size in the water column. In such a case, the particle concentration and size distribution do not change with depth as particles of different sizes continuously settle through the water column (this applies to an infinitely long water column where the effects of the production and sedimentation boundaries are not considered). The occurrence of additional sinks for the settling particles would generally affect their concentration and (or) shape of the particle-size distribution. A sink for the settling particles of any given size class means that particles are removed from that size class, and this may take place by growth or increase in size, dissolution or decrease in size, or scavenging of some fraction of the material out of suspension. The sinks include the following processes:

Dissolution (or growth) of mineral particles
Oxidation or decomposition of organic particles
Agglomeration and attachment
Disaggregation or fragmentation
Ingestion by organisms

Of the five sinks listed, agglomeration and attachment are likely to be of importance only in the areas of high concentration of suspended materials, such as in certain lakes, coastal waters, estuaries, and near mouths of rivers. Of disaggregation and fragmentation, little can be said with reference to particles in the water column. However, fragmentation of skeletal particles in sediments, where it is promoted by dissolution (Chapters 5 and 7), and breakdown of coccospheres in the water column (cells bearing coccoliths platelets, Honjo, 1975) suggest that disaggregation may play a role as a sink for some types of settling particles.

Dissolution. A constant rate of dissolution of a spherical particle is equivalent to a constant rate of mass loss per unit area of the surface. In such a diffusion-controlled process the rate of decrease in the particle radius with time $-dr/dt$ is given in equation 5.80 as

$$-\frac{dr}{dt} = \varepsilon \quad (\text{cm sec}^{-1})$$

The settling velocity of a spherical particle obeying the Stokes law is

$$\frac{dz}{dt} = Br^2 \quad (\text{cm sec}^{-1}) \tag{6.8}$$

where z denotes the vertical distance coordinate, positive and increasing downward from some arbitrary reference depth level $z=0$. From the preceding two equations, the rate of the particle radius change with depth is

$$\frac{dr}{dz} = -\frac{\varepsilon}{Br^2} \tag{6.9}$$

Integration gives the particle radius as a function of depth z and dissolution rate ε,

$$r = \left(r_0^3 - \frac{3\varepsilon z}{B} \right)^{1/3} \quad (\text{cm}) \tag{6.10}$$

where r_0 is the initial particle radius, at depth $z=0$. As the particle settles and dissolution goes on, it dissolves completely (that is, r becomes 0) at the depth

$$z_{r=0} = \frac{Br_0^3}{3\varepsilon} \quad (\text{cm}) \tag{6.11}$$

Because of the r_0^3 dependence of the dissolution depth, the survival distances of the smaller particles are very short in comparison to the larger particles, under the conditions of the constant rate of dissolution and Stokes settling. For example, at the radial dissolution rate of $\varepsilon = 1 \ \mu m \ yr^{-1}$ and the Stokes settling parameter value of $B = 10^4 \ cm^{-1} \ sec^{-1} = 3 \times 10^{11} \ cm^{-1} \ yr^{-1}$ (see equation 6.4), a particle of initial radius $r_0 = 3 \ \mu m$ disappears at the depth

$$z_{r=0} = \frac{3 \times 10^{11} \times (3 \times 10^{-4})^3}{3 \times 1 \times 10^{-4} \times (100 \ cm \ m^{-1})} = 270 \ m$$

The distance of 270 m is the depth through which the particle settles counting from some level $z = 0$ where dissolution begins to take place. A particle of initial radius $r_0 = 9$ μm would settle through a distance 27 times longer, about 7300 m. This means that the particle would effectively reach the bottom between 3 and 4 km depth, where it may or may not continue to dissolve, depending on the rate of sedimentation and other factors.

The settling velocity of a spherical particle diminishes as it dissolves and its radius decreases. Over a settling distance z ($z < z_{r=0}$), the mean settling velocity $\bar{U}_s$ of a particle of initial radius r_0 is

$$\bar{U}_s = \frac{1}{z} \int_0^z Br^2 \, dz$$

$$\bar{U}_s = 0.6 Br_0^2 \cdot \frac{z}{z_{r=0}} \left[1 - \left(1 - \frac{z}{z_{r=0}} \right)^{5/3} \right] \qquad (6.12)$$

where $z_{r=0}$ is the distance to complete dissolution defined in equation 6.11, and r is given in 6.10. Over the entire survival distance, the mean settling velocity is 60% of the initial velocity, or

$$\bar{U}_s = 0.6 Br_0^2$$

A particle-size spectrum changes with depth owing to dissolution as follows. If the size spectrum initially, at depth $z = 0$, is a Pareto-type distribution (Section 5.2.3),

$$dN_0 = A r_0^{-b} \, dr_0 \qquad \text{(particles cm}^{-3}\text{)} \qquad (6.13)$$

then at depth $z > 0$ the size distribution dN becomes

$$dN = A \left(r^3 + \frac{3\varepsilon z}{B} \right)^{-b/3} dr \qquad \text{(particles cm}^{-3}\text{)} \qquad (6.14)$$

At $z = 0$, the two preceding equations are identical. Derivation of the latter was given in Lal and Lerman (1975). The change in the size of settling particles with depth owing to dissolution has also been dealt with theoretically in the papers by Okubo (1954, 1956) and Brun-Cottan (1976). At successively increasing depths, a plot of $\log dN / dr$ against $\log r$ looks like a series of curves drawn in Figure 5.11 for the case of dissolution without settling. Instead of successively increasing values of time t, the increasing depth z produces a similar decrease in the number of smaller particles, as drawn in Figure 5.11. Equation 6.14 is a modified Pareto distribution equation 5.41. At any fixed depth, for particles smaller than $r^3 \ll 3\varepsilon z / B$, the particle-number concentration dN/dr is nearly constant and independent of the particle size: in Figure 5.11, these are the nearly horizontal sections to which the curves labeled $t = 10$ and 100 yr tend.

Removal of Settling Particles by a First-Order Process. A process that removes settling particles from water in proportion to their abundance is analogous to a first-order chemical reaction or radioactive decay. In a well-mixed closed volume of water where a particle assemblage is kept in suspension, removal of an aliquot from suspension withdraws a sample of the particle assemblage that has, in principle, the same size distribution as in the bulk suspension. The remaining mass of particles in suspension is by some amount smaller, but the relative proportions of the different particle sizes are not affected by the withdrawal of the small-volume aliquots (excluding the statistical effects of sampling). In steady-state settling through a water column, continuous removal of particles in proportion to the abundance of each size fraction leads to a different result: both the mass and the shape of the size distribution change with depth. The change will be demonstrated for the following simplified case. A particle-number distribution at depth $z = 0$ is a power-law spectrum of the type used in the preceding section,

$$dN \equiv N_{r, z=0} = Ar^{-b}dr \qquad (\text{cm}^{-3}) \qquad (6.15)$$

where the notation N_r is introduced instead of the previously used form dN, and both denote the number of particles of size between r and $r + dr$ in a unit volume of water. The downward flux of particles settling with the velocity U_s is $U_s N_r$. Removal of particles at a rate that is proportional to their concentration N_r affects the concentration but not the particle size. The particle-number flux $U_s N_r$ changes with depth according to the following mass balance equation:

$$\frac{d(U_s N_r)}{dz} = -kN_r \qquad (\text{cm}^{-3}\,\text{yr}^{-1}) \qquad (6.16)$$

where k is a first-order rate constant (yr^{-1}). As the particle radii do not change with depth, the settling velocity U_s is independent of depth z. If the Stokes settling velocity $U_s = Br^2$ is inserted in the above equation, it integrates to

$$N_r = N_{r,z=0}\exp\left(-\frac{kz}{Br^2}\right) \quad (\text{cm}^{-3}) \tag{6.17}$$

Using the explicit form of $N_{r,z=0}$ from equation 6.15 and returning to the dN notation (that is, $dN \equiv N_r$ cm^{-3}), the decrease in the particle number with depth is

$$dN = A\exp\left(-\frac{kz}{Br^2}\right)r^{-b}dr \quad (\text{cm}^{-3}) \tag{6.18}$$

The equation shows that the particle concentration decreases with depth as a function of both depth and the particle radius. The smaller the radius, the more pronounced is the decrease in particle concentration. The particle-number spectrum in equation 6.18 has a maximum at the value of the particle radius r, which is

$$r = \left(\frac{2kz}{Bb}\right)^{1/2} \quad (\text{cm}) \tag{6.19}$$

The peak value of dN/dr is therefore

$$\frac{dN}{dr} = Ae^{-b/2}\left(\frac{2kz}{Bb}\right)^{-b/2} \quad (\text{cm}^{-4}) \tag{6.20}$$

The latter equation shows that the particle concentration at the peak decreases with depth in direct relation to $1/z^{b/2}$. The value of $b \simeq 4$ is commonly encountered in suspended materials (Table 6.4). Thus if the suspensoid is removed from water by a first-order process, the particle-concentration peak decreases with depth in direct relation to $1/z^2$. The occurrence of the concentration maximum makes the particle distribution curve similar in shape to the curve for atmospheric aerosols shown in Figure 5.6.

Removal of Particles by Other Mechanisms. Filtration of suspended material is a process in which particles are removed from the fluid in proportion to their number and cross-sectional areas. Experimental work on filtration of suspensions flowing through porous beds shows that the efficiency of retention on the filter having a certain distribution of pore diameters is related to the cross-sectional areas of particles and the particle-number concentration in the inflowing suspensions (Yao et al., 1971). Although it is not clear to what extent, if at all, the bigger

Table 6.4 Parameter b of the Particle-Size Distribution
Equation $dN/dr = Ar^{-b}$ (cm^{-4}) for Suspended Materials in the Oceans

b	Range of particle radius (μm)	Range of water depths (m)	Type of material and location
4.5 ± 0.3	ca. 0.3–30	Surface waters	Foraminifera and diatoms, Indian Ocean [a]
4.2 ± 0.3	0.7–10	300–800	Calcareous material abundant, Guinea Basin, East Atlantic [b]
4.0 ± 0.4	0.7–10	300–800	Western Mediterranean [b]
3.6 ± 0.2	0.5–50	200–4500	Northwestern Atlantic [c]
4.0 ± 0.3	1–7	30–5100	Mostly organic, minor clays and carbonate, North Equatorial Atlantic [d]

[a]Lal and Lerman (1975), based on data of Lisitzin (1972).
[b]Brun-Cottan (1976).
[c]Data of R. W. Sheldon cited by McCave (1975).
[d]Lerman et al. (1977).

zooplankton can filter particles in a manner analogous to porous filters, or to what extent the bigger mucous aggregates of organic material can collect the smaller particles, a removal mechanism similar to filtration is worth considering briefly, because it leads to results observable in particulate materials in the ocean.

In the particle removal process, where the settling particles are removed in proportion to their number concentration *and* cross-sectional areas, the result is that the mass concentration of suspended material decreases with depth, but the shape of the size distribution remains unchanged. Derivation of the equations demonstrating these conclusions follows.

The conservation condition for the particle-number flux, as given in equation 6.16, is

$$\frac{d(U_s N_r)}{dz} = -k_s r^2 N_r \qquad (\text{particles cm}^{-3}\ \text{yr}^{-1}) \qquad (6.21)$$

where U_s, N_r, r, and z retain their meaning, and k_s is a rate parameter $(cm^{-2}\ yr^{-1})$. The term $-k_s r^2 N_r$ describes a removal mechanism where the small cross-sectional area of the smaller particles can be compensated for by their occurrences in large numbers, whereas for the bigger particles occurring at low concentrations N_r, the removal rate is compensated for by their relatively large cross-sectional areas. The settling velocity of the particles $U_s = Br^2$ does not change with depth as long as the particle radius remains constant (i.e., particles neither grow nor dissolve). With the Stokes settling velocity, equation 6.21 integrates to

$$dN = A \exp\left(-\frac{k_s z}{B}\right) r^{-b} dr \qquad \text{(particles cm}^{-3}) \qquad (6.22)$$

where $dN \equiv N_r$ is the particle concentration, and the particle size distribution is initially, at depth $z=0$, the power-law spectrum of equation 6.15. Comparing the results for the filtration-type of removal, equation 6.22, and first-order removal, equation 6.18, it is evident that in the filtration-type scavenging process the particle concentration decreases exponentially with depth, but the rate of decrease is independent of the particle size (the exponential is a function of z but not a function of r, as was the case in the first-order removal model). A log-log plot of equation 6.22 has the form

$$\log\frac{dN}{dr} = \log A - \frac{k_s z}{2.3B} - b\log r \qquad (6.23)$$

A plot of $\log dN/dr$ against $\log r$ is a straight line with the negative slope $-b$. The slope does not change with depth, the dN/dr values plot as parallel lines, and the decrease in the particle-number concentration with depth is determined by the z-containing term only. The mass concentration of particles dM (mass volume^{-1}) also decreases exponentially with depth. For spherical particles, the mass concentration is the following function of depth when the particle-number concentration is given by equation 6.22:

$$dM = \tfrac{4}{3}\pi\rho_s A \exp\left(-\frac{k_s z}{B}\right) r^{3-b} dr \qquad \text{(g cm}^{-3}) \qquad (6.24)$$

The equations for the number and mass concentration of settling particles are discussed in more detail in Section 6.2.5.

6.2.5 Fine Particles in the Ocean

Concentration of suspended materials in the open ocean generally decreases from the surface down, the decrease being more or less exponential with depth. Most of the decrease in concentration takes place within a few hundred meters near the surface, whereas within the deeper 3–4 km concentrations are generally low. In the surface layer, 100 or so meters thick, concentrations of suspended materials are in the range of tens to a few hundred micrograms dry weight per liter of water. By volume, fresh (wet) material amounts to between 0.1 and 0.5 ppm (μl l^{-1}). In the intermediate and deeper ocean waters, mass concentrations (dry weight) are typically $\lesssim 20 \ \mu g \ l^{-1}$, and the fresh volume fractions drop by a factor of 5–10, into the range of 10^{-2}–$10^{-1} \ \mu l \ l^{-1}$. Over a distance of a few hundred meters above the ocean bottom in some areas, concentrations of suspended material increase, forming a nepheloid layer. Mass concentrations in the nepheloid layers are in the tens of $\mu g \ l^{-1}$ range, and the origins of such layers have been attributed to resuspension from the bottom due to a stronger turbulence (Brewer et al., 1976; Sheldon et al., 1972; Biscaye and Eittreim, 1974). A considerable amount of data on the distribution of suspended materials in ocean water that appears in the oceanographic literature is based on light-scattering measurements, which give only relative changes in concentration and which cannot always be translated into concentration in units of mass per volume.

A decrease with depth in the particle-mass and number concentration is shown in Figure 6.4, based on the data from an area in the north equatorial Atlantic Ocean, east and west of the Mid-Atlantic Ridge. In the surface water layer, down to about 200 m depth, mass concentration decreases rapidly by a factor of 4–5. Below the 200-m depth, the decrease is much slower, and most of the removal of the mass takes place within the surface layer. The particle-size distributions from one of the vertical sections plotted in Figure 6.4b show that the suspended material consists mostly of fine particles, of radius between 1 and 7 μm. (These are the sizes of the most abundant material. Larger particles of organic and inorganic origins may easily remain undetected because of their low abundance, particularly when water samples of only a few liters in volume are taken. The lower particle-size limit of about 1 μm mean radius is an instrumental limit for the data plotted in the figure.) Although the total mass of suspended material decreases with depth, the shapes of the particle-size spectra—straight lines in log-log coordinates—remain virtually constant. The negative slope of the size distribution lines, $\log dN/dr = \log A - b \log r$, varies within relatively narrow limits: b is between 3.7 and 4.3. However, the parameter A decreases with depth, reflecting the smaller number and mass concentrations of particles. Mean values of A and b for different depths at the four stations are listed in Table 6.5.

The mass-concentration data in Figure 6.4a can be described by an equation relating concentration M (g cm^{-3}) to depth z,

$$M = M_\infty + (M_0 - M_\infty) e^{-\beta z} \qquad (\text{g cm}^{-3}) \qquad (6.25)$$

where M_∞ is a steady concentration at depth, M_0 is concentration at the layer top, and β is a constant (units of reciprocal distance). In the surface layer, between depths of 0 and 200 m, mass concentration M changes with

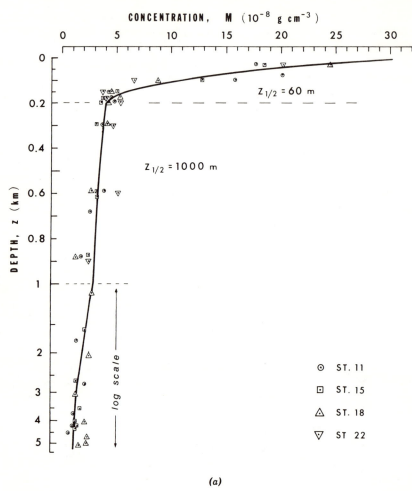

(a)

Figure 6.4 Suspended material in the water column of the central equatorial Atlantic. (*a*) Decrease in mass concentration with depth. (*b*) Particle-size distributions in one of the vertical profiles (from Lerman et al., 1977). Reproduced by permission of Elsevier Scientific Publishing Company, Amsterdam, The Netherlands.

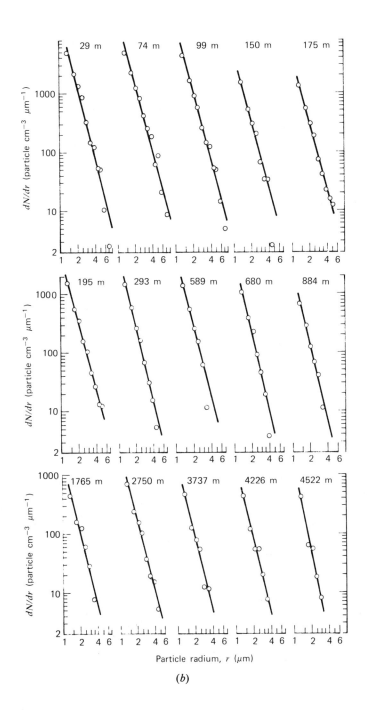

(b)

Table 6.5   Parameters A and b of the Particle-Size Distribution Equation $dN/dr = Ar^{-b}$ (cm^{-3} μm^{-1}) for Suspended Materials in North Equatorial Atlantic. Mean Values for Four Stations [a]

Water depth (m)	A (cm^{-3} μm^{b-1})	b
30	24360	4.00
74	19550	3.73
98–100	11720	3.83
147–150	4860	3.89
172–175	5010	3.94
195–200	5060	3.91
290–300	5110	4.06
588–600	5390	4.13
617–680	3900	4.11
876–900	3380	4.38
1600–1770	1870	3.91
2060–2750	2060	3.82
3050–3740	1740	4.06
4030–4710	2030	4.20
5010–5110	1830	3.73
Mean $b \pm$ standard deviation [b]		4.0 ± 0.3

[a] Data from Lerman et al. (1977); Stations at 11°N, between longitudes 29.5°W and 48°W.

[b] Mean from 53 individual samples, all stations and depths.

depth as

$$M = 1 \times 10^{-8} + 30 \times 10^{-8} \exp(-1.16 \times 10^{-2} z) \quad \text{(g cm}^{-3}) \quad (6.26)$$

where z is in meters. In the deep-water layer, depths greater than 200 m, the particle-mass concentration decreases with depth as

$$M = 1 \times 10^{-8} + 3.1 \times 10^{-8} \exp[-6.9 \times 10^{-4}(z - 200)] \quad \text{(g cm}^{-3})$$

$$(6.27)$$

where z is in meters. The depth at which concentration is halfway between the values for the layer top and bottom is the half-concentration depth $z_{1/2}$. From equation 6.25, the quotient of concentration differences is $\frac{1}{2}$,

$$\frac{M - M_\infty}{M_0 - M_\infty} = \frac{1}{2}$$

at the depth $z_{1/2}$, equal to

$$z_{1/2} = \frac{\ln 2}{\beta} \qquad (6.28)$$

In the surface water layer, the half-concentration depth is $z_{1/2} \simeq 60$ m. In the deep layer, the half-concentration is attained at depth $z_{1/2} \simeq 1000$ m below the layer top.

Transport and Removal of Particles.   The settling velocities of small and light particles can be affected by water turbulence. The downward flux F of settling particles can be considered as made of the flux due to eddy diffusional transport and the flux due to gravitational settling (Chapter 2), approximately,

$$F \simeq -\overline{\kappa} \frac{d(M)}{dz} + \overline{U}_s M \qquad (\text{g cm}^{-2}\,\text{yr}^{-1}) \qquad (6.29)$$

where $\overline{\kappa}$ is a mean eddy diffusion coefficient independent of the particle size, and $\overline{U}_s$ is the mean settling velocity of particles in the size range between some lower-limit radius r_1 and upper limit r_2. The quantity (M) is mass concentration, identical to M defined in equations 6.25 through 6.26; the notation of the distance derivative of mass $d(M)/dz$ refers to the change with depth in the total mass of suspended particles, in the size range $r_1 < r < r_2$, and should not be confused with the particle-mass spectrum dM, referring to particles of size between r and $r + dr$, as in equation 6.24 and in other chapters.

For suspended material that is being removed from the water column by a first-order removal process, the balance condition for the mass flux is

$$\frac{dF}{dz} = -k(M - M_\infty) \qquad (\text{g cm}^{-3}\,\text{yr}^{-1}) \qquad (6.30)$$

which is analogous to equation 6.16 for the gradient of the particle-number flux. The removal rate term $k(M - M_\infty)$ assumes that some fraction of the settling material, concentration M_∞, escapes the k-removal mechanism. The derivative of F from equation 6.29 in combination with the preceding gives

$$\overline{\kappa} \frac{d^2(M)}{dz^2} - \overline{U}_s \frac{d(M)}{dz} = k(M - M_\infty) \qquad (\text{g cm}^{-3}\,\text{yr}^{-1}) \qquad (6.31)$$

The latter equation is a diffusion-advection-chemical reaction equation of the type also discussed in Chapters 2 and 8. The derivatives of mass concentration M with respect to distance z can be obtained by differentiating M in equation 6.25, and the explicit result can be used in equation 6.31. This gives a relationship between the removal rate parameter k and other constant parameters,

$$k = \beta\left(\bar{\kappa}\beta + \bar{U}_s\right) \quad (yr^{-1}) \tag{6.32}$$

The term $\bar{\kappa}\beta$ represents the contribution of vertical eddy diffusion to transport and removal of particles, and $\bar{U}_s$ represents the contribution of the settling velocity. If $\bar{\kappa}\beta \ll \bar{U}_s$, then the settling velocity is the main transport mechanism and the removal rate parameter k becomes

$$k = \beta\bar{U}_s \quad (yr^{-1}) \tag{6.33}$$

To evaluate the removal rate constant k for the suspended material in the deep and surface layers, the values of β given previously (Figure 6.4a), can be combined with the mean settling velocity and mean vertical eddy diffusion coefficient. This is done in the next section, whereas here, in conclusion, the concept of a mean settling velocity of a particle assemblage will be discussed briefly.

The mean settling velocity $\bar{U}_s$ can be an average of the settling velocities of the mass fractions dM, number fractions dN, or of particles of different radii r, each mean being somewhat different from the others. The mean settling velocity of a dissolving particle is given in equation 6.12. A mean with respect to the mass fractions dM is

$$\bar{U}_s = \frac{\int_{r_1}^{r_2} U\,dM}{\int_{r_1}^{r_2} dM} \quad (cm\ yr^{-1}) \tag{6.34}$$

where dM is the mass of dN particles of radius between r and $r+dr$, as defined in equations 5.5 and 6.15. The integration is carried out for all particle sizes, from the lower-limit radius r_1 to the upper limit r_2. For spherical particles, Stokes settling, and the case of $b \neq 4$, the mean settling velocity is

$$\bar{U}_s = \frac{(4-b)\left[1-(r_1/r_2)^{6-b}\right]}{(6-b)\left[1-(r_1/r_2)^{4-b}\right]} \cdot Br_2^2 \quad (cm\ yr^{-1}) \tag{6.35}$$

For the case of $b = 4$, the mean settling velocity is

$$\bar{U}_s = \frac{1-(r_1/r_2)^2}{2\ln(r_2/r_1)} \cdot Br_2^2 \quad (cm\ yr^{-1}) \tag{6.36}$$

A somewhat simpler equation for the mean settling velocity is obtained by averaging the Stokes settling velocities Br^2 over the particle radii,

$$\bar{U}_s = \frac{1}{r_2 - r_1} \int_{r_1}^{r_2} Br^2 \, dr$$

$$= \frac{1}{3} \left[1 + \frac{r_1}{r_2} + \left(\frac{r_1}{r_2} \right)^2 \right] Br_2^2 \quad (\text{cm yr}^{-1}) \tag{6.37}$$

By either method of averaging, the mean settling velocity is independent of concentration of suspended matter, but it depends on the size range and especially on the upper size limit r_2 of the particle assemblage. The Stokes-law dependence of the settling velocity on r^2 is preserved in the mean velocity equations, such that the choice of the upper limit radius r_2 affects the estimate of the mean settling velocity.

Residence Times of Suspended Matter. The mean residence time of particles in a water layer of thickness h through which they settle with velocity U_s is (Section 1.2)

$$\tau_U = \frac{h}{U_s} \quad (\text{yr}) \tag{6.38}$$

For particles that are removed from water by a first-order k-removal process, the mean residence time is

$$\tau_k = \frac{1}{k} \quad (\text{yr}) \tag{6.39}$$

Mean settling velocities in the range of

$$\bar{U}_s = 35 \text{ to } 47 \quad (\text{m yr}^{-1})$$

apply to particles of sizes between $r_1 = 1$ μm and $r_2 = 5$ μm (mean velocity from equation 6.36, with the Stokes settling parameter $B = 1.43 \times 10^3$ cm^{-1} sec^{-1}. The latter is taken from equation 6.4; for viscosity of sea water $\eta = 0.015$ g cm^{-1} sec^{-1}, and the density difference $\rho_s - \rho = 0.1$ g cm^{-3}). The range of settling velocities is from equations 6.36 and 6.37. The residence times of fine particles, the size distributions of which are shown in Figure 6.4b, are estimated below for the deep and surface layers.

Deep Layer ($>$200 m). The vertical eddy diffusion coefficient in the thermocline region of the ocean is characterized by values of $\bar{K} \simeq$ 0.2 cm^2 sec^{-1} and in the deeper ocean, $\bar{K} \simeq 1$ cm^2 sec^{-1} (see Chapter 2,

Figure 2.7). Contribution of the vertical eddy diffusion to transport of the smaller particles is not significant in the deeper water, insofar as the product $\overline{K}\beta$ appearing in equation 6.32 is smaller than the mean Stokes settling velocity $\overline{U}_s$: using $\overline{K} = 1$ cm^2 sec^{-1} and $\beta = 6.9 \times 10^{-6}$ cm^{-1} as given in equation 6.27, we have

$$\overline{K}\beta \simeq 7 \times 10^{-6} \qquad (\text{cm sec}^{-1})$$

$$\overline{U}_s \simeq 1 \times 10^{-4} \qquad (\text{cm sec}^{-1})$$

and therefore

$$\overline{K}\beta \ll \overline{U}_s$$

Consequently, equation 6.33 can be used to compute the removal rate constant k for suspended material in the deep layer. Its residence time τ_k with respect to the k-removal process is

$$\tau_k = \frac{1}{(6.9 \times 10^{-4}\ \text{m}^{-1})(40\ \text{m yr}^{-1})} \simeq 40 \qquad (\text{yr})$$

In the preceding section it is pointed out that the values of the mean settling velocity $\overline{U}_s$ are sensitive to the upper size limit of the settling particles r_2 and the Stokes settling parameter B that depends on the particle density. A slightly smaller value of r_2 (4 instead of 5 μm) and a slightly smaller density difference between the particles and sea water ($1.09 - 1.03 = 0.06$ instead of 0.1 g cm^{-3}) give mean settling velocities a factor of 2 lower and, correspondingly, the residence time of suspended material with respect to the k-process a factor of 2 higher, about 80 yr. Thus an estimate of the residence time τ_k for the deeper water layer is

$$\tau_k = 60 \pm 20 \qquad (\text{yr})$$

The mean Stokes settling velocity of 30 ± 10 m yr^{-1} for particles in a water layer about 4000 m thick gives the mean residence time with respect to settling

$$\tau_U = \frac{4000\ \text{m}}{(30 \pm 10\ \text{m yr}^{-1})} \simeq 150 \pm 50 \qquad (\text{yr})$$

The first-order k-removal is faster than the Stokes settling of the small particles.

Surface Layer (0–200 m). The vertical eddy diffusivity that would affect the transport of small particles, the Stokes settling velocities of which are $1-1.5 \times 10^{-4}$ cm sec^{-1}, as computed in a preceding section, is

$$\bar{K} \simeq \frac{\bar{U}_s}{\beta} \simeq \frac{1.2 \times 10^{-4} \text{ cm sec}^{-1}}{1.16 \times 10^{-4} \text{ cm}^{-1}} \simeq 1 \text{ cm}^2 \text{ sec}^{-1}$$

The coefficient of vertical eddy diffusion in surface ocean water can be one or more orders of magnitude higher than $K_z = 1$ cm^2 sec^{-1} (Figure 2.7). An estimated value of $\bar{K} = 16 \pm 8$ cm^2 sec^{-1} (Lerman et al., 1977) requires that the vertical eddy diffusion be taken into account when the residence time of suspended matter with respect to the k-removal process is evaluated by equations 6.32 and 6.39. The removal rate constant k is

$$k = 1.16 \times 10^{-4} \left[(16 \pm 8) \times 1.16 \times 10^{-4} + 1 \times 10^{-4} \right]$$

$$\times (3.16 \times 10^7 \text{ sec yr}^{-1}) = 7 \pm 3 \text{ yr}^{-1}$$

The residence time τ_k is therefore

$$\tau_k = 0.2 \pm 0.1 \qquad \text{(yr)}$$

The short residence time implies a rapid removal rate, comparable to the regeneration rates of phytoplankton in the biological production zone of surface waters. Thus one may think of scavenging by the larger plankton in a process analogous to filtration, as suggested by the fast removal rates and the constancy of the particle-size distribution slope, discussed with Figure 6.4b.

If the eddy diffusional contribution to the vertical transport of particles were ignored, then the residence time with respect to the k-removal would be longer than in the preceding estimate, from equations 6.33 and 6.39,

$$\tau_k \simeq \frac{1}{(1.16 \times 10^{-2} \text{ m}^{-1})(30 - 45 \text{ m yr}^{-1})} = 2.3 \pm 0.5 \text{ yr}$$

The residence times with respect to the Stokes settling out of the 200-m-thick layer are

$$\tau_U = \frac{200 \text{ m}}{30 \text{ to } 45 \text{ m yr}^{-1}} = 4 - 7 \text{ yr}$$

As in the case of the deep layer, the k-removal process is faster than the sedimentation sink.

6.2.6   Three Mechanisms of Coagulation

Particles of micron and submicron dimensions can either remain as individual particles in suspension or they may coalesce and form cohesive aggregates, depending on their concentration and the nature of electrostatic forces acting on them. The forces of attraction or repulsion between particles are conditioned by the electrical potential developing on the particle surfaces in the presence of the hydrogen and metal ions in solution. A suspension is considered stable when no aggregates form, and one speaks of the stabilizing or destabilizing effects of various ionic species in solution that can be adsorbed on the particle surfaces, changing the surface potential and, consequently, determining the attraction or repulsion between the particles. Cohesive aggregates may sink faster or slower than individual particles, depending on their bulk densities and sizes. The formation of aggregates is referred to variably as aggregation, agglomeration, coagulation, or flocculation—terms that are for practical purposes synonymous. (Flocculation is commonly used as a term describing the formation of flocs in solution. Aggregation and agglomeration are, if anything, less specific than the term coagulation.)

On the physical scale of particle dimensions, coagulation is a process producing larger from smaller units, irrespective of the nature of the molecular or ionic forces that cause the particles to adhere to each other. The random Brownian motion of particles in water and air is one of the mechanisms that leads to collisions between particles, and a successful collision produces a floc or an aggregate that can in turn collide with another particle. Velocity gradients in the fluid are another mechanism that causes collisions between particles: particles moving at different velocities can overtake one another, collide, and form aggregates. In laboratory experiments, velocity gradients in a vessel can be produced by stirring or, more generally, by shear stresses. In rivers and open channel flow, velocities decrease from the flow axis toward the banks and the bottom, and practically in any situation where flow or turbulence occur, velocities vary, and more or less pronounced velocity gradients exist. The third coagulation mechanism is capture by settling: a larger particle settling faster can collide and capture smaller and slower particles. The situation is similar to interparticle collisions in a velocity gradient, but with one principal difference: in gravitational settling the larger particles are the ones that move faster, provided all particles are of the same density and shape. In a flow that has a gradient of velocity in the direction perpendicular to the flow direction (such as a vertical gradient of velocity in a horizontally flowing stream), two particles of the same radius can collide if one moves in a slightly faster section of the stream and the other in a slower section, within the distance equal to the radius of either particle. In capture by settling, the collisions take place between particles of unequal radius only, according to the definition of the process.

In a suspension, the probability of two particles colliding with each other depends on their cross-sectional areas and on their number concentrations. For a suspension containing two kinds of particles, of radius r_1 and r_2, at concentrations N_1 and N_2 (particles cm^{-3}), the rate J at which aggregates form by collisions in a unit of suspension volume is

$$\text{J} = f\mathcal{P}(r_1, r_2)N_1N_2 \qquad (cm^{-3}\,sec^{-1}) \qquad (6.40)$$

where f is the fraction of collisions that are successful (that is, collisions resulting in cohesion of particles), and $\mathcal{P}(r_1, r_2)$ is a function that defines the probability of collisions between particles of radii r_1 and r_2. The collision frequency function is characteristic of each collision mechanism and generally depends on the particle sizes, the temperature and viscosity of the fluid, and the velocities of the fluid and particles, but it does not depend on the number concentrations of particles. The function $\mathcal{P}$ is also known as the kernel of the coagulation equation. The rate of coagulation J can be equated to the rate of decrease in the number of particles with time, because in a closed volume, the particle concentrations N_1 and N_2 decrease as coagulation progresses. Thus $\text{J} \equiv -dN/dt$. In a more general case of a system where sources and sinks for particles exist, particles of some radius r form by coagulation of the smaller sizes and are removed by collisions with particles of other sizes. In such a process, the material balance equation for the rate of change in the number of particles of size r, $dN(r)/dt$, has in a general case the form of an integral-differential equation. An extensive summary of the kernel functions $\mathcal{P}$ and integral-differential equations describing production and coagulation of particles, mostly with reference to atmospheric aerosols, can be found in a review article by Drake (1972).

Coagulation of particles in aqueous solutions of different ionic compositions can be treated as an electrochemical problem. The composition of the solution and the solid is the basic information that helps to determine whether coagulation will or will not take place under given conditions. This is a physical-chemical approach to coagulation, aimed at the fundamental mechanisms behind the attraction or repulsion of particles. Hahn and Stumm (1968) and Stumm et al. (1970) dealt with the chemical aspects of coagulation in solutions and under model conditions resembling those of natural waters.

A purely physical approach to coagulation, inaugurated by von Smoluchowski (1918), is based on the probability of collisions between two particles in a fluid. The equations for the rate of coagulation due to the Brownian motion of particles and due to the velocity gradients in the fluid were originally developed by von Smoluchowski. Modern treatment and derivations of the coagulation equations, mostly in application to aerosols, can be found in an article by Hidy (1973) and in the book by Friedlander (1977). Coagulation in aqueous suspensions, with examples from chemical engineering systems, has been dealt with by Weber (1972), and coagulation in natural waters is the subject of the articles by Hahn and Stumm (1970), Edzwald et al. (1974), and Stumm (1977), who also refer to additional publications in this field.

In this section, the equations for the rates of coagulation due to the three mechanisms—Brownian motion, velocity gradients, and scavenging by settling—are presented without derivation. Their physical significance is made clear by the discussion of each equation and by the examples given at the end of the section.

Coagulation by Brownian Motion.   For particles immersed in a fluid and which are much bigger than the free path of the fluid molecules (as is the case for micron-size particles in water and air), the collision probability function for particles in the Brownian motion is

$$\mathcal{P}_B = \frac{2kT}{3\eta} \cdot \frac{(r_1+r_2)^2}{r_1 r_2} \qquad (cm^3\ sec^{-1}) \qquad (6.41)$$

The equation applies to spherical particles, where k is the Boltzmann constant (Appendix D), T is temperature in degrees Kelvin, and η is the viscosity of the fluid. The rate at which individual particles of radii r_1 and r_2 are removed due to successful collisions in a unit of fluid volume is, from equations 6.40 and 6.41,

$$J_B = \frac{2kT}{3\eta} \cdot \frac{(r_1+r_2)^2}{r_1 r_2} f N_1 N_2 \qquad (cm^{-3}\ sec^{-1}) \qquad (6.42)$$

For particles of the same size ($r_1 = r_2$), the collision probability function $\mathcal{P}_B$ becomes

$$\mathcal{P}_B = \frac{8kT}{3\eta} \qquad (6.43)$$

and the removal rate is

$$J_B = \frac{4kT}{3\eta} f N^2 \qquad (cm^{-3}\ sec^{-1}) \qquad (6.44)$$

The latter equation shows that the rate of coagulation is directly related to the second power of the particle concentration. This is analogous to a second-order chemical reaction. The term J_B can be replaced by the rate of particle removal $-dN/dt$ and (6.44) can be integrated to give the particle concentration N decreasing as a function of time from the initial value N_0,

$$N = \frac{N_0}{1 + N_0 k_2 t} \quad (\text{cm}^{-3}) \tag{6.45}$$

where the rate constant k_2 is $k_2 = 4kTf/3\eta$ cm^{-3} sec^{-1}. The rate of coagulation J_B is independent of the particle size.

Coagulation in a Velocity Gradient. A picture easy to visualize is a horizontal flow in the x direction, and horizontal flow velocity U_x changing in value in the vertical z direction. The shearing stress causes the velocity gradient G to be perpendicular to the flow,

$$G \equiv \frac{dU_x}{dz} \quad (\text{sec}^{-1}) \tag{6.46}$$

The vertical gradient will make the suspended particles move at differnet velocities, depending on their location along the vertical distance coordinate in the flow. The collision frequency function $\mathscr{P}_G$ for collisions in a velocity gradient is

$$\mathscr{P}_G = \tfrac{4}{3} G(r_1 + r_2)^3 \quad (\text{cm}^3 \text{ sec}^{-1}) \tag{6.47}$$

A different form of the collision frequency function has been derived from considerations of the rate of dissipation of mechanical energy in a fluid. The rate of dissipation of mechanical energy by turbulence ϵ (cm^2 sec^{-3}, dimensions of energy per unit mass of fluid per unit of time) and the kinematic viscosity of the fluid $\nu = \eta/\rho$ (cm^2 sec^{-1}) enter in the equation for the function $\mathscr{P}_G$ in the form of the quotient $(\epsilon/\nu)^{1/2}$, which has the dimensions of the velocity gradient G (Saffman and Turner, 1956). The probability of collisions of particles of radii r_1 and r_2 is

$$\mathscr{P}_G = 1.3(r_1 + r_2)^3 \left(\frac{\epsilon}{\nu}\right)^{1/2} \quad (\text{cm}^3 \text{ sec}^{-1}) \tag{6.48}$$

The rate of removal of particles due to successful collisions is

$$J_G = \tfrac{4}{3} G(r_1 + r_2)^3 f N_1 N_2 \quad (\text{cm}^{-3} \text{ sec}^{-1}) \tag{6.49}$$

For particles of the same size ($r_1 = r_2$), the coagulation rate is

$$J_G = \tfrac{16}{3} Gr^3 f N^2 \qquad (\text{cm}^{-3} \text{ sec}^{-1}) \qquad (6.50)$$

The coagulation rate depends on the particle size r as well as the particle-number concentration N. In a closed system, the product $\tfrac{4}{3}\pi r^3 N$ is the total volume of solids in a unit volume of water, and because the total volume is conserved

$$V = \frac{4}{3}\pi r^3 N = \text{const}$$

Using V in equation 6.50, the coagulation rate J_G or $-dN/dt$ is

$$J_G = \frac{4}{\pi} GVf N \qquad (\text{cm}^{-3} \text{ sec}^{-1}) \qquad (6.51)$$

The latter is a first-order removal process, where the rate of removal is directly related to the particle-number concentration N, and is determined by the rate constant $k = 4GVf/\pi$ sec^{-1}. The particle concentration N decreases with time from the initial value N_0 as

$$N = N_0 e^{-kt} \qquad (\text{cm}^{-3}) \qquad (6.52)$$

Scavenging by Settling Particles.   Collisions between settling particles can occur only for those that settle with unequal velocities. Thus for particles of the same density and shape, collisions can take place only between particles of unequal size. For particles of different densities, the constraint of unequal size presumably does not apply, but the settling velocities of any two particles must be different. The collision frequency function $\mathcal{P}_S$ for settling particles of two sizes r_1 and r_2 is

$$\mathcal{P}_S = \pi E (r_1 + r_2)^2 |U_{s1} - U_{s2}| \qquad (\text{cm}^3 \text{ sec}^{-1}) \qquad (6.53)$$

where U_{s1} and U_{s2} are the settling velocities. of particles of radii r_1 and r_2, and E is a collision efficiency factor that depends on the particle sizes. When a relatively large particle settles and approaches smaller particles settling more slowly in its path, collision and capture do not necessarily take place, because very small particles can evade the larger particle by flowing around it with the streamlines of the fluid that follows the surface of the faster moving particle. Because of this flow-by phenomenon, collision efficiency is generally poor for particles of very dissimilar sizes (for particles settling in air, the theory predicts the collision efficiency of about

1 when the ratio of the particle radii is about $r_1/r_2 \simeq 0.6$ and the larger particle is $r_2 \gtrsim 20$ μm). Explicit equations for the collision efficiency factor E exist for particles in the atmosphere, although they are complicated. A fairly simple approximation for the radius ratio $r_1/r_2 \lesssim 0.5$ is the following (Hidy, 1973):

$$E = 1 - \frac{3}{2(1+r_1/r_2)} + \frac{1}{2(1+r_1/r_2)^3} \qquad (6.54)$$

As the particle sizes become very dissimilar and the ratio r_1/r_2 tends to zero, so does the collision efficiency E.

The removal rate of particles J_S is

$$J_S = \pi E (r_1 + r_2)^2 |U_{s1} - U_{s2}| f N_1 N_2 \qquad (\text{cm}^{-3} \text{ sec}^{-1}) \qquad (6.55)$$

where E is defined in (6.54). The product $\pi(r_1+r_2)^2$ in equations 6.53 and 6.55 is the circular area within which a settling particle of radius r_2 can collide with a smaller particle of radius r_1.

Disregarding momentarily the collision efficiency and success factors (E and f), the rate of removal by scavenging in equation 6.55 can be written in a simpler form. For particles differing in size by a factor of 2 (that is, $r_1 = 0.5 r_2$) and settling with the Stokes velocity, the scavenging rate under these conditions would be

$$J_S \simeq 5.3 r_2^4 B N_1 N_2 \qquad (\text{cm}^{-3} \text{ sec}^{-1})$$

This is an upper limit for the rate of coagulation, as it assumes that all the geometrically possible collisions take place and all are successful ($E=1$ and $f=1$). In effect, however, the collision factors are $E \simeq 0.15$ and $f < 1$, such that the real rate of scavenging would be much smaller.

When a settling particle collides and captures in succession smaller particles in its path, the size of an aggregate grows. The mean radius of an aggregate r_n, formed by a series of collisions between a larger particle (r_2) and n smaller particles (r_1), is

$$r_n = r_2 \left[1 + n \left(\frac{r_1}{r_2} \right)^3 \right]^{1/3} \qquad (\text{cm})$$

In a layer of water h cm thick, containing a particle concentration N_1 (cm^{-3}), there are initially $h N_1^{1/3}$ particles in a linear path. If all collisions were successful, a larger particle (r_2) could scavenge $n = h N_1^{1/3}$ smaller particles, forming an aggregate of mean radius r_n,

$$r_n \leqslant r_2 \left[1 + h N_1^{1/3} \left(\frac{r_1}{r_2} \right)^3 \right]^{1/3} \qquad (cm)$$

Thus a long settling path h or a high concentration N_1 of the smaller particles can produce aggregates of a radius significantly larger than the initial radius r_2 of the scavenging particle. The settling distances of the order of 10^3–10^5 cm, as in lakes and oceans, can alone be responsible for aggregation by scavenging among the settling particles of different sizes (see the example at the end of this section). A conclusion that can be drawn from the preceding discussion is that the formation of aggregates by particle scavenging in the water column can be an efficient process if the settling distances are long; a significant fraction of the settling material can reach the sediment surface in an aggregated state.

Comparison of the Three Coagulation Mechanisms. The relative efficiency of each of the three coagulation processes—Brownian motion, velocity gradient (shear), and scavenging by settling—is measured by the values of the collision frequency function $\mathcal{P}$ and by the frequency f of successful collisions. For collisions between two particles, one of radius $r_1 = 1$ μm and the other in the size range $1 < r_2 < 20$ μm, the collision frequency function $\mathcal{P}$ for each of the coagulation mechanisms is plotted in Figure 6.5. For coagulation by the Brownian motion, equation 6.41, the values of $\mathcal{P}_B$ are symmetrical about the point of equal radii ($r_1 = r_2$). Because of the dependence of $\mathcal{P}_B$ on temperature and viscosity, the collision frequencies are higher by about 50% in warmer fresh water than in colder ocean water.

Coagulation by shear in a velocity gradient is more effective than coagulation by Brownian motion for velocity gradients greater than 0.5 sec^{-1}. The value of $G = 0.5$ sec^{-1} corresponds to a change in the velocity of flow of 0.5 cm sec^{-1} over a distance of 1 cm. For small-scale currents in lakes and oceans, with the velocities of the order of 1 ± 10 cm sec^{-1}, velocity gradients of the orders of 10^{-1}–10^0 sec^{-1} are a reasonable ball-park figure. Another estimate of the velocity gradients that gives comparable values of G for natural waters is based on the energy dissipation function $(\epsilon/\nu)^{1/2}$ appearing in equation 6.48. The kinematic viscosity of water ν is of the order of 10^{-2} cm^2 sec^{-1}; the rate at which mechanical energy is dissipated by turbulence in the ocean is to an order of magnitude between 10^{-3} and 10^{-2} cm^2 sec^{-3} (Levich, 1962; Kraus and Turner, 1967; Niiler, 1977). Thus the quotient $(\epsilon/\nu)^{1/2}$ equal to the velocity gradient G is somewhat less than 1. The values of the collision frequency function $\mathcal{P}_G$

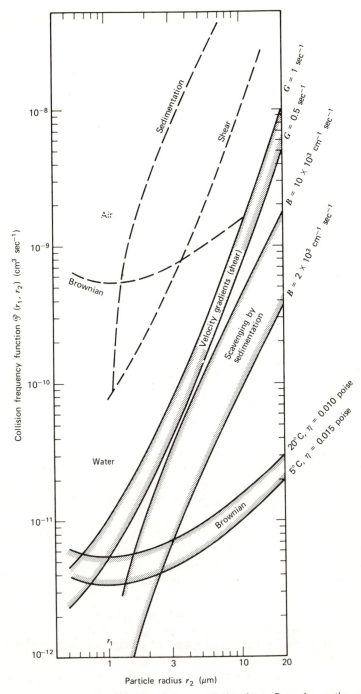

Figure 6.5 Collision frequency function $\mathscr{P}$ for coagulation due to Brownian motion, velocity gradient, and scavenging by settling in water. The diagram for collision frequencies in air (dotted lines) is shown for comparison (adopted from Hidy, 1973; Friedlander, 1977).

for coagulation by shear, shown in Figure 6.5, are bracketed by the velocity gradients of 0.5 and 1 $\sec^{-1}$.

The frequency of collisions between particles of radius $r_1 = 1$ μm and $r_2 \gtrsim 3$ μm due to settling is greater than the frequency due to the Brownian motion. The settling velocities of particles, on which the frequency of collisions depends, as given in equation 6.53, were taken as the Stokes velocities for the curves drawn in Figure 6.5. The parameter B in the Stokes settling velocity defined in equation 6.4 determines the settling velocity of a particle of any given radius r. For the lower curve in the figure, the value of $B = 2 \times 10^3$ cm^{-1} $\sec^{-1}$ represents light particles in fresh water at 20°C. For the upper curve, the value of $B = 10 \times 10^3$ cm^{-1} $\sec^{-1}$ represents somewhat heavier particles. A fivefold increase in the value of B and the settling velocity can also be caused by a relatively small increase in the particle density from 1.1 to 1.5 g cm^{-3}. The fields labeled velocity gradients and scavenging can be made to overlap for smaller velocity gradients and faster settling, such that the difference between the relative efficiency of the two coagulation mechanisms is not great.

Diagrams of the type shown in Figure 6.5 have been used in the analysis of coagulation of atmospheric aerosols. For the air, the relative positions of the three coagulation curves differ from their positions in the $\mathcal{P}$-r field for particles in water, as shown in the figure. In the atmosphere, scavenging by sedimentation is generally faster than coagulation by shear; the two mechanisms become more efficient than the Brownian coagulation for particles of radius greater than about 3 μm. The frequency of collision between settling particles in air is higher than in water because the Stokes settling velocities in air are about 100 times greater (Figure 6.1). The frequency of collisions in the Brownian motion is inversely related to the viscosity of the fluid: the viscosity of air is about 100 times lower than the viscosity of water, and the frequency of Brownian collisions in air is therefore about 100 times higher. The position of the curves for the Brownian coagulation and for the scavenging by settling relative to one another is about the same for air and water: the reasons are in the factor of 100 differences between the settling velocities and viscosities that displace the curves in the $\mathcal{P}$-r field upward by about the same factor.

The curve for the collision frequency by shear or velocity gradient in the atmosphere drawn in Figure 6.5 is somewhat higher than for the conditions in water. The atmospheric curve corresponds to the velocity gradient $G = 7$ $\sec^{-1}$, a value cited for the atmosphere at a height of about 100 m above the ground. Near the ground, within 1 m of the surface, the velocity gradients can be much stronger, of the order of 100 $\sec^{-1}$, and coagulation by shear can become more effective than scavenging by settling particles (Hidy, 1973; Friedlander, 1977; Saffman and Turner, 1956).

The fraction of the collisions that are successful, denoted f in the equations for the coagulation rate J_B, J_C, and J_S, can be highly variable. For clays (0.5–2 μm in diameter) and colloidal suspensions in general, the factor f depends on the ionic strength and the H^+-ion concentration in the solution. Data from the literature, summarized in part by Edzwald et al. (1974), give the factor f in the range between 0.01 and 0.2 for clay minerals and silica colloids in solutions of brackish waters, NaCl, and hydrolized alumina. Other values of f for coagulation of polysterene latex particles in NaCl solution are between 0.34 and 0.45. These data show that the percentage of collisions that are successful is generally less than 100%, and it can be as low as between 1 and 20%.

Examples. Estimate the rate of removal of particles J_B and J_S by Brownian motion and by scavenging in settling. Two cases will be considered: a very dilute suspension containing 20 μg l^{-1} solids (concentration comparable to deep ocean water) and a more concentrated suspension of 1 mg l^{-1}.

In the first case, a suspension contains 20 μg l^{-1} of solids, made of two kinds of particles: smaller particles of radius $r_1 = 1$ μm and larger particles of radius $r_2 = 5$ μm, present in equal proportions. For particles of density $\rho_s = 1.5$ g cm^{-3}, the suspension contains 25 particles cm^{-3} of the smaller size and 25 particles cm^{-3} of the bigger size. The equations for the rates of coagulation, equations 6.42, 6.49, and 6.55, give the initial rates in a closed volume containing initially 25 particles cm^{-3} of each kind. As an approximation to a steady-state process, where concentrations of the particles of two sizes are taken as constant, the equations give a measure of the rate of formation of aggregates made by two-particle collisions. As a third possibility, concentration of the larger particles N_2 can be considered constant, and the equations can be used to estimate approximately how fast the smaller particles N_1 are swept out of suspensions by collisions with the larger particles.

For coagulation by the Brownian motion, the value of the collision frequency function $\mathscr{P}_B$ from the graph in Figure 6.5 is about $\mathscr{P}_B = 8 \times 10^{-12}$ cm^3 sec^{-1}. The fraction of successful collisions f will be arbitrarily taken as 10% or $f = 0.1$. Using the above values of N_1, N_2, $\mathscr{P}_B$, and f in equation 6.42, the rate of coagulation J_B is

$$J_B = 8 \times 10^{-12} \times 0.1 \times 25 \times 25 = 5 \times 10^{-10} \quad cm^{-3} \ sec^{-1}$$

Thus particles of *either* size are removed from suspension through the formation of aggregates at the rate of 5×10^{-10} cm^{-3} sec^{-1}. The residence time of particles is, approximately,

$$(25 \text{ cm}^{-3})/(5 \times 10^{-10} \text{ cm}^{-3} \text{ sec}^{-1}) = 5 \times 10^{10} \text{ sec} = 1700 \text{ yr.}$$

The suspension is so dilute, that removal by coagulation plays virtually no role: particles of radii 1 and 5 μm settle through a distance of 1 cm in 3 hours and in 7 minutes, respectively (Stokes settling, $B = 10 \times 10^3$ cm^{-1} sec^{-1} given in Figure 6.5). With reference to the oceans, Arrhenius (1963) pointed out that coagulation by Brownian motion plays no role in speeding up the removal of particles from the deep ocean.

The rate of scavenging by settling, using the same numbers as above in equation 6.55 and the value of the collision frequency function from Figure 6.5, $\mathcal{P}_S = 1 \times 10^{-10}$ cm^3 sec^{-1}, is

$$J_S = 1 \times 10^{-10} \times 0.1 \times 25 \times 25 = 6 \times 10^{-9} \quad \text{cm}^{-3} \text{ sec}^{-1}$$

Although the scavenging by settling is ten times faster than coagulation by the Brownian motion, it is also ineffective as a removal mechanism in a dilute suspension containing only 20 μg l^{-1} of solids, when the settling distance is short.

The rate of particle removal by either mechanism of coagulation increases strongly with an increasing particle concentration. A suspension containing 1 mg l^{-1} of particles of the same two sizes and proportions as in the preceding example, has $N_1 = 1250$ and $N_2 = 1250$ particles cm^{-3}. The mass and number concentrations are 500 times higher. The value of 1 mg l^{-1} is a fairly low measure of concentration of suspended matter in lakes and estuaries. The rates of coagulation due to the Brownian motion and scavenging by settling in the thicker suspension are 500^2 times faster. Thus

$$J_B = 1.25 \times 10^{-4} \quad \text{cm}^{-3} \text{ sec}^{-1}$$

$$J_S = 1.56 \times 10^{-3} \quad \text{cm}^{-3} \text{ sec}^{-1}$$

The residence times for removal by the Brownian coagulation and scavenging are 0.3 yr and 0.03 yr, respectively. Although the residence times are relatively short, they are still higher than the residence time with respect to settling out of a 1-cm-thick water layer. For particles in the size range 1–5 μm radius, an additional improvement in the rate of coagulation by one order of magnitude can be achieved if velocity gradients of the order of $G = 1$–10 sec^{-1} are present in the fluid (see Figure 6.5 for the values of the collision frequency function, $\mathcal{P}_G$.)

Another question that should be asked is how efficient are 5-μm particles in removing by settling 1-μm particles from water? Clays and bacteria fall in the class of particles of about 1 μm in linear dimensions. Concentration of bacteria at the thermocline of a lake shown in Figure 6.6a is about 3×10^5 cells cm^{-3}. This is not a very thick suspension, as there are on the average 67 cells in a linear section 1 cm long. Allowing 2 μm for the length of a cell, the average distance between the cells in water is about 150 μm. If the concentration of the bigger particles N_2 is constant, the rate of removal of the smaller particles $-dN_1/dt$ is identical to J_S in equation 6.55. Therefore, the rate of removal becomes a first-order process,

$$-\frac{dN_1}{dt} = \mathcal{P}_S f N_2 N_1 \qquad \text{(particles cm}^{-3}\text{ sec}^{-1}\text{)}$$

As the parameters $\mathcal{P}_S$, f, and N_2 are constants, the equation integrates to the following form:

$$N_1 = N_{1,t=0} \exp(-kt) \qquad \text{(particles cm}^{-3}\text{)}$$

where k is a rate constant,

$$k = \mathcal{P}_S f N_2 \qquad \text{(sec}^{-1}\text{)}$$

The data used previously are: collision frequency for settling particles $\mathcal{P}_S = 1 \times 10^{-10}$ cm^3 sec^{-1}, $f=0.1$, and $N_2 = 1250$ particles cm^{-3}. These values give the rate constant k and the residence time $1/k$ of the 1 μm particles with respect to scavenging,

$$k = 0.4 \text{ yr}^{-1} \qquad \text{and} \qquad 1/k = 2.5 \text{ yr}$$

Particles somewhat bigger than 5 μm in radius scavenge the 1-μm particles three to four times faster (see Figure 6.5), such that the overall residence time of the small particles is likely to be closer to 0.5 yr. Note that these are residence times with respect to scavenging by settling, whereas the residence times with respect to ingestion by plankton or respiration may be shorter.

6.2.7 Concentration Peaks of Suspended Matter

The vertical distribution of suspended matter in lakes and oceans often shows concentration peaks at the thermoclines and pycnoclines. Water layers, from centimeters to meters thick, with a higher content of suspended material also occur at the boundaries between oceanic water masses and in estuaries, within the layer of the density gradient between the less saline surface waters and saline deeper waters. The term turbidity

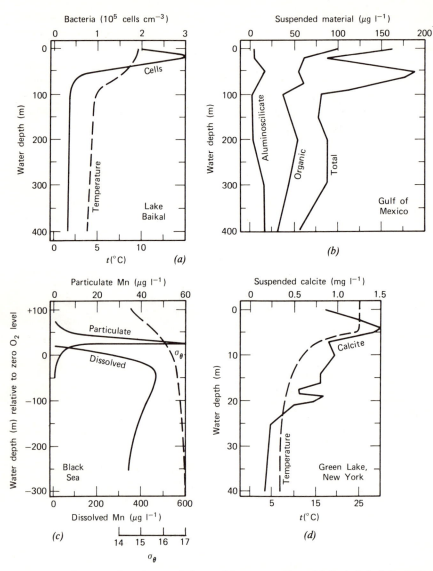

Figure 6.6 Concentration of suspended materials at pycnoclines. (*a*) Bacteria in Lake Baikal (Kuznetsov, 1975). (*b*) Suspended matter in the Gulf of Mexico (Feely et al., 1971). (*c*) Particulate and dissolved manganese in Black Sea. σ_θ is a function of sea water density ρ_θ at the potential temperature $\theta°C$: $\sigma_\theta = (\rho_\theta - 1) \times 1000$ (Spencer and Brewer, 1971). (*d*) Suspended calcite in Green Lake, near Syracuse, New York (data of G. J. Brunskill in Lerman et al., 1974).

maxima is used to describe the peaks in concentration of suspended matter, particularly in the literature dealing with the measurements of light scattering *in situ* and transparency of water. Figure 6.6 shows the vertical distribution of different suspended materials in the sections of the water column of a big lake, a small lake, the Black Sea, and the Gulf of Mexico. Concentration peaks occur in the region of the density gradient (pycnocline). The peaks of suspended manganese oxide particles in the Black Sea and of calcite nucleating in the surface waters of Green Lake are examples of inorganic materials accumulating in the pycnocline. The peaks of bacteria in Lake Baikal, and of organic and inorganic particles in the Gulf of Mexico are additional examples of the general phenomenon. In the surface waters of the ocean, the occurrence of concentration peaks of the phytoplankton and bacterial cells at some depths below the surface has been known since long ago (Sverdrup et al., 1942).

Suspended material rains down from the surface waters where most of it is produced biologically. The occurrence of concentration peaks much below the zone of photosynthesis implies that the settling material encounters resistance. In a steady-state settling process, the residence time of suspended material in a unit volume of the layer where peak concentration occurs is longer than in a unit volume of the water column above and below it. The longer residence time in water bears on the uptake or release of various chemical species (Section 6.3), and on the type and rates of sediment accumulation on the bottom.

Conditions of Peak Formation. Concentration of settling particles within a water layer can increase if the particles are slowed down in comparison to their settling velocities higher up in the water column. However, for a concentration peak to form, the particle concentration below the layer of accumulation must decrease. A decrease can be achieved by a faster transport or by other sinks to which particles are removed from water. The differences in the density of water layers, such as occur in the ocean or across the seasonal thermoclines of fresh water lakes, are small, and they alone do not explain the occurrence of the turbidity peaks. (Density differences across the pycnoclines in stratified saline lakes can be large, and the settling velocities of the lighter particles can be significantly affected at the interfaces between brackish waters and saline brines. For the effect of water density on the settling velocity see Section 6.2.2.) An increase in the viscosity of water due to a decrease in water temperature is percentwise significantly greater than an increase in density, and changes in viscosity across the thermoclines have a greater effect on the Stokes settling velocities. In seasonally stratified lakes, the temperature below the thermocline is lower than in the surface layer. In a steady-state settling, the velocities would decrease below the thermocline and the effect would be a higher concentration of suspended material in the colder water layer, although this would not produce a concentration peak in the water column at the thermocline.

The settling flux of particles in the upper water layer can be written as $F_1 = N_1 U_1$ (particles cm^{-2} sec^{-1}). N_1 is the particle-number concentration (cm^{-3}) and U_1 is the settling velocity, which may be a combination of gravitational settling and other transport mechanisms, such as up or down water currents and eddy diffusion. In the deeper layer, the effective settling velocity U_2 may be different because of a lower temperature and higher viscosity, and the flux of particles is, by analogy with upper layer, $F_2 = N_2 U_2$. If the particles neither grow nor shrink, at the steady state the flux conservation requires that the fluxes must be equal, $F_1 = F_2$. Therefore, the concentration of particles N_2 in the deeper layer is

$$N_2 = \frac{N_1 U_1}{U_2} \qquad (\text{cm}^{-3}) \qquad\qquad (6.56)$$

Concentration N_2 is either greater or smaller than N_1, depending on whether the particles in the deeper water layer settle more slowly or faster than in the surface layer. An increase in the viscosity of water by a factor of 1.5, when the water temperature decreases from about 25 to 10°C, would make the gravitational settling velocity U_2 smaller and the particle concentration N_2 higher by the same factor.

The flux conservation condition that applies to two layers with different settling velocities can be extended to a water column subdivided into a number of layers. In such a model, particles enter the top layer at a steady rate and are transported downward with velocities U_i that may vary from one layer to another. When the settling velocity U_i decreases and then increases over a number of thin water layers, concentration maxima of the settling particles are obtained in the layers of the small U_i, and concentration minima form in the layers where U_i is greater than the average. A more sophisticated approach to the problem of concentration maxima and minima is a one-dimensional stochastic model in which the particles have certain probabilities to move from any one of the layers to others above and below it. As the net settling direction is downward, the probabilities of moving to the lower layers are greater than to the layers above. The probability of a particle leaving a given water layer is analogous to its residence time or velocity within the layer at some depth in the water column. Concentration peaks occur in the model in those layers that are characterized by relatively low probabilities of exit (Lerman et al., 1974).

Instead of a water column subdivided into a number of thin layers or boxes, a one-dimensional transport model for particles, which treats the water column as a continuous space, can be constructed for the case of settling, eddy diffusional transport, and chemical source or sink reactions (Chapter 2). The following equation describes the vertical transport of particles:

$$\frac{\partial}{\partial z}\left(\kappa \frac{\partial N}{\partial z}\right) - \frac{\partial}{\partial z}\left[(U_s + U_a)N\right] \pm \mathfrak{R} = 0 \qquad (\text{cm}^{-3}\,\text{sec}^{-1}) \qquad (6.57)$$

where N is concentration (particles cm^{-3}), z is the vertical distance coordinate, positive and increasing downward, κ is the vertical eddy diffusion coefficient ($\text{cm}^2\,\text{sec}^{-1}$, identical to κ_z in the notation used in Chapter 2), U_s is the gravitational settling velocity ($\text{cm}\,\text{sec}^{-1}$), U_a is the vertical advective velocity of water (it can be either zero, positive if directed downward, or negative in the case of upwelling), and $\mathfrak{R}$ denotes any of the chemical or biochemical processes that may affect the particle concentration or mean size. For suspended material made of particles of different sizes, the eddy diffusion coefficient κ and gravitational settling velocity U_s have the significance of mean quantities (Section 6.2.5). Rearrangement of the derivatives in equation 6.57 gives

$$\kappa \frac{\partial^2 N}{\partial z^2} - \left(U_s + U_a - \frac{\partial \kappa}{\partial z}\right)\frac{\partial N}{\partial z} - N\frac{\partial U_s}{\partial z} \pm \mathfrak{R} = 0 \qquad (6.58)$$

In a water column where the eddy diffusion coefficient κ and gravitational settling velocity U_s remain constant with depth, the production $(+\mathfrak{R})$ or consumption $(-\mathfrak{R})$ of the settling material can alone control the occurrence and position of a concentration peak. In some oceanic sections, concentrations of the phytoplankton increases from the water surface to the base of the euphotic zone at about 50 m depth. This is due to the fact that the organisms are produced within the 50-m-thick layer, and those from the top of the layer sink and mix with those in the lower part of the euphotic layer. Below the base of the layer, where photosynthesis is no longer active, concentration of biogenic particles decreases with depth. The increase downward to the 50-m depth level and subsequent decrease result in a concentration peak (compare Figure 6.6a). A mathematical model of such a peak was developed for the case of production of the phytoplankton within the euphotic zone, as given by the term $+\mathfrak{R}$ in equation 6.58, and its destruction below the euphotic zone, represented by a negative production rate $-\mathfrak{R}$ (Riley et al., 1949; Riley, 1963).

Without the biological production and consumption, variation in the transport velocity of particles with depth can produce a concentration peak, as explained at the beginning of this section. The temperature and viscosity of water usually do not change with depth in a manner that creates minima in the gravitational settling velocities of particles. Therefore, the reasons for the occurrence of layers with low net velocity of transport have been sought in the vertical profiles of eddy diffusivity (Jerlov, 1959). The coefficients of vertical eddy diffusion are at a minimum in the zone of strong density gradients, such as exist at the thermoclines (Chapter 2). From the surface to the zone of the maximum density gradient, the coefficient of vertical eddy diffusion can be thought of as decreasing with depth, and from pycnocline downward it increases again, although not to the high values characteristic of the upper water layer exposed to winds. If the vertical eddy turbulence is the main mechanism that transports particles through the water column, then the low eddy turbulence near a pycnocline is analogous to a layer of slow transport, in comparison to the faster transport above and below it.

A change in the coefficient of vertical eddy diffusion κ with depth indicates that the gradient $d\kappa/dz$ is nonzero. $d\kappa/dz$ has the dimensions of velocity and enters in the advective term of equation 6.58. A change with depth in the value of $d\kappa/dz$ affects the advective term even if the gravitational settling velocity of particles U_s is constant. Some typical values of $d\kappa/dz$ and U_s given below show that the diffusion gradient is comparable in magnitude to the settling velocity. Therefore, the vertical eddy diffusion can be an important mechanism in the transport of settling materials under certain conditions. Above the thermocline, based on the studies of the vertical eddy diffusion in the epilimnion of a large lake, Lake Huron (Csanady, 1973, p. 107), the gradient $d\kappa/dz$ is in the range

$$-2\times10^{-3}<\frac{d\kappa}{dz}<-50\times10^{-3}\qquad(\text{cm sec}^{-1})$$

The higher value of about -0.05 applies to the upper several meters of the water column and the lower value applies to the depths extending down to the thermocline depth near 25 m. In lakes, below the pycnocline (from sources listed in Figure 2.12), the gradients of the coefficient of vertical eddy diffusion are positive because the coefficient increases somewhat with depth. The gradient $d\kappa/dz$ is in the range

$$0.02\times10^{-3}<\frac{d\kappa}{dz}<1\times10^{-3}\qquad(\text{cm sec}^{-1})$$

The gravitational settling velocities of small particles in water, from Figure 6.1, are between 10^{-4} and 10^{-2} cm sec^{-1}, for particles of radius between 1 and 10 μm. The graph in Figure 6.1 applies to quartz spheres whose net density in water is $2.65-1.0=1.65$ g cm^{-3}. Lighter particles containing organic matter can have a net density of about 0.1 g cm^{-3}, and these would settle about 15 times slower. An overall range of settling velocities for the smaller particles is

$$10^{-5} < U_s < 10^{-2} \quad (\text{cm sec}^{-1})$$

Thus the ranges of the settling velocities and gradients of the diffusion coefficient can overlap in natural waters, making both mechanisms important to the vertical transport of suspended materials.

An interesting mechanism for an increase in concentration of particles within a water layer was proposed by Stommel (1949): particles can be trapped in convection cells such that concentration of particles of certain sizes may increase within a layer where vertical convection takes place. A simple model of convection cell, such as the Langmuir convection cell, has water circulating in a vertical plane. The cell consists schematically of an upward directed flow, which bends to a horizontal flow, and the latter turns downward and horizontally in the opposite direction, closing the cell. Particles, whose gravitational settling velocities are smaller than the upward velocity of water, ascend with the flow and fall out of it toward the core of the convection cell when the streamlines bend from the vertical into the horizontal direction. Some particles may leave the cell but some are picked up again by ascending flow. In the notation of equation 6.57, a negative difference between the velocities of a settling particle and upward flow of water, $U_s - U_a < 0$, causes the particle to rise. Entrapment of the smaller particles within the convection cells makes them, according to the theoretical analysis, move along closed contour stream lines within the cells. Because of some water exchange that can take place between adjacent convection cells in the horizontal direction by eddy turbulence, concentrations of the trapped particles can appear as a more or less continuous horizontal layer. If a sharp density change occurs below the level of convection, downward transport of the trapped particles by eddy turbulent exchange with the underlying water would be restricted. The result is a layer of higher concentration of particles or a concentration peak in a vertical profile of suspended material.

Chemical Sources and Sinks. Peaks and minima in concentration of dissolved or particulate material can form in a vertical concentration profile if there are sources and sinks, respectively, occurring within a water layer. For example, minima in concentration profiles of oxygen in water are related to the consumption of oxygen by organic matter, and maxima in concentration of silica in pore waters of sediments are often related to the abundance of diatom skeletons within sediment layers. Peaks in the abundance of bacteria in water, as in Figure 6.6a, may be caused by the availability of live and dead organic material on which bacteria thrive. A peak in concentration of particulate manganese oxide in the Black Sea, shown in Figure 6.6c, occurs a short distance (20–30 m) above the level of zero concentration of dissolved oxygen. Oxidation of dissolved manganese, transported upward from the deeper water, causes the precipitation of manganese oxide. As the particles settle toward and below the zero oxygen level, manganese is regenerated to solution. The rate of regeneration or the rate of dissolved manganese production J in the water column has been given as (Spencer and Brewer, 1971) $J = 35 \exp(-0.038z)$ μg kg^{-1} yr^{-1}, where z is in meters. The rate at which suspended manganese oxide is consumed by reduction and dissolution falls off with depth from the value of about $J = 35$ μg kg^{-1} yr^{-1} at a level 20 m above the zero oxygen depth ($z = 0$). The rate decreases by a factor of 2 over a distance of 20 m. The decrease with depth in the rate of production of dissolved manganese reflects the fact that most of the particulate material is regenerated to solution within a few tens of meters below its concentration peak. A similar concentration peak of suspended iron-containing material that settles and dissolves has also been described for the Black Sea (Brewer and Spencer, 1974).

Concentration profiles of dissolved species, such as the profile of manganese shown in Figure 6.6c, that form by chemical production or consumption reactions can be described by one-dimensional vertical transport models that include eddy diffusion, vertical advection of water, and a constant or a depth-dependent chemical rate term. Such one-dimensional transport models for reactive chemical species (oxygen, carbon) and radioactive tracers in the ocean have been dealt with by Wyrtki (1962), Munk (1966), and Craig (1969) for steady-state vertical concentration profiles. Solutions for transient states and certain sets of boundary conditions in one-dimensional columns of water have been given by Lerman (1971). A three-dimensional model based on diffusion, advection, and production, applying to the regeneration of silica in the Pacific Ocean, has been dealt with by Fiadeiro (1975).

Along a vertical concentration profile of suspended material, such as the profiles in Figures 6.6b and 6.6d, the condition of the flux conservation can hold *either* for the number *or* mass of particles in a unit volume of water. For particles that are transported downward by gravitational settling only, the particle-number flux is NU (particles cm^{-2} yr^{-1}) and the particle-mass flux is MU (g cm^{-2} yr^{-1}), using the notation of the preceding sections for the particle-number concentration N and mass concentration M. If particles grow by chemical precipitation or dissolve below the layer of a concentration maximum, then the particle-number flux is conserved (as long as the particles have not completely dissolved). However, if particles coagulate or break into fragments, then the number of particles changes but the mass is conserved. Either of the two conservation conditions leads to certain relationships between the sizes of particles and their concentrations within and below the layer of the maximum.

The particles-number fluxes, as given in equation 6.56, in the layer of high concentration $N_1 U_1$ and in the deeper layer $N_2 U_2$ are equal when the flux is conserved. The particle-number and particle-mass fluxes are interrelated. The mass concentration M of N spherical particles, each of radius r, is $M = \frac{4}{3} \pi \rho_s r^3 N$ (g cm^{-3}) and their Stokes settling velocity is $U = Br^2$. Using these in the flux conservation equation 6.56, we obtain

$$\frac{M_2}{M_1} = \frac{r_2}{r_1} < 1 \qquad (6.59)$$

The suspended matter concentration below the peak is M_2, which is smaller than the concentration in the peak layer M_1, and this means that the particle size below the peak layer becomes smaller. A smaller size implies dissolution and an increase in the particle concentration N_2 (particles cm^{-3}) owing to the slower settling velocity. However, the occurrence of the turbidity maxima is usually detected by light-scattering techniques as an indication of a zone of high concentration of particles. Thus equation 6.59 does not explain a decrease in the number of particles below the peak layer, but it applies to general conditions for settling particles that settle through a water layer and dissolve in the layer below (see Section 6.3.2).

The mass-flux conservation between the peak layer and deep layer is

$$M_1 U_1 = M_2 U_2 \qquad (\text{g cm}^{-2} \text{ yr}^{-1}) \qquad (6.60)$$

For the Stokes-settling particles, the relationship between the suspended matter concentrations and particle radii becomes

$$\frac{M_2}{M_1} = \left(\frac{r_1}{r_2}\right)^2 < 1 \qquad (6.61)$$

Because the mass concentration below the peak is smaller than within the peak layer, we have $M_2 < M_1$, and the particle radius in the deep layer is greater, $r_2 > r_1$. Growth by chemical precipitation on particles is ruled out in this case, because the particle mass is conserved by the conditions of equation 6.60. Another process that can lead to an increase in the particle radius is coagulation, which forms aggregates of an average radius greater than the individual particles. For example, the concentration profile of suspended calcite in Figure 6.6d shows a decrease from about 1.5 mg l^{-1}

at the concentration maximum to 0.25 mg l^{-1} at depth. A decrease in concentration by a factor of 2 corresponds to the mass concentrations ratio $M_2/M_1 = 0.5$. From equation 6.61, the average particle radius in the deep layer is

$$r_2 = 2^{1/2} r_1 = 1.41 r_1$$

A 40% increase in radius can be achieved if three particles of equal size form a floc: the volume of a floc is not less than the sum of the volumes of individual particles of radius r_1. The mean radius of a three-particle floc r_2 is

$$r_2 \simeq (3r_1^3)^{1/3} = 1.44 r_1$$

In terms of this model of particle aggregation, the bigger flocs settle faster and maintain the concentration peak in the overlying layer to which material is supplied by slower settling from above.

Retardation at Density Interfaces. In stratified solutions where a sharp interface is present between the lighter upper layer and denser lower layer, flocs of suspended material settling through the upper layer can stop at the interface zone. As opposed to flocculated materials, single particles (or compact flocs) go through the interface, as expected, without stopping, and sink through the lower layer at a slower velocity that depends on the differences in viscosity and density between the two layers. For flocs, a schematic picture of retardation of sinking is shown in Figure 6.7a. Such materials as goethite (α-FeOOH) in needle crystals of dimensions $10^{-1} \times 10^0$ μm, and amorphous silica powder (SiO_2) in spherules of diameters between 70 and 400 Å (7×10^{-3}–40×10^{-3} μm) flocculate readily in water and NaCl solutions, and form flocs of dimensions from microscopic to about 1 mm, observable by naked eye.

An analog of a density-stratified water column that is easy to make is a layer of pure water or dilute salt solution on top of a more concentrated solution layer. One method of making a stratified column is to pour the less dense fluid first into a vessel, and to let the heavier fluid flow slowly through a pipette held near the wall of the vessel and almost touching the bottom. If the density difference between the two solutions is appreciable and the flow of the heavier fluid to the bottom of the vessel is controlled with care, a fairly sharp interface zone can be obtained. Some mixing of the two solutions is unavoidable, such that the interface zone is effectively a thin layer within which salt and water concentration gradients exist. An easier method of making a sharp interface is to pour the heavier fluid in the vessel first, place on top of the fluid a floating porous disc (cut out of

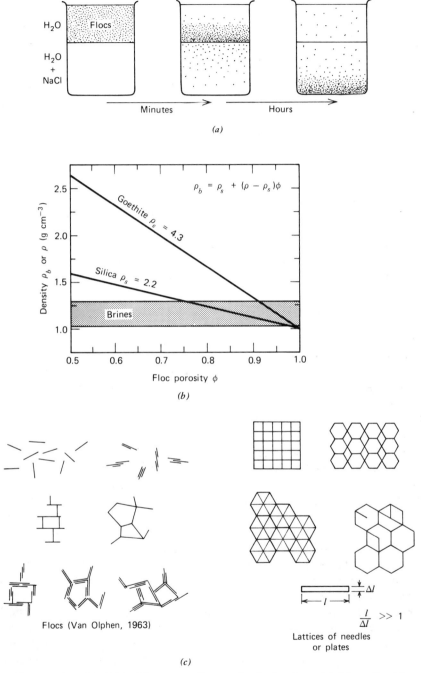

Figure 6.7 Flocculated materials at pycnoclines or density interfaces. (a) Schematic picture of retardation of settling at density interface. (b) Bulk density of flocs ρ_b as a function of their porosity or water content by volume ϕ. (c) Shapes of two-dimensional flocs (from van Olphen, 1963, reproduced by permission), and models of two- and three-dimensional lattices.

thin foam rubber or similar material), and to pipette the lighter solution on the disc, allowing it to flow through the pores and build an upper layer. More than two layers of different density can be easily made by this method.

Flocs dispersed in the upper layer settle to the interface zone and accumulate there without sinking into the deeper layer (Figure 6.7a). Within minutes to tens of minutes individual flocs begin to sink and eventually all of the material settles on the bottom of the deeper layer. Goethite and silica, mentioned earlier, aggregate into bigger flocs in the interface zone; in experiments with kaolinite in aqueous NaCl solutions, the flocs settling below the interface are much smaller but still discernable by eye. The retardation phenomenon is not of an all-or-none type, as some particles and flocs from the upper layer sink through, while a large fraction of the flocculated material is retarded.

Large interfacial tension that exists at the water-air interface is sometimes sufficient to retard the settling of nonwettable particles of density higher than water. The interfacial tension σ for the water-air interface at 20°C is $\sigma = 73$ dyn cm^{-1}. For a saturated NaCl solution at the same temperature, the interfacial tension at the solution-air interface is somewhat higher, $\sigma = 83$ dyn cm^{-1}. The difference between the two tensions is a measure of the interfacial tension between a concentrated salt solution and pure water, $\sigma \simeq 10$ dyn cm^{-1}. Such an interface is thermodynamically unstable, as diffusion tends to equalize the concentration differences of salt and solvent that exist across the interface. The floating of dust and pollen particles at the water surface, often attributable to the surface tension effect of the water-air interface, has another aspect to it: a cohesive floc of particles containing air in the interparticle space can have a density about equal to the density of water. For example, the bulk density of a floc made by volume of 40% particles (density 2.5 g cm^{-3}) and 60% air (density 0.001 g cm^{-3}) is about 1.0 g cm^{-3}. The same floc filled with water would have the density of 1.6 g cm^{-3}, which would make it denser than water. The time required for water to penetrate into the pore space of a floc generally depends on the mechanical cohesiveness of the floc, sizes of the pores, and wettability of the material.

Flocs of clay and other mineral particles in water are irregularly shaped aggregates forming more or less flat or framework lattices. In such flocs as those illustrated schematically in Figure 6.7c, the linear dimensions of the spaces between the individual particles are comparable to the particle dimensions. In flocs made of clay mineral particles or needle-shape crystals, the longer dimensions of the particles and the interparticle distances are of the order of 1 μm, or smaller if the floc collapses. The bulk density of a floc is a weighted sum of the densities of the solid and solution in the interparticle space (as long as the floc maintains its shape),

$$\rho_b = \rho_s - (\rho_s - \rho)\phi \qquad (\text{g cm}^{-3}) \qquad\qquad (6.62)$$

where ϕ is the volume fraction of the floc filled with solution of density ρ, and ρ_s is the density of the solid material. For two solids of dissimilar densities—goethite of density 4.3 g cm^{-3} and amorphous silica of density 2.2 g cm^{-3}—the bulk densities of flocs are plotted as a function of their porosity in Figure 6.7b. The bulk density approaches the density of the filling water ($\rho = 1$) as the floc porosity tends to unity. Flocs containing 75–95 vol % water, have densities around 1.15–1.3 g cm^{-3}, similar to some of the more saline natural brines. This suggests that porous flocs filled with fresh or brackish water would not immediately sink into the brine if the brine were covered by a layer of lighter solution, as happens during the periods of flooding in saline lakes.

The porosity of flocs can be estimated for two- and three-dimensional arrangements of particles, and the porosities of some of the simpler arrangements are the following. Spheres in a random packing have the porosity of about 40% or $\phi \simeq 0.4$. Spheres in a cubic lattice arrangement have the porosity of about 50% or $\phi \simeq 0.5$. More porous structures, with porosities over 90%, can be formed from rods and needles, making two-dimensional grids of squares and hexagons, and three-dimensional frameworks of cubes and hexagonal prisms. With reference to the drawings of flat flocs in Figure 6.7c, the porosity in cross section or in space depends on the length/width ratio of the rods or plates making the floc. Some cross-section porosities for grids made of rods of the length/width ratio of 20 are: grids of hexagons and squares, $\phi = 0.8$–0.9; and grids made of equilateral triangles, $\phi = 0.45$–0.8, depending on whether all triangles are intact (lower porosity) or whether some lose their laterals and retain their bases forming hexagons (higher porosity). Simple regular three-dimensional arrangements are lattices made of cubes and hexagonal prisms, the edges of which are rods of a length/width ratio of 20. For such flocs, the porosities are from $\phi = 0.97$ to over 0.99. Flocs made of solid particles of density 2.2 g cm^{-3} and having porosity of between 0.97 and 0.99, have bulk densities of 1.03–1.01 g cm^{-3}, comparable to or less than the mean density of sea water. A sinking floc that encounters an interface at a layer of density higher than the floc would be stopped by buoyancy. Settling can continue either if the floc breaks down or coalesces into denser units, or if the lighter solution in the interparticle space is replaced by the denser solution from the underlying water layer. When the density differences are caused by salt concentration, diffusion of salt through the interface zone and into the interparticle space of the floc can raise the bulk density of the floc, enabling it to sink. The time required for the salt concentration within the floc space to increase due to diffusion from below can be estimated at $t = L^2/D$, where L is the floc length and D is the molecular diffusion coefficient of salt. For lengths L between 0.1 and 1 mm, and diffusion coefficient between 10^{-6} and 10^{-5} cm^2 sec^{-1}, the times are $t = 10^2 – 10^3$ sec. The times of tens of minutes agree with the observations of goethite, silica, and kaolinite flocs at the interfaces in aqueous salt solutions. In a density gradient induced by a temperature gradient, thermal equilibrium between the interior of a floc and a colder exterior can be attained much faster because the diffusion of heat in water is faster than diffusion of solutes (10^{-3} compared to 10^{-5} cm^2 sec^{-1}).

The abundance of flocculated particles in lakes, estuaries, and coastal ocean waters does not imply that the retardation of their settling by the mechanism described in this section is primarily responsible for the turbidity maxima observable at the pycnoclines. The first-order effects in surface waters are variations in the settling velocities caused by the vertical eddy turbulence, and by the changes with depth in the relative importance of production and regeneration of biogenic materials.

6.3 REGENERATION OF SUSPENDED MATERIALS

6.3.1 Biological Production and Regeneration

The end result of the biological production in waters is the building of organic and mineral materials from the dissolved chemical species in solution. The phytoplankton is the primary producer of organic material on which the zooplankton and higher members of the ecological pyramid depend. Regeneration is the return of the biologically fixed materials to water. Some materials are regenerated to water by exchange across the cell walls and excretion from living organisms, but the net consumed by an organism during its lifetime is regenerated to ocean water in a series of steps involving physical and chemical breakdown and decomposition. This section, as its heading implies, is aimed at the final stage of the process that begins with the biological production and ends in the return of materials to solution in lake and ocean waters. The questions of how much is regenerated in water, and how fast, are the main questions dealt with in this section. The background information and discussion of regeneration center around the biologically fixed carbon in organic matter and in calcium carbonate, as examples of organic and mineral materials involved in the cycle of biological production and regeneration.

Insofar as the photosynthetic activity is limited to a surface water layer where a sufficient amount of light is available, a general model of the cycle of biological productivity and regeneration drawn in Figure 6.8 shows a water column divided into two sections, a surface layer and a deep layer. A more detailed discussion of this simple box model will be given in Section 6.3.2. The major steps of the cycle are summarized below for the three reservoirs: surface layer, deep layer, and the bottom. Each a item is a source and b item a sink for a given reservoir, following the pattern introduced in Chapter 1.

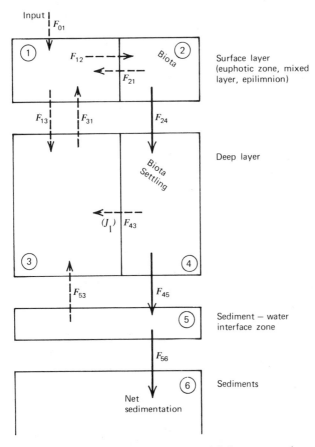

Figure 6.8   A model for regeneration of biogenic materials in a water column and on the bottom.

1. Surface water layer
    a. Production of organic and mineral material. Availability of nutrients, light, and temperature control the production. Source materials supplied from the atmosphere, by river inflow, and by water exchange with the deep water layer. Source materials include carbon dioxide, nitrogen in different forms, phosphorus, sulfur, trace metals, and for skeletal construction, primarily calcium and silicon (magnesium and strontium enter as minor components of skeletal materials).
    b. Sinks of the biologically produced material in the surface layer include regeneration (oxidation, dissolution, respiration), ingestion by higher organisms, and settling out of the layer.
2. Deep water layer
    a. Settling of biogenic materials out of the surface layer.
    b. Regeneration to water by chemical and biochemical reactions involving oxidation, bacterial decomposition, and dissolution.
3. Bottom
    a. Deposition at the sediment-water interface.
    b. Regeneration to overlying water until the reactive material has been either exhausted or covered by a sufficiently thick layer of accumulating sediment.

Table 6.6   Rates of Biological Production and Standing Crop of Organic Carbon in Lakes

Net primary production rates (mol C m^{-2} yr^{-1})[a]	
Tropical	2.5–200
Temperate	0.2–80
Alpine	<0.1–8
Arctic	<0.1–3
Antarctic	<0.1–1
Standing crop of organic carbon	
Particulate C	5–500 μmol l^{-1}
Dissolved C[b]	100–1700 μmol l^{-1}
Removal rate constants[c]	
Mineralization (oxidation)	1–10 yr^{-1}
Sedimentation	0.5–50 yr^{-1}

[a]Likens (1975). Net=gross assimilation-respiration. High values for tropical and temperate lakes represent extremely eutrophic conditions.

[b]Livingstone (1963). Carbon taken as fraction $\frac{12}{30}=0.4$ of organic matter.

[c]Imboden and Lerman (1978). Rate constant×particulate concentration = removal rate per unit volume.

The chemical (or biochemical) reactivity of the settling material in water controls ultimately how much is regenerated in the different parts of the water column. However, the length of the water column and the rates of sediment accumulation also affect the total amount of materials regenerated. As long as oxygen dissolved in water is not a limiting factor in decomposition of organic materials, a greater fraction of the total amount regenerated may be oxidized in a longer water column. In shallow lakes the settling times can be sufficiently short, such that a significant fraction of the biogenic material produced in surface water reaches the bottom. For organic carbon, such information as the rates of biological production, amounts of the standing crop, and the rates of removal from water in lakes and oceans are summarized in Tables 6.6 and 6.7. With the exception of the highly eutrophic and unusually productive lakes in the tropical and

Table 6.7   Rates of Biological Production and Standing Crop of Organic Carbon in Oceans

Mean net primary production rates (mol C m^{-2} yr^{-1}) and areas (km^2)[a]		
Oligotrophic subtropical central waters	2.1	(148×10^6)
Transitional waters between subtropical and polar zones	4.2	(83×10^6)
Equatorial divergence and subpolar waters	6.1	(86×10^6)
Inshore waters	10.3	(39×10^6)
Neritic waters	30.4	(11×10^6)
Ocean mean and total area	5.2	(367×10^6)
Standing crop of organic carbon		
Surface ocean 0–100 m:		
Particulate C[b]	2.5–4.6 μmol l^{-1}	
Dissolved C[c, d]	50–90 μmol l^{-1}	
Regeneration rate constant for respiration[e]	5–120 yr^{-1}	
Deep ocean 500-5000 m:		
Particulate C[b, c]	0.25–0.7 μmol l^{-1}	
Dissolved C[c]	25–60 μmol l^{-1}	
Regeneration rate constant for oxidation of dissolved C[f]	$(0.5–9) \times 10^{-3}$ yr^{-1}	

[a] Koblentz-Mischke et al. (1970).
[b] Wangersky (1976); Romankevich (1977).
[c] Menzel and Ryther (1970).
[d] Holm-Hansen et al. (1966).
[e] Parsons and Takahashi (1973).
[f] Riley (1963); Munk (1966); Craig (1971).

temperate zones, the net rates of the biological productivity in the lakes and oceans are comparable. In the oceans, 3% of the surface area taken up by neritic waters (depths not greater than 200 m) produce as much organic material as the 40% of the surface allocated to the oligotrophic waters of the central parts of the oceans. The standing crop of particulate and dissolved organic carbon in lakes is generally higher than in the surface layer of the oceans (0–100 m). The more productive and eutrophic lakes are commonly only a few tens of meters deep, and the shallowness accounts for the higher values at the upper end of the concentration range of the particulate and dissolved organic carbon.

The rates of regeneration through oxidation or mineralization (a term used to denote conversion of organics to dissolved inorganic or mineral material) are commonly given in the literature in units of mass of carbon oxidized or oxygen consumed per unit of dry mass of plankton per unit of time $(MM^{-1}T^{-1})$. Also in common use is the rate of regeneration in units of mass per unit volume of water per unit of time $(ML^{-3}T^{-1})$. A rate constant for removal process can be defined as a ratio of the mass removal rate to concentration: $k = (ML^{-3}T^{-1})/(ML^{-3})$. This gives rate constants in units of time^{-1}, as listed in Tables 6.6 and 6.7 for the different removal processes. In surface ocean waters, the higher values of the rate constant for regeneration near 120 yr^{-1} reflect the fast turnover in tropical coastal waters, whereas the lower values between 10 and 20 yr^{-1} are more representative of the global averages.

6.3.2 Production and Regeneration: a Model

The Major Fluxes. An outline of the main steps of the production and regeneration cycle was given in the preceding section with reference to Figure 6.8. The fluxes indicated by broken arrows represent the fluxes of dissolved material. These include input to the surface layer from outside (F_{01}), uptake and release of dissolved material by the living and dead biota $(F_{12}$ and $F_{21})$, mixing of the surface and deep waters $(F_{13}$ and $F_{31})$, and regeneration from the settling (F_{43}) and settled biogenic material (F_{53}). The fluxes of the solids are the settling flux from the surface to the deep layer (F_{24}), deposition on the bottom (F_{45}), and inclusion in the sediment as the sediment column grows (F_{56}). A modification and improvement of the model shown in the figure can include, for example, an additional reservoir for the sediments in contact with the surface water layer. In the ocean, such an additional box would represent the settling and regeneration of material in shallow water. For lakes, it would include the interactions between the waters and sediments of the epilimnion, where the chemical processes taking place under oxidizing conditions can be very different from those in the anoxic environment of the hypolimnion during the period of thermal stratification. The two reservoirs labeled *biota* have different sizes and fluxes for different biological and chemical species. Each of the regeneration fluxes F_{21}, F_{43}, or F_{53} can vary from zero up, depending on the settling velocities and chemical reactivity of the material.

At a steady state, input of a chemical species into the surface layer, flux F_{01}, is balanced by its removal into sediments, flux F_{56}. Other mass balance equations for the individual reservoirs can be written in the form as explained in Chapter 1. In the model of Figure 6.8, the amount of material leaving the surface layer by settling, flux F_{24}, can be estimated from the following mass-balance condition:

$$F_{24} = F_{01} + F_{31} - F_{13} \qquad \left(\text{mass cm}^{-2}\ \text{yr}^{-1} \right) \qquad (6.63)$$

The fluxes on the right hand side of the equation are

$$F_{01} = F_{56} \qquad (6.64)$$

$$F_{31} = k_d h_d C_d \qquad (6.65)$$

$$F_{13} = k_s h_s C_s \qquad (6.66)$$

where subscripts d and s denote the deep and surface layers, $1/k$ is the residence time of water within the layer (yr), h is the thickness of the layer (cm^3 cm^{-2}), and C is concentration of the chemical species in solution that is being taken up into the biologically produced solids. Concentrations C_d and C_s are mean concentrations in each layer. A mass-balance condition for water exchange between the surface and deep layers connects the rate constants k and the layer volumes h,

$$k_d h_d = k_s h_s \qquad \left(\text{cm yr}^{-1} \right) \qquad (6.67)$$

The water flux kh and the water residence time $1/k$ in each layer depend on the thickness of the layer. For an "average ocean,"

$$k_s h_s = k_d h_d \simeq 3 \pm 1 \quad \text{m yr}^{-1}$$

The latter figure is based on the residence time of deep ocean water of about 1000 yr and a volume of between 3 and 4 km.

For seasonally stratified lakes, estimates of the water residence time with respect to mixing across the thermocline can be obtained from the following ball-park figures. The coefficient of vertical eddy diffusion in the thermocline layer, about 0.05 cm^2 sec^{-1} (Figure 2.13), and the thermocline thickness of about 5 m, give the rate of water transport (0.05/500 cm) as

$$k_s h_s = k_d h_d \simeq 30 \quad \text{m yr}^{-1}$$

The seasonal thermocline separating the epilimnion from the hypolimnion occurs rarely at depth greater than 25 m. Taking $h_s = 20$ m and $h_d = 50$ m, the residence time in the epilimnion is $1/k_s = 0.7$ yr, and in the hypolimnion $1/k_d = 1.7$ yr. The residence times can be longer or shorter, depending on the sharpness of the thermocline layer, vertical eddy diffusivity in the thermocline layer, the degree of mixing within the hypolimnion, and the depth of the lake. Overall, the residence times are somewhat longer than the length of time the lakes remain seasonally stratified in temperate regions. For lengths of time that are shorter than the residence times in either of the two water layers, lake models often treat the epilimnion and hypolimnion as two boxes isolated one from the other during the period of stratification.

Regeneration of Biogenic Silica in the Ocean. The rates of removal of biologically fixed silica from the surface layer of the ocean and its rate of regeneration are estimated using the flux equations of the preceding section with the following information on the rates of silica deposition and concentrations in the Pacific Ocean water.

$$F_{56} = F_{01} = 2 \quad \mu\text{mol cm}^{-2}\text{ yr}^{-1} \quad \text{(Burton and Liss, 1973; Heath, 1974)}$$

$$C_d = 140 \pm 20 \quad \mu\text{mol l}^{-1} \qquad \text{(Fiadeiro, 1975)}$$

$$C_s = 15 \pm 5 \quad \mu\text{mol l}^{-1} \qquad \text{(Fiadeiro, 1975)}$$

$$k_s = 3 \times 10^{-3} \quad \text{yr}^{-1}, \qquad k_d = 1 \times 10^{-3} \quad \text{yr}^{-1}$$

$$h_s = 100 \quad \text{l cm}^{-2}, \qquad h_d = 350 \pm 50 \quad \text{l cm}^{-2}$$

Note that the thickness of the upper water layer h_s is taken as 1 km and the lower layer h_d as between 3 and 4 km. This makes the water column between 4 and 5 km thick, instead of the mean depth of the ocean, 3.8 km. However, the area of the world ocean that lies between the depths 1 and 6 km covers 87% of the total area of the ocean (Menard and Smith, 1966), such that the 4–5 km thick water column is a valid model for the pelagic ocean. The settling flux of biogenic silicate particles leaving the surface layer is, from equations 6.63 through 6.66,

$$F_{24} = 10^{-3} \times (350 \pm 50) \times (140 \pm 20) + 2 - 3 \times 10^{-3} \times 100 \times (15 \pm 5)$$

$$= 47 \pm 12 \quad \mu\text{mol cm}^{-2}\text{ yr}^{-1}$$

The total amount of silica regenerated annually to the deep water layer is the sum of the amounts regenerated in the water column and at the sediment surface,

$$F_{43} + F_{53} = F_{24} - F_{56}$$

$$= 47 - 2 = 45 \quad \mu\text{mol cm}^{-2}\,\text{yr}^{-1}$$

Additional information is needed to estimate one of the two regeneration fluxes, either F_{43} in the water column or F_{53} at the bottom. Regeneration in the water column is

$$F_{43} = 23 \pm 4 \quad \mu\text{mol cm}^{-2}\,\text{yr}^{-1}$$

One estimate of F_{43} is based on a numerical model of three-dimensional distributions of dissolved silica concentrations in the Pacific Ocean (Fiadeiro, 1975). The other estimate, independent of the concentration field of dissolved silica, is based on a one-dimensional model of settling and dissolution (Lerman and Lal, 1977), also discussed with reference to calcium carbonate in the ocean in Section 6.3.4. Using the above value for the rate of regeneration in the water column, the amount of silica regenerated at the sediment-water interface is

$$F_{53} = (45 \pm 12) - (25 \pm 2) = 20 \pm 12 \quad \mu\text{mol cm}^{-2}\,\text{yr}^{-1}$$

The amounts of SiO_2 regenerated in the water column and on the bottom are comparable. In those areas of the ocean where the abundance of the silica secreting organisms is either very low or very high, the local rates of regeneration can be very different from the averages derived in this section. A summary of rates of regeneration of silica and calcium carbonate in different parts of the oceanic water column and in sediments is given in Table 7.1.

6.3.3 Dissolution of Calcium Carbonate in the Ocean

The bulk of modern calcium carbonate sediments in the oceans is of biogenic origin. The skeletal materials of planktonic foraminifera and coccoliths (calcite), and pteropods (aragonite) are the main components of calcareous oozes in the open ocean (see Plate I). In the shallower areas near continents and islands, organisms living on the bottom and secreting skeletal parts of calcium carbonate can form massive deposits locally. Coral reefs, shell beds, and accumulations of calcareous algae are examples of biogenic carbonates formed by benthonic organisms. In the geologic past, echinoderms and brachiopods played an important role as builders of calcareous beds, in a manner resembling the younger beds made of the shells of bivalve molluscs. Inorganically formed calcium carbonate is conspicuous as cement precipitated in the form of calcite or low-magnesium calcite in calcareous sediments, where it often forms at the expense of the more soluble aragonite or high-magnesium calcites. Another type of an inorganic calcareous sediment, beds made of oolites, forms in shallow water environments, where the degree of supersaturation of sea water with respect to $CaCO_3$ is generally high.

The fact that calcium carbonate forming in surface waters of the ocean dissolves at depth has been known for about a century, following the work done during the *HMS Challenger* expedition in 1870s (Murray and Renard, 1891). The amount of calcium carbonate in deep ocean sediments that are in contact with ocean water generally decreases with an increasing water depth from about 3500 m down, indicating dissolution. At about 3500 m depth first signs of dissolution of calcite shells appear, and this depth zone in the Pacific Ocean, where more or less significant dissolution begins, was given the name of *lysocline* (Berger, 1967). The depth where calcium carbonate in sediments is reduced to a few percent is known as the compensation depth. The lysocline and compensation depth are indicated in Figure 6.9. The surface layer of the ocean is supersaturated with respect to calcite, and the approximate depth of saturation is also indicated in the figure. At saturation, the product of the calcium- and carbonate-ion activities (ion-activity product, IAP) is equal to the thermodynamic dissociation constant for calcite (K_{eq}), and the ratio IAP/K_{eq} is equal to unity, as shown in the figure (see also Section 7.7.5). The lysocline lies much below the saturation level, and the rates of dissolution of calcite between the two depths are slow, whereas from the lysocline down the dissolution rate increases strongly. The increase in the dissolution rate has been attributed to the more acidic pH of ocean water relative to the pH value for the equilibrium with calcite under the *in situ* conditions (Morse and Berner, 1972): experimental work has shown that the rate of calcite dissolution increases with the increasing difference $\Delta pH = pH(equilibrium) - pH$. At the depth of the lysocline, the ΔpH values are between 0.14 and 0.16, and the departure from the calcite–sea water equilibrium pH becomes greater at greater depths. The more recent literature on the dissolution of calcium carbonate in the ocean is extensive. The chemical thermodynamic relationships in the system sea water–$CaCO_3$ and dissolution at the ocean floor have been treated by the following authors: Revelle and Fairbridge (1957), Bramlette (1961), Turekian (1965), Pytkowicz (1965), Peterson (1966), Berger (1967, 1971), Heath and Culberson (1970), Edmond and Gieskes (1970), Edmond (1974), Morse and Berner (1972),

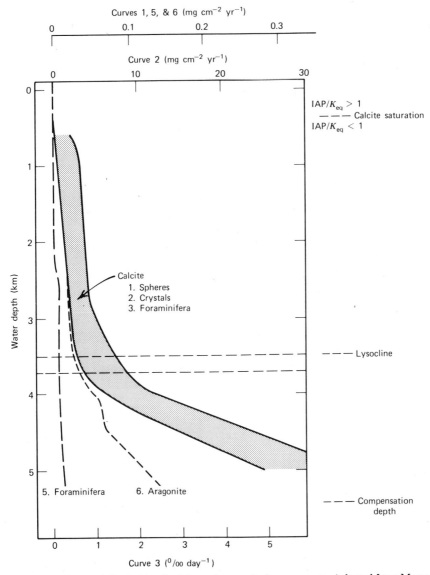

Figure 6.9 Rates of dissolution of calcite and aragonite in ocean water (adopted from Morse and Berner, 1972). Curves 5 and 6 from Milliman (1975, 1977). See also Table 6.8. Note the differences in the rates of dissolution of different calcitic materials.

Broecker and Broecker (1974), Ben-Yaakov et al. (1974), Honjo (1975), Takahashi (1975), and Berner (1976).

The rates of dissolution of calcite plotted in Figure 6.9 include different forms of calcite used for the *in situ* and laboratory experiments. The general picture is that the rate of dissolution, measured as a loss of mass per unit area of the solid per unit of time, is low and nearly constant between 1000 and 3500 m depth. Below 3500 m, the rate increases by a factor of about 10 between the depths of 3500 and 5000 m. The band of dissolution rates shown in the figure includes four sets of results, two from an *in situ* experiment in the Pacific Ocean (calcite spheres and foraminifera) and two from a laboratory study (calcite crystals and calcareous sediment). The rates of dissolution determined on the samples kept in the Pacific Ocean are considerably lower than the laboratory rates. An *in situ* experiment on dissolution of calcitic foraminifera carried out by Milliman (1975) in the Sargasso Sea shows very little dissolution of calcite and virtually no change with depth in the rate of dissolution. The rate of dissolution of aragonite shows a noticeable increase from depth 2500 m down. The rates of dissolution from the different experiments are summarized in Table 6.8 for the oceanic water column above the lysocline, where the rates are nearly constant with depth. The rate of dissolution,

Table 6.8   Dissolution Rates of Calcite ($CaCO_3$) in ocean water, Depths between 1000 and 3500 m

Calcitic material	Particle diameter or length $2r$ (μm)	Mean dissolution rate or mass loss[a] $-\dfrac{dM}{dt}$ (mg cm^{-2} yr^{-1})	$-\dfrac{100}{M}\cdot\dfrac{dM}{dt}$ (% yr^{-1})	Equivalent radial dissolution rate[a] ε (μm yr^{-1})
Calcite spheres[b]	2000–11,000	0.03		0.11
Calcite crystals[c]	1–20	1–3		7.5±3.5
Foraminifera shells[d]	250–500		1±1	0.6±0.6
Foraminifera shells[e]	125–250		30±30	9±9

[a]Radial dissolution rate of a sphere $\varepsilon = -dr/dt$ defined in equation 5.80. Radial rates from mass loss: $dr/dt=(dM/dt)/\rho_s$ or $dr/dt=(1/M)(dM/dt)$ $r/3$. Calcite density $\rho_s=2.7$ g cm^{-3}.
[b]*In situ* experiment, Pacific (Peterson, 1966).
[c]Laboratory experiment (Morse and Berner, 1972).
[d]*In situ* experiment, Sargasso Sea (Milliman, 1975, 1977).
[e]*In situ* experiment, Pacific (Berger, 1967, 1970).

measured as a constant rate of decrease of the particle radius with time ε (cm yr^{-1}), rises approximately linearly with depth below the lysocline,

$$\varepsilon_z = \varepsilon + \beta(z - z_1) \qquad (\text{cm yr}^{-1}) \qquad (6.68)$$

where β is the gradient of the dissolution rate (yr^{-1}). From Peterson's (1966) data on dissolution of calcite spheres in the Pacific Ocean, the rate of dissolution below 3500 m depth is approximately

$$\varepsilon_z = 1.1 \times 10^{-5} + 8.4 \times 10^{-10}(z - 3.5 \times 10^5) \qquad (\text{cm yr}^{-1}) \qquad (6.69)$$

where z is in cm and $z > 3.5 \times 10^5$ cm. At the depth of 5000 m, the rate of dissolution ε_z is about 1.5 μm yr^{-1}.

6.3.4   Regeneration of Settling Particles

Settling and Dissolution Model. As particles of different sizes and densities sink at different velocities, the rates at which they are regenerated and dissolved material is added to water depend on the particle size. The larger particles sinking relatively fast may loose little of their mass in transit through the water column. The rates of regeneration, expressed as the rate of addition of dissolved material to water, are relatively low for the bigger particles, both because of their fast sinking rates and low abundance. The mass flux of settling particles dF_m is

$$dF_m = \rho_s(r)v(r)U_s(r)dN(r,z) \qquad (\text{g cm}^{-2}\,\text{yr}^{-1}) \qquad (6.70)$$

The parentheses denote the parameters dependent on the particle radius r or settling distance z. The mass flux dF_m and the particle-number concentration dN refer to particles of radius between r and $r + dr$. The equation is a spectrum of fluxes. The other parameters, used previously, are the density ρ_s, volume v, and settling velocity U_s of a particle of radius r. How the particle-number concentration dN changes with depth due to dissolution was discussed in Section 6.2.4. Equation 6.70 reduces to a simple form if all particles are of the same radius r, settling with the Stokes velocity. Then the downward flux is

$$F_m = \tfrac{4}{3}\pi\rho_s BNr^5 \qquad (\text{g cm}^{-2}\,\text{yr}^{-1}) \qquad (6.71)$$

When particles dissolve in the course of settling, the mass balance condition states that the mass of material dJ dissolved and added to a unit volume of water in a unit of time is equal to the change of the flux with distance, or

$$dJ = -\frac{\partial}{\partial z}(dF_m)_r, \qquad (\text{g cm}^{-3}\,\text{yr}^{-1}) \qquad (6.72)$$

where subscript r denotes that the derivative of the flux with respect to distance is taken at a constant particle size or radius r (see also Section 2.1). The quantity dJ is the rate of regeneration of particles of radius between r and $r+dr$. If particles of radius r are removed from water by additional mechanisms, such as agglomeration or scavenging, the gradient of the flux must be balanced by the regeneration rate dJ plus other removal rate terms on the left-hand side of the equation. The particle-number concentration dN in equation 6.70 is a function of the particle radius and depth, or $dN = f(r,z)\,dr$. An explicit form of $f(r,z)$ can be used in equation 6.70, and the rate of regeneration of a settling particle assemblage, between the sizes of radius r_1 and r_2, at any depth z is

$$J = \int_{r_1}^{r_2} dJ \qquad\qquad (\text{g cm}^{-3}\,\text{yr}^{-1})$$

$$J = \int_{r_1}^{r_2} \rho_s(r)v(r)U_s(r)\frac{\partial f(r,z)}{\partial z}\,dr \qquad (\text{g cm}^{-3}\,\text{yr}^{-1}) \qquad (6.73)$$

For particles of the same initial radius r_0 and number concentration N_0, the flux conservation condition is $Nr^2 = N_0 r_0^2 = \text{const}$. The rate of regeneration J is, from equations 6.70, 6.72 and 6.9,

$$J = 4\pi\rho_s\varepsilon N_0 r_0^2 \qquad (\text{g cm}^{-3}\,\text{yr}^{-1}) \qquad (6.74)$$

The latter equation states that the rate of regeneration J due to dissolution of settling particles of the same size is constant and independent of the settling depth. The product $N_0 r_0^2$ can be replaced by Nr^2 at depths $z>0$.

The rate of regeneration for a water column between depths $z = z_1$ and $z = z_2$ (with $z_2 > z_1$) is an integrated regeneration rate J_I,

$$J_I = \int_{z_1}^{z_2} J\,dz \qquad (\text{g cm}^{-2}\,\text{yr}^{-1}) \qquad (6.75)$$

where J is as defined in the preceding two equations. The rates of regeneration J and J_I can be computed if the dependence of the other parameters on the particle radius, and the radius dependence on the settling distance are known explicitly. In the next two sections such equations are given for the cases of a constant dissolution rate ε and a depth-dependent dissolution rate ε_z. Subsequently, regeneration of the settling calcium carbonate in the ocean will be evaluated with the use of the equations for J and J_I.

Equations for J and J_I: Constant Dissolution Rate ε. The equations for the regeneration rate per unit volume of water J and per unit area of a water column J_I are given in this section for the following case: spherical particles, all of the same density, dissolve at a constant rate of the radius decrease ε; the initial particle size distribution is of the Pareto type $dN/dr = Ar^{-b}$ (Section 6.2.5). Thus the parameters that enter in equations 6.70 and 6.73 are

$$\rho_s = \text{const} \qquad (\text{g cm}^{-3})$$
$$\varepsilon = \text{const} \qquad (\text{cm yr}^{-1})$$
$$v(r) = \tfrac{4}{3}\pi r^3 \qquad (\text{cm}^3)$$
$$U_s(r) = Br^2 \qquad (\text{cm yr}^{-1})$$
$$dN = A\left(r + \frac{3\varepsilon z}{B}\right)^{-b/3} dr \qquad (\text{cm}^{-3})$$

The above parameters give the following equations for dJ and J:

$$dJ = \tfrac{4}{3}\pi\rho_s A b\varepsilon r^5\left(r^3 + \frac{3\varepsilon z}{B}\right)^{-b/3-1} dr \qquad (\text{g cm}^{-3}\,\text{yr}^{-1}) \quad (6.76)$$

$$J = \frac{4\pi\rho_s Ab\varepsilon}{3(3-b)}\left(r^3 + \frac{9\varepsilon z}{bB}\right)\left(r^3 + \frac{3\varepsilon z}{B}\right)^{-b/3}\Bigg|_{r_1}^{r_2} \qquad (\text{g cm}^{-3}\,\text{yr}^{-1}) \quad (6.77)$$

where the vertical bar denotes the difference between the r-dependent terms at the value of $r = r_2$ and their value at $r = r_1$. For two special cases, the equation for the rate of regeneration of a particle assemblage J can be simplified.

Case 1. For the bigger particles, the following inequalities apply:

$$r^3 \gg \frac{3\varepsilon z}{B} \qquad \text{and} \qquad r^3 \gg \frac{9\varepsilon z}{bB}$$

Then the equation for J reduces to a simpler form,

$$J \simeq \frac{4\pi\rho_s Ab\varepsilon}{3(b-3)}\left(\frac{1}{r_1^{b-3}} - \frac{1}{r_2^{b-3}}\right) \qquad (\text{g cm}^{-3}\,\text{yr}^{-1}) \qquad (6.78)$$

The latter equation shows that the rate of regeneration due to the larger particles is independent of the settling distance or water depth z. The particle sizes to which the equation applies can be illustrated with the aid of the numerical values of the quotient $3\varepsilon z/B$. For example, for calcite particles (Section 6.3.3) one can use $\varepsilon = 0.1$–0.5 μm yr^{-1}, $B \approx 1 \times 10^4$ cm^{-1} sec^{-1}, and $z = 2.5$ km. Then

$$\left(\frac{3\varepsilon z}{B}\right)^{1/3} = \left(\frac{3(0.1 \text{ to } 0.5) \times 10^{-4} \times 2.5 \times 10^5}{1 \times 10^4 (3.16 \times 10^7 \text{ sec yr}^{-1})}\right)^{1/3} = 3 \times 10^{-4} \text{–} 5 \times 10^{-4} \text{ cm}$$

Thus particles significantly bigger than 3–5 μm in radius settle sufficiently fast and the rate of their regeneration remains constant with depth.

Case 2. The size limits of particles in equation 6.77 for the regeneration rate J are r_1 and r_2. If r_1 is sufficiently small and r_2 large such that the following inequalities hold:

$$r_1^3 \ll \frac{3\varepsilon z}{B} \qquad \text{and} \qquad r_2^3 \gg \frac{3\varepsilon z}{B}$$

then the limits of integration can be taken approximately as $r_1 = 0$ and $r_2 \to \infty$, and the equation for J becomes

$$J_{r>0} \simeq \frac{4\pi \rho_s A \varepsilon}{b-3} \left(\frac{3\varepsilon z}{B}\right)^{-b/3+1} \qquad (\text{g cm}^{-3} \text{ yr}^{-1}) \qquad (6.79)$$

The latter equation gives the rate of regeneration per unit volume of water $J_{r>0}$ due to dissolution of all the particles of radius $r > 0$. The equation overestimates the rate of regeneration in comparison to the complete equation 6.77 by 2–7%, when the upper radius limit of particles is $r_2 = 100$ μm, the quotient $3\varepsilon z/B$ is of the orders of magnitude as given in the computational example above, and the power exponent b of the size distribution is between 4.0 and 4.2.

In a water column of thickness $z_2 - z_1$, the integrated rate of regeneration per unit area of the water column J_I is, from equations 6.75 and 6.77,

$$J_I = \frac{12\pi \rho_s A (B/3)^{b/3-1}}{(6-b)(b-3)} \left[(\varepsilon z_2)^{-b/3+2} - (\varepsilon z_1)^{-b/3+2}\right] \qquad (\text{g cm}^{-2} \text{ yr}^{-1})$$

$$(6.80)$$

Regeneration Rates J and J_I: Depth-Dependent Dissolution Rate ε_z. The equations for the rates of regeneration per unit volume of water J and per unit area of the water column J_I, in the case of the dissolution rate varying with depth, are similar in form to the equations for regeneration at a constant rate of dissolution that are given in the preceding section. Details of the derivation for the case of ε_z can be found in the paper by Lerman and Lal (1977). The rate of regeneration $J_{r>0}$ due to dissolution of all settling particles of radius greater than zero is

$$J_{r>0} = \frac{4\pi\rho_s A \varepsilon_z}{b-3} \left(\frac{3}{B} (\varepsilon z_1 + E_z) \right)^{-b/3+1} \qquad (\text{g cm}^{-3}\,\text{yr}^{-1}) \qquad (6.81)$$

where

$$E_z \equiv \int_{z_1}^{z} \varepsilon_z\, dz$$

$$= \varepsilon(z - z_1) + \frac{\beta}{2}(z - z_1)^2 \qquad (\text{cm}^2\,\text{yr}^{-1}) \qquad (6.82)$$

The rate of dissolution of calcite below the lysocline ε_z is given in equation 6.68, for $z > z_1$. The product εz_1 in equation 6.81 represents dissolution that took place while the particles settled from depth $z = 0$ to $z = z_1$. If there had been no dissolution above depth z_1 then the term εz_1 is omitted from the sum in small parentheses in equation 6.81 and the subsequent equation for J_I.

Below the depth z_1, the rate of regeneration integrated for a water column of thickness $z_2 - z_1$ is

$$J_I = \frac{12\pi\rho_s A (B/3)^{b/3-1}}{(6-b)(b-3)} (\varepsilon z_1 + E_z)^{-b/3+2} \Big|_{z_1}^{z_2} \qquad (\text{g cm}^{-2}\,\text{yr}^{-1}) \quad (6.83)$$

where the vertical bar indicates the difference between the z-dependent terms at the value of $z = z_2$ and their value at $z = z_1$.

Regeneration of Calcium Carbonate in the Ocean. The equations for the rates of regeneration of settling material derived in the preceding section are used to evaluate the rate of regeneration of biogenic calcite in the ocean. It is stated in Section 6.3.1 that dissolution of the settling material in the water column *and* dissolution of the settled material on the ocean floor both contribute to regeneration. The model behind the rates of regeneration per unit volume of water J and per unit area of the water column J_I applies to particles of a certain shape that settle and dissolve according to specified mechanisms. For other shapes, settling velocities, and dissolution mechanisms, appropriate mathematical relationships must be included in the equations for the mass flux of particles dF_m and the rates of regeneration dJ. Possible departures from the assumptions of the model include the following. Calcite particles, represented by the coccolith platelets and smaller foraminifera, can be ingested by the zooplankton and excreted as fecal pellets. In the pellets, the particles are held together by mucous material in a form of a membrane, and the sizes of pellets are sufficiently large, $10^1 \times 10^2$ μm, such that their settling times in water are short. If transport of most of the calcareous material through the water column is rapid, then most of the material is regenerated on the bottom. Because of the irregular shape and nonuniform density of biogenic particles, the settling velocities may deviate from the Stokes-law velocities, as some experimental data suggest (Figure 6.3). Finally, dissolution of porous particles may take place in a manner different from the diffusion-controlled dissolution of a sphere, such that the relationships between the loss of mass and decrease in the particle radius are different from those taken in the model (see Figure 5.9).

Data for the Regeneration Model. A particle-number spectrum of the Pareto type $dN/dr = Ar^{-b}$ contains two constants, A and b. The values of b for suspended materials in the ocean are summarized in Table 6.5. For computation of the rates of regeneration, the value of $b \simeq 4.2$ will be used. The dimensional parameter A can be evaluated from equation 5.35, which relates the cumulative number of particles $N_{>r}$ of radius greater than r to the parameters A and b of the size distribution. By numbers, coccoliths are the most abundant components of the calcareous plankton. In the bioproductivity zone of the Pacific Ocean, estimates of the abundance of coccoliths, based on samples collected over many tens of degrees of latitude in north to south transects, range from 10^{-2}–10^{-1} per cubic centimeter of ocean water to 10^0–10^1 cm^{-3} (Lisitzin, 1971; Honjo, 1975). Below the bioproductivity zone, at depths between 400 and 1000 m, the number concentration of coccoliths increases to 10^1–10^2 cm^{-3}, owing to the breakdown of the coccolithophore cells (coccospheres), which shed off individual coccolith platelets. Using the number concentration value of $N_{>r} = 50$ cm^{-3}, the lower size limit of 2–3 μm ($r_1 = 2.5 \times 10^{-4} \pm 0.5 \times 10^{-4}$ cm), and $b = 4.2$, the parameter A is

$$A = (50 \text{ cm}^{-3})(4.2 - 1)[(2.5 \pm 0.5) \times 10^{-4} \text{ cm}]^{(4.2-1)} = (5 \pm 3) \times 10^{-10} \text{ cm}^{b-4}$$

The density value of 1.5 g cm^{-3} approximates the density of porous water-filled skeletal particles containing about 50% water by volume. For $CaCO_3$, the molar density is $\rho_s = 1.5$ g cm^{-3}/100 g mol^{-1} = 0.015 mol cm^{-3}, the value used in computation of the rates of regeneration.

The Stokes settling parameter B, from equation 6.4, is

$$B = \frac{2 \times 981(1.5 - 1.0)}{2 \times 0.015} \times 3.16 \times 10^7 \text{ sec yr}^{-1} = 2.3 \times 10^{11} \quad \text{cm}^{-1} \text{ yr}^{-1}$$

A summary list of the parameters used to compute the rates of regeneration J and J_I is:

$$A = 5 \times 10^{-10} \pm 2.5 \times 10^{-10} \quad \text{cm}^{0.2}$$

$$b = 4.2$$

$$\rho_s = 0.015 \quad \text{mol cm}^{-3}$$

$$B = 2.3 \times 10^{11} \quad \text{cm}^{-1} \text{ yr}^{-1}$$

$$\varepsilon = 1.1 \times 10^{-5} \quad \text{cm yr}^{-1}$$

$$\varepsilon_z = 1.1 \times 10^{-5} + 8.4 \times 10^{-10}(z - 3.5 \times 10^5) \quad \text{cm yr}^{-1}$$

Numerical Results. Explicit relationships between the rates of regeneration J and J_I, and water depth z, are obtained by using the above parameters in equations 6.79 through 6.83. Above the lysocline, at depths between 1000 and 3500 m, the rate of $CaCO_3$ regeneration due to dissolution of all settling particles of radius $r > 0$ is

$$J_{r>0} = 1.88 \times 10^{-3} z^{-0.4} \quad (\mu\text{mol cm}^{-3} \text{ yr}^{-1})$$

where z is in centimeters in all equations.

Below the lysocline, at depths greater than 3500 m,

$$J_{r>0} = 1.04 \times 10^{-4}(2.97 \times 10^{-6} z - 1)(1 - 5.5 \times 10^{-6} z + 8.16 \times 10^{-12} z^2)^{-0.4}$$

$$(\mu\text{mol cm}^{-3} \text{ yr}^{-1})$$

The integral rates of regeneration J_I are, for the water column above the lysocline,

$$J_I = 3.13 \times 10^{-3} z^{0.6} - 3.13 \quad (\mu\text{mol cm}^{-2} \text{ yr}^{-1})$$

and below the lysocline,

$$J_I = 31.5\left[(1 - 5.5 \times 10^{-6} z + 8.16 \times 10^{-12} z^2)^{0.6} - 0.21\right] \quad (\mu\text{mol cm}^{-2} \text{ yr}^{-1})$$

where $z > 3.5 \times 10^5$ cm. Between 1 and 3.5 km depth, the integrated rate of regeneration is

$$J_I = 3.5 \quad \mu\text{mol cm}^{-2}\,\text{yr}^{-6} \quad (\pm 50\%)$$

and between 3.5 and 5 km the rate is

$$J_I = 8.4 \quad \mu\text{mol cm}^{-2}\,\text{yr}^{-1} \quad (\pm 50\%)$$

The sum of the two rates is the total mass of calcium carbonate regenerated to ocean water annually from settling material between the depths of
1 and 5 km,

$$\text{total } J_I = 12 \pm 6 \quad \text{mol cm}^{-2}\,\text{yr}^{-1}$$

The rate of regeneration $J_{r>0}$ for calcite in the ocean is shown graphically
as a function of depth in Figure 6.10, allowing for a plus or minus 50%
spread. The mean rate above the lysocline is 1.4 and below the lysocline
5.6 $\mu\text{mol cm}^{-3}\,\text{yr}^{-1}$. The higher rate of regeneration reflects a faster
dissolution in the deep ocean.

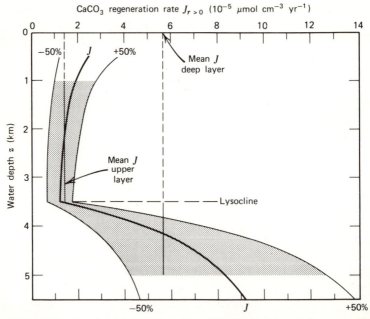

Figure 6.10 Computed rates of regeneration of $CaCO_3$ in the oceanic water column.

The integrated rate of regeneration J_I is identical to the flux F_{43} shown in Figure 6.8.

Regeneration of Organic and Inorganic Carbon. Figure 6.8 shows the flux of biogenic materials F_{24} from the surface to the deep ocean. For carbon, this flux includes organic materials and carbon in the $CaCO_3$ skeletal particles. The regeneration flux F_{43} in the deep water is also made of organic and inorganic carbon fluxes, and the same holds for the fluxes F_{53} and F_{56}. The total flux of particulate carbon F_{24}, as explained in Section 6.3.2, is the difference between the fluxes of dissolved carbon F_{31} and F_{13}, plus the input flux $F_{01} = F_{56}$. From equations 6.63 through 6.66, the flux F_{24} for total biogenic carbon is

$$F_{24} = (C_d - C_s)k_d h_d + F_{56}$$

$$= (300 \ \mu\text{mol l}^{-1})(0.3 \text{ to } 0.4 \text{ 1 cm}^{-2} \text{ yr}^{-1}) + 5$$

$$= 115 \pm 20 \quad \mu\text{mol cm}^{-2} \text{ yr}^{-1}$$

In the above computation, the difference between concentrations of total dissolved carbon in deep and surface ocean layer is taken as 300 μmol l^{-1}, based on the measured values in the Atlantic and Pacific Oceans (Li et al., 1969; the concentrations are about 2.3 and 2.0 mmol l^{-1}). The product $k_d h_d$ was given in Section 6.3.2, and the net sedimentation flux of 5 μmol cm^{-2} yr^{-1} is the sum of the organic and carbonate carbon fluxes. As each of the fluxes F_{43}, F_{53}, and F_{56} consists of the organic and inorganic component, the balance equation 6.63 can be written in the following form, where subscripts org and cal refer to the organic carbon and carbonate carbon,

$$F_{24} = F_{43,\text{org}} + F_{53,\text{org}} + F_{43,\text{cal}} + F_{53,\text{cal}} + (F_{56,\text{org}} + F_{56,\text{cal}})$$

The known fluxes are the following (in units of μmol cm^{-2} yr^{-1}):

$$F_{24} = 115$$

$$F_{43,\text{cal}} \equiv J_I = 12$$

$$(F_{43,\text{org}} + F_{53,\text{org}}) = 80 \quad \text{(see estimates below)}$$

$$F_{56,\text{org}} = 1 \quad \text{(Garrels and Mackenzie, 1972)}$$

$$F_{56,\text{cal}} = 4 \quad \text{(Turekian, 1965)}$$

Using the known fluxes in the preceding balance equation, the unknown is the flux of carbonate carbon $F_{53,\text{cal}}$ due to regeneration of $CaCO_3$ on the ocean floor, and this flux is

$$F_{53,\text{cal}} = 18 \quad \mu\text{mol cm}^{-2}\text{ yr}^{-1}$$

The sum of the fluxes $F_{43,\text{org}} + F_{53,\text{org}}$ represents the total regeneration of organic carbon in the deep ocean layer (Figure 6.8). One estimate of the sum of the two fluxes is based on the rate of regeneration of dissolved organic carbon in deep ocean water. The concentration of dissolved organic carbon is much higher than the concentration of particulate carbon (Table 6.7), and using the value of 45 ± 20 μmol 1^{-1} for concentration, a rate constant for oxidation of $(1$ to $9) \times 10^{-3}$ yr^{-1}, and water column volume of 400 1 cm^{-2}, the total regeneration rate is

$$(F_{43} + F_{53})_{\text{org}} = (1 \text{ to } 9) \times 10^{-3} \times (65 \text{ to } 25) \times 400$$

$$= 60 \pm 30 \quad \mu\text{mol cm}^{-2}\text{ yr}^{-1}$$

Another estimate can be based on the rate of production of CO_2 due to oxidation of organic matter in the deep ocean (Li et al., 1969). Concentration of dissolved CO_2 in deep ocean water, attributed to the oxidation of organic matter, is between 190 and 330 μmol 1^{-1}. The deep water column is between 300 and 400 1 cm^{-2}, and the water residence time is about 10^3 yr. Using these numbers, the rate of regeneration of organic carbon is

$$(F_{43} + F_{53})_{\text{org}} = 1 \times 10^{-3} \times (190 \text{ to } 330) \times (300 \text{ to } 400)$$

$$= 95 \pm 35 \quad \mu\text{mol cm}^{-2}\text{ yr}^{-1}$$

The mean of the two latter estimates is taken as $(F_{43} + F_{53})_{\text{org}} = 80$ gmmol cm^{-2} yr^{-1} in the preceding computation of the rate of regeneration of $CaCO_3$.

The total flux of biogenic particulate carbon F_{24} can now be partitioned into the organic and mineral components: $F_{24,\text{org}} = 81$ and $F_{24,\text{cal}} = 34$ μmol cm^{-2} yr^{-1}. The ratio of the two fluxes is lower than the ratios of carbon concentrations $C_{\text{org}}/C_{\text{cal}}$ in plankton within the euphotic layer of the surface ocean, because the fluxes F_{24} apply to the material that leaves a thicker layer (1 km), and some of the organic matter is regenerated in surface waters where calcium carbonate does not dissolve.

CHAPTER 7

Sediment-Water Interface

Two characteristic features distinguish the zone of the sediment-water interface from the water above and sediments below it: the residence times of the reactive solids and the biological activity. Sediment particles within the upper one or so centimeters of the sediment column remain in contact with the overlying water for periods of time that are generally much longer than the settling times. The activity of bacteria and benthic organisms is primarily confined to the upper few centimeters of the sediment column, where the exchange with the overlying water supplies oxygen or nutrients. The biological activity, the chemical reactions involving organic matter and minerals, and the faster inorganic reactions of certain mineral phases make the sediment-water interface a distinct zone where special conditions exist, and where diagenetic reactions peculiar to this zone take place. The nature of these reactions, and the rates of transport of dissolved and solid materials across the sediment-water interface in lakes and oceans are discussed in this chapter.

7.1 UP AND DOWN TRANSPORT

The sediment-water interface separates a mixture of solid sediment and interstitial water from an overlying body of water. Wherever sediment accumulates, growth of the sediment pile is achieved by sedimentation of solid particles and inclusion of water in the pore spaces among the particles. An observer stationed on shore would observe the sediment-water interface rising as sediments accumulate. But an imaginary observer who balances himself on the interface as sediment particles continue to arrive from above and pile up, will see the particles and pore water flow by in the downward direction. In this sense one can always speak of the fluxes of solids, water, and solutes as moving up or down, *across* the sediment-water interface.

Arrival of sediment particles and entrapment of water in the pore space between the particles are the two major fluxes of materials across the sediment-water interface. The rates of sediment deposition vary from the low values of millimeters per 1000 yr in the pelagic ocean up to centimeters per year in lakes and near-shore oceanic areas (Section 6.2.1). The porosity of sediments near the interface, measured as the water content by volume, is commonly between 70 and 90%. Upon compaction, the porosity usually decreases to 40–60% within a few tens of centimeters to a few meters below the interface. Representative values of the mass fluxes of solids and dissolved materials across the sediment-water interface can be bracketed by estimates for areas of low and high deposition rates. The slower sedimentation rates in the ocean are of the order of 5×10^{-4} cm yr^{-1} (5mm per 1000 yr) and rates in lakes and coastal sections are 5×10^{-1} cm yr^{-1}; a sediment can be thought of as made of equal volumes of water and solid particles of density 2.5 g cm^{-3} (that is, sediment volume fraction 0.5). Then the mass flux of solids to the sediment is in the range

$$6 \times 10^{-4} \text{ to } 6 \times 10^{-1} \quad g \, cm^{-2} \, yr^{-1}$$

Ocean water contains 35 g l^{-1} dissolved materials. For freshwater lakes a typical figure is 0.1 g l^{-1}. For the same range of sedimentation rates and sediment porosity as used in the preceding computation, the flux of dissolved materials across the sediment-water interface is in the range

$$0.1 \times 10^{-4} \text{ to } 0.1 \times 10^{-1} \quad g \, cm^{-2} \, yr^{-1}$$

In freshwater lakes, the dissolved material fluxes into sediments are closer to the lower value of 0.1×10^{-4} g cm^{-2} yr^{-1}.

Because the sedimentation flux brings dissolved materials in a proportion of only 1 or 2% of the flux of solids to the sediment, the dissolved material in pore water does not add much to the mass of solids. The opposite results, however, are the changes in the chemical composition of pore waters owing to reactions with solid phases in the sediment, and these are commonly observed.

In some environments, the sediment-water interface is a quiet zone where settling materials come to rest on the bottom. In some other environments, the zone of the interface is subject to a variety of disturbances: benthic organisms mix and rework the sediment within several centimeters of the interface, fluctuations in near-bottom water currents resuspend and shift the sediment, and sediments accumulating on inclines creep and slide, altering the local topographic bottom relief to a greater or lesser extent.

The settling residence times of solid particles in the water column of lakes and oceans are short in comparison to the rate of the sediment column growth, such that the materials arriving at the sediment surface may continue to react either inorganically or through bacterially mediated processes. Transport of chemical species across the sediment-water interface is caused by the advective processes of sediment column growth, as discussed above, as well as by diffusional migration in pore waters, caused by differences in concentration (or thermodynamic activity) that develop between the two sides of the interface (Chapter 8). Biogenic skeletal material made of silica and calcium carbonate phases, organic matter, and some inorganic mineral phases which are out of a thermodynamic equilibrium with the surrounding solution, will react and cause some changes in the chemical composition of pore waters. Examples of such processes include decomposition of organic matter, reduction of sulfate to sulfide, production of methane, ammonia, and carbon dioxide, and also dissolution of mineral phases of inorganic and biogenic origin. An increase or decrease in concentration of a dissolved species in sediment pore water relative to the overlying water establishes a condition for the diffusional flux that can be realized if pore water is in contact with the water space above it. Those chemical species whose concentrations increase in pore water are at least in part removed by diffusional fluxes up, across the sediment-water interface, or down the sediment pore water column. Additional removal from pore water may take place by adsorption or inorganic reactions with other solids.

The major processes responsible for the transport of chemical species across the sediment-water interface can be grouped into the following categories:

1. Sedimentation flux of solids (mineral, skeletal, and organic materials).
2. Flux of dissolved material and water into sediment, owing to the growth of the sediment column.
3. Upward flow of dissolved material and pore water flow upward, caused by hydrostatic pressure gradients of ground water in aquifers on land.
4. Molecular diffusional fluxes in pore water.
5. Mixing of sediment and water at the interface (bioturbation and water turbulence).

Processes 1–3 are advective fluxes, differing from process 4, the molecular diffusion flux. The magnitude of process 3, upward flow of pore water, is highly variable and depends to a large extent on the local configuration of a body of water and ground water aquifers. Flow of ground water into and out of lakes can amount to some fraction of the total water budget, depending on the geological conditions within the drainage basin. For the ocean, the amount of ground water inflow has been estimated as 5% of the river and glacier melt inflow (Section 6.1).

7.2 GENERAL ROLE OF THE SEDIMENT-WATER INTERFACE

The role played by the sediment-water interface in the redistribution of materials in the ocean is demonstrated in this section for three elements heavily involved in the biogeochemical cycle—carbon, calcium, and silicon. Figure 7.1 shows the major fluxes of biogenic $CaCO_3$ and SiO_2 entering the deep ocean and settling on the bottom. (The figure is a simplified version of a more complete diagram of the biogeochemical cycle given in Figure 6.8.) Flux A to the deep ocean includes all of the suspended organic material, $CaCO_3$ and SiO_2 of biogenic origin, that is, material settling as individual particles, aggregates, or fecal pellets. Some

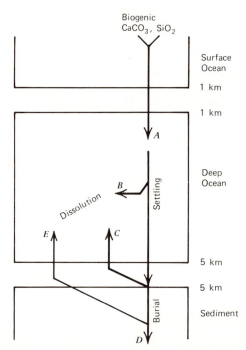

Figure 7.1 Dissolution of settling materials in the water column, at the sediment-water interface, and in sediments (from Lerman, 1978). Reproduced with permission from Annual Review of Earth & Planetary Sciences, Vol. 6. © 1978 by Annual Reviews Inc.

of the skeletal material dissolves during settling, and this dissolution flux is shown by B. Solid particles reaching the bottom remain in contact with overlying water as long as they are not covered by new sediment, and at the sedimentation rate of about 1 cm per 1000 yr this time is relatively long. Continuing dissolution at the sediment-water interface returns dissolved material to ocean water, as indicated by flux C. The material that is buried and becomes part of the sedimentary record (flux D) may continue to react with pore water, and its dissolution results in flux E, upward across the interface.

The amounts of biogenic $CaCO_3$ and SiO_2 that enter the deeper ocean with flux A are shown as 100% in Table 7.1. The other fluxes are listed as percentages of flux A and in absolute values. About 50% of $CaCO_3$ entering the deep ocean dissolves at the sediment-water interface. A smaller fraction dissolves in the water column and about 15% becomes part of the sediment record. If the amount dissolved in transit is less than shown in Table 7.1, then proportionally more dissolves at the bottom. The relatively large fraction of $CaCO_3$ that dissolves at the sediment-water interface and during settling to the bottom is the result of undersaturation of deep ocean water with respect to the $CaCO_3$ minerals calcite and aragonite, and the faster dissolution rate of calcite in the deep ocean (Section 6.3).

For biogenic SiO_2, the estimated fraction that dissolves at the sediment-water interface is about 50%.

The sum total of the fluxes $B + C + D$ in Figure 7.1 is the rate of regeneration of a chemical species in the deep ocean. The small value of the flux E carrying material out of the sediment in comparison to the dissolution flux C at the sediment-water interface can be attributed to the following factors. One, molecular diffusion responsible for the fluxes of

Table 7.1 Sedimentation and Dissolution Fluxes of Biogenic $CaCO_3$ and SiO_2 in the Ocean, Shown Schematically in Figure 7.1 [a]

Flux (Figure 7.1)	$CaCO_3$ (percentage of flux A)	(μmol cm^{-2} yr^{-1})	SiO_2 (percentage of flux A)	(μmol cm^{-2} yr^{-1})
A	100	35	100	47 ± 12
B	35 ± 17	12 ± 6	50 ± 10	23 ± 4
C	53	18	46	22
D	12	4	4	2
E	1 ± 1	0.5 ± 0.5	2 ± 2	1 ± 1

[a] From Lerman and Lal (1977).

dissolved species in pore water is generally a slow process. Two, concentrations of calcium and SiO_2 in pore waters are closer to saturation with respect to the dissolving phases, such that the rates of dissolution are slower. Three, newly formed mineral phases precipitate on the dissolving solids, reducing their total reactive area.

7.3 MODELS OF THE SEDIMENT-WATER INTERFACE

Three conceptual models of the sediment-water interface are shown in Figure 7.2, all assuming the interface to be a horizontal plane. In the simplest model of Figure 7.2a, the water above the bottom is a well-mixed homogeneous solution (no concentration gradients of dissolved species). Below the interface, concentration gradients can exist because of the slower transport in pore water. At the interface, concentrations in the bottom water and pore water are equal, but the concentration gradients are not equal in the geometric representation of Figure 7.2a. (Gas-water interface is discussed in Section 4.3.2.)

A sediment-water interface with a water boundary layer above it is shown in Figure 7.2b. Laminar flow and molecular diffusion take place within the boundary layer, as opposed to the eddy turbulent conditions above it. In the deep ocean, the thickness of the laminar boundary layer is measurable in centimeters (Wimbush and Munk, 1970). Existence of such boundary layers with concentration gradients within them, as drawn in Figure 7.2b, is potentially significant to computation of the diffusional

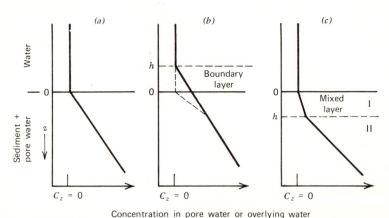

Figure 7.2 Three models of the sediment-water interface (from Lerman, 1978)[9] Reproduced with permission from Annual Review of Earth & Planetary Sciences, Vol. 6. © 1978 by Annual Reviews Inc.

fluxes at the sediment-water interface (Morse, 1974). Sampling methods of sediment pore waters are such that reliable sample withdrawals and analyses are not always possible in the uppermost 1–3 cm of the pore water column. Therefore, a concentration gradient has often to be estimated. The estimate is made between the concentration at the highest point sampled, at some distance below the interface, and the concentration in the overlying water near the bottom, taken as the concentration at the interface. The presence of the water boundary layer of centimeter thickness, comparable to the sampling distance below the interface, can introduce a factor of 2 uncertainty in the linear distance from which the concentration gradient is computed.

In Figure 7.2c is shown a diagram of the interface zone, where a sediment layer immediately below the interface is characterized by a faster transport than the deeper sediment. In near-shore carbonate sediments of Bermuda and in organic-matter-rich sediments of Long Island Sound, seasonal variations in concentrations of dissolved species have been observed within the upper few to a few tens of centimeters of sediments. Sediment zones of comparable thickness, showing the effects of mixing and reworking, have also been reported in the Caribbean deep-sea sediments and in coastal sediments of the North Sea (Thorstenson and Mackenzie, 1974; Goldhaber et al., 1977; Guinasso and Schink, 1975; Vanderborght et al., 1977). The reasons behind the formation of a mixed sediment layer are the burrowing activity of benthic organisms (bioturbation), and sporadic or continuous penetration of water turbulence near the bottom into the sediment pore space. If the vertical transport within the mixed sediment layer is regarded as analogous to the diffusional flux, equation 7.3, then conservation of the flux at the boundary $z = h$ between two layers I and II (Figure 7.2c) requires the following condition to be obeyed for dissolved species in pore water:

$$(D\,dC/dz)_{\mathrm{I}} = (D\,dC/dz)_{\mathrm{II}} \tag{7.1}$$

The coefficient D_{I} for the fast-transport layer can be computed using the observed values of the concentration gradients in the mixed layer $(dC/dz)_{\mathrm{I}}$, and in the undisturbed sediment below it $(dC/dz)_{\mathrm{II}}$, and the molecular diffusion coefficient D_{II} for the undisturbed-layer pore water. For near-shore environments, values of D_{I} have been reported in the range

$$\frac{D_{\mathrm{I}}}{D_{\mathrm{II}}} = 5\text{--}100$$

where the molecular diffusion coefficient D_{II} is of the order of 10^{-6} $cm^2\ sec^{-1}$.

For two- and three-layer models, mathematical relationships for concentrations and fluxes in pore waters, in the presence of diffusion, pore water advection, and dissolution and precipitation reactions, can be found in Lerman (1977).

7.4   FLUXES AT THE SEDIMENT-WATER INTERFACE

7.4.1   General Flux Equation

The net flux of a chemical species across the sediment-water interface $F_{z=0}$ due to diffusion in pore water, and due to advective transport of the species in pore water solution and on sediment particles (see Section 7.1 and for fluxes in general, Chapter 2) is

$$F_{z=0} = -\phi D \frac{dC}{dz} + \phi UC + \phi U_s C_s \bigg|_{z=0} \qquad (\text{g cm}^{-2}\,\text{yr}^{-1}) \qquad (7.2)$$

where the flux due to molecular diffusion is

$$F_d = -\phi D \frac{dC}{dz} \qquad (7.3)$$

the flux due to advection of pore water is

$$F_a = \phi UC \qquad (7.4)$$

and the flux due to deposition of solid particles is

$$F_s = \phi U_s C_s \qquad (7.5)$$

In the above equations, the parameters are as follows: concentrations in solution C and in solids C_s are in units of mass per cubic centimeter of *pore water*, the vertical distance coordinate z is positive and increasing downward from the sediment-water interface at $z=0$, the rates of sedimentation U_s and pore water advection U (cm yr^{-1}) are positive when the sediment and water flow downward relative to the sediment-water interface $z=0$, and the sediment porosity ϕ is a fraction, as used previously.

The magnitudes of the first two terms in equation 7.1 determine the relative importance of diffusional and advective transport in pore water (Section 2.4). If the concentration gradient is positive ($dC/dz>0$), then concentration increases with depth, indicating that the diffusional flux $-\phi D\,dC/dz$ is directed upward (the flux value is negative). In a growing sediment column, U is positive (if there is no forced flow upward), and when the two fluxes are of opposite sign, then diffusion tends to counteract the effects of transport into sediment. When C and ΔC are of comparable magnitude, then the ratio $D/\Delta z$ and U, both having the dimensions of velocity, can be compared in their relative importance. Taking the vertical distance over which the concentration gradient exists as $\Delta z \simeq 5$ cm and $D \simeq 5 \times 10^{-6}$ cm^2 sec^{-1}, the ratio is $D/\Delta z \simeq 10^{-6}$ cm sec$^{-1} = 30$ cm yr^{-1}; sedimentation rates U are of the order of <1 cm yr^{-1}. Thus we have $D/\Delta z > U$, and this indicates that the diffusional flux $D\,dC/dz$ term in the general flux equation 7.1 is almost always significant in comparison to the advective flux UC in pore waters.

7.4.2 Fluxes on Solids

If the concentration of a chemical species in water is small in comparison to the concentration on solid sediment particles ($C \ll C_s$), then the flux to the sediment is, from equation 7.5,

$$F_{z=0} = \phi U_s C_s \qquad \left(\text{g cm}^{-2}\ \text{yr}^{-1}\right)$$

A unit of concentration more convenient than C_s (mass per unit volume of pore water) is C_g, in units of mass per unit mass of solid (g g^{-1}). The relationship between the two concentrations is

$$C_s = C_g \rho_s (1-\phi)/\phi$$

where ρ_s is solid density (g cm^{-3}). The flux of a chemical species due to deposition of sediment is

$$F_{z=0} = U_s C_g \rho_s (1-\phi) \qquad \left(\text{g cm}^{-2}\ \text{yr}^{-1}\right) \tag{7.6}$$

Equation 7.6 can be generalized for lakes and near-continent oceanic sediments where the sedimentation rates U_s are of the order of 10^{-1} cm yr^{-1}. Using $\phi \simeq 0.7$ for the sediment porosity near the sediment-water interface and the solid density $\rho_s \simeq 2.5$ g cm^{-3}, the flux into the sediment is

$$F_{z=0} \simeq 10^{-1} C_g \qquad \left(\text{g cm}^{-2}\ \text{yr}^{-1}\right) \tag{7.7}$$

Concentrations of some heavy metals on sediment particles are usually measurable in parts per million to hundreds of ppm (10^{-6}–10^{-4} g g^{-1}), and such concentrations produce fluxes of the order of 0.1–10 μg cm^{-2} yr^{-1}. In Table 7.2 are summarized concentrations of four heavy metals in settling material and in solution in Greifensee, a freshwater lake near Zurich, Switzerland. The fluxes of Pb, Cd, Cu, and Zn to the sediment

Table 7.2   Heavy Metals in Water and Settling Sediments in Greifensee [a]

	Concentration		
	In lake water C	In settling material C_g	Flux to sediments $F_{z=0}$
Metal	$(\mu g \, l^{-1})$	$(\mu g \, g^{-1})$	$(\mu g \, cm^{-2} \, yr^{-1})$
Cd	0.06	4	2
Cu	0.95	34	17
Pb	0.6	77	39
Zn	4.9	134	68

[a]J. Tschopp, EAWAG, Dübendorf/Zurich (personal communication, 1977). Mean sedimentation rate $U_s = 0.85$ cm yr^{-1}; solid sediment density $\rho_s = 2.0$ g cm^{-3}; sediment porosity $\phi = 0.7$; suspended matter concentration (as dry solids) 15 ± 5 mg l^{-1}.

are almost exclusively due to deposition of these metals with settling solids. Potential concentrations of the metals in lake water, if they were completely solubilized from suspended materials, would be of the order of 1 μg l^{-1} (more if surface sediments were also contributing). Such potential concentration levels in lake water should be viewed against the currently acceptable health hazard limits for heavy-metal concentrations in natural waters, of the order of 10^0–10^1 μg l^{-1}.

7.5   UPTAKE AND ADSORPTION BY SOLIDS

7.5.1   Fluxes in Noncompacting and Compacting Sediments

Uptake is an inclusive term that refers to any combination of the processes of chemical exchange or adsorption of dissolved species by solid particles. As it is generally difficult to evaluate how several dissolved substances would distribute themselves between a solution of a given composition and a mixture of several solid phases, a simplified approach to adsorption in water-sediment systems is outlined in this section.

Distribution of a chemical species between a solution and a solid can under certain conditions obey a simple distribution law,

$$C_s = KC \tag{7.8}$$

where C and C_s are concentrations in solution and on the solid, as defined in Section 7.4.1, and K is a factor relating the concentrations to each another. In general, K is a function of temperature, solution composition, and the nature of the solid substrate. The equation also describes an equilibrium in a dilute solution, where concentrations can be used instead of the thermodynamic activities. The product KC can be substituted for C_s in the general flux equation 7.2 with the condition of equal advection velocities of sediment particles and pore water $(U = U_s)$, giving

$$F_{z=0} = -\phi D \frac{dC}{dz} + \phi U (K+1) C \Big|_{z=0} \qquad (7.9)$$

A large value of K means that adsorption is strong, and the larger K is $(K \gg 1)$, the more important is the flux of a chemical species on solid particles UKC in comparison to its flux in pore water UC.

The condition of $U = U_s$ applies to a sediment column with a porosity that is constant with depth. If compaction of sediments is significant and porosity decreases with depth, then, as explained in Section 8.1, the advective velocities of pore water and sediment particles can become equal (U_∞) only at some depth below the sediment-water interface, where porosity has attained a steady value (ϕ_∞). In the region below the interface where porosity changes with depth, the velocity terms U and U_s are

[margin note:] 8.3.3

$$U = \frac{U_\infty \phi_\infty}{\phi} \quad \text{and} \quad U_s = \frac{U_\infty (1 - \phi_\infty)}{(1 - \phi)}.$$

Using the last two relationships together with the definition of $C_s = KC$ in equation 7.1, the total flux $F_{z=0}$ in a compacting sediment becomes

$$F_{z=0} = -\phi D \frac{dC}{dz} + \phi UC \left(\frac{K \phi (1 - \phi_\infty)}{\phi_\infty (1 - \phi)} + 1 \right) \qquad (7.10)$$

Note the difference between the adsorption factor $(K+1)$ in equations 7.9 and 7.10. In the latter, taking $\phi = 0.75$ for the porosity at the sediment-water interface and $\phi_\infty = 0.5$ at depth, where compaction has stabilized, the adsorption factor is $(3K+1)$. A greater change in porosity between the sediment surface and depth (for example, 0.8 and 0.4) gives the factor of $(6K+1)$. Thus if K is large and the sedimentation flux is the main flux mechanism, the value of $F_{z=0}$ is sensitive to the porosity changes due to the sediment compaction.

7.5.2 Ion Exchange and Adsorption

The exchange reaction of two species A and B between their aqueous solution and a solid is

$$A(aq) + mB(s) = mB(aq) + A(s) \qquad (7.11)$$

where (aq) and (s) denote the species in solution and in adsorbed state on the solid, and m is a stoichiometric coefficient. The equilibrium constant K_{eq} for the reaction is

$$K_{eq} = \frac{a_{A(s)} a_{B\,(aq)}^{m}}{a_{A(aq)} a_{B\,(s)}^{m}} \qquad (7.12)$$

Using the conventional relationships between thermodynamic activity, concentration, and activity coefficient (see Section 4.3), the activities of the dissolved species are $a_{i(aq)} = C_i \gamma_i$, where γ is the molal activity coefficient and C is molal concentration; for the adsorbed species, its activity is related to its mole fraction X_i and the activity coefficient in the adsorbed state λ_i by $a_{i(s)} = X_i \lambda_i$. Introducing these relationships into equation 7.12, the concentration of A on the solid, expressed as mole fraction X_A, depends on its concentration in solution C_A according to

$$C_{A(s)} \equiv X_A = \frac{K_{eq} X_B^m \lambda_B^m \gamma_A}{C_B^m \lambda_A \gamma_B^m} \cdot C_A$$

$$= K C_A \qquad (7.13)$$

Comparing equations 7.8 and 7.13 it can be seen that the fractional term, which formally represents the distribution coefficient K, is a complicated function of concentrations in aqueous and solid phases (the activity coefficients themselves are also functions of concentration). The thermodynamic data for adsorbed species in reactions such as equation 7.11 are not widely available, and the equilibrium constants K_{eq} are known only for specific systems. A list of K_{eq} values for ion exchange between clays and aqueous solutions is given in Table 7.3. The activity coefficients of ions in solution γ can be estimated from the Debye-Hückel and related equations, but for the activity coefficients of adsorbed species λ no comparable method exists. Thus equation 7.8 or 7.13 becomes effectively an empirical relationship with a distribution coefficient K that needs to be determined for any given set of conditions.

If species A is present at a concentration much lower than B ($C_A \ll C_B$ and $X_A \ll X_B$), then small changes in the concentration of A do not affect concentrations of the major species B or the activity coefficients, and the coefficient K may be constant over some range of A concentration.

For one chemical species only, the relationship between concentrations in the adsorbed state and in solution is, from equation 7.13,

$$C_s = \frac{K' \gamma}{\lambda} C \qquad (7.14)$$

Table 7.3 Equilibrium Constants of Cation Exchange Reactions on Clays [a]

Cation exchange reaction [b]	Mineral (X)	Equilibrium constant [b] $\log K$
$Na_2X_2 \rightarrow H_2X_2$	Montmorillonite [c]	6.452
$Na_2X_2 \rightarrow Li_2X_2$	Montmorillonite [d]	−0.059
	Bentonite [d]	−0.035
	Vermiculite [d]	−1.059
$Na_2X_2 \rightarrow K_2X_2$	Bentonite [e]	0.449
	Beidellite [c]	1.598
$Na_2X_2 \rightarrow Rb_2X_2$	Bentonite [e]	0.930
	Bentonite [f]	3.400
$Na_2X_2 \rightarrow Cs_2X_2$	Montmorillonite [g, h]	3.16–3.34
	Bentonite [e]	1.585
$Na_2X_2 \rightarrow MgX_2$	Vermiculite [i]	−0.194
$Na_2X_2 \rightarrow CaX_2$	Vermiculite [i]	0.009
$Na_2X_2 \rightarrow SrX_2$	Montmorillonite [j]	0.243
	Vermiculite [j]	−0.009
$Na_2X_2 \rightarrow BaX_2$	Montmorillonite [c]	0.040
$K_2X_2 \rightarrow Li_2X_2$	Montmorillonite [c]	−1.796
$CaX_2 \rightarrow K_2X_2$	Montmorillonite [l,k]	1.06–1.82
	Vermiculite [l, m]	1.51–1.58
$CaX_2 \rightarrow SrX_2$	Bentonite [j]	0.113
	Vermiculite [j]	0.119
$BaX_2 \rightarrow MgX_2$	Vermiculite [i]	−0.271
$BaX_2 \rightarrow CaX_2$	Vermiculite [i]	−0.053
$BaX_2 \rightarrow SrX_2$	Vermiculite [i]	−0.006

[a] Compiled by R. M. Garrels, Northwestern University.
[b] Exchange reaction, using $Na_2X_2 \rightarrow Li_2X_2$ as an example, is $Na_2X_2(s) + 2Li^+(aq) = Li_2X_2(s) + 2Na^+(aq)$. K is the thermodynamic equilibrium constant based on the standard free energy change of the reaction at 25°C: $\log K = -\Delta G°/1.364$, where $\Delta G°$ is in kilocalories per mole.
[c] Truesdell and Christ (1968).
[d] Gast and Klobe (1971).
[e] Gast (1969).
[f] Kunishi and Heald (1968).
[g] Eliason (1966).
[h] Lewis and Thomas (1963).
[i] Wild and Keay (1964).
[j] Heald (1960).
[k] Laudelou et al. (1962).
[l] Rich and Black (1964).
[m] Dolcater et al. (1968).

The quotient $K'\gamma/\lambda$ depends on concentration, similar to the two-species case.

A nonlinear dependence of adsorbed concentration C_s on the solution concentration C is often represented by the Freundlich adsorption equation

$$C_s = KC^n \qquad (7.15)$$

where K and n are constants that have to be determined experimentally for any given system. Generally, values of the power exponent n lie in the range $0.1 < n < 1$.

Another power-law equation describing distribution of two exchangeable species relates the ratio of their concentrations on the solid to the ratio of their activities in solution (cited in Garrels and Christ, 1965, p. 269). Referring to reaction 7.11, the distribution equilibrium is given by the equation

$$\frac{C_{A(s)}}{C_{B(s)}} = K\left(\frac{a_{A(aq)}}{a_{B\,(aq)}^m}\right)^n \qquad (7.16)$$

where K and n are empirical constants. For various ion exchangers, K varies widely (by more than a factor of 1000), whereas the power exponent n falls in the range between about 0.2 and 1.1.

The Langmuir adsorption equation relates the amount of a species adsorbed (adsorption density Γ_A mol cm^{-2}) to its concentration in solution C_A,

$$\Gamma_A = \frac{\Gamma_T C_A}{K_i + C_A} \qquad (\text{mol cm}^{-2}) \qquad (7.17)$$

where K_i is a constant and Γ_T is the adsorption density at the full monolayer coverage of the adsorbing substrate. Equation 7.17 also follows from the exchange reaction 7.11 under the following simplifying assumptions (Parks, 1974): $m = 1$, and concentrations instead of activities are used in the equilibrium constant $K' = X_A C_B / X_B C_A$. If the maximum adsorption density $\Gamma_T = \Gamma_A + \Gamma_B$ is maintained by the sum of the adsorbed amounts of the two species (that is, $\Gamma_A = X_A \Gamma_T$ and $\Gamma_B = X_B \Gamma_T$), then

$$\Gamma_A = \frac{\Gamma_T C_A}{C_B / K' + C_A} \qquad (\text{mol cm}^{-2}) \qquad (7.18)$$

The latter is similar to the Langmuir equation 7.17 in the case when the major species concentration C_B is nearly constant. At low concentrations of A, equation 7.18 reduces to the form of

$$\Gamma_A = \frac{\Gamma_T K'}{C_B} C_A \qquad (7.19)$$

which is identical to $C_s = KC$ in equation 7.8.

7.5.3 Measures of Adsorption

Experimental measurements are certainly the most reliable way to find out the type of exchange or adsorption that takes place in a given solution-solid system under given conditions. However, estimates of distribution coefficients reported in the literature are sometimes based on chemical analyses of waters and of exchangeable ions on sediments. Some values of the distribution coefficient K for a number of chemical species in different sedimentary environments are listed in Table 7.4. The differences in the K values for individual ions, and the differences between adsorption in sea

Table 7.4 Distribution Factor $K = C_s / C$
for Some Chemical Species in Oceanic and Freshwater Sediments
(K Dimensionless, Concentration in Solids C_s and in Solution C
are in Units of Mass per Unit Volume of Solution)

Chemical species	K	Notes and sources
Ca	1.6	Sea water and clays (Li and Gregory, 1974)
Cs	3,500–10,000	^{137}Cs in freshwater sediments (Lerman and Lietzke, 1975)
Mg	10	Deep-ocean sediments (Lerman and Lietzke, 1977)
Na	0.3	Sea water and clays (Li and Gregory, 1974)
	2	Freshwater sediments (Lerman and Weiler, 1970)
$NH_4^+ - NH_3$	3	Anoxic marine sediments (Berner, 1974)
Phosphate	8	Sea water and clays (Berner, 1973, 1974)
Ra	2000	^{226}Ra in deep ocean sediments (Lerman, 1977)
Sr	6–12	^{90}Sr in Ligurian Sea sediments (Cerrai et al., 1969)
	45–120	^{90}Sr in freshwater sediments (Lerman and Lietzke, 1975)

water and fresh water (sodium and strontium) are large. High values of K are generally associated with adsorption of radioactive tracers, such as shown in the data listed in Table 7.4 and additional data for other radionuclides reported by Aston and Duursma (1973).

Another measure of adsorption is an ion exchange capacity, commonly expressed as milliequivalents of exchanged ion per 100 g of exchanger material. Solids, the ion exchange capacity of which is sought, are placed in a solution of sufficiently high concentration of an easily exchangeable ion (such as solutions of NaCl or NH_4Cl) until all the exchange sites are presumably taken by Na^+ or NH_4^+ or, in principle, any other desirable cation or anion. Subsequently, the adsorbed ion is replaced by another, such as H^+ in water at the pH of usually 7, and the amount replaced is determined in solution. The number of milliequivalents released in the replacement process is reported as the ion exchange capacity (milliequivalents per 100 g).

Table 7.5   Cation Exchange Capacities
of Natural Materials, Arranged in a Decreasing Order
of Upper-Limit Values [a]

Cation exchange capacity (milliequivalents per 100 g)	Material
350–130	Organic matter
300–100	Zeolites
260	Hydrous manganese oxide [b]
150–100	Vermiculite
100–70	Montmorillonites (typical)
60–20	Micas from soils [c]
50–40	Halloysite ($4H_2O$)
47–4	Chlorite
40–10	Illite
30–20	Palygorskite
25–4	Various soils [c]
20–11	Glauconite
15–3	Kaolinite
10–5	Halloysite ($2H_2O$)
4	Pyrophyllite
1	Feldspar, quartz
0.5	Basalt (Snake River)

[a] Unmarked data from Carroll (1970), Na^+-H^+ exchange at pH = 7.
[b] Murray and Brewer (1977), at pH = 8.
[c] Schachtschabel (1940), pH = 7, NH_4^+, Ca^{2+}, and K^+ exchange.

The ability of a solid to adsorb ions from solution, and the resulting values of the ion exchange capacity can be affected by the following. Whether a cation or an anion is adsorbed on the solid surface depends on the electric surface charge. The charge can vary and change its sign depending on the H^+-ion concentration of the solution and, at least to some extent, depending on the presence of organic coatings on solid particles. Decrease in the particle size and comminution of grains increase the solid surface area per unit of its mass and usually lead to a higher value of the exchange capacity. Specific adsorption exhibited by some ions on different substrates may result in exchange capacity values that differ from one ion to another and, significantly, differ from one pH value of the solution to another.

For some minerals, organic matter, and soils, the cation exchange capacities are listed in Table 7.5. Organic matter of soils, zeolites, and a clay mineral vermiculite top the list as substances with a high cation exchange capacity.

Saturation of the cation exchange capacity of a sediment by adsorption from solution can amount to a few percent of the sediment weight. For example, when a K^+ ion occupies the exchangeable sites of a sediment, the exchange capacity of which is 50 milliequivalents/100 g, then the concentration of K^+ in the solid is:

$$(50 \times 10^{-5} \text{ equivalents g}^{-1}) \times (39 \text{ g equivalent}^{-1}) = 0.02 \text{ g g}^{-1} \text{ or } 2 \text{ wt \%}.$$

7.6 SILICA IN THE INTERFACE ZONE

7.6.1 Deep Sea

Concentration profiles of silica in pore water of the eastern Caribbean Sea sediments, within several tens of centimeters below the sediment-water interface, are shown in Figure 7.3. The silica concentration increases pronouncedly with depth in sediment, from the ocean water value of 28 μmol SiO_2 l^{-1} near the bottom into the 200–300 μmol l^{-1} range. The siliceous shells of diatoms are a likely source of dissolved silica in sediments. The positive concentration gradients within the upper 5 cm of the sediment pore water column show that silica concentration increases six- to tenfold over a short distance. From about 5 cm depth down, some of the profiles remain nearly constant with depth, and some show a slight decrease in concentration, suggesting the presence of mineral sinks in sediments.

The flux of silica across the sediment-water interface can be evaluated using equations 7.1 or 7.9. A reasonable value for the sediment porosity at the interface is $\phi = 0.7$, and for the diffusion coefficient of SiO_2 in pore

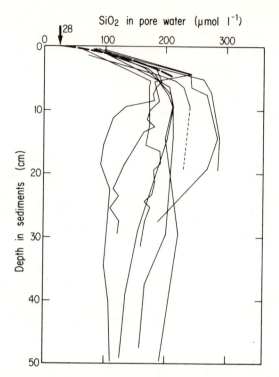

Figure 7.3 Dissolved silica in pore waters of sediments in the eastern Caribbean Sea. Heavy dashed line shows a generalized concentration gradient below the sediment-water interface (from unpublished data of K. A. Fanning). By permission of Prof. K. A. Fanning, St. Petersburg, Florida.

water $D \simeq 3 \times 10^{-6}$ cm^2 sec$^{-1} = 100$ cm^2 yr^{-1}; the concentration gradient can be read off Figure 7.3. The molecular diffusional flux is

$$F_d = -\phi D \frac{\Delta C}{\Delta z} = \frac{-0.7 \times 100 \times (200 \text{ to } 300) \times 10^{-3}}{5}$$

$$= -2.4 \quad \mu\text{mol cm}^{-2} \text{yr}^{-1}$$

The diffusional flux is out of the sediment and its magnitude is comparable to the value of the silica flux E listed in Table 7.1.

The advective flux into the sediment, owing to inclusion of dissolved silica in the growing pore water column, at the rates of sedimentation between 1 and 20 cm per 1000 yr, is

$$F_a = \phi U C_{z=0} = 0.7 \times (1 \text{ to } 20) \times 10^{-3} \times 28 \times 10^{-3}$$

$$= 0.2 \times 10^{-4} \text{ to } 4 \times 10^{-4} \quad \mu\text{mol cm}^{-2} \text{yr}^{-1}$$

In this case, the advective flux in pore water is negligibly small in comparison to the diffusional flux out of the sediment, such that an effective mean value for $F_{z=0}$ is essentially equivalent to the diffusional flux, about 2.4 μmol cm^{-2} yr^{-1}. If adsorption of silica takes place on sediment particles, only very strong adsorption can counteract the diffusional flux of dissolved silica out of the sediment. To make the value of the flux F_a comparable in magnitude to F_d, the adsorption coefficient K should be of the order of magnitude $K \simeq 10^4$ (see equation 7.9, term UKC). In the presence of such a strong adsorption, the net flux out of sediment would be zero. If silica is adsorbed, the question should be asked, how much is taken up by the solid sediment over some period of time? An adsorption flux of the order of 2 μmol cm^{-2} yr^{-1} adds about 100 mg SiO$_2$ to solids in 1 cm^3 of bulk sediment in 1000 yr. The mass of solids in 1 cm^3 of bulk sediment is between 500 and 1000 mg. Thus the mass of silica added would not affect much the solid weight fraction or the porosity of the sediment.

7.6.2   Near-Shore Sediments

A concentration profile of dissolved silica in a near-shore sediment, similar to the profiles in deep-sea pore waters, is shown in Figure 7.4a. The sediment is made of a mixture of biogenic carbonates and detrital aluminosilicates transported from soils and partly weathered volcanic rocks of the vicinity of Kaneohe Bay, Island of Oahu, Hawaii.

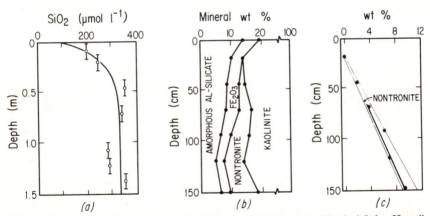

Figure 7.4   Diagenesis of silica in recent sediments of Kaneohe Bay, Island of Oahu, Hawaii (Ristvet, 1977, and personal communication, 1975). (a) Dissolved silica concentration in pore water. The computed curve is $C = 0.34 - 0.24 \exp(-0.55z)$ mmol^{-1}, where z is in centimeters. (b) Weight % of minerals in the fine size fraction of sediment ($<0.2\mu$m) amounting to approximately 0.5 wt % of dry sediment. (c) Weight % of nontronite in the fine fraction. Thin lines bracket the range of data. Thicker line is a computed increase with depth (from Lerman, 1978). Reproduced with permission from Annual Review of Earth & Planetary Sciences, Vol. 6. © 1978 by Annual Reviews Inc.

The mineralogical composition of the fine fraction of sediment (particle size < 0.2 μm, Figure 7.4b) shows that the aluminosilicate mineral nontronite appears at a depth of about 20 cm and its relative abundance increases with an increasing depth. A possible reaction between kaolinite, dissolved silica, and other cations (in sea water or in solid phases) producing nontronite can be written as

$$Al_2Si_2O_5(OH)_4 + 12H_4SiO_4 + cations \rightarrow 2(cations)Fe_4AlSi_5O_{20}(OH)_4$$
$$\text{Kaolinite} \qquad\qquad\qquad\qquad\qquad \text{Nontronite}$$

Combination of the data in Figures 7.4a and 7.4b suggests that nontronite grows at silica concentrations higher than about 250 μmol l^{-1} and constitutes a sink for silica in pore water. The removal of silica from pore water at depths $z > 20$ cm will be considered as a first-order precipitation reaction, and the removal rate is $k(C - C_{z=20})$ μmol cm^{-3} yr^{-1}. The length of time precipitation has taken place at any depth z is $(z-20)/U_s$ yr, where U_s is the sedimentation rate (cm yr^{-1}). The amount of silica removed from pore water into solids of density ρ_s (g cm^{-3}) is

$$\frac{\phi k(C - C_{z=20})(z-20)}{(1-\phi)\rho_s U_s} \qquad \text{(mol SiO}_2 \text{ in 1 g of solids)}$$

Silica concentration in pore water below 75 cm depth is nearly constant and the difference $C - C_{z=0}$ is therefore also constant with depth. If porosity does not change appreciably below 75 cm, then doubling of the distance $(z-20)$ cm in sediment corresponds to doubling of the amount of silica precipitated. When all the precipitated silica goes into the formation of nontronite, then the nontronite fraction in sediment should increase linearly with depth below 75 cm. This is nearly the case, as shown by the few points plotted in Figure 7.4c: below 75 cm depth, the abundance of nontronite increases approximately linearly with depth. The fraction of nontronite in sediment, 4% at $z = 75$ cm, increases by a factor of 2, to 8% at $z = 130$ cm, and this is an increase of a factor of 2 in the distance measured from the 20-cm depth level (that is, from 55 to 110 cm).

The occurrence of nontronite in the very fine size fraction of the sediment and its low concentration in bulk sediment point to the general difficulties that are encountered in identification of the mineral sinks for dissolved chemical species in pore waters. The presence or absence of mineral sinks controls the magnitude of diffusional fluxes of reactive species in or out of sediments, even though the mass of the mineral sink can be very small in comparison to the bulk of solid phases making the sediment.

7.6.3   Lake Sediments

In lake sediments, diatoms are as an important source of dissolved silica as in the ocean. A dissolved silica profile, shown in Figure 7.5, approaches the solubility of opal near 20 cm below the sediment-water interface (25 mg Si $l^{-1} \simeq 900$ μmol l^{-1}). Over a period of one year, there is little variation in the silica concentrations in sediments and water: the range of variation, shown by the hatched area between the curves, amounts to only a few percent. Concentrations of Si in the sediment pore water can be approximated by a curve

$$C = 25 - 8.9 \exp(-0.135z) \qquad (\text{mg Si } l^{-1})$$

where z is in cm. A model of silica diagenesis given in equation 8.46, is based on dissolution and precipitation, and diffusional transport in pore water. The power exponent 0.135 for the silica concentration profile can be equated to the exponent term $(k/D)^{1/2}$ to obtain the reaction rate con-

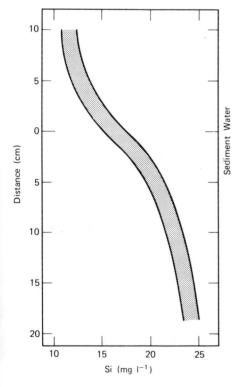

Figure 7.5   Dissolved silica in the sediment-water interface zone of a freshwater lake (Ursee, Germany) (from von Tessenow, 1975).

stant k. Taking the diffusion coefficient of silica in pore water as $D = 3 \times 10^{-6}$ cm^2 sec$^{-1} = 95$ cm^2 yr^{-1}, we have

$$\left(\frac{k}{D}\right)^{1/2} = 0.135 \quad \text{cm}^{-1}$$

$$k = 1.7 \quad \text{yr}^{-1}$$

Sedimentation, supplying the silica to sediments, can be neglected as long as the inequality $U \ll (4kD)^{1/2}$ is maintained, according to equation 8.41. The quantity $(kD)^{1/2}$ is $(1.7 \times 95)^{1/2} = 25$ cm yr^{-1}. Rates of sedimentation U in lakes are usually about 10 times lower, so that no significant error is introduced by computing the rate constant k without the sedimentation rate term. The high value of k for silica in the sediments of Lake Ur is comparable to the high rate constants of silica dissolution in laboratory experiments (Table 8.1).

The flux of silica across the sediment-water interface can be evaluated, using equation 7.25, as

$$F_{z=0} = \frac{0.75 \times (25 \text{ cm yr}^{-1})(16 - 25 \text{ mg Si l}^{-1})}{(10^3 \text{ cm}^3 \text{ l}^{-1})(28 \times 10^{-3} \text{ mg Si } \mu\text{mol}^{-1})}$$

$$= -6 \ \mu\text{mol cm}^{-2} \text{ yr}^{-1}$$

In comparison to the silica fluxes in oceanic sediments, the flux $6 \ \mu\text{mol cm}^{-2} \text{ yr}^{-1}$ is of the same order of magnitude. Because of the high concentrations of dissolved silica in sediments, this value is more characteristic of the lake sediments containing soluble silicate material.

7.7 FLUXES AND DIAGENETIC MODELS

To evaluate the flux across the sediment-water interface using equation 7.2, the concentration gradient dC/dz can be estimated from a profile of concentration plotted against depth, as was done in the section on the fluxes of silica in the deep sea. A more general approach is to have an analytic expression for the concentration gradient dC/dz and use it in the flux equation. Explicit equations for the gradient dC/dz can be derived if concentration is known as a function of depth z, transport parameters (D and U), and the chemical reaction rate terms. Explicit relationships for concentration C are obtained as solutions of the differential equations, which serve as models of diagenesis in sediments and which are discussed in Chapter 8 (Sections 8.3.1, 8.4.2, and 8.5.1). Conceptually, the mathematical models of diagenesis are based on the combinations of chemical sources, sinks, and transport processes in sediments, as was elaborated on at the beginning. The equations for the fluxes at the sediment-water interface that are given in this section fall into two categories. One, fluxes in the presence of fast adsorption by solids, or instantaneous removal; and two, fluxes in the presence of minerals sinks acting as first-order precipitation reactions, making the removal time dependent.

7.7.1 Production and Adsorption

A model of diagenesis that describes production of a chemical species in pore water due to decomposition of organic matter in sediment and adsorption from pore water onto sediment particles is given for a one-dimensional sediment column in equation 8.28. The parameters of the model are explained in Chapter 8, and here they will be restated only briefly, to introduce the flux equations. The model can be applied to the production of phosphate and ammonia from organic matter in sediments under certain conditions. A numerical example using the equations is given at the end of the section.

$J_* e^{-\beta z}$ is the rate of production of the chemical species (μmol in 1 cm^3 of pore water per year). β is a constant of dimensions of reciprocal distance. The rate of production decreases exponentially with distance if $\beta > 0$. Alternatively, the rate of production is constant with depth if $\beta = 0$.

U is the rate of pore water advection relative to the sediment-water interface at $z = 0$ (cm yr^{-1}). The rates of pore water and solid sediment advection are taken as equal ($U = U_s$) if the porosity of the sediment is constant with depth. Then U or U_s is the mean rate of sedimentation.

D is the diffusion coefficient of the chemical species in pore water, K is the adsorption or exchange coefficient defined in equation 7.8, ϕ is the sediment porosity, and $C_{z=0}$ is concentration in pore water at the sediment-water interface.

The flux $F_{z=0}$ is

$$F_{z=0} = \frac{-\phi D J_*}{D\beta + U(K+1)} + \phi U(K+1) C_{z=0} \qquad (7.20)$$

The first term on the right-hand side of the equation is negative but the second term is positive. Production in the sediment can be balanced by the rate of the sediment column growth and adsorption on solids. If the rate of sedimentation is high and adsorption is strong (U and K large) then $F_{z=0}$ is positive, which means that the flux is directed downward into the sediment. Alternatively, if the rate of sedimentation is low and adsorption is weak (U and K small), the negative term dominates $F_{z=0}$ and the flux is out of the sediment into the overlying water. Strong adsorption retards diffusion in pore water because it lowers concentration in solution. In equation 7.20, the effect of large adsorption coefficient K is equivalent to the very low value of the diffusion coefficient D: either makes the first term small. The second term of the equation is the sum of the two advective fluxes $F_a + F_s$, as defined in equations 7.4 and 7.5.

7.7.2 Production and First-Order Removal

First-order precipitation reactions and radioactive decay are examples of the first-order removal processes. The equations for concentration in pore water C, 8.9 and 8.28, include the rates of production $J_* e^{-\beta z}$ and dissolution J (both in units of mol cm^{-3} yr^{-1}), and the rate constant k (yr^{-1}) for a first-order removal process. More than one physical interpretation of the parameters J_*, J, and k is possible. For example, the ratio J/k (mol cm^{-3}) can be interpreted as a constant concentration in pore water to which the concentration profile tends with depth.

For this model, a general equation of the flux $F_{z=0}$ at the sediment-water interface is

$$F_{z=0} = \frac{-\phi D J_*}{D\beta^2 + U\beta - k}\left(\beta + \frac{U}{2D} - R\right)$$

$$-\phi D\left(C_{z=0} - \frac{J}{k}\right)\left(\frac{U}{2D} - R\right) + \phi U C_{z=0} \qquad (7.21)$$

where R is defined as

$$R = \left(\frac{U^2}{4D^2} + \frac{k}{D}\right)^{1/2}$$

$$= \frac{U}{2D}\left(1 + \frac{4kD}{U^2}\right)^{1/2} \qquad (\text{cm}^{-1}) \qquad (7.22)$$

The flux can be either out of the sediment or into the sediment, depending on the values of the individual parameters that enter in the equation. A number of particular cases lead to simpler forms of the flux $F_{z=0}$.

Slow Precipitation Reaction.   When the rate of removal of a dissolved species from a unit of pore water volume by precipitation is slower than its removal by advection and diffusion, the following inequality applies:

$$k \ll \frac{U^2}{4D} \quad (\text{yr}^{-1})$$

In this case we have $U/2D - R \simeq 0$, and the flux equation 7.21 reduces to

$$F_{z=0} = \frac{-\phi DJ_*}{D\beta + U} + \phi U C_{z=0} \tag{7.23}$$

No Reactions of Organic Matter.   The production rate term $J_* e^{-\beta z}$ can be equated to the rate of decomposition of organic matter decreasing exponentially with depth (Chapter 8). The case of no organic matter reactions corresponds to the condition of $J_* = 0$. The flux equation 7.21 becomes

$$F_{z=0} = -\phi D\left(C_{z=0} - \frac{J}{k}\right)\left(\frac{U}{2D} - R\right) + \phi U C_{z=0} \tag{7.24}$$

Equation 7.24 is the flux equation for a semi-infinite pore water column within which concentration approaches a steady value at depth $C_{z=\infty} = J/k$, and the concentration gradient tends to zero. Concentration profiles of this type are shown in Figures 7.3 and 7.4.

Zero Sedimentation Rate.   If the sediment column does not grow and there is no forced flow of pore water, then $U = 0$. The flux equation 7.24 becomes

$$F_{z=0} = \phi(kD)^{1/2}\left(C_{z=0} - \frac{J}{k}\right) \tag{7.25}$$

If the concentration at depth is greater than at the sediment-water interface, then the condition $J/k > C_{z=0}$ exists and the flux is out of the sediment ($F_{z=0} < 0$).

No Precipitation Reaction.   The case of no precipitation corresponds to the condition of the reaction rate constant zero: $k = 0$. The rate of dissolution may be nonzero and, according to the terms of the model, the rate of dissolution J is a constant or a zeroth order reaction. The flux equation 7.24 becomes, with $k = 0$,

$$F_{z=0} = -\frac{\phi DJ}{U} + \phi U C_{z=0} \tag{7.26}$$

As the first term on the right-hand side is negative and the second term is positive, a condition of zero flux can exist. Dissolution builds up concentration in pore water, from which it is removed by diffusional flux upward. A sufficiently fast rate of sedimentation U may counteract the upward flux.

No Mineral Reactions. The case of no mineral-pore water reactions corresponds to the condition of $J=0$ and $k=0$. In the absence of any chemical reactions within the sediment, there are no concentration gradients in pore water and the flux across the sediment-water interface is due only to the growth of the sediment pore water column. The flux $F_{z=0}$ is

$$F_{z=0} = \phi U C_{z=0} \tag{7.27}$$

7.7.3 Example: The Flux of Phosphorus

Concentrations of dissolved phosphorus in pore water of a sediment core from the deep Timor Trough in the southwestern Pacific Ocean are shown in Figure 7.6. Concentration increases from the sediment-water interface down to about 50 m depth, from where it decreases further down in the core. The concentration maximum near 50 m depth suggests that decomposition of organic matter may be responsible for the higher phosphorus concentration. The concentration gradients above and below the peak indicate that molecular diffusion should be taking place upward and downward in the pore water column. For the downward flux, the sink is

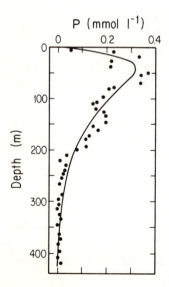

Figure 7.6 Dissolved phosphorus in pore waters of a deep sea sediment (from data of Cook, 1974).

likely to be uptake by solid particles of the sediment. For the upward flux, the sink is ocean water even if uptake by solids takes place uniformly throughout the sediment column. The smooth curve drawn through the points gives the phosphorus concentration in pore water as a function of depth,

$$C=0.746e^{-0.011z} - 0.726e^{-0.042z} \quad (\text{mmol } 1^{-1}) \quad (7.28)$$

where the distance below the sediment-water interface z is in meters. The explicit form of C is obtained by inserting the numerical values of the parameters listed below in equation 8.32. The reasons for the choice of this model equation, as compared to other possible models for the concentration profiles with maxima as in Figure 7.6, are explained in Chapter 8. The numerical values of the transport and chemical reaction parameters will be used to evaluate the flux across the sediment-water interface according to equations 7.20 through 7.27. This procedure makes clear the role of the individual processes and their contributions to the flux at $z=0$. The parameters for the flux equations are (from Toth and Lerman, 1977):

Diffusion coefficient	$D=1\times10^{-6}$ cm^2 sec$^{-1}=32$ cm^2 yr^{-1}
Sedimentation rate	$U=13$ cm per 1000 yr $=1.3\times10^{-2}$ cm yr^{-1}
Sediment porosity at $z=0$	$\phi=0.7$
Concentration at $z=0$	$C_{z=0}=0.02$ μmol cm^{-3}
Concentration at $z=\infty$	$J/k=8.4\times10^{-6}$ μmol cm$^{-3}\simeq0$
Removal rate constant	$k=1.9\times10^{-6}$ yr^{-1}
Production rate at $z=0$	$J_*=6.5\times10^{-6}$ μmol cm^{-3} yr^{-1}
	$\beta=4.15\times10^{-4}$ cm^{-1}

The value of the adsorption coefficient K in equation 7.20 is $K=8$, from Table 7.4. The fluxes from the individual models are:

Equation 7.21	$F_{z=0}=-4.7\times10^{-3}$	(μmol cm^{-2} yr^{-1})
Equation 7.23	$F_{z=0}=-5.4\times10^{-3}$	
Equation 7.24	$F_{z=0}=+0.23\times10^{-3}$	
Equation 7.25	$F_{z=0}=+0.11\times10^{-3}$	
Equation 7.26	$F_{z=0}=+0.18\times10^{-3}$	
Equation 7.27	$F_{z=0}=+0.18\times10^{-3}$	
Equation 7.20	$F_{z=0}=+0.50\times10^{-3}$	

The first result, -4.7×10^{-3} μmol cm^{-2} yr^{-1}, is based on the more complete model of equation 7.21. The same result could be obtained by taking the concentration derivative dC/dz from equation 7.28 and using it with other parameters in the flux equation $F_{z=0} = -DdC/dz + UC$. The next value of the flux, -5.4×10^{-3}, is based on a model that ignores the mineral sinks, and the flux is therefore somewhat higher than in the preceding example. All the other values of the flux on the above list are positive, indicating transport from the water into the sediment column. The models of equations 7.24–7.27 ignore the production of phosphorus due to decomposition of organic material as expressed in the J_* term. Therefore, they underestimate the phosphorus gradients in pore water and give incorrect fluxes. The model of equation 7.20 also gives a positive value of the flux, despite the fact that the model includes the production term J_*. The relatively strong adsorption of phosphorus by sediments, as represented by the adsorption coefficient of about $K \simeq 8$, hinders the diffusional flux out of the sediment.

7.7.4 Fluxes and the Rates of Sedimentation

The fluxes across the sediment-water interface of six ionic species—sulfate, ammonia, phosphate, calcium, magnesium, and bicarbonate—are plotted in Figure 7.7 against the rates of sedimentation at each site represented by a dot in the graph. The sediments cover a range of environments from near shore to the pelagic ocean, and a range of sedimentation rates.

Straight lines drawn through the points have the slopes of approximately unity, indicating that there is a nearly linear relationship between the flux across the sediment-water interface and the mean sedimentation rate,

$$F_{z=0} \propto U \qquad\qquad (7.29)$$

The straight lines in the log-log graphs of Figure 7.7 are given by the following equations for the flux as a function of the sedimentation rate (U is in cm yr^{-1}, and $F_{z=0}$ is in μmol cm^{-2} yr^{-1});

$$F_{S,z=0} = 81\, U^{0.98 \pm 0.15}$$

$$F_{P,z=0} = -0.31\, U^{1.29 \pm 0.58}$$

$$F_{N,z=0} = -18\, U^{1.32 \pm 0.54}$$

$$|F|_{Ca,z=0} = 88\, U^{1.27 \pm 0.17}$$

$$F_{Mg,z=0} = 48\, U^{0.86 \pm 0.09}$$

$$F_{HCO_3,z=0} = -310\, U^{1.22 \pm 0.52}$$

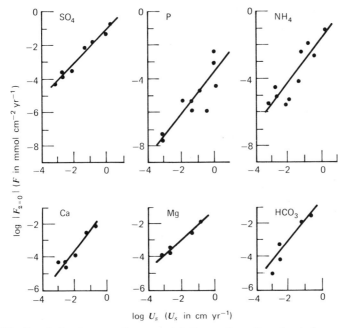

Figure 7.7 Correlation between the computed fluxes of reactive chemical species at the sediment-water interface $F_z = 0$ and the rates of sedimentation U_s (from Toth, 1976; Toth and Lerman, 1977).

The fluxes of phosphorus, ammonia, and bicarbonate are out of the sediments because of their higher concentrations in pore waters due to the decomposition of organic materials. The fluxes of sulfate are into the sediments where sulfate is being reduced in pore waters. Magnesium is also transported from ocean water into the sediments, likely because of its uptake by clays (Drever, 1977). The fluxes of calcium are out of the sediment at some of the localities, and into the sediment at some other sites. The behavior of calcium in pore waters is sensitive to the mineralogical composition of sediments and the mineral-pore water reactions taking place in the presence or in the absence of organic matter.

At first glance, a positive correlation between the flux $F_{z=0}$ out of the sediment and the rate of sedimentation seemingly contradicts the discussion following equation 7.2, where it was pointed out that sedimentation can counteract the effects of the diffusional transport upward. However, rapid sedimentation usually characterizes near-shore environments, where relatively large amounts of organic materials are deposited with the inorganic sediments. A greater amount of reactive organic materials raises concentrations of such chemical species as nitrogen and phosphorus, and is

also reponsible for the stronger concentration gradients than would other-
wise develop in organic-matter-poor sediments. At a higher rate of sedimentation, more reactive organic materials are supplied to the sediments,
and their reactions with pore waters overcome the effect of the higher
sedimentation rate. The net result is a positive correlation between the flux
(computed from the model) and the rate of sedimentation. For sulfate and
magnesium, the diffusional fluxes in pore water are directed downward,
because of their lower concentrations in pore water in comparison to the
overlying ocean water. Therefore, the higher rate of sedimentation en-
hances the flux to the sediments.

The direct proportionality relationship 7.29 arises from the following. In
Chapter 8 it is shown that the parameter β (cm^{-1}) appearing in the
production rate term $J \cdot e^{-\beta z}$ correlates positively with the rate of sedimentation U, that is, $\beta \propto U$. Also, the parameter R defined in equation 7.22 is
approximately proportional to the first power of U, or $R \propto U$. Next, if a
constant $\times U$ is substituted for R in equation 7.21, and a *constant* $\times U$ is
substituted for β, then it can be verified by inspection that the relationship
$F_{z=0} \propto U$ is nearly valid.

7.7.5 Dissolution and Saturation in Pore Waters

The reactions of silicates in the zone of the sediment-water interface,
discussed in Section 7.6, usually produce such concentration gradients of
dissolved SiO_2 in pore waters that the flux direction is out of the sediments
into the overlying water. The existing concentration gradients reflect the
fact that the solids in homogeneous sediments are not in a thermodynamic
equilibrium with pore water solutions. The mass of dissolved solids that
can bring a pore water solution to saturation is estimated in this section for
opal and calcite. A state of saturation or near saturation may exist (1) if
there is no removal of dissolved species by transport or chemical processes
from pore water, or (2) if dissolution is much faster than removal.

The solubility of amorphous silica, taken as an analog of opal, at the
temperatures of the earth's surface is about 100 mg SiO_2 l^{-1}, or
0.1 mg cm^{-3} (from data in Figure 8.7). The mass of solid SiO_2 contained in
a 1-cm-thick layer of fresh sediment is between 400 and 1000 mg cm^{-3}.
When about 0.1 mg SiO_2 is removed from 400–1000 mg of solid to 1 cm^3
of pore water, the mass loss is negligible. The undersaturation with respect
to amorphous silica in pore waters of opal-containing sediments is an
indication of other sinks for dissolved silica, as well as of the flux out of
sediments.

Calcite or aragonite dissolving in deep-ocean water (Section 6.3.3) can
bring a closed volume of solution to saturation, which can be determined

from the solubilty product K_{sp} of the solid phase, written as

$$K_{sp} = ([Ca^{2+}] + x)([CO_3^{2-}] + x) \quad (mol^2\, l^{-2}) \quad\quad (7.30)$$

where the brackets denote concentrations of total dissolved calcium and carbonate ions in ocean water (mol l^{-1}), and x is the number of moles of $CaCO_3$ that should be added to solution to bring it to an equilibrium with the solid $CaCO_3$ phase. Concentration of Ca^{2+} is nearly conserved in ocean water, but CO_3^{2-} is controlled by the equilibria between the dissolved carbonate species (Sections 1.5.3 and 4.4.3). For an approximate computation of how much $CaCO_3$ is needed to bring deep-ocean water to saturation, only the solubility product, equation 7.30, will be considered. Representative concentrations of the calcium and carbonate ions in the deep ocean are

$$[Ca^{2+}] = 10 \times 10^{-3} \quad mol\, l^{-1}$$

$$[CO_3^{2-}] = 0.075 \times 10^{-3} \quad mol\, l^{-1}$$

The solubility product K_{sp} for calcite depends on temperature, chlorinity of sea water, and pressure, according to the two following equations (Edmond and Gieskes, 1970). At atmospheric pressure, K_{sp} is a function of chlorinity and temperature,

$$K_{sp} = (0.1614 + 0.02892\, Cl - 0.0063t) \times 10^{-6} \quad\quad (7.31)$$

where Cl is the chlorinity (‰ or g kg^{-1}) and t is temperature (°C). The equation is valid in the range $0 < t < 40°C$ and $15 < Cl < 25‰$. The dependence of K_{sp} on pressure is

$$\log \frac{(K_{sp})_P}{(K_{sp})_1} = -\frac{\Delta V}{2.303 RT}(P-1)$$

$$= \frac{35.4 - 0.23t}{188.9(273.15 + t)}(P-1) \quad\quad (7.32)$$

where P is total pressure, in atmospheres. The average chlorinity of sea water is $Cl = 19‰$, and at a depth of 5000 m the pressure is $P \simeq 500$ atm and the temperature is $t \simeq 5°C$. Using these values in equations 7.31 and 7.32, the solubility product of calcite is

$$K_{sp} = 1.437 \times 10^{-6} \quad (mol^2\, l^{-2})$$

(The solubility products of calcite and aragonite, reported by Edmond and Gieskes (1970), are based on concentration units of moles per kilogram. For sea water, the differences between the concentration units of moles per kilogram, moles per 1000 g H_2O, and moles per liter are sufficiently small to be neglected in this computation.) The degree of saturation of deep-ocean-like water, containing the calcium and carbonate ions at concentrations given above, is

$$\frac{[Ca^{2+}][CO_3^{2-}]}{K_{sp}} = \frac{10 \times 10^{-3} \times 0.075 \times 10^{-3}}{1.437 \times 10^{-6}}$$

$$= 0.52 \quad \text{or} \quad 52\%$$

The mass of $CaCO_3$ needed to bring 1 liter of deep-ocean-like water to saturation with respect to calcite is, using equation 7.30,

$$(10 \times 10^{-3} + x)(0.075 \times 10^{-3} + x) = 1.437 \times 10^{-6}$$

The latter can be solved for x, giving

$$x = 6.77 \times 10^{-5} \text{ mol l}^{-1} \simeq 7 \text{ μg cm}^{-3}$$

The mass of calcite needed to saturate 1 cm^3 of deep-ocean water is only 7 μg, and this corresponds to dissolution of one calcite crystal, a cube of edge 140 μm.

Most of the calcareous material dissolves in the deep-ocean water and at the bottom. Little can dissolve after the material is covered by accumulating sediments and becomes part of the sedimentary column. This point is illustrated by the size distribution of particles in a calcareous sediment, as shown in Figure 7.8. The sediment, at 3850 m depth in the west equatorial Pacific Ocean, shows signs of corrosion and dissolution. However, the shape of the particle-size distribution in the upper 1 cm of the sediment is essentially the same as at 20 cm depth. The smaller particles do not dissolve faster than the larger particles, as might have been anticipated if they behaved as dissolving spheres of different sizes (see Figure 5.11). Slow dissolution of porous and fragmented particles can take place at the surfaces exposed to solution without affecting the overall particle dimensions. In such a case, or in a case of no dissolution because of saturation, the particle-size distribution can remain nearly constant with time, at least as long as the particles have not lost a major fraction of their mass.

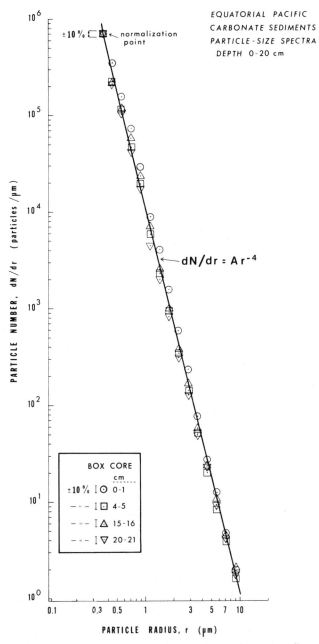

Figure 7.8 Particle-size distributions at four depths in a calcareous sediment from west equatorial Pacific Ocean, 1°6.0′S, 161°36.6′E, box core no. 136, water depth 3848 m (sediments courtesy of Prof. W. H. Berger, Scripps Institution of Oceanography).

7.8   TRANSPORT OUT OF THE SEDIMENTS

Sedimentary materials that dissolve, decompose, or are degraded biologically near the sediment-water interface are regenerated to the pore waters
and to the overlying body of water. The existence of a positive concentration gradient (see Figure 7.2) in pore water is a necessary but not sufficient
condition for the flux out of the sediment: strong adsorption on sediment
particles or fast rates of sedimentation can counteract the diffusional
fluxes directed upward, as explained in the preceding section. Direct
observations of the fluxes out of sediments are possible through measurement of concentrations in vertical profiles above the sediment: a
concentration gradient of a chemical species in water, with higher concentration near the bottom, usually indicates transport up from the sediments. In seasonally stratified eutrophic lakes it has been recognized long
ago from the concentration profiles in the water column that phosphorus,
nitrogen, iron species, and hydrogen sulfide are released from the bottom.
The shape of a vertical concentration profile of a chemical species in water
can yield information on its rate of transport (flux) and the rates of
chemical reactions producing or removing it. Because of the complex
nature of the chemical reactions involving the nutrient species, such as
those mentioned, they are not convenient tracers of the fluxes from
sediments. The vertical eddy diffusion of water is at least in part responsible for the fluxes of various chemical species from the sediments up. A
convenient tracer of the vertical eddy diffusion is the radioactive isotope of
radon, ^{222}Rn, of half-life of about 4 days (radon is a noble gas at the
atmospheric conditions). The isotope ^{222}Rn is the decay daughter of the
radium isotope ^{226}Ra that decays by alpha emission with a half-life of 1620
yr. Owing to the universal occurrence of ^{226}Ra in rock minerals, sediments,
and waters, its daughter ^{222}Rn also occurs in waters, atmosphere, and in
the pore space of water-filled sediments and air-filled soils. Because of the
short half-life of radon, its occurrence at higher concentrations in ground
waters can be indicative of the proximity to uranium-containing ores.
Higher rates of release of radon to the atmosphere in earthquake-prone
areas have been considered as a tool in earthquake forecasting, based on
the release of radon from deep fissures undergoing contraction or expansion.

In waters near the sediment-water interface, concentrations of ^{222}Rn are
higher than what can be supported by the decay of ^{226}Ra in solution. This
so-called excess radon forms from ^{226}Ra in the sediments and diffuses into
the water column. A concentration profile of radon in a freshwater lake is
shown in Figure 7.9. Because the concentration of ^{226}Ra in water is
uniform with depth, the concentration profile of ^{222}Rn supported by the

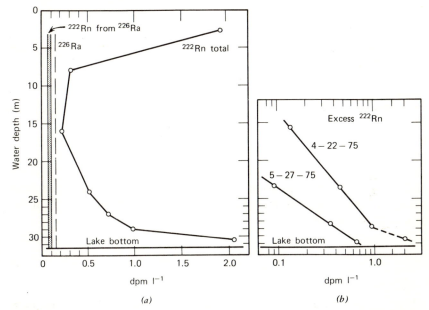

Figure 7.9 Radon and radium in water (dpm = disintegrations per minute). (a) Vertical distribution of ^{226}Ra and ^{222}Rn in a freshwater lake, Greifensee. (b) Vertical distribution of excess ^{222}Rn at the same site, plotted for determination of K_z (from unpublished data of D. M. Imboden). By permission of Dr D. M. Imboden, Dübendorf, Switzerland.

decay of ^{226}Ra would be uniform in the deeper water, and it could show a decline near the water-air interface, caused by diffusion from water into the atmosphere. The actual concentration profile shows an increase toward the lake bottom, due to diffusion of radon from the sediments, and an increase toward the surface, caused by inflow of river water.

The logarithm of radon concentration, plotted as a function of depth, makes a straight line for the lower half of the water column. Concentration that decreases exponentially with distance above the bottom is the result of radioactive decay and transport by vertical eddy diffusion. The one-dimensional transport model of radon can be written as the following differential equation for a steady state:

$$\kappa_z \frac{d^2C}{dz^2} - \lambda C = 0 \tag{7.33}$$

where κ_z is the coefficient of vertical eddy diffusion (cm^2 sec^{-1}), and λ (sec^{-1}) is the decay rate constant for ^{222}Rn. The solution of equation 7.33

is

$$C = C_{z=0}\, \exp\left[-z\left(\frac{\lambda}{\kappa_z}\right)^{1/2} \right] \qquad (7.34)$$

where $C_{z=0}$ is a constant concentration at the lower boundary of the water column and z is a positive distance above the boundary. The decay rate constant λ is related to the half-life $t_{1/2}$ as (see Table 1.3).

$$\lambda = (\ln 2)/t_{1/2}$$

The vertical eddy diffusion coefficient κ_z can be determined from equation 7.34, using two concentrations differing by a factor of 2, at two points of the profile, $z_{1/2}$ meters apart. κ_z is given by

$$\kappa_z = \frac{z_{1/2}^2}{t_{1/2}\ln 2} \qquad (\text{cm}^2\,\text{sec}^{-1}) \qquad (7.35)$$

where $z_{1/2}$ is the distance over which the concentration of ^{222}Rn decreases by a factor of 2. From the concentration profile measured in April 1975 (Figure 7.9), concentration decreases by a factor of 2 every 4 m. Using the half-life of ^{222}Rn $t_{1/2} = 3.80$ days, or 3.283×10^5 sec, the vertical eddy diffusion coefficient is

$$\kappa_z = \frac{(4 \times 10^2)^2}{3.28 \times 10^5 \times 0.69} = 0.7 \quad \text{cm}^2\,\text{sec}^{-1}$$

About one month later, concentration of radon near the bottom became lower, and the concentration gradient steeper. The vertical eddy diffusion coefficient is about $\kappa_z = 0.3-0.4$ cm^2 sec^{-1}. This difference shows some short-term variability of the transport conditions in a small lake. Applications of the radon method to the determination of the vertical eddy diffusivity in the deep ocean are also discussed in Section 2.5.

Migration, Reactions, and Diagenesis in Sediments

8.1 BACKGROUND PROBLEMS

Diagenesis is the sum total of changes—physical, chemical, and mineralogical—taking place in sediments during and after their deposition. The term diagenesis is reserved for the processes which operate under conditions not too different from those of the original depositional environment. The effects caused by the higher temperatures and pressures in deeply buried sediments fall in the domain of metamorphism and other substantive transformations associated with the endogenic geochemical cycle. Processes of either predominantly chemical or physical nature that produce changes in sediments and their interstitial waters can be classified as chemical or physical diagenesis. The distinction between the two is not always sharp, just as the distinction between the chemical and physical weathering, discussed in Chapter 5, can be blurred in many of the important phenomena.

About one-quarter of the material eroded from land enters the oceans in solution, and the remaining three quarters are delivered to the oceans as a suspended load. (Airborne materials amount to 5–20% of the suspended load of rivers, see Section 5.4.1.) For ocean water to remain of constant composition, the dissolved load in rivers must find its way back to the solids in sediments. This means that about one-quarter of the global mass of sediments are materials taken out of solution in ocean water. The main processes that remove dissolved materials from solution and transfer them into solids are the chemical precipitation and adsorption reactions and the biological productivity of the plankton. An additional source of materials added to sediments is the oceanic crust. The older basalts of the ocean floor react with pore waters of sediments above them, supplying or removing such chemical species as silica, calcium, magnesium, potassium, and sodium. Hot solutions circulating in the tectonically active areas and spreading zones of the ocean floor carry a dissolved load that is eventually added to the sediments. In the latter category are the occurrences of the manganese oxide crusts and nodules, and the higher concentrations of iron and other heavy metals that have been traced to the active zones of spreading. The seven major elements that account for more than 99% of the earth's crust are, in order of a decreasing abundance (excluding the most abundant, oxygen), Si, Al, Fe, Ca, Na, Mg, and K (Table 1.1). Their relative abundance in sediments is similar to the crustal abundance, with some differences in the abundance of iron (less in sediments than in crustal rocks) and calcium (more in sediments). The chemical diagenesis in sediments involves the main elements, as well as many minor or trace elements, which add to the mineralogical and chemical character of the sediments. Carbon, nitrogen, phosphorus, and sulfur—the building blocks of organic matter—participate in inorganic and organic reactions, wherever organic materials are deposited with sediments. The subject of the chemical reactivity of sedimentary materials, or why certain solid phases in sediments are more reactive than others, is discussed in Section 7.1.

This last chapter of the book covers the early stages of chemical diagenesis of sediments, which involve a number of inorganic and biogeochemical reactions. A general theoretical treatment of diagenesis is followed by some case studies of oceanic sediments, exposing the role of chemical reactions and migration of the reactive species in the post-depositional history of the sediment.

8.2   GENERAL MODEL OF SEDIMENT-PORE WATER SYSTEM

Sediments accumulating in a body of water are made of solids and aqueous solutions filling the interstitial pore space between the solid particles. A diagrammatic illustration of a sediment–pore water system is shown in Figure 8.1: growth of the sediment column takes place by deposition of solid material and entrapment of water in the interparticle pore space. The sediment-water interface is a convenient level of reference, relative to which particles and water move downward in a growing sediment column, or move upward in a column that is being eroded. (An analogy to an observer who balances himself at the interface, jumping instantaneously from his position to a position on top of a newly arrived particle, is drawn at the beginning of Chapter 7.)

The pore-water space of sediment is a communication channel to the overlying body of water. The organic and inorganic sediment particles may react in pore water, thereby changing the concentrations of various dissolved species. Molecular diffusion taking place in pore water (D in Figure

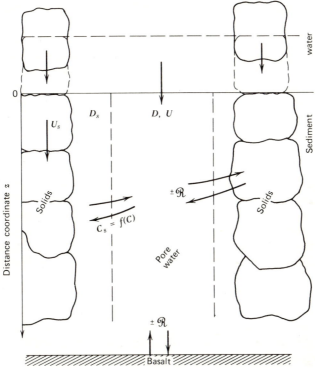

Figure 8.1 General model of transport processes and diagenetic reactions in a sediment column.

8.1) and, possibly, in the adsorbed water layer of molecular dimensions adjacent to the solid surfaces (D_s) would transport dissolved and adsorbed material in response to the gradients of thermodynamic activity generated by the changes in concentration. In a steadily growing sediment column, solid particles and pore water move downward relative to the sediment-water interface with velocities U_s and U, as shown in the figure. The chemical reactions, indicated by arrows between pore water and solids, can be dissolution and precipitation reactions and adsorption or desorption of dissolved materials. For the solutes in pore water, precipitation and adsorption are sinks, and dissolution, desorption, and decomposition of organics are sources of dissolved material. The one-dimensional vertical picture of a sediment column is both a convenience and a necessity for modeling. The convenience is the mathematics of one-dimensional differential equations, in comparison to the system of equation describing transport and chemical reactions in two or three dimensions. The use of one-dimensional models is justified for many sedimentary environments where the vertical changes in the profile are more pronounced than horizontally. In the ocean, there may be little variability in the composition of sediments over substantial areas of the ocean floor. However, in the areas of strong bottom relief and in lakes, where horizontal variability of sediments may be significant, the one-dimensional approach becomes a cruder approximation. Scarceness of data often makes it a necessity.

Considering the major sources and sinks for the individual chemical elements, the following generalizations can be made. The seven major elements—Si, Al, Fe, Ca, Na, Mg, and K—are generally involved in the reactions between pore waters of sediments and basalts, volcanic materials, and clays. The reactive solids include aluminosilicate minerals of fairly simple and well-known composition, such as feldspars, as well as mineral phases of very complex stoichiometry. In Table 8.1 are listed three types of volcanic glasses, two zeolites, and three clay minerals that commonly occur in oceanic sediments. Volcanic glass is a source phase, at whose expense zeolites and/or clays can form. It is easy to visualize how the kinetics of the chemical reactions leading from a volcanic glass to a zeolite or clay can present some difficulties to interpretation. Individual elements can react with other more or less abundant species and produce minerals characteristic of an aqueous environment. For example, iron forms iron sulfides and, more rarely, iron phosphates in reducing environments; hydrated iron oxides appear as coatings and crusts in oxidizing environments.

The presence of multiple sources or sinks for a given chemical element in sediments is a common feature. Sulfur, occuring as sulfate in ocean

Table 8.1 Mean Chemical Composition of Volcanic Glasses and Some Aluminosilicate Minerals as Reactive Sources and Sinks in Oceanic Sediments[a]

Glasses	
Rhyolitic volcanic glass	$(Na_{0.14}K_{0.20})(Ca_{0.22}Mg_{0.06}Fe^{2+}_{0.05})(Fe^{3+}_{0.10}Al_{1.35}Si_{7.85})O_{18.375}$
Sideromelane	$(Na_{0.40}K_{0.10})(Ca_{1.08}Mg_{1.12}Fe^{2+}_{0.91})(Fe^{3+}_{0.20}Al_{1.81}Ti^{4+}_{0.10}Si_{4.45})O_{15.475}$
Palagonite	$(Na_{0.40}K_{0.20})(Ca_{0.77}Mg_{0.87}Fe^{2+}_{0.08})(Fe^{3+}_{1.29}Al_{1.83}Si_{4.56})O_{15.82}$
Zeolites	
Clinoptilolite	$(Na_{3.0}K_{1.6})(Ca_{0.5}Mg_{0.2})Al_6Si_{30}O_{72} \cdot 24H_2O$
Phillipsite	$KAlSi_2O_6 \cdot 2H_2O$
Clays	
Montmorillonite (smectite)	$K_{0.40}(Mg_{0.34}Fe^{2+}_{0.04}Fe^{3+}_{0.17}Al_{1.50})(Al_{0.17}Si_{3.83})O_{10}(OH)_2$
Palygorskite	$Mg_2Al_2Si_8O_{20}(OH)_2 \cdot 4H_2O$
Sepiolite	$Mg_2Si_3O_6(OH)_4 \cdot 1.5H_2O$

[a]Toth (1976).

$$Mn^{2+} \longrightarrow MnS$$
$$MnCO_3$$
$$MnO_2 \ by \ bac$$

water, can be reduced to sulfide by bacterial activity, and the sinks for sulfide can be either oxidation back to sulfate or a reaction with iron. Mineral sinks for sulfate are the minerals $CaSO_4$, $SrSO_4$, and $BaSO_4$, occurring in the evaporitic environments and in oceanic sediments. Silicon, in addition to its major role in the formation of aluminosilicate minerals, is to a great extent involved in the biological production in surface waters. The formation and deposition of the siliceous tests of diatoms and radiolarians is a major sink for dissolved silica. Similarly, carbon and phosphorus are involved in inorganic and organic reactions. Inorganic carbon in sediments is associated with calcium and, to a lesser extent, magnesium in $CaCO_3$, magnesian calcites, and dolomite. Organic carbon is oxidized and released as CO_2 through oxidation of organic matter. Phosphorus released from organic matter can be adsorbed by sediments or it may react with calcium forming apatite.

The preceding brief account introduces the broad classes of the chemical diagenetic reactions occurring in sediments. Some of these will be discussed in more detail in the subsequent sections.

8.3   MATHEMATICAL MODELS

8.3.1   Basic Equations

A one-dimensional differential equation describing the dependence of concentration of a chemical species in solution, in the presence of diffusion, water flow (advection), and chemical reactions, is discussed in Sections 2.4 and 3.1. For a sediment made of solid particles and pore water, as shown schematically in Figure 8.1, the rate of change in concentration C of a given chemical species at any point z in pore water can be written as a mass balance equation based on Fick's second law of diffusion,

$$\frac{\partial(\phi C)}{\partial t} = -\frac{\partial F}{\partial z} \pm \phi \mathcal{R} \qquad (\text{g cm}^{-3}\,\text{yr}^{-1}) \qquad (8.1)$$

where concentration C has the dimensions of *mass per unit volume of pore water*, ϕ is the sediment porosity (fraction), F is the flux in the vertical direction, z is the vertical distance coordinate increasing downward from $z = 0$ at the sediment-water interface, and $\pm \mathcal{R}$ denotes all the chemical supply (+) or removal (−) processes affecting the dissolved species. Like other terms, $\mathcal{R}$ has the dimensions of mass per unit time *per unit volume of pore water*.

A relationship similar to equation 8.1 can be written for the same chemical species occurring in the solid phase at concentration C_s (mass *per unit volume of pore water*).

$$\frac{\partial(\phi C_s)}{\partial t} = -\frac{\partial F_s}{\partial z} \pm \phi \mathcal{R}_s \qquad (\text{g cm}^{-3}\, \text{yr}^{-1}) \qquad (8.2)$$

The flux due to diffusion and advection in pore water is

$$F = -\phi D \frac{dC}{dz} + \phi UC \qquad (\text{g cm}^{-2}\, \text{yr}^{-1}) \qquad (8.3)$$

and a similar relationship can be written for the solid, provided diffusion can take place in the adsorbed layer characterized by a diffusion coefficient D_s (Figure 8.1),

$$F_s = -\phi D_s \frac{dC}{dz} + \phi U_s C_s \qquad (\text{g cm}^{-2}\, \text{yr}^{-1}) \qquad (8.4)$$

Substitution of the relationships for F and F_s in equations 8.1 and 8.2 gives

$$\frac{\partial(\phi C)}{\partial t} = \frac{\partial}{\partial z}\left(\phi D \frac{\partial C}{\partial z}\right) - \frac{\partial(\phi UC)}{\partial z} \pm \phi \mathcal{R} \qquad (8.5)$$

$$\frac{\partial(\phi C_s)}{\partial t} = \frac{\partial}{\partial z}\left(\phi D_s \frac{\partial C_s}{\partial z}\right) - \frac{\partial(\phi U_s C_s)}{\partial z} \pm \phi \mathcal{R}_s \qquad (8.6)$$

Equation 8.5 relates the rate of change in concentration at any fixed position z in pore water to the diffusional and advective transport, and to chemical supply or removal reactions. Equation 8.6 does the same for concentration on the solids. If the concentration on the solid C_s is a known function of concentration in pore water C,

$$C_s = f(C)$$

then one concentration can be substituted for the other and the two equations can be combined into one. For example, a linear distribution of a species between pore water and solids $C_s = KC$, discussed in equation 7.8, can be used in equations 8.5 and 8.6, giving one equation in C,

$$\frac{\partial C}{\partial t} = \frac{1}{\phi(K+1)} \frac{\partial}{\partial z}\left[\phi(D + KD_s)\frac{\partial C}{\partial z}\right] - \frac{1}{\phi(K+1)} \frac{\partial}{\partial z}\left[\phi(U + KU_s)C\right]$$

$$\pm \frac{\mathcal{R} + \mathcal{R}_s}{K+1} - \frac{C}{\phi}\frac{\partial \phi}{\partial t} \qquad (\text{g cm}^{-3}\, \text{yr}^{-1}) \qquad (8.7)$$

In the last equation, the distribution factor K is treated as a constant independent of concentration and depth. This condition is far from universal, and other forms of the exchange between solutions and solids are discussed in Section 7.5. Many one-dimensional equations describing transport and reactions in sediments are derivable from the more general equation 8.7. For example, if the porosity does not change with time, and there are no adsorption, advection, and chemical reactions (that is, the terms K, U, and $\mathcal{R}$ are zero), then the equation reduces to the simple case of Fick's second law of diffusion: $\partial C/\partial t = (\partial/\partial z)(D\,\partial C/\partial z)$.

The term diagenetic equation has been used for the one-dimensional diffusion-advection-reaction equation describing the chemical diagenesis and migration of reactive species in sediments (Berner, 1971).

At a steady state, concentration and sediment porosity are constant at any depth z such that the time-dependent terms $\partial C/\partial t$ and $\partial \phi/\partial t$ cancel. Equation 8.7 then becomes

$$\frac{\partial}{\partial z}\left[\phi(D+KD_s)\frac{\partial C}{\partial z}\right] - \frac{\partial}{\partial z}\left[\phi(U+KU_s)C\right] \pm \phi(\mathcal{R}+\mathcal{R}_s) = 0 \quad (8.8)$$

Transient-state models based on equation 8.7 (that is, $\partial C/\partial t \neq 0$ and C is a function of both t and z) apply to those situations where the conditions within the system or at its boundaries vary with time. For example, this is the case of man-induced chemical perturbations in natural waters where the events have been relatively recent and the entire sediment-water system has not yet adjusted to the imposed change, even if the new inputs have in certain cases come to a stable level. Transient-state models of diagenesis, applying to different situations resembling those in natural waters and sediments, have been described by Lerman (1971), Imboden (1975), and Lasaga and Holland (1976).

Steady-state models based on equation 8.8 have the appeal of mathematically simpler solutions (no time-dependent terms) and a conceptual justification of near constancy of the conditions prevailing in the ocean when viewed on a time scale of millions to tens of millions of years. On a shorter time scale, even such changes as took place in the ocean during the glacial periods, and changes in the kind and rate of sediment deposition in different sections of the ocean were likely to result in changes in the chemical composition of sediment pore waters. On a longer time scale of millions of years, pore waters could have adjusted to the new conditions: diffusional redistribution of dissolved materials, characterized by a diffusion coefficient of the order of magnitude $D \simeq 3 \times 10^{-6}$ cm^2 sec^{-1} = 10^2 cm^2 yr^{-1}, is effective within a 100-m long pore water column on a time scale of

$$t = \frac{(10^4 \text{ cm})^2}{10^2 \text{ cm}^2 \text{ yr}^{-1}} = 10^6 \text{ yr}$$

Thus steady-state models can be acceptable approximations in those cases where the systems have not been perturbed for a period of at least 10^6 yr, or where perturbations did not affect a particlar diagenetic process.

Some concentration against depth profiles to which the steady-state diagenetic equation 8.8 applies are schematized in Figure 8.2. Figure 8.2*a* shows a steady-state concentration profile in a semi-infinite column of sediment: the profile on the left, at time t_1, and the profile on the right, at a later time t_2, are identical in the sense that at any depth z below the sediment-water interface concentration does not change with time. If the reference level $z = 0$ were set on shore or at the water level, then a point at any depth z would be moving farther away from the sediment-water interface, as the sediment column continued to grow.

Figure 8.2*b* shows a sediment layer near the top of the sediment column. Concentrations in pore water within the layer and below it may be controlled by different mineral phases. At the steady state, the thickness of the layer does not change, although it may either move up with the rising sediment-water interface or it may move down relative to the interface. If the layer moves up with the interface, this implies that the processes responsible for its boundaries take place at such a rate that some of the material in the layer is transformed into the material below the layer,

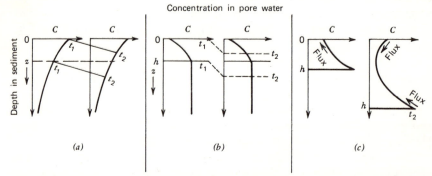

Figure 8.2 Vertical concentration profiles in sediments. Concentration C, distance z, and time t. (*a*) Steady state in a semi-infinite sediment column. (*b*) Steady state in a sediment layer with boundaries that move upward as the sediment accumulates. (*c*) A growing sediment layer with sources at the upper and lower boundaries, and reaction sinks within the layer.

allowing the layer boundaries to keep pace with the sedimentation rate. Another type of a layer in sediments, such as a layer of volcanic ash, may have fixed boundaries and its distance relative to the sediment-water interface will be increasing with time.

Figure 8.2c shows a growing sediment layer. The layer boundaries recede from each other with time as sedimentation adds material at the top. If concentrations in pore water at the two boundaries remain constant, then at any depth within the layer, concentration changes with time and its gradient also changes. A curvilinear concentration profile, drawn in the figure, may be caused by precipitation or uptake reactions acting as sinks within the sediment column. In a thin layer, at time t_1, the concentration profile is slightly convex to the left and the concentration gradient near the upper boundary is positive, indicating that the flux of dissolved material is upward, out of the layer. In a thicker layer, at time t_2, the concentration profile is more pronouncedly curved, owing to the fact that diffusional transport of dissolved material from the sources located at the boundaries cannot keep pace with the removal of material to the precipitation sink. The shape of the concentration profile shown in the figure is maintained by diffusional fluxes downward from the upper boundary and upward from the lower boundary, toward the point of the minimum concentration. Thus the direction of the flux at the upper boundary at time t_2 is reversed relative to its direction at an earlier time t_1. Application of a steady-state model to a growing layer can be justified only if the layer thickness is large in comparison to the thickness added over a period of about 10^6 yr. For example, at the sedimentation rate of 1 cm per 1000 yr, a section 10 m thick is built up in 10^6 yr; relative to a 100- to 200-m-thick layer within which a concentration profile is observed, an additional 10 m may not change the picture substantially.

The physical significance of the individual parameters in the diagenetic equations 8.7 and 8.8 will be discussed in the following sections.

8.3.2 Diffusional Terms

The diffusion coefficient D for a dissolved species in pore water includes any of the effects of the porosity and tortuosity of the sediment that may affect molecular diffusivity in solution (Section 3.2).

The possibility of diffusion of chemical species within a thin layer at the particle surface that contains adsorbed ions (Figure 8.1) has been suggested by the studies of electrical conductance and ionic diffusion in clay-water mixtures (van Olphen, 1957, 1963; van Schaik and Kemper, 1966; Ellis et al., 1970). Some fraction of the ionic current flowing through an electrolyte solution in a capillary or a porous medium is carried by the ions in the adsorbed layer. Such a diffusional flux, analogous to the electric current flux, is represented by the diffusion coefficient D_s in the first and second Fink's law equations 8.4 and 8.2.

If the diffusion coefficients D and D_s are independent of the distance coordinate z (that is, they are independent of concentration that varies along the coordinate z), then the diffusional term in equation 8.7 is

$$\frac{D + KD_s}{K + 1} \cdot \frac{\partial^2 C}{\partial z^2}$$

The quotient $(D + KD_s)/(K + 1)$ cm^2 sec^{-1} is an effective diffusion coefficient, when adsorption and diffusion in the adsorbed layer take place in addition to diffusion in solution. In the case of strong adsorption $(K \gg 1)$, the quotient becomes

$$D\left(\frac{1}{K} + \frac{D_s}{D}\right) \qquad (\text{cm}^2 \text{ sec}^{-1})$$

For diffusion in the adsorbed layer to add significantly to diffusion in solution, the ratio D_s/D should be of the same order of magnitude or greater than $1/K$. For a value of the exchange coefficient $K \simeq 100$ (see Section 7.4), the ratio D_s/D should be $\geqslant 0.01$ for the effect to be significant. Adsorption retards diffusion in pore water, because an effective diffusion coefficient in the presence of adsorption is $D/(K + 1)$ or D/K. Diffusion in the adsorbed layer could add something to the diffusional flux, although no independent estimates of D_s are available for sediment-pore water systems.

8.3.3 Advective Terms

The rates of movement of sediment particles U_s (cm yr^{-1}) and pore water U (cm yr^{-1}) relative to the sediment-water interface are the same if the sediment porosity does not change with depth. In most sediments, however, the porosity decreases due to compaction of sediments from the top of the sediment column downward. A generalized porosity against depth profile in oceanic sediments is shown in Figure 8.3. In oceanic and lake sediments, the porosity at the sediment-water interface is commonly in the range between 70 and 90% $(0.70 < \phi < 0.85)$, and its value decreases significantly within the upper several tens to hundreds of centimeters of the sediment column. Typical values for unconsolidated noncemented sediments are between 40 and 60% $(0.4 < \phi < 0.6)$, after compaction has reached a steady value. In a growing column of sediment, the changing

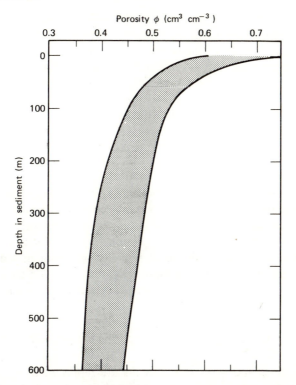

Figure 8.3   Porosity ϕ as a function of depth in oceanic sediments (adopted from von Huene and Piper, 1973).

proportions of water and sediment should obey the flux conservation condition for solids and pore water when the sediment column is in a steady state, that is, when the rates of sedimentation and compaction are constant. The flux conservation conditions are

$$\frac{\partial(\phi U)}{\partial z}=0 \quad \text{and} \quad \frac{\partial\left[(1-\phi)U_s\right]}{\partial z}=0 \qquad (8.9)$$

Deeper in the sediment, where the porosity gradient is essentially zero, pore water and sediment particles do not move relative to each other, and their advective velocity relative to the sediment-water interface is the same: U_∞. Then the following conditions apply from equation 8.9

$$\phi U=\phi_\infty U_\infty \qquad (8.10)$$

$$(1-\phi)U_s=(1-\phi_\infty)U_\infty \qquad (8.11)$$

The last two relationships show that in the region where compaction takes place and porosity decreases with depth, sediment particles move faster than pore water relative to the sediment-water interface: $U_s > U$ as long as $\phi > \phi_\infty$. For such typical values of porosity as $\phi_\infty = 0.5$ at depth and $\phi = 0.7$ near the top of the sediment column, the rate of pore water advection near the column top is $U = 0.7 U_\infty$, and the rate of sediment advection is $U_s = 1.7 U_\infty$. The difference between the two advection rates is more than a factor of 2.

A porosity gradient term $\partial \phi / \partial z$ enters in the steady-state equation 8.8. Many empirical relationships dealing with the compaction of sediments are available in the literature (Athy, 1930; Emery and Rittenberg, 1952; Engelhardt, 1960; Hamilton, 1971; Rieke and Chilingarian, 1974), most of them representing porosity as an exponentially decreasing or as some power function of depth. Depending on the mathematical form of the function $\phi(z)$, explicit solutions of equations 8.7 and 8.8 may or may not exist. In general, the dependence of porosity on depth in sediment may have to be taken into account in those cases that deal with the region close to the sediment-water interface, where the porosity gradients are most pronounced.

8.3.4   Chemical Terms: Decomposition of Organic Matter

Chemical sources and sinks in sediments are represented by the reaction rate terms $\pm \mathfrak{R}$ (g cm^{-3} yr^{-1}) in equations 8.7 and 8.8. Organic matter in sediments is a source of a number of chemical species that can be restored to the environment through inorganic oxidation or organic decomposition. A major portion of the organic material produced in the photosynthesis zone of surface waters is decomposed and regenerated before settling to the bottom. Decomposition of the settled material is generally understood as a process mediated by different bacteria, the end results of which can be written as a chemical reaction (the Redfield reaction),

$$(CH_2O)_{\alpha_C}(NH_3)_{\alpha_N}(H_3PO_4)_{\alpha_P} + \tfrac{1}{2}\alpha_C SO_4^{2-} \rightarrow$$

$$\alpha_C HCO_3^- + \alpha_N NH_3 + \alpha_P H_3PO_4 + \tfrac{1}{2}\alpha_C H_2S \quad (8.12)$$

where CH_2O stands for the organic matter and α_C, α_N, and α_P are the stoichiometric coefficients of C, N, and P in the sediment organic matter. The reaction represents the reduction of sulfate and production of bicarbonate, ammonia, phosphate, and sulfide. In the absence of sulfate, decomposition of organic matter can lead to production of methane CH_4

and other organic residues (Martens and Berner, 1974; Barnes and Goldberg, 1976). For the oceanic plankton, the ratio $\alpha_C:\alpha_N:\alpha_P$ is about
106:16:1, although significant departures from this ratio have also been
reported for different sections of the ocean. For other groups of the biota
and some derived organic materials the ratios $C:N:P$ are given on
pp. 23 and 33.

Reduction of sulfate is the first step accompanying bacterial decomposition of organic matter. The bacterial activity is primarily confined to the
uppermost few centimeters of the sediment column, and Figure 8.4 shows
an example of a rapid decrease in the number of bacteria with depth.
Organic matter is a mixture of many different compounds the resistance of
which to the bacterial attack is nonuniform. The fraction of the total
organic matter that reacts with sulfate according to reaction 8.12 decreases
with depth, as less sulfate remains in pore water. The rate of decrease in
the amount of reactive organic matter has been equated with the rate of
sulfate reduction (Berner, 1974) and represented by a relationship of the

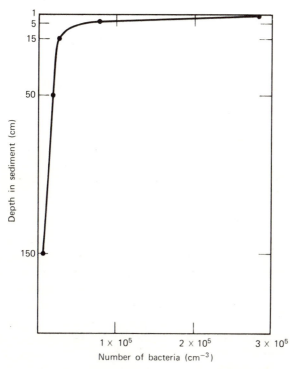

Figure 8.4   Decrease with depth in the number concentration of bacteria in sediment, Persian
Gulf (from Rheinheimer, 1974 reproduced by permission).

type

$$\mathcal{R} = J_* e^{-\beta z} \qquad (\text{g cm}^{-3} \text{ yr}^{-1}) \qquad (8.13)$$

where $\mathcal{R}$ is the production or consumption rate decreasing exponentially with depth from its value J_* at the sediment-water interface. The parameter J_* is

$$J_* = \kappa_i \alpha_i C_{org}^{\circ} \qquad (\text{g cm}^{-3} \text{ yr}^{-1}) \qquad (8.14)$$

where κ_i is the reaction rate parameter (yr^{-1}), α_i is the stoichiometric coefficient, as in reaction 8.12, indicating the number of mols of a species produced or consumed by decomposition of 1 mol of organic matter, and C_{org}° is the concentration of reactive organic matter at the sediment-water interface (mol cm^{-3} of pore water). The parameter β is the ratio of the decomposition reaction rate constant κ_i and the sedimentation rate U supplying organic matter to the sediment,

$$\beta = \frac{\kappa_i}{U} \qquad (\text{cm}^{-1}) \qquad (8.15)$$

The reaction rate constant κ_i and the mean rate of sedimentation U are interrelated: higher values of the rate constant κ_i occur in the sediments of higher rates of sedimentation. The correlation between the two parameters shown in Figure 8.5 follows a relationship

$$\kappa_i \propto U^2$$

or   (8.16)

$$\beta \propto U$$

Explicit equations for the reduction of sulfate (κ_S), and the production of ammonia (κ_N) and phosphate (κ_P) as a function of the sedimentation rate U are given in the figure. The data shown represent areas of a wide range of sedimentation rates in the deep Caribbean, Indian Ocean, Santa Barbara Basin off California, Long Island Sound, Sommes Sound in Maine, and Saanich Inlet in British Columbia. The rates of sedimentation, reported in the literature independently of the chemical information on pore waters composition, are generally based on paleontological and radioactive dating techniques. The values of the rate constant κ_i, however, are determined from the concentrations of dissolved sulfate, ammonia, and

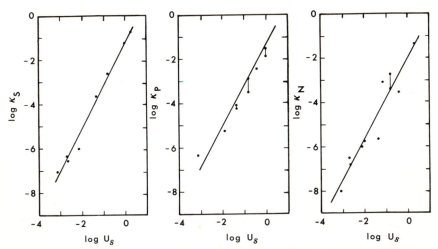

Figure 8.5 Correlation of the sedimentation rate U_s (cm yr^{-1}) and first-order reaction rate constant κ_i (yr^{-1}) for the reduction of sulfate (κ_S), and production of phosphate (κ_P) and ammonia (κ_N) in near-shore and pelagic oceanic sediments (from Toth and Lerman, 1977). Reproduced by permission of the *American Journal of Science.*

phosphate in pore waters (Section 8.4). The direct relationship between κ_i and U^2 reflects the nature and past history of the reactive organic matter reaching the sediment surface. The areas of the more rapid rates of sedimentation receive a greater flux of organic matter that carries a greater fraction of reactive materials. In the areas where the rates of sedimentation are slow, relatively little organic material reaches the ocean floor, and the settled material has already reacted to some extent, either bacterially or inorganically. The fraction entering the sediment consists of more resilient components of organic matter, which continue to react, but only slowly. The near-shore and hemipelagic sections of the ocean are characterized by higher rates of sedimentation in comparison to the pelagic areas, and the κ_i and U values are higher for the former.

8.3.5   Chemical Terms: Dissolution and Precipitation

Coupled Reactions. In a closed system, the rate of dissolution or precipitation of a solid phase can be measured by the rate of change of concentration in solution dC/dt (Section 5.5). The chemical reaction term $\pm \mathcal{R}$ can be equated to dC/dt in equations 8.7 and 8.8, and the following situations should be considered. A dissolving phase supplies material to pore water and a precipitating phase removes it from pore water. Such simultaneous first-order dissolution and precipitation reactions form a couple,

Dissolution $\qquad \dfrac{dC}{dt} = k_1(C_{s,1} - C) \qquad (C_{s,1} > C) \qquad\qquad$ (8.17)

Precipitation $\qquad \dfrac{dC}{dt} = k_2(C_{s,2} - C) \qquad (C_{s,2} < C) \qquad\qquad$ (8.18)

where $C_{s,1}$ is the saturation concentration for the dissolving phase, $C_{s,2}$ is the saturation concentration for the precipitating phase, and k_1 and k_2 are the reaction rate constants (yr^{-1}). In a system with no inflow and outflow at a steady state, the rates of dissolution and precipitation are equal,

$$k_1(C_{s,1} - C) = k_2(C - C_{s,2}) \qquad (g\ cm^{-3}\ yr^{-1}) \qquad\qquad (8.19)$$

A steady-state concentration $C = C_{ss}$ follows from the latter equality,

$$C_{ss} = \frac{k_1 C_{s,1} + k_2 C_{s,2}}{k_1 + k_2} \equiv \frac{J}{k} \qquad (g\ cm^{-3}) \qquad\qquad (8.20)$$

where the notation of J and k is introduced for brevity,

$$J = k_1 C_{s,1} + k_2 C_{s,2} \qquad (g\ cm^{-3}\ yr^{-1}) \qquad\qquad (8.21)$$

$$k = k_1 + k_2 \qquad (yr^{-1}) \qquad\qquad (8.22)$$

The parameter J has the dimensions of a rate of reaction and k the dimensions of a reaction rate constant. Note that the equality of the reaction rates in 8.19 is not a condition of a chemical equilibrium between two phases. The equality applies to a nonequilibrium situation where one solid phase dissolves and another precipitates. If only one reaction, either dissolution or precipitation, takes place, then the steady-state concentration C_{ss} takes one of the two solubility values, $C_{s,1}$ or $C_{s,2}$.

In an open system at a steady state, a chemical species is supplied to pore water by dissolution, and it is removed by precipitation, diffusion, and advection of pore water. For such a system, the material balance condition is

$$k_1(C_{s,1} - C) - k_2(C - C_{s,2}) = \text{diffusional and advective transport}$$

$$(8.23)$$

With the J and k notation given above, the material balance equation 8.23

becomes

$$J - kC = \text{diffusional} + \text{advective transport} \qquad (8.24)$$

The latter contains only two reaction rate parameters instead of four, which makes their evaluation easier from observed concentration profiles in pore waters. More than one interpretation of the parameters J and k is possible. The rate term J can be considered as a zeroth-order dissolution reaction representing a source, and the term $-kC$ as representing a precipitation sink. If there is only one reaction taking place, either dissolution or precipitation, the rate of which is written as $\pm k(C_s - C)$, then the terms $kC_s \equiv J$ and $-kC$ can be viewed as a source and sink pair. Thus the solid phase precipitating because of a concentration difference $C_s - C$ can be formally treated as a source-sink reaction, while there is a net removal of dissolved material from solution to the solid. The values of J and k for a dissolved species can be evaluated from a concentration against depth profile (Section 8.5). Then, provided the nature and solubilities of the dissolving and precipitating mineral phases are known, the rate constants k_1 and k_2 can be computed from the equations

$$k_1 = \frac{J - kC_{s,2}}{C_{s,1} - C_{s,2}} \qquad (\text{yr}^{-1}) \qquad (8.25)$$

$$k_2 = \frac{kC_{s,1} - J}{C_{s,1} - C_{s,2}} \qquad (\text{yr}^{-1}) \qquad (8.26)$$

The physical significance and numerical magnitudes of the individual rate constants will be illustrated below.

Reaction Rates in Sediments: Silica. A concentration profile of silica in pore waters at a drilling site in the Indian Ocean is shown in Figure 8.6. In the upper 100 m, the profile is curved slightly to the left, toward lower silica concentrations, indicating that silica is being removed from pore waters. A steady-state situation can be maintained by diffusional fluxes of silica down the sediment-water interface and up from depths greater than 100 m. The removal of silica, regarded as a first-order precipitation reaction, proceeds with the first-order rate constant $k = 2 \times 10^{-6}$ yr^{-1} in the upper 100 m of the sediment pore water column. The residence time of silica with respect to its removal by precipitation is approximately $1/k = 5 \times 10^5$ yr.

Below 100 m, the curvature of the profile toward higher silica concentrations suggests dissolution from a source present throughout the sediment column, and precipitation that removes part of the dissolved silica.

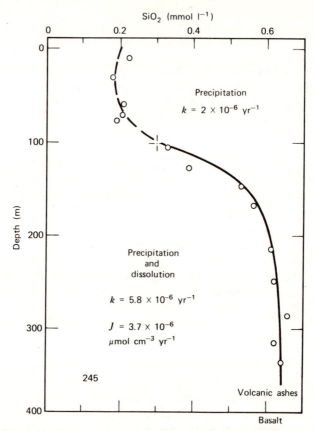

Figure 8.6 Dissolved silica in pore waters of Deep Sea Drilling Project site 245, southern Madagascar Basin. Control of the concentration profile by diagenetic reactions is shown in the figure (data from Schlich et al., 1974; Gieskes, 1974).

Volcanic ashes and basalt (also possible sources of dissolved silica) underlie the pore water column. In the lower part, between 100 and 350 m, the silica concentration profile gives $k = 6 \times 10^{-6}$ yr^{-1} and $J = 4 \times 10^{-12}$ mol SiO$_2$ cm^{-3} yr^{-1}. The steady-state concentration at depth approaches the value of $C = J/k = 0.65 \times 10^{-6}$ mol cm^{-3} or 0.65 mmol l^{-1}. As both dissolution and precipitation are indicated by the profile, the terms J and $-kC$ cannot be identified individually with the rates of dissolution and precipitation of the source and sink phases.

Amorphous SiO$_2$ (opal) of diatom and radiolarian shells is a common source of dissolved silica in pore waters of sediments. The sinks can be clay minerals, zeolites or, as an extreme of a low-solubility mineral, quartz.

For a computational example, two source and sink pairs are considered: opal and quartz, and opal and a clay mineral. Solubilities of pure SiO_2 phases at different temperatures are shown in Figure 8.7.

For the opal-quartz pair, the solubility of opal will be taken as $C_{s,1} \simeq$ 1.5 mmol 1^{-1} and the solubility of quartz $C_{s,2} \simeq 0.1$ mmol 1^{-1} (see Figure 8.7). These solubilities and the J and k values given above can be used in equations 8.25 and 8.26 to compute the dissolution (k_1) and precipitation (k_2) rate constants

$$k_1 = \frac{4 \times 10^{-12} - 6 \times 10^{-6} \times 0.1 \times 10^{-6}}{1.5 \times 10^{-6} - 0.1 \times 10^{-6}} = 2.4 \times 10^{-6} \quad yr^{-1}$$

$$k_2 = \frac{6 \times 10^{-6} \times 1.5 \times 10^{-6} - 4 \times 10^{-12}}{1.5 \times 10^{-6} - 0.1 \times 10^{-6}} = 3.6 \times 10^{-6} \quad yr^{-1}$$

The difference between the solubilities of opal and quartz is extreme, but the rate constants of the individual dissolution and precipitation reactions turn out to be of the same order of magnitude as the combined rate parameter $k = 6 \times 10^{-6}$.

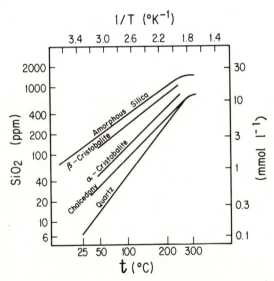

Figure 8.7 Solubility of the polymorphic forms of SiO_2 in water as a function of temperature (from Fournier, 1973). By permission of Dr. R. O. Fournier, Menlo Park, California.

An alternative sink of silica in sediments may be a mineral with a solubility in the vicinity of $C_{s,2} = 0.3$ mmol l^{-1}, comparable to the solubilities of some clays. Then, as in the computational example above, the rate constants are $k_1 = 1.8 \times 10^{-6}$ yr^{-1} and $k_2 = 4.2 \times 10^{-6}$ yr^{-1}. The difference between these values and those computed for the opal-quartz pair is not great, although the solubilities of the sink minerals differ by a factor of 3 in the two examples. The reason behind the similarity of the results is the relatively large difference between $C_{s,1}$ (source) and $C_{s,2}$ (sink). As the solubilities of the source and sink come closer one to another, the difference between the precipitation and dissolution rate constants becomes greater. However, both k_1 and k_2 are positive quantities, and this fact places certain limits on the magnitudes of the sink and source solubilities. The condition that must be obeyed is

$$C_{s,1} > \frac{J}{k} > C_{s,2} \qquad (8.27)$$

which states that the solubility of the source mineral is greater than the steady-state concentration J/k (concentration in the region of zero concentration gradient, lower part of the profile in Figure 8.6), and the steady-state concentration is greater than the solubility of the sink.

From the preceding numerical examples it can be concluded that when opal is a source of SiO_2 in pore water, the value of the rate parameter k may be used at least as an approximation for the precipitation rate constant. For dissolution of silica from different silicate phases, some values of the rate parameters are listed in Table 8.2. In laboratory experiments, the rates of equilibration of solutions with silicate minerals are fast: for clays, zeolites, pure SiO_2 phases, and shells of radiolarians, the rate parameters are large. For siliceous sediments, the results derived from steady-state models and diagenetic equations give values of k that are several orders of magnitude lower. Had the rates of reactions in sediments been as fast as the values obtainable in laboratory dissolution experiments, there would have been a much faster and a much more pronounced change with depth in the bulk mineralogical composition of sediment cores due to dissolution and precipitation.

The kinetic parameters J and k are constant and independent of depth in a sediment to which the model of the coupled first-order reactions applies. Constant J and k imply that the *reacting* surface areas of the solid phases do not change with depth. An explanation for this lies in the relationship between the rate constant k and the surface area of the solid S, as given in equation 5.68: $k = k_s S/V$. If the rate parameter k_s (cm sec^{-1}) and volume of the solution V (cm^3) do not change, then k can be constant only if the reactive surface area of the solid also remains

Table 8.2   First-Order Dissolution Reaction Rate Constants of SiO$_2$ for Silicate Minerals and Sediments[a]

Silicate material	k (yr^{-1})	k_s (cm yr^{-1})
Siliceous sediments[b]		
Deep ocean	$0.2 \times 10^{-5} - 10 \times 10^{-5}$	
Near shore	0.02–0.5	
Clays[c]	2.4–21	
Zeolites[c]	4–12	
Quartz	23[c]	0.41 ± 0.06[d]
Tridymite[d]		0.36
Cristobalite[d]		0.43
Amorphous SiO$_2$[d]		0.64
Radiolarian oozes[e]		
26°C		1–11
3°C		0.1–0.6

[a] Rate constants k (yr^{-1}) and k_s (cm yr^{-1}) are interrelated: $k = k_s S / V$, where S/V (cm^2 cm^{-3}) is surface area of the solid per unit volume of solution, equation 5.68.

[b] From diffusion-reaction models of pore waters (Anikouchine, 1967; Lerman, 1975, 1977)

[c] Size fraction < 62 μm, sea water, room temperature. Clays: kaolinite, montmorillonite, illite, bentonite, glauconite. Zeolites: analcite, phillipsite, prehnite (Lerman, Mackenzie, Bricker, 1975).

[d] In sea water, 25°C, pH = 8.0–8.4 (Wollast, 1974).

[e] In surface sea water, shells acid cleaned (Hurd and Theyer, 1975).

constant. In a long column of pore water, changes in temperature, the solution pH, and other characteristics of the solution may affect the rate parameter k_s as well as the rate parameter k. Therefore, the model based on the constant kinetic parameters k and J is only an approximation, even if the two parameters provide results that agree well with the available data.

The total surface areas of sedimentary materials, determined by gas adsorption techniques, do not necessarily represent the total reactive surface area. The surface that reacts with the solution may be smaller than the total measured, as the smallest pores and joints may be not accessible to exchange with the surrounding solution. The presence of organic films or adsorbed ions on the surfaces may hinder their reactivities such that the effective rate of a reaction may be considerably slower than for a clean surface. Some representative values of the surface areas of solids in oceanic sediments are listed in Table 8.3. The finer is the material, the greater is its specific surface area, measured in square meters per gram of dry solid. Among the calcareous materials, the surface areas generally increase from coarse foraminiferal shells to small coccolith platelets. Among the detrital materials, clays have greater specific surface areas than sands and volcanic ash, and sediments on the continental slope have greater surface areas than the sediments on the continental shelves.

In shallow-water siliceous sediments the silicate reactions proceed faster than in the pelagic sediments, as the data in Table 8.2 show: the rate parameter k is of the order of $10^{-2}-10^{-1}$ yr^{-1} for near-shore and continental-shelf sediments. The abundance of siliceous skeletons in the sediments of the Eastern Atlantic (Figure 8.8) decreases rapidly with the increasing sediment age. The number of siliceous particles decreases by a factor of about 2 every 500 yr, within the youngest 3000 yr-old layer of sediment. From these data, the rate of dissolution parameter k is about $k \simeq 1/500 = 2 \times 10^{-3}$ yr^{-1}, a value within range of near-shore and deep-ocean data in Table 8.2.

Table 8.3   Specific Surface Areas (S_s) of Oceanic Sediments

Sediment type	S_s (m^2 g^{-1})
Calcareous[a]	
Coarse foraminifera	1–5
Pteropods	3–6
Small foraminifera	5–10
Fine coccolith material	$\leqslant 40$
Siliceous	
Diatom ooze[a]	18–32
Radiolarian ooze[a]	13–70
$< 10^7$ yr old[b]	10–260
$> 10^7$ yr old[b]	2–10
Detrital and mixed[a]	
Fine sand	1–5
Volcanic ash	5–10
Foraminiferal grey clay	10–30
Continental shelf sediments	2–20
Continental slope sediments	10–45
Red clay, deep ocean	25–90

[a]Weiler and Mills (1965).
[b]Acid cleaned tests, Hurd and Theyer (1975).

Number of SiO$_2$ skeletons per gram of sediment (10^3)

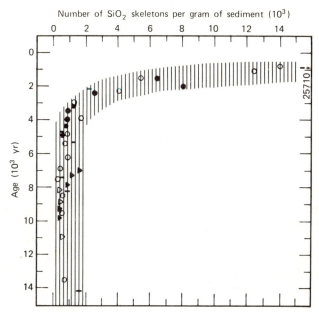

Figure 8.8 Decrease in the abundance of siliceous skeletal material with depth in sediments on the Iberian-African continental slope and abyssal plane (from Schrader, 1972 *b*).

The faster reactions indicated by the steeper concentration gradients in pore waters of near-shore sediments and the abundance gradients of the siliceous skeletons, such as those shown in Figure 8.8, point to the faster rates of diagenesis that take place during the earlier stages of sediment deposition. The smaller values of *k*, typical of slower reactions, are averages for longer, older sediment sections.

8.4 NUTRIENTS AND RELATED SPECIES IN SEDIMENTS

8.4.1 Pore Water Solutions

The final products of degradation of organic materials in sediments, discussed in Section 8.3.4, are ammonia, phosphates, and carbon dioxide. The reduction of sulfate to sulfide is a concomitant phenomenon of the bacterial activity, where sulfate is the metabolic source for the sulfate reducing bacteria. In near-shore marine environments and in eutrophic lakes, the amount of organic material in fresh sediments can be high, on the level of a few percent, measured as weight percent of organic carbon.

In such environments, decomposition of organic matter and the resulting changes in the concentrations of the sulfur, nitrogen, and phosphorus species in pore waters have been treated extensively in the literature (for example, Berner, 1974; Goldhaber et al., 1977; Golterman, 1977).

For deep-ocean sediments, vertical distributions of dissolved sulfate, ammonia, and phosphate are shown in Figures 8.9–8.11. The two profiles from the Caribbean, one from the Cariaco Trench and the other from the Venezuelan Basin, are characterized by very different rates of sedimentation: 0.75 cm per 1000 yr in the Venezuelan Basin (Figure 8.10), as compared to 46 cm per 1000 yr in the Cariaco Trench (Figure 8.9). Concentrations in pore water change faster with depth in the more rapidly accumulating sediments of the Cariaco Trench. This illustrates the positive correlation between the rates of sedimentation and the rates of reaction of organic matter (Section 8.3.4) that ultimately determines the shape of the concentration profiles and steepness of the concentration gradients in pore water. A decrease in the sulfate concentration with depth, and the increase in the concentrations of the other two species is the main observational evidence that the bacterial decomposition of organic materials is responsible for the picture. A fraction of the organic material that is reactive is responsible for the increase in the concentrations of phosphate and ammonia over some distance below the sediment-water interface. It is

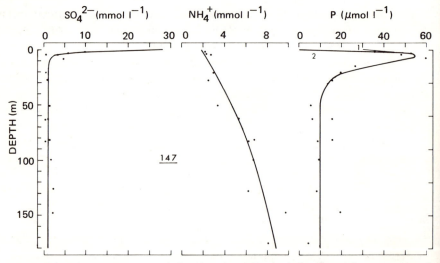

Figure 8.9 Sulfate, ammonia, and phosphate (total dissolved phosphorus) in pore waters of the Cariaco Trench, Deep Sea Drilling Project site 147 (data from Sayles et al., 1973; Gieskes, 1973; computed curves from Toth and Lerman, 1977). Reproduced by permission of the *American Journal of Science*.

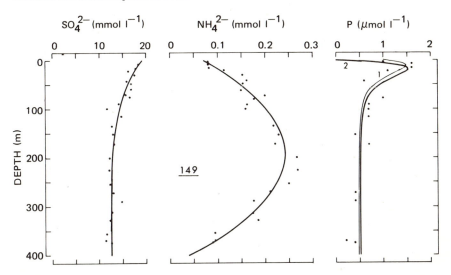

Figure 8.10 Sulfate, ammonia, and phosphate in pore waters of the central Venezuelan Basin, Deep Sea Drilling Project site 149 (data from Sayles et al., 1973; Gieskes, 1973; computed curves from Toth and Lerman, 1977). Reproduced by permission of the *American Journal of Science.*

conceivable that detrital minerals, such as apatite, and clays may release phosphate to solution under some conditions of the bacterial activity in sediments. Deeper in the sediment, the reduction of sulfate ceases, either at an exhaustion level or at some higher concentration: no further removal of sulfate is indicated by the zero gradient of concentration (concentration stays constant with an increasing depth). Phosphate (Figures 8.9 and 8.10) and ammonia (Figure 8.11) show a similar behavior: concentration gradients become zero at depth, within the uncertainty of the scatter of the data. The occurrence of the concentration peaks on the phosphate and ammonia profiles indicates that diffusional fluxes in pore water can carry each species in the upward *and* downward direction from the peak. Thus there are sinks deeper in the sediment that are responsible for the removal of phosphate and ammonia from pore water. Uptake by solid particles is a likely sink for the two species.

Concentrations of ammonia, drawn in Figure 8.9 and 8.10, continue to rise with depth after the reduction of sulfate has stopped. This suggests that ammonia may continue to form in sediments independently of the sulfate reduction, under the conditions that fall outside the scheme of reaction 8.12.

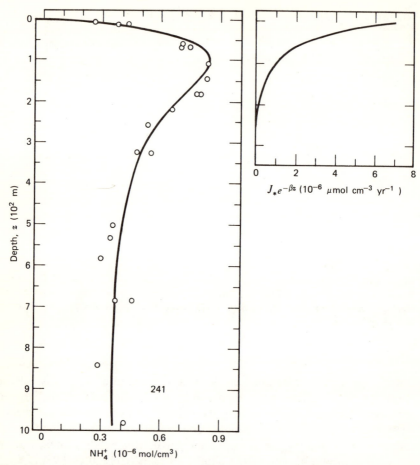

Figure 8.11 Ammonia in pore waters of sediments at the Deep Sea Drilling Project site 241, Indian Ocean. Two differnet concentrations at one depth indicate analyses of pore water extracted at 5° and 23°C (data from Gieskes, 1974; computed profile from Lerman, 1977).

Oxidation of organic carbon in sediments is reflected in a rise in concentrations of dissolved CO_2. In pore waters of oceanic and freshwater sediments, the main species of dissolved CO_2 is the bicarbonate ion HCO_3^-, in the pH range between 7 and 8.5. However, a rise in the bicarbonate-ion concentration in pore waters is not always due to oxidation of organic carbon only. Dissolution of calcium carbonate minerals can contribute additional amounts of HCO_3^- to pore waters, with the result that the net increase in HCO_3^- concentration can be greater than expected from the stoichiometric proportions $C:N:P$ given in reaction 8.12.

8.4.2   Information from the Models

The processes of production, removal, and transport in pore waters can be combined in a model that follows from the one-dimensional steady-state equation 8.8, and the equations for the chemical reactions 8.13 and 8.24. For phosphate and ammonia the following form of the diagenetic equation takes into account organic matter sources and mineral sinks

$$D\frac{\partial^2 C}{\partial z^2} - U\frac{\partial C}{\partial z} - kC + J + J_* e^{-\beta z} = 0 \qquad (8.28)$$

where the diffusion coefficient D and pore water advection rate U are taken as constant and independent of the vertical distance coordinate z. (This is valid for a sediment column of a constant porosity.) For sulfate that is being consumed bacterially, the equation without the mineral sinks is

$$D\frac{\partial^2 C}{\partial z^2} - U\frac{\partial C}{\partial z} - J_* e^{-\beta z} = 0 \qquad (8.29)$$

The concentration profiles of sulfate and phosphate, and the ammonia profile in Figure 8.11 approach a zero-concentration gradient with depth. Thus the boundary conditions for such profiles can be specified as a constant concentration at the sediment water interface $(z=0)$ and a constant concentration deeper in the sediment $(z\rightarrow\infty)$, that is,

$$\text{at}\quad z=0: \qquad C=C_{z=0} \qquad (8.30)$$

$$\text{at}\quad z\rightarrow\infty: \qquad C=C_{ss}\equiv\frac{J}{k} \qquad (8.31)$$

With these boundary conditions the solution of equation 8.28 is derived in a step-by-step procedure, shown in Appendix B. The solution is

$$C=\frac{J}{k} - \frac{J_* e^{-\beta z}}{D\beta^2 + U\beta - k} + \left(C_{z=0} - \frac{J}{k} + \frac{J_*}{D\beta^2 + U\beta - k}\right)\exp\left(\frac{Uz}{2D} - Rz\right)$$

$$(8.32)$$

where R is, as defined in equation 7.22,

$$R = +\left(\frac{U^2}{4D^2} + \frac{k}{D}\right)^{1/2} \quad (\text{cm}^{-1}) \qquad (8.32a)$$

For sulfate,

$$C = C_{z=0} - \frac{J_*}{D\beta^2 + U\beta}(1 - e^{-\beta z}) \qquad (8.33)$$

If sulfate concentration decreases to zero at depth, then the consumption rate $-J_*$ and concentration at the sediment-water interface $C_{z=0}$ are interrelated,

$$C_{z=0} = \frac{J_*}{D\beta^2 + U\beta}$$

Equation 8.32 contains three constant coefficients and two negative exponentials, and can be written in a simpler form suitable for evaluation of the individual terms as

$$C = A_1 + A_2 \exp(-\beta z) + A_3 \exp\left(\frac{Uz}{2D} - Rz\right) \qquad (8.34)$$

The equation contains four chemical parameters that have to be evaluated: J, J_*, k, and κ_i (κ_i appears in $\beta = \kappa_i/U$ and k enters in the parameter R). The rate of sedimentation or pore water advection U and the molecular diffusion coefficient D can be obtained from sources independent of the concentration profile. One way of estimating the unknown parameters is by a trial-and-error fitting of a concentration against depth profile to a plot of C against z. Another method is to obtain with a computer the best-fit values of the parameters A_1, A_2, A_3, β, and R, using some initially specified values. (Computer programs for exponential curve fitting are routinely available to users of computational centers.) Then, when U and D are known, the four chemical rate parameters J, J_*, k, and κ_i can be immediately computed from the best-fit values of A_1, A_2, A_3, and β, as follows from equations 8.32 and 8.34.

Concentrations of dissolved species in long sections of sediments, such as those shown in Figures 8.9–8.11, are usually based on samples taken several meters or even tens of meters apart. A thin impermeable layer or a layer of distinctly different mineralogical composition acting as a source or sink, not detected in the sampling, would make a semi-infinite model for the whole sediment column not valid. In the concentration profiles drawn in Figures 8.9–8.11, the physical continuity of the pore water column is only suggested by the concentration trends. In other cases, such as in the case of the silica profile in Figure 8.6, mineralogical differences between the upper and lower sections of the core are clearly suggested by the shape

Table 8.4 Parameters for the Computed Concentration Profiles of Sulfate (SO_4^{2-}), Total Dissolved Phosphorus (P), and Ammonia (NH_4^+) in Pore Waters, Drawn in Figures 8.9–8.11[a]

Parameter	Units	SO_4^{2-}		P		NH_4^+
		DSDP[b] 149	DSDP 147	DSDP 149	DSDP 147	DSDP 241
$C_{z=0}$	$\mu mol\ cm^{-3}$	19.4	28.0	0 (0.001)[c]	0 (0.030)[c]	0
J_*	$10^{-3}\ \mu mol\ cm^{-3}\ yr^{-1}$	0.0066	52.8	0.000072 (0.000045)	0.03 (0.017)	0.0075
β	$10^{-4}\ cm^{-1}$	1.187	52.17	6.0 (6.0)	17.17 (13.04)	2.1
J/k	$10^{-3}\ \mu mol\ cm^{-3}$	—	—	0.518 (0.509)	10.0 (9.1)	350.
k	$10^{-6}\ yr^{-1}$	—	—	11. (11.)	170. (110.)	0.7
κ	$10^{-6}\ yr^{-1}$	0.09	240.	0.45 (0.45)	79. (60.)	0.6
U	$10^{-3}\ cm\ yr^{-1}$	0.75	46.	0.75	46.	3.
D	$cm^2\ yr^{-1}$	63.2	63.2	31.6	31.6	94.8

[a] From Toth and Lerman (1977).

[b] Deep Sea Drilling Project site number.

[c] Numbers in parentheses refer to the second concentration profile of phosphorus in the figures.

Table 8.5 Equations for Concentration Against Depth Profiles Drawn in Figures 8.9 through 8.11

Dissolved species	DSDP site no.	Equation [a]	Units of C	Text equation
Sulfate (SO_4^{2-})	149	$C = 19.4 - 6.7(1 - e^{-0.012z})$	mmol l^{-1}	8.33
	147	$C = 28.2 - 27.0(1 - e^{-0.522z})$	mmol l^{-1}	8.33
Phosphate (total P)	149	$C = 0.518 - 87.17e^{-0.06z}$ $+ 86.65e^{-0.0578z}$	μmol l^{-1}	8.32
	147	$C = 10 - 1.402 \times 10^4 e^{-0.1717z}$ $+ 1.401 \times 10^4 e^{-0.1703z}$	μmol l^{-1}	8.32
Ammonia (NH_4^+)	241	$C = 0.35 - 1.825e^{-0.021z}$ $+ 1.475e^{-0.7155z}$	mmol l^{-1}	8.32

[a]Depth in sediment z is in meters.

of the dissolved silica profile. Other features related to the varying mineralogical composition of sediments are dealt with in Section 8.5.

The curves drawn through the data points for sulfate and phosphorus in Figure 8.9 and 8.10, and for ammonia in Figure 8.11, can be computed by using the parameters listed in Table 8.4 and equations 8.33 and 8.32. Explicit equations for the curves are listed in Table 8.5. For the ammonia profiles shown in Figures 8.9 and 8.10, the model behind equation 8.32 could be used only if it were assumed that concentrations tend to a steady value with depth. Because neither of the profiles approaches a zero gradient of concentration, a model based on equation 8.28 for a sediment layer can be applied. The solution of 8.28 with the boundary conditions for a sediment layer, such as the boundary conditions given in equations 8.39 and 8.40, is an algebraically more involved form than the solution for a semi-infinite sediment column that was used for the sulfate and phosphorus profiles. The solution and the statistical method of computation of the individual parameters were described by Toth and Lerman (1977).

8.5 MINERAL-PORE WATER REACTIONS

8.5.1 Model Equations

For the chemical reactions between mineral phases and pore water that can be approximated to the coupled dissolution and precipitation, the diagenetic equation 8.28 reduces to the form

$$D\frac{\partial^2 C}{\partial z^2} - U\frac{\partial C}{\partial z} - kC + J = 0 \qquad (8.35)$$

For other than first-order reactions, the reaction rate term will have a different functional form and the resulting equation may have to be solved by numerical methods. Equation 8.35 is identical to the mass balance equation 8.24. As long as the transport (D, U) and chemical (k, J) coefficients in the differential equation are constant and independent of concentration C and depth in the sediment z, the steady-state equation 8.35 and its transient-state analog (that is, the right-hand side is $\partial C/\partial t$ instead of 0) can be solved explicitly for a variety of initial and boundary conditions. In this section, mathematical solutions for a steady-state will be given for two types of boundary conditions more commonly encountered in sediments. One is a semi-infinite column of sediment, defined only by one boundary, such as the sediment-water interface or a reference horizon within the sediment. The other type is a sediment layer having two boundaries (see Figure 8.2).

Semi-Infinite Column of Sediment. The solution of equation 8.35 for a semi-infinite column of sediment is equation 8.32 with the rate of production term J_* taken as zero. The solution is (for details, see Appendix B)

$$C = \frac{J}{k} + \left(C_{z=0} - \frac{J}{k} \right) \exp\left(\frac{Uz}{2D} - Rz \right) \qquad (8.36)$$

A plot of concentration C against depth z, computed from the preceding equation, gives a curve for C that either rises or declines with depth from the value $C_{z=0}$, approaching asymptotically the value $C = J/k$ at depth. The derivative of the curve dC/dz does not change sign, and neither concentration maxima nor minima occurring in a vertical concentration profile can be described by equation 8.36. Compare the sulfate concentration profiles in Figures 8.9 and 8.10, and the latter equation with equation 8.33.

The two chemical parameters J and k can be evaluated as follows. The quotient J/k is concentration at depth, in the region of the zero gradient, and it can be read directly from a plotted profile. The midpoint of the concentration range is concentration $C_{1/2}$ that satisfies the following condition:

$$\frac{C_{1/2} - J/k}{C_{z=0} - J/k} = \frac{1}{2} \qquad (8.37)$$

The depth $z_{1/2}$ where the midpoint of the concentration range occurs can also be read off a plotted concentration profile. Then the kinetic parameter k can be determined from equation 8.37 and 8.36 with $C = C_{1/2}$ and

$z = z_{1/2}$ as

$$k = \left(U + \frac{D \ln 2}{z_{1/2}} \right) \frac{\ln 2}{z_{1/2}} \qquad (\text{yr}^{-1}) \qquad (8.38)$$

The deeper in the sediment the half-concentration point is, the larger the value of $z_{1/2}$ is, and, consequently, the smaller the value of k is. A large $z_{1/2}$ means that the concentration gradient is weak and a small k means that the mineral diagenetic reactions are slow. Differences in the gradients of dissolved species, such as those of the silica profiles in oceanic and lake sediments shown in Section 7.6, illustrate the point about the half-concentration depth.

Sediment Layer. The boundary conditions for a sediment pore water layer of thickness h (cm) are

$$\text{at} \quad z = 0: \qquad C = C_{z=0} \qquad\qquad (8.39)$$

$$\text{at} \quad z = h: \qquad C = C_{z=h} \qquad\qquad (8.40)$$

With these boundary conditions, the solution of equation 8.35 is

$$C = \frac{J}{k} + \frac{(C_{z=0} - J/k)e^{Uz/2D} \sinh R(h-z)}{\sinh Rh}$$

$$+ \frac{(C_{z=h} - J/k)e^{U(z-h)/2D} \sinh Rz}{\sinh Rh} \qquad (8.41)$$

The choice of the upper or the lower boundary for $z = 0$ is immaterial, although setting $z = 0$ at the upper boundary is compatible with the frame of reference for the processes below the sediment-water interface.

For large h (thick layer) or for $z \ll h$ (section near the upper boundary), equation 8.41 reduces at the limit $h \to \infty$ to the form of equation 8.36. The significance of this limiting relationship is that the mathematically simpler models for a semi-infinite sediment column may be used for the upper part of a sediment layer. In a thick layer, the diagenetic reactions and transport of dissolved species in pore water are, as an approximation, not affected by the processes taking place deeper within the layer.

In a growing column of sediment, the thickness h of a layer may be increasing with time, as discussed in connection with Figure 8.2. If h is a function of time, then the steady-state equation 8.41 strictly does not apply. However, the rates of sedimentation in the pelagic ocean are slow in comparison to diffusional fluxes in pore water, and they are also slow in comparison to the redistribution of dissolved species by the chemical source and sink reactions. Numerical computations of concentration profiles in pore water of a growing sediment layer show that transient-state profiles are very close to steady-state profiles for a layer of a given thickness h, of the order of 10^2 m, as long as the rates of the layer growth are of the order of centimeters per 1000 yr. More generally, if an increment of sediment thickness added to a layer over a period of 10^6 yr amounts to only a few percent of the layer thickness, then a steady-state model is a valid approximation to a growing layer.

The relationships between the mean rate of sedimentation U, the layer thickness h, and the chemical reaction rate parameter k determine the relative importance of some of the terms in the steady-state equation 8.41 for a layer. The equation can be simplified if the following two conditions are met:

$$\frac{U^2}{4D} \ll k \qquad\qquad (8.42)$$

$$\frac{Uh}{2D} \ll 1 \qquad\qquad (8.43)$$

In the former case, the term R (defined in equation 8.32a) reduces to $(k/D)^{1/2}$; in the latter case, the exponential terms containing $Uh/2D$ and $Uz/2D$ (with $z < h$) can be omitted. Representative values of the two quotients can be based on the sedimentation rate of the order of $U \leqslant 1$ cm per 1000 yr $= 1 \times 10^{-3}$ cm yr^{-1}, molecular diffusion coefficient in pore water $D = 3 \times 10^{-6}$ cm^2 sec$^{-1} = 100$ cm^2 yr^{-1}, and sediment column thickness $h = 100$ m $= 10^4$ cm. Then

$$\frac{U^2}{4D} \leqslant 2 \times 10^{-9} \text{ yr}^{-1} \qquad \text{and} \qquad \frac{Uh}{2D} \leqslant 0.05$$

The values of the reaction rate parameter k derived from application of steady-state models to concentration profiles, such as those listed in Table 8.6, all are much greater than 2×10^{-9} yr^{-1}. Thus condition 8.42 is commonly obeyed for oceanic sediments. The value of the quotient $Uh/2D \simeq 0.05$ is sufficiently small to be neglected, because $e^{0.05} \simeq 1.05 \simeq 1$. But for the sedimentation rates or pore water advection rates relative to the sediment layer boundaries that are appreciably greater than 10^{-3} cm yr^{-1}, the terms containing $Uh/2D$ must be retained in the equation. When both $U^2/4D$ and $Uh/2D$ can be neglected, equation 8.41 reduces to a simpler form,

$$C = \frac{J}{k} + \frac{(C_{z=0}-J/k)\sinh(h-z)(k/D)^{1/2}+(C_{z=h}-J/k)\sinh z(k/D)^{1/2}}{\sinh h(k/D)^{1/2}}$$

(8.44)

In a slowly growing sediment layer, profiles near steady-state at different times can be computed by equating the layer thickness with the product of the sedimentation rate and time (Lerman and Lietzke, 1977),

$$h = Ut \qquad (\text{cm}) \tag{8.45}$$

Substitution of Ut for h in equation 8.44 makes concentration C time dependent; however, it should be remembered that the procedure is an approximation to a concentration profile that is close to a steady-state profile, within a layer of a fixed thickness h and fixed boundary concentrations $C_{z=0}$ and $C_{z=h}$.

After long times (large t), or near the upper boundary of the layer ($z \ll h$ or $z \ll Ut$), the concentration profile approaches a profile in a semi-infinite column,

$$C = \frac{J}{k} + \left(C_{z=0} - \frac{J}{k}\right)\exp\left[-z\left(\frac{k}{D}\right)^{1/2}\right] \tag{8.46}$$

The latter equation is identical to 8.41 with the pore water advection or sedimentation rate U equal zero.

8.5.2   Diagenesis in a Carbonate-Silicate Sediment

The major changes in the geography of the continental and oceanic plates that have been taking place during the last 100 million years have left their imprint in the oceanic sediments. Variation in the mineralogical composition and in the abundances of inorganic or biogenic materials in vertical sections of sediments are local reflections of the global evolution of the geological conditions. An example of a change in the sediments is a vertical section from the Eastern Caribbean Sea, site 149 of the Deep Sea Drilling Project, the main mineralogical features of which are presented in Figure 8.12. The reactions between organic matter and pore waters in the same sedimentary column are discussed in Section 8.4.

A coarse subdivision of the sediment column into three units, shown in the figure, is based on the gross mineralogical composition of the sediments. The upper unit, about 100 m thick, contains terrigenous clays

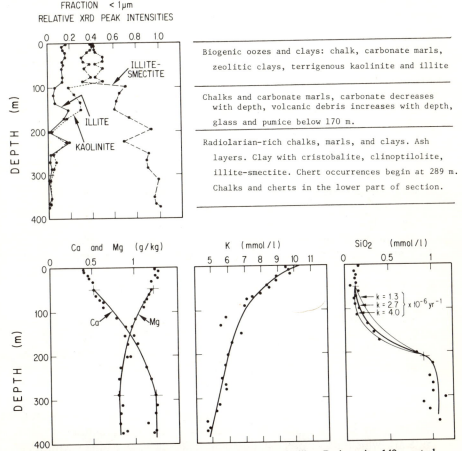

FRACTION < 1μm
RELATIVE XRD PEAK INTENSITIES

Biogenic oozes and clays: chalk, carbonate marls,
zeolitic clays, terrigenous kaolinite and illite

Chalks and carbonate marls, carbonate decreases
with depth, volcanic debris increases with depth,
glass and pumice below 170 m.

Radiolarian-rich chalks, marls, and clays. Ash
layers. Clay with cristobalite, clinoptilolite,
illite-smectite. Chert occurrences begin at 289 m.
Chalks and cherts in the lower part of section.

Figure 8.12 Diagenesis in sediments at the Deep Sea Drilling Project site 149, central Venezuelan Basin. Mineralogical composition of the small-size sediment fraction, and concentrations of calcium, magnesium, potassium, and silica in pore waters (data from Edgar et al., 1973; Gieskes, 1973; Sayles et al., 1973; Perry et al., 1976). Copywrited portions of this figure reproduced by permission of Pergamon Press, Ltd., Oxford, England.

(kaolinite) and other silicates, and calcareous plankton, which amounts to up to 50% of the sediment mass. The second unit, between about 100 and 200 m depth, is made of chalks and clayey oozes, with volcanic debris, glass, and pumice becoming more abundant deeper in the sediment. The third unit, from 200 m down, is characterized by abundant radiolarian skeletons and volcanic materials; chert layers appear at 289 m and occur in the bottom part of the section.

The changes in the chemical composition of pore waters—shown in Figure 8.12 for calcium, magnesium, potassium, and silica—are in good qualitative agreement with the changes in the mineralogy observable in the vertical profile. The shape of the concentration profile of calcium in pore water indicates dissolution within the sediment column and flux upward, judging by the concentration gradient. (Whether there is a net flux out of the sediment into the overlying ocean water is determined by the profile immediately below the sediment-water interface. The scale of sampling in the sediments at site 149 is too coarse, and it does not allow refinements on the scale of less than a meter, at the best. Note that some samples are tens of meters apart.) The deduced dissolution and upward flux of calcium are compatible with the abundance of biogenic calcium carbonate and calcium silicates in the volcanic debris in the sediments.

The concentration profile of magnesium indicates that a flux downward may exist. The curvature of the profile to the left, toward the lower concentration values, can be due to a precipitation reaction removing magnesium from pore water to solids. Such a reaction may cause the formation of the mixed-layer illite–smectite, the abundance of which increases from 100 m down. The decrease in dissolved potassium with an increasing depth, and the decrease in dissolved silica concentration from 200 m upward also agree qualitatively with the neoformation of clay minerals, which act as precipitation sinks. The curvature of concentration profiles—such as in the profiles of calcium, magnesium, and silica—should be interpreted with caution. In a pore water column, in which a dissolved species is transported by diffusion only, a linear concentration gradient will exist if there are no other sources or sinks within the column. The concentration profile can become curved due to any of the following factors: the diffusion coefficient varies with distance; the porosity or temperature vary in the sediment; advection of pore water is significant in comparison to the diffusional transport; and chemical sources or sinks exist within the layer (Lerman, 1977). The shape of a concentration profile alone does not allow one to determine whether it is uniquely due to a single factor only: either the chemical reactions or such physical factors as the gradients of porosity and temperature. In the case of the concentration profiles drawn in Figure 8.12, an unusual combination of physical factors would be needed to account for the different curvature of the calcium and magnesium profiles. However, an explanation that the curvature reflects chemical reactions is compatible with the changes in the mineralogical composition of the sediments.

The sum of the molal concentrations of calcium and magnesium at any depth is approximately constant. At a first glance, this feature suggests that for each mole of magnesium removed from pore water, one mole of calcium is released. A logical choice for such an exchange is the formation of dolomite in a reaction between calcite and magnesium in pore water. However, neither dolomite nor another potential sink for magnesium, the clay mineral sepiolite, have been detected in the sediments. The mixed-layer illite–smectite, forming by reactions between the calc-alkaline volcanic debris and magnesium, silica, and other ions in pore water, is a more likely sink for the reactive species in solution (Perry et al., 1976).

Below 220 m, concentrations of silica in pore water show no clear gradient. Concentrations approach the solubility of opal (Figure 8.7), and the abundance of the radiolarian skeletons in the lower part of the sediment section is a likely source of dissolved silica, in addition to the volcanic debris. In an uninterrupted column of pore water, the high concentrations of dissolved silica in the deeper sediment are a partial source of silica for the younger sediments. The curves drawn through the concentration points for the four dissolved species in Figure 8.12 were computed using the equations given in the text and the parameter values listed in Table 8.6. Explicit equations for each concentration profile are given in Table 8.7.

Table 8.6 Parameters for the Concentration Profiles of Calcium, Magnesiun, Potassium, and Silica in Pore Waters, Drawn in Figure 8.12

Parameter	Units	Ca	Mg	K	Si	
Depth in core	m	50–270	50–270	0–370	50–210	210–370
h	m	220	220	∞	160	∞
	$g \, l^{-1}$	0.52	1.22			
$C_{z=0}$						
	$mmol \, l^{-1}$	13.0	50.2			
	$g \, l^{-1}$	1.23	0.80			
$C_{z=h}$						
	$mmol \, l^{-1}$	30.7	32.9	5.2	0.90	1.05
	$g \, cm^{-3} \, yr^{-1}$	7.0×10^{-10}	0			
J						
	$\mu mol \, cm^{-3} \, yr^{-1}$	1.7×10^{-5}	0	1.6×10^{-5}	0	2.2×10^{-5}
k	$10^{-6} \, yr^{-1}$	0.57	0.18	3.1	2.7	21.0

Table 8.7   Equations for Concentrations Against Depth Curves Drawnin Figure 8.12. Data from Table 8.6

Dissolved species	Equation	Units of C	Range of z (m)	Text equation
Ca	$C=1.23-0.734\exp(-7.37\times10^{-3}z)$ $+0.024\exp(8.16\times10^{-3}z)$	g l^{-1}	$0<z<220$	8.44
Mg	$C=1.10\exp(-3.98\times10^{-3}z)$ $+0.12\exp(4.47\times10^{-2}z)$	g l^{-1}	$0<z<220$	8.44
K	$C=5.24+5.02\exp(-1.09\times10^{-2}z)$	mmol l^{-1}	$0<z<\infty$	8.46
Si	$C=0.074\exp(-1.65\times10^{-2}z)$ $+0.056\exp(1.73\times10^{-2}z)$	mmol l^{-1}	$<z<160$	8.44
	$C=1.05-0.15\exp(-4.71\times10^{-2}z)$		$0<z<\infty$, from 160 m down	8.46

Precipitation Rates of Silica and Magnesium. The amounts of silica and magnesium precipitated from pore water at any depth in sediment can be estimated from the chemical kinetic parameters given in Table 8.6. The precipitation rate is approximately $-kC$ (g cm^{-3} yr^{-1}), and the amount precipitated at any given depth z during the time period Δt is

$$\frac{\phi k C \Delta t}{1-\phi} \quad \text{(g in 1 cm}^3 \text{ of solids)}$$

For a sediment of porosity 50% ($\phi=0.5$), and age Δt of about 30 million years, a representative SiO$_2$ concentration of 0.5 mmol l^{-1} gives the mass precipitated,

$$(2.7\times10^{-6}\ \text{yr}^{-1})(0.5\times10^{-6}\ \text{mol cm}^{-3})(60\ \text{g mol}^{-1})(3\times10^7\ \text{yr})$$

$$=2.4\times10^{-3}\ \text{g SiO}_2 \text{ in 1 cm}^3 \text{ solids}$$

For Mg between the depths 50 and 200 m, using a representative concentration of 1 g l^{-1}, the mass precipitated is

$$(1.8\times10^{-7}\ \text{yr}^{-1})(1\times10^{-3}\ \text{g cm}^{-3})(3\times10^7\ \text{yr})$$

$$=5.4\times10^{-3}\ \text{g Mg in 1 cm}^3 \text{ solids}$$

The estimates of the precipitated masses would be larger if precipitation took place over a longer period of time. Conversely, they would be smaller if the rates of precipitation were computed from the equation $k(C-C_s)$ instead of kC. For silica, the lower estimate of the precipitation rate can be obtained from the difference term $C_s-C=0.5-0.1$ mmol l^{-1}. For magnesium, the driving concentration difference can be taken as $C_s - C = 1.0 - 0.7$ g l^{-1} (see Figure 8.12).

According to the computation given above, the total mass of such species as silica, magnesium, and potassium, precipitated out of a unit of volume of pore water over a period of about 30 million years, amounts to an order of magnitude figure of 10 mg per 1 cm^3 of solids or about 5 mg per 1 g of solids. The mass precipitated is only 0.5% by weight of the bulk dry sediment. If the precipitated phases are associated with the smallest particles in the sediment, then the mass precipitated can be percentwise greater in the small size fraction of the sediment. For example, if the precipitated material is associated with a particle-size fraction that accounts for 1 wt % of the dry sediment, then the newly formed material adds diagenetically 50% to the small size fraction. In the sediment section from the Eastern Caribbean, the newly formed illite–smectite is associated with the size fraction <1 μm. The fine fraction contains the following proportions of silica, magnesium, and potassium (Perry et al., 1976): SiO_2, 47–50%; MgO, 1.5–3.5%; and K_2O, 1–3%. Some of the materials in the fine-size fraction are probably altered residues of the volcanic debris that acts as a precursor for the growth of clay minerals. Leaching of some of the cations leaves behind a phase enriched in silica (Section 5.5), which may react with other ions in pore water. The global abundance of smectites in ocean sediments has been attributed to alteration of volcanic ash (Section 6.2).

Opal and Chert.   In oceanic sediments, amorphous silica (opal) transforms to disordered crystobalite and tridymite in microcrystalline aggregates, and finally to microcrystalline quartz, forming the chert (Heath and Moberly, 1971; von Rad and Rösch, 1972; Berger and von Rad, 1972). Although chert in the form of layers, cement, fillings, and nodules is ubiquitously abundant in the older sediments, there are apparently no chert layers younger than 25×10^6 yr. The more common occurrences are in sediments of early Tertiary age and older than 70×10^6 yr. For a layer of chert to form out of an unconsolidated layer of biogenic opal, the pore space volume of about 50% must be reduced to almost zero. The rates of dissolution and precipitation of silica in oceanic sediments that are discussed on the preceding pages can be used to estimate the amount of reduction of the sediment porosity that can take place during the reactions between opal and pore waters.

In the East Caribbean sediment section pictured in Figure 8.12, the high concentration of dissolved SiO_2 in pore water coincides with the occurrence of radiolarian skeletons and chert layers in the lower part of the sediment column, below 270 m level. The rate constant for the dissolution of opal in a first-order reaction is $k_1 = 1.4 \times 10^{-5}$ yr^{-1}, and the rate constant for a precipitation reaction is $k_2 = 0.7 \times 10^{-5}$ yr^{-1}. (These values can be obtained as shown in the computation example on pp. 385-387, using $k = 2.1 \times 10^{-5}$ yr^{-1} and $J = 2.2 \times 10^{-11}$ mol cm^{-3} yr^{-1} from Table 8.6.) Precipitation of SiO_2 takes place from a pore water solution containing 1 mmol l^{-1} of dissolved silica. Over a period of 50 million years, the mass of silica dissolved per unit volume of solids is (porosity $\phi = 0.5$)

$$\frac{\phi k_1 (C_{s,1} - C)\Delta t}{1 - \phi} \simeq (1.4 \times 10^{-5} \text{ yr}^{-1})(2-1) \times 10^{-6} \text{ mol cm}^{-3}$$

$$\times (60 \text{ g mol}^{-1})(50 \times 10^6 \text{ yr})$$

$$\simeq 4 \times 10^{-2} \text{ g SiO}_2 \text{ in 1 cm}^3 \text{ solids}$$

Similarly, the mass of SiO_2 precipitated during 50×10^6 yr is

$$\frac{\phi k_2 (C - C_{s,2})\Delta t}{1 - \phi} \simeq 0.7 \times 10^{-5}(1 - 0.1) \times 10^{-6} \times 60 \times 50 \times 10^6$$

$$\simeq 2 \times 10^{-2} \text{ g SiO}_2 \text{ in 1 cm}^3 \text{ solids}$$

Basically, a mass of about 1×10^{-2} g SiO_2 added to 1 cm^3 of solids reduces the pore volume fraction by about 1%. To reduce the porosity of a layer of opal from 50% to near zero, within 50 million years, the rates of precipitation should be about 100 times higher, precipitating a mass of the order of 1 g SiO_2 in 1 cm^3 of solids. Such a mass of silica can be derived over a period of 50 million years from a column of sediment 0.5–2 m thick (based on the dissolution and precipitation rates of the order of 10^{-5} yr^{-1}, as used in the computation given above). The mass, precipitated within a layer 1 cm thick, would be equivalent to an impermeable layer of chert. The preceding discussion and computation show that, in principle, thin chert layers of the Cenozoic age might have formed by the mechanism of local sinks for silica. The amount of dissolution of the adjacent SiO_2 sediments is volumetrically insignificant when they act as sources for centimeter-thin layers of chert. For thick chert beds occuring over large areas, such as the Eocene chert in the Atlantic Ocean sediments and the Cretaceous cherts of the Mediterranean region, other sources must have existed. The possible mechanisms include *in situ* recrystallization of opaline sediments. Also, recirculation of ocean water through the volcanically and tectonically active zones of the oceanic crust might have raised hot solutions of higher silica concentration, from which SiO_2 precipitated during the ascent and cooling.

APPENDIX A

The Error Function

The error function $\mathrm{erf}\, x$ and the complement of the error function $\mathrm{erfc}\, x$ are mentioned in Chapters 2 and 5. The error function also appears often in solutions of partial differential equations describing diffusion and chemical reactions, such as those given in Appendix B. The error function is defined as the integral

$$\mathrm{erf}\, x = \frac{2}{\sqrt{\pi}} \int_0^x e^{-y^2} dy \tag{A.1}$$

where y is an integration variable and x is a number, positive or negative. The limiting values of $\mathrm{erf}\, x$ are

$$\mathrm{erf}\, 0 = 0, \quad \mathrm{erf}\, \infty = 1, \quad \text{and} \quad \mathrm{erf}(-\infty) = -1$$

The complement of the error function is defined as

$$\mathrm{erfc}\, x = 1 - \mathrm{erf}\, x \tag{A.2}$$

or as an integral

$$\mathrm{erfc}\, x = \frac{2}{\sqrt{\pi}} \int_x^\infty e^{-y^2} dy \tag{A.3}$$

For positive x, the limits of $\mathrm{erfc}\, x$ are (see Table A.1)

$$\mathrm{erfc}\, 0 = 1 \quad \text{and} \quad \mathrm{erfc}\, \infty = 0$$

For negative x, the following relationships hold:

$$\mathrm{erfc}(-x) = \begin{cases} 1 - \mathrm{erf}(-x) \\ 1 + \mathrm{erf}\, x \\ 2 - \mathrm{erfc}\, x \end{cases} \tag{A.4}$$

409

Table A.1   The Error Function erf and the Error Function Complement erfc for Negative and Positive Values of the Argument x^a

x	erf x	erfc x	x	erf x	erfc x
$-\infty$	-1.0	2.0	0	0	1.0
-3.0	-1.0	2.0	0.1	0.1125	0.8875
-2.8	-0.9999	1.9999	0.2	0.2227	0.7773
-2.6	-0.9998	1.9998	0.3	0.3286	0.6714
-2.4	-0.9993	1.9993	0.4	0.4284	0.5716
-2.2	-0.9981	1.9981	0.5	0.5205	0.4795
-2.0	-0.9953	1.9953	0.6	0.6039	0.3961
-1.8	-0.9891	1.9891	0.7	0.6778	0.3222
-1.6	-0.9763	1.9763	0.8	0.7421	0.2579
-1.4	-0.9523	1.9523	0.9	0.7969	0.2031
-1.2	-0.9103	1.9103	1.0	0.8427	0.1573
-1.0	-0.8427	1.8427	1.2	0.9103	0.0897
-0.9	-0.7969	1.7969	1.4	0.9523	0.0477
-0.8	-0.7421	1.7421	1.6	0.9763	0.0237
-0.7	-0.6778	1.6778	1.8	0.9891	0.0109
-0.6	-0.6039	1.6039	2.0	0.9953	0.0047
-0.5	-0.5205	1.5205	2.2	0.9981	0.0019
-0.4	-0.4284	1.4284	2.4	0.9993	0.0007
-0.03	-0.3286	1.3286	2.6	0.9998	0.0002
-0.2	-0.2227	1.2227	2.8	0.9999	0.0001
-0.1	-0.1125	1.1125	3.0	1.0	0
0	0	1.0	∞	1.0	0

^{a}From Gautschi (1964); Carslaw and Jaeger (1959); Crank (1956).

Thus the limit of erfc(x) at $x = -\infty$ is

$$\text{erfc}(-\infty) = 2$$

The tabulated values of erf x and erfc x for positive and negative x (Table A.1) show that the functions approach their limiting values fast, when $|x| > 3$. Detailed tabulations of the functions, their integrals and derivatives can be found in the references listed under Table A.1.

To compute erfc x, the following series can be used (Gautschi, 1964). For $0 \leqslant x < 3$,

$$\text{erfc}\, x = \frac{1}{\left(1 + a_1 x + a_2 x^2 + a_3 x^3 + a_4 x^4 + a_5 x^5\right)^8} \tag{A.5}$$

where the constants a_i are

$$a_1 = 0.14112821 \qquad a_4 = -0.00039446$$

$$a_2 = 0.08864027 \qquad a_5 = 0.00328975$$

$$a_3 = 0.002743349$$

For $x \geqslant 3$,

$$\operatorname{erfc} x = \frac{e^{-x^2}}{\sqrt{\pi}} \left(\frac{1}{x} - \frac{1}{2x^3} + \frac{3}{2^2 x^5} - \frac{15}{2^3 x^7} + \frac{105}{2^4 x^9} - \frac{945}{2^5 x^{11}} + \frac{10365}{2^6 x^{13}} \right)$$

$$(A.6)$$

For large x, the error function complement $\operatorname{erfc} x$ tends to zero faster than e^{-x^2}, as follows from equation A.6. Therefore, the product $e^{x^2} \operatorname{erfc} x$ tends to zero when x increases indefinitely, and for large x it becomes approximately

$$e^{x^2} \operatorname{erfc} x \simeq \frac{1}{x \sqrt{\pi}}$$

Solution of Differential Equations

In this Appendix are given two solutions of differential equations that appear in Chapters 2, 3, and 8. The equations apply to a one-dimensional system where diffusion, advection, and chemical reactions take place. The case of a steady state corresponds to an *ordinary* differential equation, where the dependent variable (concentration C) is a function of the distance coordinate only, but not a function of time. In transient state, concentration is a function of both distance and time, and this leads to a *partial* differential equation.

B.1 STEADY STATE

An ordinary differential equation in which C is a function of z,

$$\frac{\partial^2 C}{\partial z^2} - A_1 \frac{\partial C}{\partial z} - A_2 C = f(z) \tag{B.1}$$

is a linear equation of second order with constant coefficients A_1 and A_2. The right-hand side is a function of z, $f(z)$. A general solution of equation B.1 is (Margenau and Murphy, 1956, p. 54)

$$C = \frac{1}{r_1 - r_2} \left(e^{r_1 z} \int e^{-r_1 z} f(z)\, dz - e^{r_2 z} \int e^{-r_2 z} f(z)\, dz \right)$$

$$+ K_1 e^{r_1 z} + K_2 e^{r_2 z} \tag{B.2}$$

where K_1 and K_2 are constants, and r_1 and r_2 are the roots of the auxiliary equation. The auxiliary equation contains all the C-dependent terms on the left-hand side of (B.1), and its roots are

$$r_1 = \frac{A_1}{2} + \left[\left(\frac{A_1}{2}\right)^2 + A_2\right]^{1/2} \tag{B.3}$$

$$r_2 = \frac{A_1}{2} - \left[\left(\frac{A_1}{2}\right)^2 + A_2\right]^{1/2} \tag{B.4}$$

The constants K_1 and K_2 are determined by the boundary conditions of the problem.

A steady-state diagenetic equation given in Chapter 8 includes diffusion, advection, and mineral and organic chemical reactions. The equation

$$D\frac{\partial^2 C}{\partial z^2} - U\frac{\partial C}{\partial z} - kC + J + J_* e^{-\beta z} = 0 \tag{B.5}$$

has constant coefficients D, U, and k, and the terms independent of C can be transferred to the right-hand side, to give $f(z)$ in the form

$$f(z) = -\frac{J}{D} - \frac{J_* e^{-\beta z}}{D} \tag{B.6}$$

Equation B.5 can be written in a form similar to B.1,

$$\frac{\partial^2 C}{\partial z^2} - \frac{U}{D}\frac{\partial C}{\partial z} - \frac{k}{D}C = f(z) \tag{B.7}$$

The roots of the auxiliary equation are

$$r_1 = \frac{U}{2D} + \left(\frac{U^2}{4D^2} + \frac{k}{D}\right)^{1/2} \qquad (r_1 > 0) \tag{B.8}$$

$$r_2 = \frac{U}{2D} - \left(\frac{U^2}{4D^2} + \frac{k}{D}\right)^{1/2} \qquad (r_2 < 0) \tag{B.9}$$

The product of the two roots is

$$r_1 r_2 = -\frac{k}{D}$$

Substitution from equations B.6, B.8, and B.9 into the general solution, equation B.2, gives

$$C = \frac{J}{k} - \frac{J_* e^{-\beta z}}{D\beta^2 + U\beta - k} + K_1 e^{r_1 z} + K_2 e^{r_2 z} \qquad (B.10)$$

Because r_1 is positive, the term $K \exp(r_1 z)$ increases indefinitely with the increasing distance z. If concentration tends to a finite value at depth, then the constant K_1 must be zero. The boundary conditions for a semi-infinite column, where concentration at the upper boundary is constant and concentration at depth approaches a steady value are

$$\text{at} \quad z = 0: \qquad C = C_{z=0} \qquad (B.11)$$

$$\text{at} \quad z \to \infty: \qquad \frac{dC}{dz} = 0 \qquad (B.12)$$

The boundary condition B.11 is sufficient to determine the constant K_2, if K_1 is taken as zero. The constant is

$$K_2 = C_{z=0} - \frac{J}{k} + \frac{J_*}{D\beta^2 + U\beta - k} \qquad (B.13)$$

Substitution from equation B.13 for K_2 in equation B.10 gives the final form of C as a function of z and other constant parameters,

$$C = \frac{J}{k} - \frac{J_* e^{-\beta z}}{D\beta^2 + U\beta - k} + \left(C_{z=0} - \frac{J}{k} + \frac{J_*}{D\beta^2 + U\beta - k} \right)$$

$$\times \exp\left[\frac{zU}{2D} - z\left(\frac{U^2}{4D^2} + \frac{k}{D} \right)^{1/2} \right] \quad (B.14)$$

Equation B.14 is identical to 8.32.

B.2   TRANSIENT STATE

If the left-hand side of equation B.5 is equal to the time derivative of C instead of zero, the equation becomes a partial differential equation with constant coefficients,

$$\frac{\partial C}{\partial t} = D\frac{\partial^2 C}{\partial z^2} - U\frac{\partial C}{\partial z} - kC + J + J_* e^{-\beta z} \qquad (B.15)$$

Numerical methods of solving partial differential equations include various computational routines (Carnahan et al., 1969) and such specialized techniques as the finite difference method, the relaxation method, and the finite element method.

One-dimensional cases to which equation B.15 with certain initial and boundary conditions applies can be solved explicitly. Solutions pertaining to heat flow, and methods of solution, are covered extensively in Carslaw and Jaeger (1959). A convenient method of obtaining specific solutions of equation B.15 is the method of the Laplace transformation. The method allows to reduce a partial differential equation in t and z to an ordinary differential equation in z only. Instead of a dependent variable $C(t,z)$, the equation contains a transformed variable $\overline{C}(z)$, which can in many cases be solved by the method of Section B.1. The result is an explicit equation for $\overline{C}$ as a function of z and other constant parameters. To convert the solution for $\overline{C}$ back into an explicit function of t and z, the inverse of the Laplace transformation is applied to the solution. The inversion can be done either using the tables of the Laplace transformation or algebraic or numerical techniques. For equations of the type of equation B.15, tables of the commonly used Laplace transformations can be found as appendices in the books by Carslaw and Jaeger (1956) and Crank (1959). The following Laplace transformations are needed for the solution of equation B.15 in a semi-infinite medium.

The Laplace transformation of a function of time and other variables $C(t,z)$ is defined as

$$\mathcal{L}\{C(t,z)\} \equiv \overline{C} = \int_0^\infty e^{-pt} C(t,z)\, dt \tag{B.16}$$

where p is treated as a constant. As mentioned previously, tables of many transformed functions are available, and those needed for a solution of B.15 are given below. The Laplace transformation of a constant or of a function that does not depend on t is

$$\mathcal{L}\{\text{const}\} = \frac{\text{const}}{p} \tag{B.17}$$

For example, the term $J_* e^{-\beta z}$ depends on z only, and its transform is

$$\mathcal{L}\{J_* e^{-\beta z}\} = \frac{J_* e^{-\beta z}}{p}$$

The Laplace transform of $\partial C / \partial t$ is

$$\mathcal{L}\left\{\frac{\partial C}{\partial t}\right\} = p\overline{C} - C_{t=0} \tag{B.18}$$

where $C_{t=0}$ is the initial value of C.

The Laplace transform of a partial derivative of C with respect to z is

$$\mathcal{L}\left\{\frac{\partial^n C}{\partial z^n}\right\} = \frac{\partial^n \overline{C}}{\partial z^n} \tag{B.19}$$

which states that the transform of a derivative is the same derivative of the transformed function. Using the definitions from equations B.17 through B.19, the Laplace transformation of the individual terms in equation B.15 gives

$$\frac{\partial^2 \overline{C}}{\partial z^2} - \frac{U}{D}\frac{\partial \overline{C}}{\partial z} - \frac{p+k}{D}C = -\frac{C_{t=0}}{D} - \frac{J}{pD} - \frac{J_* e^{-\beta z}}{pD} \tag{B.20}$$

The last equation is an ordinary differential equation, and can be solved for $\overline{C}$ using the procedure outlined under equations B.1 through B.14. The boundary conditions for a semi-infinite column at $t > 0$ are

$$\text{at} \quad z = 0: \quad \overline{C} = \frac{C_{z=0}}{p} \tag{B.21}$$

$$\text{at} \quad z \to \infty: \quad \frac{d\overline{C}}{dz} = 0 \tag{B.22}$$

The latter two boundary conditions are the Laplace transformations of C and dC/dz in equations B.11 and B.12. The initial conditions are

$$\text{at} \quad t = 0, \, z \geqslant 0: \quad C = C_{t=0}$$

The solution of equation B.20 is

$$\overline{C} = \frac{C_{t=0}}{p+k} + \frac{J}{p(p+k)} + \frac{J_* e^{-\beta z}}{p(p-a_2)} + \left(\frac{C_{z=0}}{p} - \frac{C_{t=0}}{p+k}\right.$$

$$\left. - \frac{J}{p(p+k)} - \frac{J_*}{p(p-a_2)}\right)e^{r_2 z} \tag{B.23}$$

The solution contains three constants, r_2, a_1, and a_2, that include the constant coefficients of the differential equation. r_2 is the root of the auxiliary equation, from the left-hand side of equation B.20,

$$r_2 = \frac{U}{2D} - \left(\frac{U^2}{4D^2} + \frac{p+k}{D}\right)^{1/2}$$

$$= \frac{U}{2D} - \left(\frac{p+a_1}{D}\right)^{1/2} \tag{B.24}$$

and the constant a_1 is defined as

$$a_1 = k + \frac{U^2}{4D} \tag{B.25}$$

and the constant a_2 is

$$a_2 = D\beta^2 + U\beta - k \tag{B.26}$$

To obtain C as an explicit function of t and z, all the inverse Laplace transformations of the terms on the right-hand side of equation B.23 must be found. Examples for two types of the terms are given below.

The inverse of $J/p(p+k)$. This term can be written as an algebraic sum of two fractions

$$\frac{J}{p(p+k)} = \frac{J}{k}\left(\frac{1}{p} - \frac{1}{p+k}\right)$$

The inverse Laplace transformation is

$$\mathcal{L}^{-1}\left\{\frac{J}{p(p+k)}\right\} = \frac{J}{k}\mathcal{L}^{-1}\left\{\frac{1}{p}\right\} - \frac{J}{k}\mathcal{L}^{-1}\left\{\frac{1}{p+k}\right\}$$

$$= \frac{J}{k} - \frac{J}{k}e^{-kt}$$

In the preceding, use was made of the following rule of the Laplace transformation: if the Laplace transformed function is $f(p+k)$, where k is a constant, then its inverse is a product of e^{-kt} and the inverse of $f(p)$,

$$\mathcal{L}^{-1}\{f(p+k)\} = e^{-kt}\mathcal{L}^{-1}\{f(p)\} \tag{B.27}$$

In the example given above, $1/(p+k)$ is the function $f(p+k)$, and $1/p$ is the function $f(p)$.

The inverse of $e^{r_2 z}/p(p-a_2)$. The last term on the right-hand side of equation B.23 can be written in the following form, suitable for taking the inverse Laplace transformations:

$$\frac{J_*e^{r_2 z}}{p(p-a_2)} = \frac{J_*e^{zU/2D}}{a_2}\left(\frac{e^{-z[(p+a_1)/D]^{1/2}}}{p-a_2} - \frac{e^{-z[(p+a_1)/D]^{1/2}}}{p}\right) \tag{B.28}$$

By the rule given in equation B.27, the inverse Laplace transformation of the term containing p in the denominator is

$$\mathcal{L}^{-1}\left\{\frac{e^{-z[(p+a_1)/D]^{1/2}}}{p}\right\} = e^{-a_1 t}\left(\mathcal{L}^{-1}\left\{\frac{e^{-z[p/D]^{1/2}}}{p-a_1}\right\}\right)$$

$$= e^{-a_1 t}\left[e^{a_1 t}\Psi(a_1)\right] \qquad (B.29)$$

The last product in brackets $[e^{a_1 t}\Psi(a_1)]$ is the inverse of the Laplace transformation enclosed in large braces. The transformation is given in Carslaw and Jaeger (1959, p. 495, transform 19). The function $\Psi(a_1)$ contains exponentials and error function terms,

$$\Psi(a_1) = \frac{1}{2}\left[e^{-z\sqrt{a_1/D}}\,\text{erfc}\left(\frac{z}{2(Dt)^{1/2}} - (a_1 t)^{1/2}\right)\right.$$

$$\left. + e^{z\sqrt{a_1/D}}\,\text{erfc}\left(\frac{z}{2(Dt)^{1/2}} + (a_1 t)^{1/2}\right)\right] \qquad (B.30)$$

The inverse of the Laplace transformations for all other terms in equation B.23 can be written in a manner similar to equation B.29. The final result, after the individual terms have been rearranged and assembled, is

$$C = \frac{J}{k} - \frac{J_* e^{-\beta z}}{a_2} + \left(C_{z=0} - \frac{J}{k} + \frac{J_*}{a_2}\right)e^{zU/2D}\Psi(a_1)$$

$$+ e^{-kt}\left(C_{t=0} - \frac{J}{k}\right)\left[1 - e^{zU/2D}\Psi(a_1 - k)\right]$$

$$+ \frac{J_*}{a_2}e^{a_2 t}\left[e^{-\beta z} - e^{zU/2D}\Psi(a_1 + a_2)\right] \qquad (B.31)$$

where the constants a_1 and a_2 are defined in B.25 and B.26. Although equation B.31 is not appealing to the eye, the physical significance of the individual terms is not difficult to perceive with the aid of the limiting values of the function $\Psi(a)$ in equation B.30. The limits of $\Psi(a)$ follow from the limiting values of erfc$x$ at $+\infty$ and $-\infty$, as defined in Appendix A, and they are

at $t \to \infty$: $\quad \Psi(a) = e^{-z(a/D)^{1/2}}$ $\qquad$ at $z \to \infty$: $\quad \Psi(a) = 0$

at $t = 0$: $\quad \Psi(a) = 0$ $\qquad\qquad\qquad$ at $z = 0$: $\quad \Psi(a) = 1$

Initially, at time $t=0$, concentration is uniform throughout the column $z \geqslant 0$, and is $C = C_{t=0}$. In response to a newly imposed boundary concentration $C_{z=0}$, concentration against distance profiles at different times can be computed using equation B.31. After a long time, as t increases indefinitely, the terms on the second and third line of the equation cancel. The term premultiplied by e^{-kt} tends to zero if k is positive and $t \to \infty$, for obvious reasons. The term on the third line, premultiplied by $e^{a_2 t}$, also tends to zero as shown below, even if the constant a_2 is positive. When t increases, the function $\Psi(a)$ tends to its limit faster than the exponential $e^{a_2 t}$, because of the rapid approach of erfc to its limit. The sum $a_1 + a_2$ in the term $\Psi(a_1 + a_2)$ is

$$a_1 + a_2 = \frac{(D\beta + U/2)^2}{D}$$

The product of $e^{a_2 t}$ and the difference in brackets in B.31 tends to zero when $t \to \infty$ as follows:

$$e^{a_2 t}(e^{-\beta z} - e^{zU/2D}e^{-z[(a_1 + a_2)/D]^{1/2}})$$

$$= e^{a_2 t}(e^{-\beta z} - e^{zU/2D}e^{-\beta z - zU/2D}) = 0$$

On the first line of equation B.31, the product $e^{zU/2D}\Psi(a_1)$ tends to the following limit as time increases indefinitely:

$$e^{zU/2D}\Psi(a_1) = e^{zU/2D - z[(a_1/D)]^{1/2}}$$

$$= \exp\left[\frac{zU}{2D} - z\left(\frac{U^2}{4D^2} + \frac{k}{D} \right)^{1/2} \right]$$

Thus finally equation B.31 reduces to the form

$$C = \frac{J}{k} - \frac{J_* e^{-\beta z}}{a_2} + \left(C_{z=0} - \frac{J}{k} + \frac{J_*}{a_2} \right) \exp\left[\frac{zU}{2D} - z\left(\frac{U^2}{4D^2} + \frac{k}{D} \right)^{1/2} \right]$$

The latter is identical to equation B.14 for the steady state.

APPENDIX C

Settling Velocities of Particles

The net gravitational force F_G acting on a particle of volume v_s and density ρ_s immersed in a fluid of density ρ is

$$F_G = g v_s (\rho_s - \rho) \qquad (\text{g cm sec}^{-2}) \qquad \text{(C.1)}$$

where g is the acceleration due to the force of gravity. A settling particle experiences a resisting or drag force F_D, and when the gravitational and drag forces are equal, the particle settles with a constant velocity. The constant or terminal settling velocity U_s can be derived from the condition of the equality of the forces,

$$F_G = F_D \qquad \text{(C.2)}$$

For particles of different shapes, explicit equations of the settling velocities U_s are derived from the equality of the forces F_G and F_D in the subsequent sections of this appendix. All equations apply to the laminar flow regime where the Stokes law of settling holds. Monograph-level treatment of the theory of drag forces acting on settling particles or on stationary frameworks of packed beds is given by Lamb (1945) and Happel and Brenner (1973). Mathematical relationships between the drag forces and certain shapes of settling particles have also been summarized in the books by Allen (1968) and Taylor (1968). Additional references to the literature are given under the specific cases discussed in this appendix.

C.1 SOLID SPHERE

The net gravitational force acting on a sphere is

$$F_G = \tfrac{4}{3} \pi r^3 g (\rho_s - \rho) \qquad (\text{dyn}) \qquad \text{(C.3)}$$

The Stokes drag on a sphere is

$$F_D = 6\pi\eta r U_s \quad \text{(dyn)} \qquad \text{(C.4)}$$

where η is the viscosity of the fluid, r is the sphere radius, and U_s is the settling velocity. From the equality of the forces $F_G = F_D$, the settling velocity is

$$U_s = \frac{2g(\rho_s - \rho)}{9\eta} r^2$$

$$U_s = Br^2 \quad (\text{cm sec}^{-1}) \qquad \text{(C.5)}$$

The parameter B is defined as

$$B = \frac{2g(\rho_s - \rho)}{9\eta} \qquad (\text{cm}^{-1} \text{ sec}^{-1}) \qquad \text{(C.6)}$$

and will appear in the equations for the settling velocities of other particles. B is independent of the particle size and shape, but it depends on the nature of the solid and fluid medium, through its dependence on the density and viscosity.

The original derivation of the drag force and velocity, as given in equations C.4 and C.5, was worked out by G. G. Stokes (1851) for the case of a pendulum swinging slowly in a viscous fluid and, as a generalization, for flow past a sphere. Stokes emphasized that the equations apply to the conditions of slow motion, when the resisting force can be approximated to a form that depends on the first power of velocity only. For faster motion and flow, the resisting force becomes proportional to the second power of the velocity. Referring to the earlier experiments on disc-shaped pendulums, carried out by the French physicist C. A. de Coulomb, Stokes had noticed that some of the reported velocites could be explained by drag forces falling in the region between the first and second power-dependence. In present terminology, this is the region between the Stokes and impact-law settling (see Figure 6.1).

Bigger or heavier spheres settle sufficiently fast to generate turbulence in their wake such that the drag force departs from the Stokes equation C.4. The sum of the Stokes and impact drag is

$$F_D = 6\pi\eta r U_s + \rho\pi r^2 U_s^2 \qquad \text{(C.7)}$$

From the equality of F_G to F_D, the settling velocity is (Rubey, 1933)

$$U_s = \left(6\nu Br + \frac{9\nu^2}{r^2} \right)^{1/2} - \frac{3\nu}{r}$$

$$= \frac{3\nu}{r} \left[\left(1 + \frac{2Br^3}{3\nu} \right)^{1/2} - 1 \right] \tag{C.8}$$

where ν is the kinematic viscosity of the fluid: $\nu = \eta/\rho$ cm^2 sec^{-1}. The settling velocities of large particles increase as the power $r^{1/2}$, whereas in the region of the Stokes settling the velocity is proportional to the power r^2. For small r, the settling velocity in equation C.8 merges with the Stokes settling law. Rubey's (1933) experiments on the impact-law settling included measurements of the settling velocities of quartz (SiO_2) and galena (PbS). A more recent experimental study of the settling of quartz particles produced results that fell between the impact law and Stokes law, equations C.8 and C.5 (Gibbs et al., 1971).

For very small particles in a gas, when the linear dimensions of particles are comparable to the free path of the gas molecules, a phenomenon known as slippage occurs: the particles fall faster in the space between the gas molecules than predicted by the Stokes law, which treats the fluid as a continuum. The equation for the faster settling velocity of small particles in air includes the Cunningham correction term to the Stokes law,

$$U_s = \left(1 + \frac{a\lambda}{2r} \right) Br^2 \tag{C.9}$$

where a is a constant parameter and λ is the free path of the gas molecules. The dimensionless quotient $\lambda/2r$ is called the Knudsen number. When the particle radius is sufficiently large in comparison to the molecular free path, the Cunningham correction factor $(1 + a\lambda/2r)$ tends to unity and the settling velocity equation reduces to the Stokes-law form. The following values of the Cunningham correction factor for spherical particles in air at 20–21°C were given by Lapple (1951) and Davies (1973): 2.88 ($r = 0.05$ μm), 1.68 ($r = 0.125$ μm), 1.33 ($r = 0.25$ μm), 1.16 ($r = 0.5$ μm), and 1.02 ($r = 5$ μm). Thus for particles of radius between 0.25 and 5 μm, their settling velocities in air are between 2 and 30% higher than give by the Stokes-law equation C.5.

The settling velocities of spherical particles in water and air are shown graphically in Figure 6.1.

C.2 HOLLOW SPHERE

A thin-walled sphere, the interior volume of which is filled with fluid, will be called a hollow sphere. When the filling fluid is the same as the exterior environment through which the hollow sphere settles, then the bulk density of the hollow sphere ρ_b depends on the wall thickness and the densities of the wall material and the fluid as

$$\rho_b = \rho_s + (\rho - \rho_s)\left(\frac{R_i}{R}\right)^3 \qquad (\text{g cm}^{-3})$$

where ρ_s is the solid material density, ρ is the fluid density, R_i is the interior wall radius, and R is the outer wall radius. If the wall thickness ΔR is small in comparison to the sphere radius, then it follows from the relationship between the radii and the wall thickness

$$R_i = R - \Delta R$$

that the ratio R_i/R is, approximately,

$$\left(\frac{R_i}{R}\right)^3 = 1 - \frac{3\Delta R}{R}$$

The difference $\rho_b - \rho$ between the densities of the particle and fluid appears in the definition of the parameter B in equation C.6. This difference for a hollow sphere is

$$\rho_b - \rho = (\rho_s - \rho) \cdot \frac{3\Delta R}{R}$$

Thus the settling velocity of a hollow sphere, of radius R and wall thickness ΔR, when the interior of the sphere is filled with the external fluid, is

$$U_s = \frac{3\Delta R}{R} BR^2 \qquad (C.10)$$

If, for example, the wall thickness is 10% of the sphere radius ($\Delta R/R \simeq 0.1$), then the settling velocity of such a sphere is only 0.3 of the settling velocity of a solid sphere.

In comparing the settling velocities of particles of different shapes, it is convenient to refer to the settling velocity of a solid sphere, the volume of which is equal to the other particle. Such a sphere is called the *equivalent sphere*, defined in equation 6.7. The volume of the wall material of a hollow sphere of radius R is equal to the volume of a solid sphere of radius r according to the following:

$$\tfrac{4}{3}\pi\left[R^3-(R-\Delta R)^3\right]=\tfrac{4}{3}\pi r^3$$

From this, the relationship between the radii R and r, when the wall of the hollow sphere is thin such that $\Delta R/R \ll 1$, is

$$R=\left(\frac{3\Delta R}{R}\right)^{-1/3}r \qquad\qquad (C.11)$$

Radius r is the radius of the equivalent sphere. Using equation C.11 in C.10, the settling velocity of the hollow sphere is expressed in terms of the radius of equivalent sphere,

$$U_s=\left(\frac{3\Delta R}{R}\right)^{1/3}Br^2 \qquad (\text{cm sec}^{-1}) \qquad\qquad (C.12)$$

As long as the wall thickness changes in proportion to the hollow sphere diameter, the ratio $\Delta R/R$ is a constant, and the hollow sphere settles according to the Stokes law, as shown by equations C.10 and C.12. Graphically, the settling velocity of a hollow sphere is shown in Figure 6.2. Note that a hollow sphere of diameter equal to a solid sphere is much lighter and therefore settles slower.

If the wall thickness of a hollow sphere remains constant while its diameter increases, then the settling velocity of such a particle is, from equation C.10,

$$U_s=3\Delta RBR \qquad (\text{cm sec}^{-1}) \qquad\qquad (C.13)$$

This shows that the settling velocity is a function of the first power of radius R, rather than a function of R^2 as in the Stokes law. The settling velocities of certain planktonic organisms in sea water, discussed in Section 6.2.4, are functions of a power of R that lies between 1 and 2. Such settling velocities fall in the range between the hollow sphere and solid sphere Stokes equations.

The settling velocity of a thin-walled sphere ($\Delta R/R=0.1$) is plotted in Figure 6.2a.

C.3 POLYHEDRAL PARTICLES

Polyhedra that are spherically isotropic and of uniform density can be placed in any orientation in a liquid, and they should theoretically settle without rotation or change in their orientation. Such spherically isotropic bodies include, obviously, the sphere and the simple regular polyhedra (all faces equal): cube, octahedron, tetrahedron, dodecahedron (surface made of twelve pentagons), and icosahedron (surface made of twenty equilateral triangles). Certain combinations of regular polyhedra also fall in the class of spherically isotropic bodies, for example, the cube-octahedron, or the cube with corners that are symmetrically cut off at an angle of 45°. (The rhombohedron is not a spherically isotropic body, despite the fact that all six of its faces are equal rhombs.) For spherically isotropic particles, an empirical equation that relates the settling velocity to the particle linear dimensions is the Stokes-law equation for a sphere with a shape factor α,

$$U_s = \alpha B r^2 \quad (\text{cm sec}^{-1}) \tag{C.14}$$

where r is the radius of the sphere whose volume is equal to the volume of the polyhedral particle (that is, the radius of the equivalent sphere). The shape factor α for regular polyhedra is given by the equation (Pettyjohn and Christiansen, 1948; Heiss and Coull, 1952)

$$\alpha = 1.00 + 0.843 \log R_s \tag{C.15}$$

where R_s is the ratio

$$R_s = \frac{\text{surface area of equivalent sphere}}{\text{surface area of particle}} \tag{C.16}$$

The term sphericity is used for the quotient R_s. For some spherically isotropic polyhedra, the sphericity R_s and shape factor α are listed in Table 6.3. The reduction in the settling velocity of a tetrahedron amounts to 15%, in comparison to the settling velocity of an equivalent sphere.

C.4 PROLATE AND OBLATE ELLIPSOIDS

The shape of an ellipsoid of revolution with the equatorial radius (semiaxis) a and polar semiaxis c can be characterized by a ratio parameter p,

$$p = \frac{c}{a} \tag{C.17}$$

For a sphere, the semiaxes are equal and $p=1$. In a prolate ellipsoid, elongation is in the polar direction and, therefore, $c>a$ and $p>1$. In an oblate ellipsoid, the polar axis is shorter than the equatorial diameter, and we have $c<a$ and $p<1$.

The settling velocity of ellipsoids of revolution depends on their orientation. A prolate ellipsoid settling in the direction parallel to its polar axis will encounter less resistance and will settle faster than an oblate ellipsoid of the same mass and the same orientation. For ellipsoids that deviate slightly from the shape of the sphere, the resisting drag forces for the two settling orientations are (Brenner, 1964):

Settling direction parallel to polar axis ($\|c$)

$$F_D = 6\pi\eta U_s a(0.8-0.2p) \tag{C.18}$$

Settling direction perpendicular to polar axis ($\perp c$)

$$F_D = 6\pi\eta U_s a(0.6-0.4p) \tag{C.19}$$

The volume of an ellipsoid of revolution is $4\pi a^2 c/3$ or, using the notation $p=c/a$, the volume is $4\pi a^3 p/3$. The net gravitational force acting on an ellipsoid is

$$F_G = \tfrac{4}{3}\pi a^3 p g(\rho_s-\rho) \tag{C.20}$$

Equating F_G to each of the two relationships for the drag force F_D, the settling velocities can be expressed in terms of the ellipsoid shape p, the Stokes settling parameter B, and the equatorial radius a,

$$\|c: \qquad U_s = \frac{5p}{4+p}\, Ba^2 \tag{C.21}$$

$$\perp c: \qquad U_s = \frac{2.5p}{1.5+p}\, Ba^2 \tag{C.22}$$

The radius r of the equivalent sphere is related to the ellipsoid length/width ratio p and the equatorial radius a through

$$a = p^{-1/3}r \tag{C.23}$$

Substitution of the latter equation for a in the equations for U_s gives the settling velocities of ellipsoids in terms of the equivalent sphere radius r,

$$\|c: \qquad U_s = \frac{5p^{1/3}}{4+p} Br^2 \qquad\qquad \text{(C.24)}$$

$$\perp c: \qquad U_s = \frac{2.5p^{1/3}}{1.5+p} Br^2 \qquad\qquad \text{(C.25)}$$

The velocity of a prolate ellipsoid settling parallel to its long axis ($p = 1.5$) is slightly higher than the settling velocity of the equivalent sphere, as shown in Figure 6.2a. An oblate ellipsoid settling parallel to its shorter polar axis ($p = 0.75$) settles slower than the equivalent sphere. However, in the range of the length/width ratio p between 0.7 and 1.5, the settling velocities given by equations C.24 and C.25 differ very little from the settling velocity of the sphere.

C.5 NEEDLES

A prolate ellipsoid with a large length/diameter ratio approaches the shape of a needle. For settling in the direction either parallel or perpendicular to the long axis ($2c$) of the needle, the drag forces are (Happel and Brenner, 1973)

$$\|c: \qquad F_D = \frac{4\pi\eta U_s c}{\ln 2p - 0.5} \qquad\qquad \text{(C.26)}$$

$$\perp c: \qquad F_D = \frac{8\pi\eta U_s c}{\ln 2p + 0.5} \qquad\qquad \text{(C.27)}$$

where p is the ratio of the polar to equatorial semiaxes, $p = c/a > 1$. The volume of the needle and the net gravitational force acting on it are as given in equation C.20. The settling velocities U_s, written in terms of the needle half-length c and length/width ratio p, are

$$\|c: \qquad U_s = \frac{3(\ln 2p - 0.5)}{2p^2} Bc^2 \qquad\qquad \text{(C.28)}$$

$$\perp c: \qquad U_s = \frac{3(\ln 2p + 0.5)}{4p^2} Bc^2 \qquad\qquad \text{(C.29)}$$

The needle half-length c and the radius r of the equivalent sphere are related through

$$c = p^{2/3}r \tag{C.30}$$

Using the latter equation in the equations for U_s, the settling velocities written as functions of the equivalent sphere radius r become

$$\|c: \qquad U_s = \frac{3(\ln 2p - 0.5)}{2p^{2/3}} Br^2 \tag{C.31}$$

$$\perp c: \qquad U_s = \frac{3(\ln 2p + 0.5)}{4p^{2/3}} Br^2 \tag{C.32}$$

The resistance drag on a settling needle is greater than the drag on the sphere, and the needle settles slower than the sphere of equal volume, as shown in Figure 6.2a (length/width ratio of the needle is $p = 10$). A needle, the length of which is equal to the diameter of a sphere ($2c = 2r$), settles much slower than the sphere (Figure 6.2b), because the volume and mass of the needle are much smaller than those of the sphere.

C.6 DISCS

The shape of a disc can be approximated by a flattened oblate ellipsoid in which the ratio of the polar to equatorial semiaxes c/a is small. For an oblate ellipsoid of revolution settling face down (direction parallel to the shorter polar axis c) or edge down (settling direction perpendicular to c), the resisting drag forces are (Lamb, 1945; Happel and Brenner, 1973)

$$\|c: \qquad F_D = \frac{16\pi\eta U_s a}{2p/(1-p^2) + \left[2(1-2p^2)/(1-p^2)^{3/2}\right]\tan^{-1}\left[(1-p^2)^{1/2}/p\right]}$$

$$= \frac{16\pi\eta U_s a}{f_1(p)} \tag{C.33}$$

$$\perp c: \qquad F_D = \frac{16\pi\eta U_s a}{-p/(1-p^2) - \left[(2p^2-3)/(1-p^2)^{3/2}\right]\sin^{-1}(1-p^2)^{1/2}}$$

$$= \frac{16\pi\eta U_s a}{f_2(p)} \tag{C.34}$$

where $f_1(p)$ and $f_2(p)$ are shorthand notations for the denominator expressions, a is the equatorial radius of the disc, and p is the polar/equatorial semiaxes ratio $p = c/a$. The limiting shape of an ideal disc is the case of $c \to 0$, and for such an infintely thin disc the drag force equations reduce to the following form:

$$\|c: \qquad F_D = 5.1\pi\eta U_s a \qquad\qquad\qquad (C.35)$$

$$\perp c: \qquad F_D = 3.4\pi\eta U_s a \qquad\qquad\qquad (C.36)$$

The downward directed force F_G is given in equation C.20. The settling velocities of the disc in two orientations, face down ($\|c$) and edge down ($\perp c$) are

$$\|c: \qquad U_S = \tfrac{3}{8}pf_1(p)Ba^2 \qquad\qquad\qquad (C.37)$$

$$\perp c: \qquad U_s = \tfrac{3}{8}pf_2(p)Ba^2 \qquad\qquad\qquad (C.38)$$

The relationship between the equatorial radius a of an ellipsoid of revolution and the radius r of the equivalent sphere is given in Section C.4. Using this relationship for a, the settling velocity of the disc of thickness-to-diameter ratio p is, in terms of the equivalent sphere radius r,

$$\|c: \qquad U_s = \tfrac{3}{8}p^{1/3}f_1(p)Br^2 \qquad\qquad\qquad (C.39)$$

$$\perp c: \qquad U_s = \tfrac{3}{8}p^{1/3}f_2(p)Br^2 \qquad\qquad\qquad (C.40)$$

The settling velocity of the disc settling face down is much smaller than the settling velocity of a sphere of equal volume (thickness-to-diameter ratio of the disc is $p = 0.1$, Figure 6.2a).

In a nonlaminar flow where the Reynolds number $Re > 0.05$, the resisting drag force on the disc settling face down (parallel to the c axis) has been given in the form of the following equation (Happel and Brenner, 1973)

$$F_D = 16U_s a\left(1 + \frac{Re}{2} + \frac{2\,Re^2}{5\pi^2}\ln\frac{Re}{2}\right) \qquad\qquad (C.41)$$

where a is the disc radius and the Reynolds number Re is defined in equation 6.6.

C.6 CIRCULAR CYLINDERS

For a straight circular cylinder, approximate relationships between the drag force and the physical dimensions of the cylinder are similar to the drag equations for settling ellipsoids. The equations given below apply to long cylinders, with large height-to-base-radius ratio h/a. The drag force acting on a cylinder settling in the direction either parallel or perpendicular to its height h is (Happel and Brenner, 1973)

$$\|h: \qquad F_D \simeq \frac{2\pi\eta U_s h}{\ln 2p - 0.72} \qquad (C.42)$$

$$\perp h: \qquad F_D \simeq \frac{4\pi\eta U_s h}{\ln 2p + 0.5} \qquad (C.43)$$

where the notation $2p = h/a$ is used. In comparison to a prolate ellipsoid of diameter and length equal to those of the cylinder (that is, $h = 2c$), the resistance to settling in the length direction is slightly greater for the straight circular cylinder (compare equation C.42 and C.26). For the settling direction perpendicular to the long axis, the approximate equation C.43 is identical to the equation for a long prolate ellipsoid, equation C.27. Equations for the drag forces acting on infinitely long circular and elliptical cylinders were given by Lamb (1945, pp. 616–617). The equations have been used to estimate settling velocities of cylinder-like particles in natural waters (Munk and Riley, 1952; Lerman et al., 1974). However, when the equations for infinitely long cylinders are applied to cylinders of finite length, the computed settling velocities are considerably higher than the settling velocity of the sphere of equal volume. The drag force equations given above lead to settling velocities that are lower than the velocity of the equivalent sphere.

The net gravitational force acting on a cylinder is

$$F_G = \pi r^2 h g (\rho_s - \rho) \qquad (C.44)$$

and from the equality $F_G = F_D$, the settling velocity equations are

$$\|h: \qquad U_s = \tfrac{9}{4}(\ln 2p - 0.72) Ba^2 \qquad (C.45)$$

$$\perp h: \qquad U_s = \tfrac{9}{8}(\ln 2p + 0.5) Ba^2 \qquad (C.46)$$

The relationship between the cylinder base radius a, height-to-radius ratio

$h/a = 2p$, and the radius r of the equivalent sphere is

$$a = \left(\frac{2}{3p} \right)^{1/3} r \qquad (C.47)$$

In terms of the equivalent sphere radius r, the settling velocities of the cylinders settling in the h-parallel and h-perpendicular directions are

$$\|h: \qquad U_s = \frac{1.72(\ln 2p - 0.72)}{p^{2/3}} Br^2 \qquad (C.48)$$

$$\perp h: \qquad U_s = \frac{0.86(\ln 2p + 0.5)}{p^{2/3}} Br^2 \qquad (C.49)$$

The settling velocities as a function of the equivalent sphere radius r and as a function of the cylinder length h are plotted in Figure 6.2. The cylinder has the height/diameter ratio $p = 10$.

C.7   HEMISPHERICAL CAP

A hemispherical cap (also called the calotte), oriented with the plane of its equatorial circle facing the flow or settling with the cavity down, experiences the drag force (Payne and Pell, 1960)

$$F_D = 5.58 \pi \eta U_s a \qquad (C.50)$$

where a is the radius of the hemispherical cap. To compare the hemispherical cap with a particle of similar shape, the wall of the cap may be taken as some small fraction of its radius. Then the volume of the solid material in the cap is confined between the outer surface of radius a and inner surface of radius a_i. As long as the wall thickness to radius ratio $\Delta a/a$ is small, the volume of the wall material is

$$\tfrac{2}{3}\pi \left[a^3 - (a - \Delta a)^3 \right] \simeq 2\pi a^3 \cdot \frac{\Delta a}{a} \qquad (C.51)$$

The relationship between the radius of the cap and the equivalent sphere radius r is

$$a = \left(\frac{2a}{3\Delta a} \right)^{1/3} r \qquad (C.52)$$

The settling velocity of a thin-walled cap, in terms of its radius a and wall thickness to radius ratio $\Delta a/a$, is

$$U_s = 1.61 \frac{\Delta a}{a} Ba^2 \qquad (C.53)$$

The settling velocity, in terms of the radius r of the equivalent sphere, is, by substitution from equation C.52 into C.53,

$$U_s = 1.23 \left(\frac{\Delta a}{a} \right)^{1/3} Br^2 \qquad (C.54)$$

The settling velocities are shown graphically in Figure 6.2 for a cap, the wall thickness of which is $\Delta a/a = 0.1$.

C.8   CIRCULAR RING

The resisting drag force on a circular ring and half ring were derived theoretically by Tchen (1954: Happel and Brenner, 1973). The ring and half ring were "made" of long ellipsoids of revolution bent along a circle or half circle. The cross section of such a ring is not uniform but varies from a maximum, the equatorial diameter of the ellipsoid, to zero at the ellipsoid poles, which touch one another when the ring is closed. In a long and thin, needle-shaped ellipsoid, the maximum diameter is small in comparison to its length (equal to the circumference of the ring) and the ring diameter. The ring can settle such that it lies in a horizontal plane (hole down) or in a vertical plane (edge down). For the ring and half ring, the theoretical equations for the drag forces show that the drag is not affected by the different orientations in the vertical plane, such as when the tips of the half ring (the poles of the needle bent into a ring or half ring) are oriented up, sideways, or down.

The ratio of the ring circumference to its maximum thickness is the same as the ratio of the polar to equatorial semiaxes in a prolate ellipsoid c/a. The downward directed force F_G is as given in equation C.20. The drag forces and the settling velocities, written in terms of the half circumference of the ring c and the ratio parameter $p = c/a$, are:
Half ring, lying in horizontal plane

$$F_D = \frac{8\pi\eta U_s c}{\ln 2p + 0.56} \qquad (C.55)$$

$$U_s = \frac{3(\ln 2p + 0.56)}{4p^2} Bc^2 \qquad (C.56)$$

Half ring, oriented vertically

$$F_D = \frac{6\pi\eta U_s c}{\ln 2p - 0.68} \tag{C.57}$$

$$U_s = \frac{\ln 2p - 0.68}{p^2} Bc^2 \tag{C.58}$$

Ring, lying in horizontal plane

$$F_D = \frac{8\pi\eta U_s c}{\ln 2p + 0.75} \tag{C.59}$$

$$U_s = \frac{3(\ln 2p + 0.75)}{4p^2} Bc^2 \tag{C.60}$$

Ring, oriented vertically

$$F_D = \frac{6\pi\eta U_s c}{\ln 2p - 2.09} \tag{C.61}$$

$$U_s = \frac{\ln 2p - 2.09}{p^2} Bc^2 \tag{C.62}$$

The ring settling in a horizontal orientation experiences a drag force that is slightly smaller than the drag force on a long ellipsoid (needle) settling in the same orientation, equations C.27 and C.59. As the volumes of the ring and half ring made of bent ellipsoids are identical to the ellipsoid volumes, derivation of the settling velocity equations given above follows in a straightforward fashion from the equality of the gravitational and Stokes drag forces $F_G = F_D$. In terms of the radius r of the equivalent sphere, the relationship $c = p^{2/3}r$ from equation C.30 can be substituted for c in each of the four equations for the settling velocity of the half ring and ring, giving the settling velocity as a function of r^2. For the ring settling edge down (vertical orientation), the settling velocity is

$$U_s = \frac{\ln 2p - 2.09}{p^{2/3}} Br^2 \tag{C.63}$$

and for the ring settling in horizontal orientation, the settling velocity is

$$U_s = \frac{3(\ln 2p + 0.75)}{4p^{2/3}} Br^2 \tag{C.64}$$

The circumference of the ring bent out of a long prolate ellipsoid is approximately equal to the polar axis of the ellipsoid $2c$. The diameter of the ring R is

$$R \simeq \frac{c}{\pi}$$

Thus the settling velocity of the ring, the circumference of which lies in a horizontal plane, can be expressed in terms of the ring radius R and the circumference to maximum thickness ratio p, from equation C.60, in the form

$$U_s = \frac{3(\ln 2p + 0.75)\pi^2}{4p^2} BR^2 \tag{C.65}$$

The settling velocities of the ring are shown in Figure 6.2 (the ratio of circumference to maximum thickness is $p = 15$).

Physical Constants and Units

Note. The International System of Units (SI) was amended in 1967 by the Thirteenth General Conference on Weights and Measures, and it supercedes other systems of physical units, including the cgs (centimeter-gram-second) system. The SI is based on *meter-kilogram-second* as the basic measures of length, mass, and time. Three additional basic measures are: *ampere*, for electric current; *kelvin*, for temperature; *candela*, for luminous intensity. All other units, such as pressure, force, work, power, electric charge, and quantity of heat, are derived from the six fundamental measures. Such units as atmosphere, bar, and calorie have been dropped from the SI. However, because these units of pressure and thermal energy, and the unit of time year have been, and still are used in the literature, they are preserved in the usage throughout this book. Both the cgs units and other units derived from them appear in the text.

Constants

Avogadro's number	N	6.022×10^{23}	mol^{-1}
Boltzmann's constant	k	1.381×10^{-16}	$erg\ deg^{-1}$
		1.381×10^{-23}	$joule\ deg^{-1}$
		0.330×10^{-23}	$cal\ deg^{-1}$
Faraday (average)	$\mathscr{F}$	9.65×10^{4}	$coulomb\ equivalent^{-1}$
			or $volt\ sec\ ohm^{-1}\ equivalent^{-1}$
			or $joule\ volt^{-1}\ equivalent^{-1}$
		2.306×10^{4}	$cal\ volt^{-1}\ equivalent^{-1}$
		9.65×10^{3}	electromagnetic cgs units per equivalent

Gas constant $R = kN$ 8.314×10^7 erg deg^{-1} mol^{-1}

                             8.314   joule deg^{-1} mol^{-1}

                             1.987   cal deg^{-1} mol^{-1}

                            82.057   atm cm^3 deg^{-1} mol^{-1}

                            0.0821   atm l deg^{-1} mol^{-1}

                            83.144   bar cm^3 deg^{-1} mol^{-1}

                            5.191×10^{19} eV deg^{-1} mol^{-1}

Planck's
constant h 6.626×10^{-27} erg sec

                            4.136×10^{-15} eV sec

Temperature
of ice point   0   deg C

                    273.15   deg K

Units

1 atmosphere	1	atm
	1.0133	bar
	760	torr
	1.0333	at (absolute atmosphere)

1 bar	1×10^6	dyn cm^{-2} *or* erg cm^{-3} *or* g cm^{-1} sec^{-2}
	1×10^5	pascal *or* N m^{-2}
	0.9869	atm
	1×10^{-3}	kbar

1 calorie	1	cal
	1×10^{-3}	kcal
	4.184	joule *or* J
	4.184×10^7	erg
	41.84	bar cm^3
	2.612×10^{19}	eV

SI Units

1 newton	$1 \text{ N} = 1 \text{ kg m sec}^{-2}$
1 joule	$1 \text{ J} = 1 \text{ N m}$
1 watt	$1 \text{ W} = 1 \text{ J sec}^{-1}$
1 coulomb	1 A sec (ampere second)
1 volt	$1 \text{ volt} = 1 \text{ W A}^{-1}$
1 ohm	$1 \text{ ohm} = 1 \text{ volt A}^{-1}$

References

Aitchison, J. and J. A. C. Brown, 1963. *The Lognormal Distribution.* Cambridge University Press, Cambridge, England.

Allen, T., 1968. *Particle Size Measurement.* Chapman and Hall, London, England.

Andren, A. W. and R. C. Hariss, 1973. Methylmercury in estuarine sediments. *Nature,* **245,** 256–257.

Anikouchine, W. A., 1967. Dissolved chemical substances in compacting marine sediments, *J. Geophys. Res.,* **72,** 505–509.

Archie, G. E., 1942. The electrical resistivity log as an aid in determining some reservoir characteristics. *Am. Inst. Mech. Eng. Trans.,* **146,** 54–61.

Arrhenius, G., 1963. Pelagic sediments. In *The Sea,* Vol. 3., M. N. Hill, Ed. Wiley-Interscience, New York, pp. 655–727.

Aston, S. R. and E. K. Duursma, 1973. Concentration effects on ^{137}Cs, ^{65}Zn, ^{60}Co, and ^{106}Ru sorption by marine sediments, with geochemical implications. *Neth. J. Sea Res.,* **6,** 225–240.

Athy, L. F., 1930. Density, porosity, and compaction of sedimentary rocks. *Am. Assoc. Petrol. Geol. Bull.,* **14,** 1–24.

Auluck, F. C. and D. S. Kothari, 1965. Random fragmentation. *Z. Astrophys.,* **63,** 9–14.

Austin, L. G. and R. R. Klimpel, 1968. Statistics of random fracture. *Am. Inst. Mech. Eng. Trans.,* **241,** 219–224.

Bacastow, R. and C. D. Keeling, 1973. Changes from A.D. 1700 to 2070 as deduced from a geochemical model. In *Carbon and the Biosphere,* G. M. Woodwell and E. V. Pecan, Eds. U.S. Atomic Energy Commission, Washington, D.C., pp. 86–135.

Back, W. and B. B. Hanshaw, 1971. Rates of physical and chemical processes in a carbonate aquifer. *Adv. Chem. Ser.,* **106,** 77–93.

Baker, C. W., 1977. Mercury in surface waters of seas around the United Kingdom. *Nature (Phys. Sci.),* **270,** 230–232.

Bakker, J. P., 1954. Ueber den Einfluss von Klima, jungerer Sedimentation und Bodenprofilentwicklung auf die Savannen Nord-Surinams (Mittelguayana). *Erdkunde,* **8,** 89–112.

Barnes, R. O. and E. D. Goldberg, 1976. Methane production and consumption in anoxic marine sediments. *Geology,* **4,** 297–300.

Bé, A. W. H., J. W. Morse, and S. M. Harrison, 1975. Progressive dissolution and ultrastructural breakdown in planktonic foraminifera. In *Dissolution of Deep-Sea Carbonates,* W.V. Sliter, A. W. H. Bé, and W. H. Berger, Eds. Cushman Foundation for Foraminiferal Research Special Publication 13, Ithaca, N.Y., pp. 27–55.

Bear, J., 1972. *Dynamics of Fluids in Porous Media.* American Elsevier, New York.

Beard, D. C. and P. K. Weyl, 1973. Influence of texture on porosity and permeability of unconsolidated sand. *Am. Assoc. Petrol. Geol. Bull.*, **57**, 349–369.

Beck, R. E. and J. S. Schultz, 1970. Hindered diffusion in microporous membranes with known pore geometry. *Science*, **170**, 1302–1305.

Benedek, G. B. and E. M. Purcell, 1954. Nuclear magnetic reasonance in liquids under high pressure. *J. Chem. Phys.*, **22**, 2003–2012.

Bennett, J. G., 1936. Broken coal. *J. Inst. Fuel*, **10**, 22–39.

Ben-Yaakov, S., E. Ruth, and I. R. Kaplan, 1974. Carbonate compensation in depth: Relation to carbonate solubility in ocean waters. *Science*, **184**, 982–984.

Berger, W. H., 1967. Foraminiferal ooze: Solution at depths. *Science*, **156**, 383–385.

Berger, W. H., 1970. Planktonic foraminifera: Selective solution and the lysocline. *Mar. Geol.*, **8**, 111–138.

Berger, W. H., 1971. Sedimentation of planktonic foraminifera. *Mar. Geol.*, **11**, 325–358.

Berger, W. H., 1974. Deep-sea sedimentation. In *The Geology of Continental Margins*, C. A. Burk and C. L. Drake, Eds. Springer-Verlag, New York, pp. 213–241.

Berger, W. H., 1976. Biogenous deep-sea sediments: Production, preservation and interpretation. In *Chemical Oceanography*, Vol. 4, 2nd ed. J. P. Riley and R. Chester, Eds. Academic, London, England, pp. 265–387.

Berger, W. H. and U. von Rad, 1972. Cretaceous and cenozoic sediments from the Atlantic Ocean. In *Initial Reports of the Deep Sea Drilling Project*, Vol. 14. U.S. Government Printing Office, Washington, D.C., pp. 787–954.

Berner, R. A. , 1971. *Principles of Chemical Sedimentology*. McGraw-Hill, New York.

Berner, R. A., 1973. Phosphate removal from sea water by adsorption on volcanogenic ferric oxides. *Earth Plant. Sci. Lett.*, **18**, 77–86.

Berner, R. A., 1974. Kinetic models for the early diagenesis of nitrogen, sulfur, phosphorus, and silicon in anoxic marine sediments. In *The Sea*, Vol. 5, E. D. Goldberg, Ed. Wiley-Interscience, New York, pp. 427–450.

Berner, R. A., 1976. The solubility of calcite and aragonite in seawater at atmospheric pressure and 34.5‰ salinity. *Am. J. Sci.*, **276**, 713–730.

Berner, R. A. and G. R. Holdren, Jr., 1977. Mechanism of feldspar weathering: Some observational evidence. *Geology*, **5**, 369–372.

Bett, K. E. and J. B. Cappi, 1965. Effect of pressure on the viscosity of water. *Nature*, **207**, 620–621.

Biscaye, P. E. and S. L. Eittreim, 1974. Variations in benthic boundary layer phemonena: Nepheloid layer in the North American Basin. In *Suspended Solids in Water*, R. J. Gibbs, Ed. Plenum, New York, pp. 227–260.

Boerboom, A. J. H. and G. Kleyn, 1969. Diffusion coefficients of noble gases in water. *J. Chem. Phys.*, **50**, 1086–1087.

Bolin, B., 1970. The carbon cycle. *Sci. Am.*, **223**, 124–135.

Bolin, B., 1977. Changes of land biota and their importance for the carbon cycle. *Science*, **196**, 613–615.

Bolin, B. and W. Bischof, 1970. Variations of the carbon dioxide content of the atmosphere in the Northern Hemisphere. *Tellus*, **22**, 431–442.

Bolin, B. and E. Eriksson, 1959. Changes in the carbon dioxide content of the atmosphere and sea due to fossil fuel combustion. In *Rossby Memorial Volume*, B. Bolin, Ed. Rockefeller, New York, pp. 130–142.

Bolin, B. and H. Rodhe, 1973. A note on the concepts of age distribution and transit time in natural reservoirs. *Tellus*, **25**, 58–62.

Bolt, G. H. and M. G. M. Bruggenwert, 1976. *Soil Chemistry*, Elsevier, Amsterdam, The Netherlands.

Bowen, H. J. M., 1966. *Trace Elements in Biochemistry*, Academic, London, England.

Bradbury, J. P., 1975. Diatom stratigraphy and human settlement in Minnesota. *Geol. Soc. Am. Spec. Pap.*, **171**, 1–74.

Braitsch, O., 1971. *Salt Deposits, Their Origin and Composition*. Springer-Verlag, Berlin, Germany.

Bramlette, M. N., 1961. Pelagic sediments. In *Oceanography*, M. Sears, Ed. Am. Assoc. Adv. Sci. Publication 67, Washington, D.C., pp. 345–366.

Brenner, H., 1964. The Stokes resistance of a slightly deformed sphere. *Chem. Eng. Sci.*, **19**, 519–539.

Brewer, P. G. and D. W. Spencer, 1974. Distribution of some trace elements in Black Sea and their flux between dissolved and particulate phases. In *The Black Sea—Geology, Chemistry, and Biology*, Amer. Assoc. Petrol. Geol., Mem. 20, Tulsa, Ok, pp. 137–143.

Brewer, P. G., D. W. Spencer, P. E. Biscaye, A. Hanley, P. L. Sachs, C. L. Smith, S. Kadar, and J. Fredericks, 1976. The distribution of particulate matter in the Atlantic Ocean. *Earth Planet. Sci. Lett.*, **32**, 393–402.

Broecker, W. S. and S. Broecker, 1974. Carbonate dissolution on the western flank of the East Pacific Rise. In *Studies in Paleo-Oceanography*. Soc. Econ. Paleontologists and Mineralogists, Special Publication 20, Tulsa, Ok., pp. 44–57.

Broecker, W. S., Y.-H. Li, and T.-H. Peng, 1971. Carbon dioxide—man's unseen artifact. In *Impingement of Man on the Oceans*, D. W. Hood, Ed. Wiley-Interscience, New York, pp. 287–324.

Broecker, W. S. and T.-H. Peng, 1974. Gas exchange rates between air and sea. *Tellus*, **26**, 21–35.

Brun-Cottan, J.-C., 1976. Stokes settling and dissolution rate model for marine particles as a function of size distribution. *J. Geophys. Res.*, **81**, 1601–1606.

Bryant, W. R., A. P. Deflache, and P. K. Trabant, 1974. Consolidation of marine clays and carbonates. In *Deep Sea Sediments: Physical and Mechanical Properties*, Vol. 2, A. L. Interbitzen, Ed. Plenum, New York, pp. 209–244.

Burns, J. H. and M. A. Bredig, 1956. Transformation of calcite to aragonite by grinding. *J. Chem. Phys.*, **25**, 1281–1282.

Burton, J. D. and P. S. Liss, 1973. Processes of supply and removal of dissolved silicon in the oceans. *Geochim. Cosmochim. Acta*, **37**, 1761–1774.

Busenberg, E. and C. V. Clemency, 1976. The dissolution kinetics of feldspars at 25°C and 1 atm CO_2 partial pressure. *Geochim. Cosmochim. Acta*, **40**, 41–49.

Butler, J. C., G. M. Greene, and E. A. King, Jr., 1973. Grain size-frequency distributions and modal analyses of Apollo 16 fines. *Geochim. Cosmochim. Acta, Suppl. 4*, Proc. Fourth Lunar Sci. Conf., **1**, 267–278.

Butler, J. C. and E. A. King, Jr., 1974. Analyses of the grain size-frequency distributions of lunar fines. *Geochim. Cosmochim. Acta, Suppl. 5*, Proc. Fifth Lunar Sci. Conf., **1**, 829–841.

Byers, H. R., 1965. *Elements of Cloud Physics*. University of Chicago Press, Chicago.

Cadle, R. D., 1973. Particulate matter in the lower atmosphere. In *Chemistry of the Lower Atmosphere*, S. I. Rasool, Ed. Plenum, New York, pp. 69–120.

Cadle, R. D., 1975. *The Measurement of Airborne Particles*. Wiley-Interscience, New York.

Carder, K. L., G. F. Beardsley Jr., and H. Pak, 1971. Particle size distributions in the Eastern Equatorial Pacific. *J. Geophys. Res.*, **76**, 5070–5077.

Carnahan, B., H. A. Luther, and J. O. Wilkes, 1969. *Applied Numerical Methods*. Wiley, New York.

Carroll, D., 1970. *Rock Weathering*. Plenum, New York.

Carslaw, H. S. and J. C. Jaeger, 1959. *Conduction of Heat in Solids*, 2nd ed. Oxford University Press, London, England.

Cerrai, F., M. G. Mezzandri, and C. Triulzi, 1969. Sorption experiments of strontium, cesium, promethium, and europium on marine sediment samples. *Energ. Nucl.* (Milan), **16**, 378–385.

Chapman, C. R., 1974. Asteroid size distribution: Implications for the origin of stony-iron meteorites. *Geophys. Res. Lett.*, **1**, 341–344.

Charnley, T. and H. A. Skinner, 1951. Measurements of the vapour pressures of some alkylmercury halides. The latent heats of sublimation of methyl- and ethyl-mercury halides. *J. Chem. Soc.* (London), 1921–1924.

Chen, C.-T. and F. J. Millero, 1977. Precise equation of state of seawater for oceanic ranges of salinity, temperature and pressure. *Deep-Sea Res.* **24**, 365–369.

Churchman, G. J., R. N. Clayton, K. Sridhar, and M. L. Jackson, 1976. Oxygen isotopic composition of aerosol-size quartz in shales. *J. Geophys. Res.*, **81**, 381–386.

Clayton, R. N., R. W. Rex, J. K. Syers, and M. L. Jackson, 1972. Oxygen isotope abundance in quartz from Pacific pelagic sediments. *J. Geophys. Res.*, **77**, 3907–3915.

Clelland, D. W. and P. D. Ritchie, 1952. Nature and regeneration of the high solubility layer on siliceous dusts. *J. Appl. Chem.*, **2**, 42–48.

Cook, P. J., 1974. Geochemistry and diagenesis of interstitial fluids and associated calcareous oozes, Deep Sea Drilling Project, Leg 27, Site 262, Timor Trough. In *Initial Reports of the Deep Sea Drilling Project*, Vol. 27. U.S. Government Printing Office, Washington, D.C., pp. 463–480.

Correns, C. W., 1971. The experimental chemical weathering of silicates. *Clay Min. Bull.*, **4**, 249–265.

Correns, C. W. and W. von Engelhardt, 1939. Neue Untersuchungen über die Verwitterung des Kalifeldspates. *Chem. Erde*, **12**, 1–22.

Craig, H., 1969. Abyssal carbon and radiocarbon in the Pacific. *J. Geophys. Res.*, **74**, 5491–5506.

Craig, H., 1971. The deep metabolism: Oxygen consumption in abyssal ocean water. *J. Geophys. Res.*, **76**, 5078–5086.

Craig, H. and L. I. Gordon, 1965. Deuterium and oxygen-18 variations in the ocean and the marine atmosphere. In *Stable Isotopes in Oceanographic Studies and Paleotemperatures*. Laboratorio di Geologia Nucleare, University of Pisa, Pisa, Italy, pp. 9–130.

Crank, J., 1956. *The Mathematics of Diffusion*. Oxford University Press, London, England.

Criado, J. M. and J. M. Trillo, 1975. Effects of mechanical grinding on the texture and structure of calcium carbonate. *J. Chem. Soc.*, *Faraday Trans. I*, **7**, 961–966.

Csanady, G. T., 1973. *Turbulent Diffusion in the Enviroment*. Reidel, Dordrecht, The Netherlands.

Currie, J. A., 1960. Gaseous diffusion in porous media. Part 2—Dry granular materials. *Brit. J. Appl. Phys.*, **11**, 318–324.

Cyberski, J., 1973. Accumulation of debris in water storage reservoirs of Central Europe. In *Man-Made Lakes: Their Problems and Environmental Effects*, Geophysical Monograph 17, W. C. Ackermann, G. F. White, and E. B. Worthington, Eds. American Geophysical Union, Washington, D. C., pp. 359–363.

Danckwerts, P. V., 1953. Continuous flow systems. *Chem. Eng. Sci.*, **2**, 1–13.

Dapples, E. C., 1975. Laws of distribution applied to sand sizes. *Geol. Soc. Am. Mem.*, **142**, 37–61.

Darcy, H., 1856. *Les Fontaines Publiques de la Villa de Dijon. Exposition et Application des Principles à Suivre et des Formules à Employer dans les Questions de Distribution d'Eau.* Victor Dalmont, Paris, France.

Davies, C. N., 1973. *Air Filtration*. Academic, London, England.

Davies, T. A., W. H. Hay, J. R. Southam, and T. R. Worsley, 1977. Estimates of Cenozoic oceanic sedimentation rates. *Science*, **197**, 53–55.

Dayhoff, M. O., E. R. Lippincott, R. V. Eck, and G. Nagarajan, 1967. *Thermodynamic Equilibrium in Prebiological Atmospheres of C, H, O, N, P, S, and Cl* (NASA SP-3040). National Aeronautics and Space Administration, Washington, D.C.

Deevey, E. S., Jr., 1973. Sulfur, nitrogen, and carbon in the biosphere. In *Carbon and the Biosphere*, G. M. Woodwell and E. V. Pecan, Eds. U.S. Atomic Energy Commission, Washington, D.C., pp. 182–190.

Degens, E. T., 1969. Biogeochemistry of stable carbon isotopes. In *Organic Geochemistry*, G. Eglinton and M. J. T. Murphy, Eds. Springer-Verlag, Berlin, Germany, pp. 304–329.

Delany, A. C., D. W. Parkin, J. J. Griffin, E. D. Goldberg, and B. E. F. Reimann, 1967. Airborne dust collected at Barbados. *Geochim. Cosmochim. Acta*, **31**, 885–909.

Delwiche, C. C. and G. E. Likens, 1977. Biological response to fossil fuel combustion products. In *Global Chemical Cycles and Their Alterations by Man*, W. Stumm, Ed. Dahlem Konferenzen, Berlin, Germany, pp. 73–88.

Dempster, P. B. and P. D. Ritchie, 1953. Physicochemical studies on dusts. V. Examination of finely ground quartz by differential thermal analysis and other physical methods. *J. Appl. Chem.*, **3**, 182–192.

Dendy, F. E., W. A. Champion, and R. B. Wilson, 1973. Reservoir sedimentation surveys in the United States. In *Man-Made Lakes: Their Problems and Environmental Effects*, Geophysical Monograph 17, W. C. Ackermann, G. F. White, and E. B. Worthington, Eds., American Geophysical Union, Washington, D.C., pp. 349–357.

Diffusion Data, 1971. Vol. 5, No. 1, F. H. Wöhlbier, Ed. Trans Tech Publications, Bay Village.

Diffusion Data. 1974. Vol. 8, F. H. Wöhlbier, Ed. Trans Tech Publications, Bay Village.

Dolcater, D. L., E. G. Lotse, J. K. Syers, and M. L. Jackson, 1968. Cation exchange selectivity of some clay-sized minerals and soil materials. *Soil Sci. Soc. Am. Proc.*, **32**, 795–798.

Drake, R. L., 1972. A general mathematical survey of the coagulation equation. In *Topics in Current Aerosol Research*, Part 2, G. M. Hidy and J. R. Brock, Eds. Pergamon, Oxford, England, pp. 201–376.

Drever, J. I., 1974. The magnesium problem. In *The Sea*, Vol 5, E. D. Goldberg, Ed. Wiley-Interscience, New York, pp. 337–357.

Drozdova, V. M., O. P. Petrenchuk, E. S. Selezneva, and P. F. Svistov, 1964. *Khimicheskii Sostav Atmosfernykh Osadkov na Evropeiskoi Territorii SSSR*. Gidrometeorologicheskoye Izdat., Leningrad, USSR.

Duce, R. A., 1969. On the source of gaseous chlorine in the marine atmosphere. *J. Geophys. Res.*, **74**, 4597–4599.

Dufour, L. and R. Defay, 1963. *Thermodynamics of Clouds.* Academic, New York.

Duguid, H. A. and J. F. Stampfer Jr., 1971. The evaporation rates of small, freely falling water drops. *J. Atm. Sci.*, **28**, 1233–1243.

Eaton, G. P., 1964. Windborne volcanic ash: A possible index to polar wandering. *J. Geol.*, **72**, 1–35.

Eagleson, P. S., 1970. *Dynamic Hydrology.* McGraw-Hill, New York.

Eberling, C., W. Geins, F. Slemer, and W. Seiler, 1976. A method for measurements of mercury in the atmosphere and some results of the global Hg-distribution. *Proceedings of the APOMET Meeting*, Gothenburg, Sweden, October 1976.

Eckart, C., 1960. *Hydrodynamics of Oceans and Atmospheres.* Pergamon, Oxford, England.

Edgar, N. T., J. B. Saunders, et al., 1973. Site 146/149. In *Initial Reports of the Deep Sea Drilling Project*, Vol. 15. U.S. Government Printing Office, Washington, D.C., pp. 17–168.

Edmond, J. M., 1974. On the dissolution of silicate and carbonate in the deep sea. *Deep-Sea Res.* **21**, 455–480.

Edmond, J. M. and J. M. T. M. Gieskes, 1970. On the calculation of the degree of saturation of seawater with respect to calcium carbonate under *in situ* conditions. *Geochim. Cosmochim. Acta*, **34**, 1261–1291.

Edzwald, J. K., J. B. Upchurch, and C. R. O'Melia, 1974. Coagulation in estuaries. *Env. Sci. Tech.*, **8**, 58–63.

Eisenberg, D. S. and W. Kauzmann, 1969. *The Structure and Properties of Water.* Oxford University Press, New York.

Ekdahl, C. A., Jr. and C. D. Keeling, 1973. Quantitative deductions from the records at Mauna Loa Observatory and at the South Pole. In *Carbon and the Biosphere*, G. M. Woodwell and E. V. Pecan, Eds. U.S. Atomic Energy Commission, Washington, D.C., pp. 51–85.

Eliason, J. R., 1966. Montmorillonite exchange equilibria with strontium-sodium-cesium. *Am. Miner.*, **51**, 324–335.

Ellis, J. H., R. I. Barneisel, and R. E. Phillips, 1970. The diffusion of copper, manganese, and zinc as affected by concentration, clay mineralogy, and associated anions. *Soil Sci. Soc. Am. Proc.*, **34**, 866–870.

Emery, K. O. and S. C. Rittenberg, 1952. Early diagenesis of California Basin sediments in relation to origin of oil. *Am. Assoc. Petrol. Geol. Bull.*, **36**, 735–806.

Engelhardt, W. von, 1960. *Der Porenraum der Sedimente.* Springer-Verlag, Heidelberg, Germany.

Eugster, H. P. and L. A. Hardie, 1978. Mineralogy of saline brine lakes. In *Lakes: Chemistry, Geology, Physics*, A. Lerman, Ed. Springer-Verlag, New York, pp. 237–293.

Evans, R. B., G. M. Watson, and E. A. Mason, 1961. Gaseous diffusion in porous media at uniform pressure. *J. Chem. Phys.*, **35**, 3076–2083.

Fanning, K. A. and M. E. Q. Pilson, 1974. The diffusion of dissolved silica out of deep sea sediments. *J. Geophys. Res.*, **79**, 1293–1297.

Feely, R. A., W. M. Sackett, and J. E. Harris, 1971. Distribution of particulate aluminum in the Gulf of Mexico. *J. Geophys. Res.*, **76**, 5893–5902.

Ferrell, R. T. and D. M. Himmelblau, 1967. Diffusion coefficients of nitrogen and oxygen in water. *J. Chem. Eng. Data*, **12**, 111–11.

Fiadeiro, M. E., 1975. Numerical modeling of tracer distribution in the deep Pacific Ocean. Unpublished doctoral dissertation. University of California, San Diego.

Fick, A., 1855. Ueber Diffusion. *Ann. Phys.* (Leipzig), **170**, 59–86.

Fitts, D. D., 1962. *Nonequilibrium Thermodynamics.* McGraw-Hill, New York. 173 p.

Fitzgerald, W. F. and C. D. Hunt, 1974. Distribution of mercury in the surface microlayer and in subsurface waters of the northeast Atlantic Ocean. *J. Rech. Atm.*, **8**, 629–637.

Fleischer, M., 1970. Summary of the literature on the inorganic geochemistry of mercury. *U.S. Geol. Surv. Prof. Pap.*, **713**, 6–13.

Foland, K. A., 1974. Argon-40 diffusion in homogeneous orthoclase and an interpretation of argon diffusion in potassium feldspars. *Geochim. Cosmochim. Acta*, **38**, 151–166.

Fournier, R. O., 1973. Silica in thermal waters: laboratory and field investigations. In *Proceedings of the International Symposium on Hydrochemistry and Biochemistry, Tokyo*, Vol. 1, Clarke, Washington, D.C., pp. 129–139.

Friedlander, S. K., 1960. Similarity considerations for the particle-size spectrum of a coagulating, sedimenting aerosol. *J. Meteor.*, **17**, 479–483.

Friedlander, S. K., 1977. *Smoke, Dust, and Haze.* Wiley-Interscience, New York.

Frischat, G. H., 1975. *Ionic Diffusion in Oxide Glasses.* Diffusion and Defect Monograph Series, No. 3/4. Trans Tech Publications, Bay Village.

Garrels, R. M. and C. L. Christ, 1965. *Solutions, Minerals, and Equilibria.* Harper & Row, New York.

Garrels, R. M. and P. F. Howard, 1959. Reactions of feldspar and mica with water at low temperature and pressure. In *Clays and Clay Minerals, 6th Proceedings of the National Conference on Clays and Clay Minerals*, A. Swineford, Ed. Pergamon, Oxford, England. pp. 68–88.

Garrels, R. M., A. Lerman and F. T. Mackenzie, 1976. Controls of atmospheric O_2 and CO_2: Past, present, and future. *Am. Sci.*, **64**, 306–315.

Garrels, R. M. and F. T. Mackenzie, 1971. *Evolution of Sedimentary Rocks.* Norton, New York.

Garrels, R. M. and F. T. Mackenzie, 1972. A quantitative model for the sedimentary rock cycle. *Mar. Chem.*, **1**, 27–41.

Garrels, R. M., F. T. Mackenzie, and C. Hunt, 1975. *Chemical Cycles and the Global Environment: Assessing Human Influences.* Kaufmann, Los Altos, Ca.

Gast, R. G., 1969. Standard free energies of exchange for alkali metal cations on Wyoming bentonite. *Soil Sci. Soc. Am. Proc.*, **33**, 37–41.

Gast, R. G. and W. D. Klobe, 1971. Sodium-lithium exchange equilibria on vermiculite at 25° and 50°C. *Clays Clay Miner.*, **19**, 311–319.

Gaudin, A. M. and T. P. Meloy, 1962a. Model and a comminution distribution equation for single fracture. *Am. Inst. Mech. Eng. Trans.*, **223**, 40–43.

Gaudin, A. M. and T. P. Meloy, 1962b. Model and a comminution distribution equation for repeated fracture. *Am. Inst. Mech. Eng. Trans.*, **223**, 43–50.

Gautschi, W., 1964. Error function and Fresnel integrals. In *Handbook of Mathematical Functions*, M. Abramowitz and I. A. Stegun, Eds. Applied Mathematics Series 55. National Bureau of Standards, Washington, D.C., pp. 295–329.

Gibbs, R. J., M. D. Matthews, and D. A. Link, 1971. The relationship between sphere size and settling velocity. *J. Sed. Petrol.*, **41**, 7–18.

Gieskes, J. M., 1973. Interstitial water studies, Leg 15—alkalinity, pH, Mg, Ca, Si, PO_4, and

NH_4. In *Initial Reports of the Deep Sea Drilling Project*, Vol. 20. U.S. Government Printing Office, Washington, D.C., pp. 813–829.

Gieskes, J. M., 1974. Interstitial water studies, Leg 25. In *Initial Reports of the Deep Sea Drilling Project.*, Vol. 25. U.S. Government Printing Office, Washington, D.C., pp. 361–394.

Giletti, B. J., 1974. Studies in diffusion. I. Argon in phlogopite mica. In *Geochemical Transport and Kinetics*, A. W. Hofmann, B. J. Giletti, H. S. Yoder, Jr., and R. A. Yund, Eds. Carnegie Institute of Washington, Publ. 634, pp. 107–116.

Giletti, B. J., M. P. Semet, and R. A. Yund, 1978. Studies in diffusion—III. Oxygen in feldspars: An ion microprobe determination. *Geochim. Cosmochim. Acta*, **42**, 45–57.

Gjessing, E. T., A. Henriksen, M. Johannessen, and' R. F. Wright, 1976. Effects of acid precipitation on freshwater chemistry. In *Impact of Acid Precipitation on Forest and Freshwater Ecosystems in Norway*, F. H. Braekke, Ed. SNSF-Project 6, Oslo, Norway, pp. 64–85.

Glew, D. N. and D. A. Hames, 1971. Aqueous nonelectrolyte solutions. Part X. Mercury solubility in water. *Can. J. Chem.*, **49**, 3114–3118.

Glymph, L. M., 1973. Summary: sedimentation of reservoirs. In *Man-Made Lakes: Their Problems and Environmental Effects*, Geophysical Monograph 17, W. C. Ackermann, G. F. White and E. B. Worthington, Eds. American Geophysical Union, Washington, D.C., pp. 342–348.

Gmelins Handbuch der anorganischen Chemie, 1939. *Magnesium*. Vol. 27, Part 2. Chemie Verlag, Berlin, Germany.

Goldberg, E. D., 1971. Atmospheric dust, the sedimentary cycle and man. *Geophysics*, **1**, 117–132.

Goldhaber, M. B., R. C. Aller, J. K. Cochran, J. K. Rosenfeld, C. S. Martens, and R. A. Berner, 1977. Sulfate reduction, diffusion, and bioturbation in Long Island Sound sediments. *Am. J. Sci.*, **277**, 193–237.

Goldich, S. S., 1938. A study in rock-weathering. *J. Geol.*, **46**, 17–58.

Golterman, H. L., Ed., 1977. *Interactions Between Sediments and Fresh Water*. Junk, The Hague, The Netherlands.

Gordon, J. E., 1971. *The New Science of Strong Materials*. Penguin, New York.

Granat, L., 1972. On the relation between pH and the chemical composition in atmospheric precipitation. *Tellus*, **24**, 550–560.

Granat, L., 1976. A global atmospheric sulphur budget. In *Nitrogen, Phosphorus and Sulphur—Global Cycles*, B. H. Svensson and R. Soderlund, Eds. SCOPE Report 7, Ecol. Bull. 22, Stockholm, Sweden, pp. 102–122.

Gregor, C. B., 1968. The rate of denudation in Post-Algonkian time. *Koniinkl. Ned. Akad. Wetenschap. Proc.*, **71**, 22.

Gregor, C. B., 1970. Denudation of the continents. *Nature*, **228**, 273–275.

Gregor, C. B., 1977. Weathering rates of sedimentary and crystalline rocks. Unpublished.

Griffin, J. J., H. Windom, and E. D. Goldberg, 1968. The distribution of clay minerals in the World Ocean. *Deep-Sea Res.*, **15**, 433–459.

Guinasso, N. L., Jr. and D. R. Schink, 1975. Quantitative estimates of biological mixing rates in abyssal sediments. *J. Geophys. Res.*, **80**, 3032–3043.

Hahn, H. H. and W. Stumm, 1968. The role of absorption of hydrolyzed aluminum in the kinetics of coagulation. *Adv. Chem. Ser.*, **79**, 91–111.

Hahn, H. H. and W. Stumm, 1970. The role of coagulation in natural waters. *Am. J. Sci.*, **268**, 354–368.

Halley, E., 1687. An estimate of the quantity of vapour raised out of the sea by the warmth of the sun; derived from an experiment shown before the Royal Society at one of their late meetings. *Philos. Trans.* (no. 189, Sept.–Oct. 1687) **16**, 366–370.

Hamilton, E. L., 1971. Elastic properties of marine sediments. *J. Geophys. Res.*, **76**, 579–604.

Happel, J. and H. Brenner, 1973. *Low Reynolds Number Hydrodynamics*, 2nd rev. ed. Noordhoff, Leyden, The Netherlands.

Harris, C. C., 1968. The application of size distribution equations to multi-event comminution processes. *Soc. Min. Eng. Trans.*, **241**, 343–358.

Harris, C. C., 1971. Graphical presentation of size distribution data: An assessment of current practice. *Inst. Min. Metall. Trans. C*, **80**, 133–139.

Haul, R. A. W. and L. H. Stein, 1955. Diffusion in calcite crystals on the basis of isotopic exchange with carbon dioxide. *Trans. Faraday Soc.*, **51**, 1280–1290.

Hay, R. L. and B. F. Jones, 1972. Weathering of basaltic tephra on the Island of Hawaii. *Geol. Soc. Am. Bull.*, **83**, 317–332.

Heald, W. R., 1960. Characterization of exchange reactions of strontium or calcium on four clays. *Soil Sci. Soc. Am. Proc.*, **24**, 103–106.

Heath, G. R., 1974. Dissolved silica and deep-sea sediments. In *Studies in Paleo-Oceanography*. Society of Economic Paleontologists and Mineralogists, Special Publication 20, Tulsa, Ok., pp. 77–93.

Heath, G. R. and C. Culberson, 1970. Calcite: Degree of saturation, rate of dissolution, and the compensation depth in the deep oceans. *Geol. Soc. Am. Bull.*, **81**, 3157–3160.

Heath, G. R. and R. Moberly, Jr., 1971. Cherts from the Western Pacific, Leg 7, Deep Sea Drilling Project. In *Initial Reports of the Deep Sea Drilling Project*, Vol. 7. U.S. Government Printing Office, Washington, D.C., pp. 991–1007.

Heimann, R. B., 1975. *Auflösung von Kristallen*. Springer-Verlag, Vienna, Austria.

Heiss, J. F. and J. Coull, 1952. The effect of orientation and shape on the settling velocity of non-isometric particles in a viscous medium. *Chem. Eng. Prog.*, **48**, 133–140.

Helgeson, H. C., 1971. Kinetics of mass transfer among silicates and aqueous solutions. *Geochim. Cosmochim. Acta*, **35**, 421–470.

Helgeson, H. C., J. M. Delaney, and H. W. Nesbitt, 1978. Summary and critique of the thermodynamic properties of rock-forming minerals (in press).

Helgeson, H. C. and D. H. Kirkham, 1974a. Theoretical prediction of the thermodynamic behavior of aqueous electrolytes at high pressures and temperatures. I. Summary of the thermodynamic/electrostatic properties of the solvent. *Am. J. Sci.*, **274**, 1089–1198.

Helgeson, H. C. and D. H. Kirkham, 1974b. Theoretical prediction of the thermodynamic behavior of aqueous eletrolytes at high pressures and temperatures. II. Debye-Hückel parameters for activity coefficients and relative partial molal properties. *Am. J. Sci.*, **274**, 1199–1261.

Helgeson, H. C. and D. H. Kirkham, 1976. Theoretical prediction of the thermodynamic properties of aqueous electrolytes at high pressures and temperatures. III. Equation of state for aqueous species at infinite dilution. *Am. J. Sci.*, **276**, 97–240.

Hidy, G. M., 1973. Removal processes of gaseous and particulate pollutants. In *Chemistry of the Lower Atmosphere*, S. I. Rasool, Ed. Plenum, New York, pp. 121–176.

Himmelblau, D. M., 1964. Diffusion of dissolved gases in liquids. *Chem. Rev.*, **64**, 527–550.

Holeman, J. N., 1968. The sediment yield of major rivers of the world. *Water Resources Res.*, **4**, 737–747.

Holland, H. D., 1978. *The Chemistry of the Atmosphere and Oceans.* Wiley-Interscience, New York.

Holm-Hansen, O., J. D. H. Strickland, and P. ·M. Williams, 1966. A detailed analysis of biologically important substances in a profile off Southern California. *Limnol. Oceanogr.*, **11**, 548–561.

Honjo, S., 1975. Dissolution of suspended coccoliths in the deep-sea water column and sedimentation of coccolith ooze. In *Dissolution of Deep-Sea Carbonates.*, W. V. Sliter, A. W. H. Bé, and W. H. Berger, Eds. Cushman Foundation for Foraminiferal Research, Special Publication 13, Ithaca, N.Y., pp. 114–128.

Horne, R. A. and D. S. Johnson, 1966. The viscosity of water under pressure. *J. Phys. Chem.*, **70**, 2182–2190.

Hower, J., E. V. Eslinger, M. E. Hower, and E. A. Perry, 1976. Mechanism of burial metamorphism of argillaceous sediment: 1. Mineralogical and chemical evidence. *Geol. Soc. Am. Bull.*, **87**, 725–737.

Hubbert, M. K., 1940. The theory of ground-water motion. *J. Geol.*, **48**, 785–944.

Huene, R. von and D. J. W. Piper, 1973. Measurements of porosity in sediments of the lower continental margin, deep-sea fans, the Aleutian Trench, and Alaskan Abyssal Plain. In *Initial Reports of the Deep Sea Drilling Project*, Vol. 18. U.S. Government Printing Office, Washington, D.C., pp. 889–895.

Hurd, D. C. and F. Theyer, 1975. Changes in the physical and chemical properties of biogenic silica from the Central Equatorial Pacific. *Adv. Chem. Ser.*, **147**, 211–230.

Imboden, D. M., 1975. Interstitial transport of solutes in non-steady-state accumulating and compacting sediments. *Earth Planet. Sci. Lett.*, **27**, 221–228.

Imboden, D. M. and A. Lerman, 1978. Chemical models of lakes. In *Lakes: Chemistry, Geology, Physics,* A. Lerman, Ed. Springer-Verlag, New York, pp. 341-356

International Critical Tables, 1926–1930. 7 vols, E. W. Washburn, Ed. McGraw-Hill, New York.

Jackson, M. L., D. A. Gillette, E. F. Danielsen, I. H. Blifford, R. A. Bryson, and J. K. Syers, 1973. Global dustfall during the Quaternary as related to environments. *Soil Sci.*, **116**, 135–145.

Jackson, M. L. and E. Truog, 1939. Influence of grinding soil minerals to near molecular size on their solubility and base exchange properties. *Soil Sci. Soc. Am. Proc.*, **4**, 136–143.

Jamieson, J. C., 1953. Phase equilibrium in the system calcite-aragonite. *J. Chem. Phys.*, **21**, 1385–1390.

Jamieson, J. C. and J. R. Goldsmith, 1960. Some reactions produced in carbonates by grinding. *Am. Miner.*, **45**, 818–827.

JANAF Thermochemical Tables, 1971. 2nd ed. (NSRDS-NBS 37), Clearinghouse for Federal Scientific and Technical Information, Springfield, Va.

Jassby, A. and T. Powell, 1975. Vertical patterns of eddy diffusion during stratification in Castle Lake, California. *Limnol. Oceanogr.*, **20**, 530–543.

Jerlov, N. G., 1959. Maxima in the vertical distribution of particles in the sea. *Deep-Sea Res.*, **5**, 173–184.

Johnson, N. L. and S. Kotz, 1970. *Continuous Univariate Distributions—1.* Houghton Mifflin, Boston.

Jones, B. F., H. P. Eugster and S. L. Rettig, 1977. Hydrochemistry of the Lake Magadi basin, Kenya. *Geochim. Cosmochim. Acta*, **41**, 53–72.

Jost, W., 1960. *Diffusion in Solids, Liquids, Gases*, rev. ed. Academic, New York.

Junge, C. E., 1963. *Air Chemistry and Radioactivity*. Academic, New York.

Junge, C. E., 1974. Residence time and variability of tropospheric trace gases. *Tellus*, **26**, 477–488.

Junge, C. E., 1975. Processes responsible for the trace content in precipitation. *Proceedings of the IAHS Symposion*, International Oceanographic Assembly, Grenoble, France, Sept. 1975, pp. 63–77.

Kanwisher, J., 1963. On the exchange of gases between the atmosphere and the sea. *Deep-Sea Res.*, **10**, 195–207.

Katchalsky, A. and P. F. Curran, 1965. *Nonequilibrium Thermodynamics in Biophysics*. Harvard University Press, Cambridge, Mass.

Keeling, C. D., 1973. The carbon dioxide cycle: Reservoir models to depict the exchange of atmospheric carbon dioxide with the oceans and land plants. In *Chemistry of the Lower Atmosphere*, S. I. Rasool, Ed. Plenum, New York, pp. 251–329.

Kemp, A. L. W., R. L. Thomas, C. I. Dell, and J.-M. Jaquet, 1976. Cultural impact on the geochemistry of sediments in Lake Erie. *J. Fish. Res. Board Can.*, **33**, 440–462.

Kennedy, C. S. and G. C. Kennedy, 1976. The equilibrium boundary between graphite and diamond. *J. Geophys. Res.*, **81**, 2467–2470.

Kennedy, V. C., 1971. Silica variation in stream water with time and dischange. *Adv. Chem. Ser.*, **106**, 94–130.

King, E. A., Jr., 1977. The lunar regolith: Physical characteristics and dynamics. *Philos. Trans. R. Soc. Lond. A*, **285**, 273–278.

Klimpel, R. R. and L. G. Austin, 1965. The statistical theory of primary breakage distributions for brittle materials. *Am. Inst. Mech. Eng. Trans.*, **232**, 88–94.

Klinkenberg, L. J., 1951. Analogy between diffusion and electrical conductivity in porous rocks. *Geol. Soc. Am. Bull.*, **62**, 559–563.

Klotz, I. M., 1964. *Chemical Thermodynamics: Basic Theory and Methods*. Benjamin, New York.

Koblentz-Mishke, O. I., V. V. Volkovinsky, and J. G. Kabanova, 1970. Plankton primary production of the World Ocean. In *Scientific Exploration of the South Pacific*. W. S. Wooster, Ed. National Academy of Sciences, Washington, D.C., pp. 183–193.

Koide, M., A. Soutar, and E. D. Goldberg, 1972. Marine geochronology with ^{210}Pb. *Earth Planet. Sci. Lett.*, **14**, 442–446.

Kraus, E. B. and J. S. Turner, 1967. A one-dimensional model of the seasonal thermocline. II. The general theory and its consequences. *Tellus*, **19**, 98–105.

Krishnaswami, S. and D. Lal, 1978. Radionuclide limnochronology, In *Lakes: Chemistry, Geology, Physics*, A. Lerman, Ed. Springer-Verlag, New York, pp. 153–177.

Krumbein, W. C., 1934. Size frequency distributions of sediments. *J. Sed. Petrol.*, **4**, 65–77.

Krumbein, W. C. and G. D. Monk, 1942. Permeability as a function of the size parameters of unconsolidated sand. *Am. Inst. Mech. Eng. Tech. Publ.* **1492**, 1–11.

Krumbein, W. C. and F. J. Pettijohn, 1938, *Manual of Sedimentary Petrography*. Appleton-Century-Crofts, New York.

Krumbein, W. C. and F. W. Tisdel, 1940. Size distributions of source rocks of sediments. *Am. J. Sci.*, **238**, 296–305.

Kuenen, P. H., 1959. Experimental abrasion: 3. Fluviatile action on sand. *Am. J. Sci.*, **257**, 172–190.

Kuenen, P. H. and W. G. Perdok, 1962. Experimental abrasion: 5. Frosting and defrosting of quartz grains. *J. Geol.*, **70**, 648–658.

Kullenberg, G., 1971. Vertical diffusion in shallow waters. *Tellus*, **23**, 129–135.

Kunishi, H. M. and W. R. Heald, 1968. Rubidium-sodium exchange on kaolinite and bentonite. *Soil Sci. Soc. Am. Proc.*, **32**, 201–204.

Kuznetsov, S. I., 1975. *The Microflora of Lakes and Its Geochemical Activity*. University of Texas Press, Austin, Tex.

Lal, D. and A. Lerman, 1975. Size spectra of biogenic particles in ocean water and sediments. *J. Geophys. Res.*, **80**, 423–430, 4563.

Lamb, H., 1945. *Hydrodynamics*, 6th ed. Dover, New York.

Lamd, H. H., 1970. Volcanic dust in the atmosphere; With a chronology and assessment of its meterological significance. *Philos. Trans. R. Soc. Lond. A*, **266**, 425–533.

Landolt-Börnstein, 1956. *Zahlenwerte und Funktionen. Schmelzgleichgewichte und Grenzflächenerscheinungen*, 6th ed. Vol. 2, Part 3. Springer-Verlag, Heidelberg, Germany.

Landolt-Börnstein, 1960. *Zahlenwerte und Funktionen. Gleichgewichte Dampf-Kondensat und Osmotische Phänomene*, Vol. 2, Part 2, Sect. A. Springer-Verlag, Heidelberg, Germany.

Landolt-Börnstein, 1962. *Zahlenwerte und Funktionen. Lösungsgleichgewichte I*, Vol. 2, Part 2, Sect. B. Springer-Verlag, Heidelberg, Germany.

Landolt-Börnstein, 1968. *Zahlenwerte und Funktionen. Transportphänomene I*, Vol. 2, Part 5, Sect. A. Springer-Verlag, Heidelberg, Germany.

Landolt-Börnstein, 1969. *Zahlenwerte und Funktionen. Transportphänomene II*, Vol. 2, Part 5, Sect. B. Springer-Verlag, Heidelberg, Germany.

Landolt-Börnstein, 1976. *Zahlenwerte und Funktionen. Gleichgewicht der Absorption von Gasen in Flüssigkeiten von niedrigem Dampfdruck*, 6th ed, Vol. 4, Part 4, Sect. C. Springer-Verlag, Heidelberg, Germany.

Langleben, M. P., 1954. The terminal velocity of snowflakes. *Quart. J. R. Meteor. Soc.*, **80**, 174–181.

Lapple, C. E., 1951. *Fluid and Particle Mechanics*. University of Delaware, Newark, Del.

Lasaga, A. C. and H. D. Holland, 1976. Mathematical aspects of non-steady-state diagenesis. *Geochim. Cosmochim. Acta*, **40**, 257–266.

Laudelout, H., R. van Bladel, G. H. Bolt, and A. L. Page, 1968. Thermodynamics of heterovalent cation exchange reactions in a montmorillonite clay. *Trans. Faraday Soc.*, **64**, 1477–1488.

Leak, W. B., 1975. Age distribution in virgin red spruce and northern hardwoods. *Ecology*, **56**, 1451–1454.

Leopold, L. B., M. G. Wolman, and J. P. Miller, 1964. *Fluvial Processes in Geomorphology*. Freemen, San Francisco.

Lerman, A., 1970. Chemical equilibria and evolution of chloride brines. *Miner. Soc. Am., Spec. Pap.* **3**, 291–306.

Lerman, A., 1971. Time to chemical steady-states in lakes and oceans. *Adv. Chem. Ser.*, **106**, 30–76.

Lerman, A., 1975. Maintenance of steady state in oceanic sediments. *Am. J. Sci.*, **275**, 609–635.

Lerman, A., 1977. Migrational processes and chemical reactions in interstitial waters. In *The Sea*, Vol. 6, E. D. Goldberg, I. N. McCave, J. J. O'Brien, and J. H. Steele, Eds. Wiley-Interscience, New York, pp. 695–738.

Lerman, A., K. L. Carder, and P. R. Betzer, 1977. Elimination of fine suspensoids in the oceanic water column. *Earth Planet. Sci. Lett.*, **37**, 61–70.

Lerman, A. and D. Lal, 1977. Regeneration rates in the ocean. *Am. J. Sci.*, **277**, 238–258.

Lerman, A., D. Lal, and M. F. Dacey, 1974. Stokes setting and chemical reactivity of suspended particles in natural waters. In *Suspended Solids in Water*, R. J. Gibbs, Ed. Plenum, New York, pp. 17–47.

Lerman, A. and T. A. Lietzke, 1975. Uptake and migration of tracers in lake sediments. *Limnol. Oceanogr.*, **20**, 497–510.

Lerman, A. and T. A. Liezke, 1977. Fluxes in a growing sediment layer. *Am. J. Sci.*, **277**, 25–37.

Lerman, A., F. T. Mackenzie, and O. P. Bricker, 1975. Rates of dissolution of aluminosilicates in seawater. *Earth Planet. Sci. Lett.*, **25**, 82–88.

Lerman, A., F. T. Mackenzie, and R. M. Garrels, 1975. Modeling of geochemical cycles: Phosphorus as an example. *Geol. Soc. Am. Mem.*, **142**, 205–218.

Lerman, A. and R. R. Weiler, 1970. Diffusion and accumulation of chloride and sodium in Lake Ontario sediment. *Earth Planet. Sci. Lett.*, **10**, 150–156.

Levich, V. G., 1962. *Physicochemical Hydrodynamics*. Prentice-Hall, Englewood Cliffs, N.J.

Levorsen, A. I., 1967. *Geology of Petroleum*, 2nd ed. Freeman, San Francisco.

Lewis, G. N., M. Randall, K. S. Pitzer, and L. Brewer, 1961. *Thermodynamics*, 2nd ed. McGraw-Hill, New York.

Lewis, R. J. and H. C. Thomas, 1963. Adsorption studies on clay minerals. VIII. A consistency test of exchange sorption in the systems sodium-cesium-barium montmorillonite. *J. Phys. Chem.*, **67**, 1781–1783.

Li, Y.-H., 1972. Geochemical mass balance among lithosphere, hydrosphere, and atmosphere. *Am. J. Sci.*, **272**, 119–137.

Li, Y.-H., 1973. Vertical eddy diffusion coefficient in Lake Zurich. *Schweiz. Z. Hydrol.*, **35**, 1–7.

Li, Y.-H. and S. Gregory, 1974. Diffusion of ions in sea water and in deep-sea sediments. *Geochim. Cosmochim. Acta*, **38**, 703–714.

Li, Y.-H., T. Takahashi, and W. S. Broecker, 1969. Degree of saturation of $CaCO_3$ in the oceans. *J. Geophys. Res.*, **74**, 5507–5525.

Liesegang, R. E., 1913. *Geologische Diffusionen*. Steinkopff, Dresden, Germany.

Likens, G. E., 1975. Primary production of inland aquatic ecosystems. In *Primary Productivity of the Biosphere*. H. Lieth and R. H. Whittaker, Eds. Springer-Verlag, New York, pp. 185–202.

Likens, G. E., 1976. Acid precipitation. *Chem. Eng. News*, Nov. 22, 29–44.

Lin, I. J., S. Nadiv, and D. J. M. Grodzian, 1975. Changes in the state of solids and mechano-chemical reactions in prolonged comminution processes. *Miner. Sci. Eng.*, **7**, 313–336.

Linke, W. F., Ed., 1958. *Solubilities of Inorganic and Metal-Organic Compounds*, Vol. 1, 4th ed. American Chemical Society, Washington, D.C.

Lisitzin, A. P., 1971. Distribution of siliceous microfossils in suspension and in bottom sediments. In *The Micropalaeontology of Oceans*, B. M. Funnell and W. R. Riedel, Eds. Cambridge University Press, London, pp. 173–195.

Lisitzin, A. P., 1972. *Sedimentation in the World Ocean*. Society of Econonic Paleontologists and Mineraloists, Special Publication 17, Tulsa, Ok.

Lisitzin, A. P., 1974. *Osadkoobrazovaniye v Okeanakh*. Nauka, Moscow, USSR.

Liss, P. S., 1975. The exchange of gases across lake surface. *Proceedings of the First Specialty*

Symposium, Atmospheric Contribution to the Chemistry of Lake Waters. International Assoc. Great Lakes Research, pp. 88–99.

Livingstone, D. A., 1963. Chemical composition of rivers and lakes. *U. S. Geol. Surv. Prof. Pap.*, **440-G**, 1–64.

Luce, R. W., R. W. Bartlett, and G. A. Parks, 1972. Dissolution kinetics of magnesium silicates. *Geochim. Cosmochim. Acta*, **36**, 35–50.

McCarthy, J. H., Jr., J. L. Meuschke, W. H. Ficklin, and R. E. Learned, 1970. Mercury in the atmosphere. *U.S. Geol. Surv. Prof. Pap.*, **713**, 37–39.

McCave, I. N., 1975. Vertical flux of particles in the ocean. *Deep-Sea Res.*, **22**, 491–502.

McClelland, J. E., 1950. The effect of time, temperature and particle size on the release of bases from some common soil-forming minerals of different crystal structure. *Soil Sci. Soc. Am. Proc.*, **15**, 301–307.

MacDonald, G. J. F., 1956. Experimental determination of calcite–aragonite equilibrium relations at elevated temperatures and pressures. *Am. Miner.*, **41**, 744–756.

McDonald, J. E., 1958. The physics of cloud modification. *Adv. Geophys.*, **5**, 223–303.

Machta, L., 1972. The role of the oceans and biosphere in the carbon dioxide cycle. In *The Changing Chemistry of the Oceans*, D. Dryssen and D. Jagner, Eds. Wiley-Interscience, New York, pp. 121–145.

McKeague, J. D. and M. G. Cline, 1963. Silica in soil solutions. *Can. J. Soil Sci.*, **43**, 70–96.

Mackenzie, F. T. and R. M. Garrels, 1965. Silicates: Reactivity with sea water. *Science*, **150**, 57–58.

Mackenzie, F. T., R. M. Garrels, O. P. Bricker, and F. Bickley, 1967. Silica in seawater: Control by silica minerals. *Science*, **155**, 1404–1405.

MacKenzie, R. C. and A. A. Milne, 1953a. The effect of grinding on micas. *Clay Miner. Bull.*, **2**, 57–62.

MacKenzie, R. C. and A. A. Milne, 1953b. The effect of grinding on micas. I. Muscovite. *Miner. Mag.*, **30**, 178–185.

Manheim, F. T., 1970. The diffusion of ions in unconsolidated sediments. *Earth Planet. Sci. Lett.*, **9**, 307–309.

Manheim, F. T. and L. S. Waterman, 1974. Diffusimetry (diffusion constant estimation) on sediment cores by resistivity probe. In *Initial Reports of the Deep Sea Drilling Project*, Vol. 22. U.S. Government Printing Office, Washington, D.C., pp. 663–670.

Manning, A. B., 1952. Size distribution in a randomly fractured solid, and its application to coal. *J. Inst. Fuel*, **25**, 31–32.

Manning, J. R., 1968. *Diffusion Kinetics for Atoms in Crystals.* Van Nostrand, Toronto, Canada.

Margenau, H. and G. M. Murphy, 1956. *The Mathematics of Physics and Chemistry*, 2nd ed. Van Nostrand, Princeton, N.J.

Martens, C. S. and R. A. Berner, 1974. Methane production in the interstitial waters of sulfate-depleted marine sediments. *Science*, **185**, 1167–1169.

Martens, C. S., J. J. Wesolowski, R. C. Hariss, and R. Kaifer, 1973. Chlorine loss from Puerto Rican and San Francisco Bay area marine aerosols. *J. Geophys. Res.*, **78**, 8778–8792.

Mason, B. J., 1971. *Cloud Physics.* Clarendon, Oxford, England.

Matsunaga, K., M. Nishimura, and S. Konishi, 1975. Mercury in the Kuroshio and Oyashio regions and the Japan Sea. *Nature (Phys. Sci.)*, **258**, 224–225.

Mel'nik, Yu. P., 1972. *Termodinamicheskiye Konstanty dlya Analiza Uslovii Obrazovaniya Zheleznykh Rud.* Naukova Dumka, Kiev, USSR.

Meloy, T. P. and G. D. Gumtz, 1968/69. The fracture of single, brittle, heterogeneous particles—statistical derivation of the mass distribution equation. *Powder Technol.* **2**, 207–214.

Menard, H. W. and S. M. Smith, 1966. Hypsometry of ocean basin provinces. *J. Geophys. Res.*, **71**, 4305–4325.

Menard, H. W., 1974. *Geology, Resources and Society*. Freeman, San Francisco.

Menzel, D. W. and J. R. Ryther, 1970. Distribution and cycling of organic matter in the oceans. In *Organic Matter in Natural Waters*, D. W. Hood, Ed. University of Alaska Press, College, Alaska, pp. 31–54.

Miller, D. G., 1967. Application of irreversible thermodynamics to electrolyte solutions. II. Ionic coefficients L_{ij} for isothermal vector transport processes in ternary systems. *J. Phys. Chem.*, **71**, 616–632.

Miller, W. L. and A. R. Gordon, 1931. Numerical evaluation of infinite series and integrals which arise in certain problems of linear heat flow, electrochemical diffusion, etc. *J. Phys. Chem.*, **35**, 2785–2884.

Millero, F. J., 1971. The molal volumes of electrolytes. *Chem. Rev.*, **71**, 147–176.

Millero, F. J., 1975. The physical chemistry of estuaries. In *Marine Chemistry in the Coastal Environment*, T. M. Church, Ed. American Chemical Society, ACS Symposium Series 18, Washington, D.C., pp. 25–55.

Milliman, J. D., 1975. Dissolution of aragonite, Mg-calcite, and calcite in the North Atlantic Ocean. *Geology*, **3**, 461–462.

Milliman, J. D., 1977. Dissolution of calcium carbonate in the Sargasso Sea (Northwest Atlantic). In *The Fate of Fossil Fuel CO_2 in the Oceans*, N. R. Andersen and A. Malahoff, Eds. Plenum, New York, pp. 641–653.

Miranda, H. A., Jr. and R. Fenn, 1974. Stratospheric aerosol sizes. *Geophys. Res. Lett.*, **1**, 201–203.

Mitscherlich, G., 1973. *Wald, Wachstum, und Umwelt*. Sauerlander, Frankfurt am Main, Germany.

Mohr, E. C. J., F. A. Van Baren, and J. Van Schuylenborgh, 1972. *Tropical Soils*, 3rd ed. Mouton-Ichtiar Baru-Van Hoeve, The Hague, The Netherlands.

Moore, G. S. M. and H. E. Rose, 1973. The structure of powdered quartz, *Nature*, **242**, 187–190.

Morey, G. W., R. O. Fournier, and J. J. Rowe, 1962. The solubility of quartz in water in the temperature interval from 25° to 300°C. *Geochim. Cosmochim. Acta*, **26**, 1029–1043.

Morey, G. W., R. O. Fournier, and J. J. Rowe, 1964. The solubility of amorphous silica at 25°C. *J. Geophys. Res.*, **69**, 1995–2002.

Morse, J. W., 1974. Calculation of diffusive fluxes across the sediment-water interface. *J. Geophys. Res.*, **79**, 5045–5048.

Morse, J. W. and R. A. Berner, 1972. Dissolution kinetics of calcium carbonate in sea water: II. A kinetic origin for the lysocline. *Am. J. Sci.*, **272**, 840–851.

Munk, W. H., 1966. Abyssal recipes. *Deep-Sea Res.* **13**, 707–730.

Munk, W. H. and G. A. Riley, 1952. Absorption of nutrients by aquatic plants. *J. Mar. Res.*, **11**, 215–240.

Murray, J. and A. F. Renard, 1891. Report on deep-sea deposits based on the specimens collected during the voyage of H.M.S. Challenger in the years 1872–1876. In *Reports of the Voyage of H.M.S. Challenger*, Longmans, London, England.

Murray, J. W. and P. G. Brewer, 1978. The mechanisms of removal of iron, manganese, and other trace metals from sea water. In *Marine Manganese Deposits*, G. P. Glassby, Ed. American Elsevier, New York.

Nace, R. L., 1967. Water resources: A global problem with local roots. *Env. Sci. Tech.*, **1**, 550–560.

Naumov, G. B., B. N. Ryzhenko and I. L. Khodakovskii, 1971. *Spravochnik Termodinamicheskikh Velichin*. Atomizdat, Moscow, USSR. [Translation: *Handbook of Thermodynamic Data* (PB 226 722), National Technical Information Service, U.S. Department of Commerce, Springfield, Va.]

Neukum, G. and D. U. Wise, 1976. Mars: A standard crater curve and possible new time scale. *Science*, **194**, 1381–1387

Niehaus, F., 1976. *A Nonlinear Eight Level Tandem Model to Calculate the Future CO_2 and C-14-Burden to the Atmosphere*. Research Memorandum 76–35, International Institute for Applied Systems Analysis, Laxenburg, Austria.

Niiler, P. P., 1977. One-dimensional models of the seasonal thermocline. In *The Sea*, Vol. 6, E. D. Goldberg, I. N. McCave, J. J. O'Brian, and J. H. Steele, Eds. Wiley-Interscience, New York, pp. 97–115.

Northwood, D. O. and D. Lewis, 1968. Transformation of vaterite to calcite during grinding. *Am. Miner.*, **53**, 2089–2092.

Northwood, D. O. and D. Lewis, 1970. Strain induced calcite–aragonite transformation in calcium carbonate. *Can. Miner.*, **10**, 216–224.

Noyes, A. A. and W. R. Whitney, 1897. Ueber die Auflösungsgeschwindigkeit von festen Stoffen in ihren eigenen Lösungen. *Z. Phys. Chem.*, **23**, 689–692.

Oberbeck, V. R. and M. Aoyagi, 1972. Martian doublet craters. *J. Geophys. Res.*, **77**, 2419–2432.

Oeschger, H., V. Siegenthaler, V. Schotterer, and A. Gugelmann, 1975. A box diffusion model to study the carbon dioxide exchange in nature. *Tellus*, **27**, 168–192.

Okubo, A., 1954. A note on the decomposition of sinking remains of plankton organisms and its relationship to nutrient liberation. *J. Oceanogr. Soc. Japan*, **10**, 121–132.

Okubo, A., 1956. An additional note on the decomposition of sinking remains of plankton organisms and its relationship to nutrient liberation. *J. Oceanogr. Soc. Japan*, **12**, 45–47.

Okubo, A., 1971. Oceanic diffusion diagrams. *Deep-Sea Res.*, **18**, 789–802.

Olafsson, J., 1978. Report on the ICES international intercalibration of mercury in seawater. *Mar. Chem.,* **6**, 87–95.

Othmer, D. F. and M. S. Thakar, 1953. Correlating diffusion coefficients in liquids. *Ind. Eng. Chem.*, **45**, 589–593.

Parker, P. L., 1971. Petroleum—stable isotope ratio variation. In *Impingement of Man on the Oceans*, D. W. Hood, Ed. Wiley-Interscience, New York, pp. 431–444.

Parks, G. A., 1975. Adsorption in the marine environment. In *Chemical Oceanography*, 2nd ed, Vol. 1, J. P. Riley and G. Skirrow, Eds. Academic, New York, pp. 241–308.

Parsons, T. R. and M. Takahashi, 1973. *Biological Oceanographic Processes*. Pergamon, Oxford, England.

Payne, L. E. and W. H. Pell, 1960. The Stokes flow problem for a class of axially symmetric bodies. *J. Fluid Mech.*, **7**, 529–549.

Perry, E. A., Jr., J. M. Gieskes, and J. R. Lawrence, 1976. Mg, Ca, and $^{18}O/^{16}O$ exchange in the sediment-pore water system, Hole 149, DSDP. *Geochim. Cosmochim. Acta*, **40**, 413–423.

Perry, J. H., Ed., 1963. *Chemical Engineer's Handbook*, 4th ed. McGraw-Hill, New York.

Peters-Kümmerly, B. 1973. Untersuchungen über Zusammensetzung und Transport von Schwebstoffen in einigen Schweizer Flüssen. *Geogr. Helv.*, **28**, 137–151.

Peterson, M. N. A., 1966. Calcite: Rates of dissolution in a vertical profile in the Central Pacific. *Science*, **154**, 1542–1544.

Petrović, R., 1974. Diffusion of alkali ions in alkali feldspars. In *The Feldspars*, W. S. MacKenzie and J. Zussman, Eds. Manchester University, Manchester, England, pp. 174–182.

Pettijohn, F. J., P. E. Potter, and R. Siever, 1973. *Sand and Sandstone*. Springer-Verlag, New York.

Pettyjohn, E. S. and E. B. Christiansen, 1948. Effect of particle shape on free-settling rates of isometric particles. *Chem. Eng. Prog.*, **44**, 157–172.

Plummer, L. N., 1972. Rates of Mineral-Aqueous Solution Reactions. Unpublished doctoral dissertation. Northwestern University, Evanston, Ill.

Plummer, L. N., B. F. Jones, and A. H. Truesdell, 1976. WATQF— a Fortran IV version of WATEQ, a computer program for calculating chemical equilibrium of natural waters. *U.S. Geological Survey, Water-Resources Investigations* **76–13**, Washington, D.C.

Plummer, L. N. and T. M. L. Wigley, 1976. The dissolution of calcite in CO_2-saturated solutions at 25°C and 1 atmosphere total pressure. *Geochim. Cosmochim. Acta*, **40**, 191–202.

Poldervaart, A., 1955. Chemistry of the earth's crust. *Geol. Soc. Am. Spec. Pap.*, **62**, 119–144.

Pollard, W. G. and R. D. Present, 1948. Gaseous self-diffusion in long capillary tubes. *Phys. Rev.*, **73**, 762–774.

Pritchard, D. W., R. O. Reid, A. Okubo, and H. H. Carter, 1971. Physical processes of water movement and mixing. In *Radioactivity in the Marine Environment*. National Academy of Science, Washington, D.C., pp. 90–136.

Prospero, J. M. and T. N. Carlson, 1972. Vertical and areal distribution of Saharan dust over the western equatorial North Atlantic Ocean. *J. Geophys. Res.*, **77**, 5255–5265.

Pytkowicz, R. M., 1965. Calcium carbonate saturation in the ocean. *Limnol. Oceanogr.*, **10**, 220–225.

Quay, P., 1976. An experimental study of turbulent diffusion in lakes. Unpublished doctoral dissertation. Columbia University, New York.

Rad, V. von and H. Rösch, 1972. Mineralogy and origin of clay minerals, silica and authigenic silicates in Leg 14 sediments. In *Initial Reports of the Deep Sea Drilling Project*, Vol. 14. U.S. Government Printing Office, Washington, D.C., pp. 727–751.

Radke, L. F. and P. V. Hobbs, 1976. Cloud condensation nuclei on the Atlantic seaboard of the United States. *Science*, **193**, 999–1002.

Redfield, A. C., B. H. Ketchum, and F. A. Richards, 1963. The influence of organisms on the composition of seawater. In *The Sea*, Vol. 2, M. N. Hill, Ed. Wiley-Interscience, New York, pp. 26–77.

Renkin, E. M., 1954. Filtration, diffusion, and molecular sieving through porous cellulose membranes. *J. Gen. Physiol.*, **38**, 225–243.

Revelle, R. and R. Fairbridge, 1957. Carbonate and carbon dioxide. *Geol. Soc. Am. Mem.*, **67**, 239–296.

Rex, R. W. and E. D. Goldberg, 1958. Quartz contents of pelagic sediments of the Pacific Ocean. *Tellus*, **10**, 153–159.

Rheinheimer, G., 1974. *Aquatic Microbiology*. Wiley, New York.

Rich, C. I. and W. R. Black, 1964. Potassium exchange as affected by cation size, pH, and mineral structure. *Soil Sci.*, **97**, 384–390.

Rieke, H. H. and G. V. Chilingarian, 1974. *Compaction of Argillaceous Sediments*. Elsevier, Amsterdam, The Netherlands.

Riley, G. A., 1963. Theory of food-chain relations in the ocean. In *The Sea*, Vol. 2, M. N. Hill, Ed. Wiley-Interscience, New York, pp. 438–463.

Riley, G. A., H. Stommel, and D. F. Bumpus, 1949. Quantitative ecology of the plankton of the western North Atlantic. *Bull. Bingham Oceanog. Coll.*, **12**, 1–169.

Ristvet, B. L., 1977. Reverse weathering reactions within recent nearshore marine sediments, Kaneohe Bay, Oahu. Unpublished doctoral dissertation, Northwestern University, Evanston, Ill.

Robertson, E. C., 1967. Laboratory consolidation of carbonate sediments. In *Marine Geotechnique*, A. F. Richards, Ed. University of Illinois Press, Urbana, Ill., pp. 118–123.

Robie, R. A., B. S. Hemingway, and J. R. Fisher, 1977. Thermodynamic properties of minerals and related substances at 298.15 K and one bar (10^5 pascals) pressure, and at higher temperatures. *U.S. Geol. Surv. Bull.*, **1452**, 1–456.

Robie, R. A. and D. R. Waldbaum, 1968. Thermodynamic properties of minerals and related substances at 298.15°K (25°C) and one atmosphere (1.013 bars) pressure and at higher temperatures. *U.S. Geol. Surv. Bull.*, **1259**, 1–256.

Robinson, R. A. and R. H. Stokes, 1970. *Electrolyte Solutions*, 2nd rev. ed. Academic, New York.

Rogers, R. R., 1976. *A Short Course in Cloud Physics*. Pergamon, Oxford, England.

Romankevich, E. A., 1977. *Geokhimiya Organicheskogo Veshchestva v Okeane*. Nauka, Moscow, USSR.

Rooth, C. G. and H. G. Oestlund, 1972. Penetration of tritium into the Atlantic thermocline. *Deep-Sea Res.*, **19**, 481–492.

Rosin, P. and E. Rammler, 1933. The laws governing the fineness of powdered coal. *J. Inst. Fuel*, **7**, 29–36.

Rubey, W. W., 1933. Settling velocities of gravel, sand, and silt particles. *Am. J. Sci.*, **25**, 325–338.

Saffman, P. G. and J. S. Turner, 1956. On the collision of drops in turbulent clouds. *J. Fluid Mech.*, **1**, 16–30.

Sarmiento, J. L., H. W. Feely, W. S. Moore, A. E. Bainbridge, and W. S. Broecker, 1976. The relationship between vertical eddy diffusion and buoyancy gradient in the deep sea. *Earth Planet. Sci. Lett.*, **32**, 357–370.

Satterfield, C. N., 1970. *Mass Transfer in Heterogeneous Catalysis*. MIT Press, Cambridge, Mass.

Sayles, F. L., F. T. Manheim, and L. S. Waterman, 1973. Interstitial water studies on small core samples, Leg 15. In *Initial Reports of the Deep Sea Drilling Project*, Vol. 20. U.S. Government Printing Office, Washington, D.C., pp. 783–804.

Schachtschabel, P., 1940. Untersuchungen über die Sorption der Tonmineralien und organischen Bodenkolloide, und die Bestimmung des Anteils dieser Kolloide an der Sorption im Boden. *Kolloid Beih.* **51**, 199–276.

Scheidegger, A. E., 1960. *The Physics of Flow Through Porous Media*. University of Toronto Press, Toronto, Canada.

Schlich, R., E. S. W. Simpson, et al., 1974. Site 245. In *Initial Reports of the Deep Sea Drilling Project*, Vol. 25. U.S. Government Printing Office, Washington, D.C., pp. 187–236.

Schneider, S. H. and W. W. Kellogg, 1973. The chemical basis for climate change. In *Chemistry of the Lower Atmosphere*, S. I. Rasool, Ed. Plenum, New York, pp. 203–249.

Schrader, H.-J., 1972*a*. Anlösung und Konservation von Diatomeenschalen beim Absinken, am Beispiel des Landsort-Tiefs in der Ostsee. *Nova Hedwigia Beih.*. **39**. 191–216.

Schrader, H.-J., 1972*b*. Kieselsäure-Skelette in Sedimenten des Ibero-Marokkanishen Kontinentalrandes und angrenzender Tiefsee-Ebenen. *Meteor-Forsch. Ergebn.*, Ser. C, 10–39.

Schrader, R., R. Wissing, and H. Kubsch, 1969. Surface chemistry of mechanically activated quartz. *Z. Anorg. Allg. Chem.*, **365**, 191–198.

Schrader, R. and B. Hoffmann, 1969. Mechanical activation of calcium carbonate. *Z. Anorg. Allg. Chem.*, **369**, 41–47.

Schütz, L. and R. Jaenicke, 1974. Particle number and mass distribution above 10^{-4} cm radius in sand and aerosol of the Sahara Desert. *J. Appl. Met.*, **13**, 863–870.

Schwarzbach, M., 1963. *Climates of the Past*. Van Nostrand, London, England.

Scott, D. S. and F. A. L. Dullien, 1962. Diffusion of ideal gases in capillaries and porous solids. *J. Am. Inst. Chem. Eng.*, **8**, 113–117.

Seinfeld, J. H., 1975. *Air Pollution*. McGraw-Hill, New York.

Shaw, H. R., 1974. Diffusion of H_2O in granitic liquids: Part I. Experimental data; Part II. Mass transfer in magma chambers. In *Geochemical Transport and Kinetics*. A. W. Hofmann, B. J. Giletti, H. S. Yoder, Jr., and R. A. Yund, Eds. Carnegie Institution of Washington, Publication 634, Washington, D.C., pp. 139–170.

Sheldon, R. W., A. Prakash, and W. H. Sutcliffe, Jr., 1972. The size distribution of particles in the ocean. *Limnol. Oceanogr.*, **17**, 327–340.

Sherry, H. S., 1975. Ion exchange reactions involving crystalline aluminosilicate zeolites. In *The Nature of Seawater*, E. D. Goldberg, Ed. Dahlem Konferenzen, Berlin, Germany, pp. 523–553.

Shewmon, P. G., 1963. *Diffusion in Solids*. McGraw-Hill, New York.

Siegel, S. M., B. Z. Siegel, A. M. Eshleman, and K. Bachmann, 1973. Geothermal sources and distribution of mercury in Hawaii. *Env. Biol. Med.*, **2**, 81–89.

Siegenthaler, V. and H. Oeschger, 1978. Predicting future atmospheric carbon dioxide levels. *Science*, **199**, 388–395.

Siever, R. and N. Woodford, 1973. Sorption of silica by clay minerals. *Geochim. Cosmochim. Acta*, **37**, 1851–1880.

Sillén, L. G. and A. E. Martell, 1964, 1971. *Stability Constants of Metal-Ion Complexes*. The Chemical Society, Special Publications, **17**, **25**, London, England.

Silverman, S. R., 1964. Investigations of petroleum origin and evolution mechanisms by carbon isotope studies. In *Isotopic and Cosmic Chemistry*, H. Craig, S. L. Miller, and G. J. Wasserburg, Eds. North-Holland, Amsterdam, The Netherlands, pp. 92–102.

Skinner, K. J., 1976. Nitrogen fixation. *Chem. Eng. News*, Oct. 4, 22–35.

Smayda, T. J., 1971. Normal and accelerated sinking of phytoplankton in the sea. *Mar. Geol.*, **11**, 105–122.

Smithsonian Meteorological Tables, 1958, R. J. List, Ed. Smithsonian Institution, Washington, D.C.

Smoluchowski, M. von, 1917. Versuch einer mathematischen Theorie der Koagulationskinetik kolloider Lösungen. *Z. Phys. Chem.*, **92**, 129–168.

Spencer, D. W. and P. G. Brewer, 1971. Vertical advection, diffusion and redox potentials as controls on the distribution of manganese and other trace metals dissolved in waters of the Black Sea. *J. Geophys. Res.*, **76**, 5877–5892.

Stephen, H. and T. Stephen, Eds., 1963. *Solubilities of Inorganic and Organic Compounds*, Vol. **1**, 2 parts. Pergamon, Oxford, England.

Stephen, H. and T. Stephen, Eds., 1964. *Solubilities of Inorganic and Organic Compounds*, Vol. **2**, 2 parts. Pergamon, Oxford, England.

Stöber, W., 1967. Formation of silicic acid in aqueous suspensions of different silica modifications. *Adv. Chem. Ser.*, **67**, 161–182.

Stokes, G. G., 1851. On the effect of the internal friction of fluids on the motion of pendulums. *Trans. Camb. Philos. Soc.*, **9**, 8–106. (Reprinted in Stokes, G. G., 1901. *Mathematical and Physical Papers*, Vol. 3. Cambridge University Press, Cambridge, England, pp. 1–141.)

Stokes, R. H. and R. Mills, 1965. *Viscosity of Electrolytes and Related Properties*. Pergamon, Oxford, England.

Stommel, H., 1949. Trajectories of small bodies sinking slowly through convection cells. *J. Mar. Res.*, **8**, 24–29.

Stookey, S. D. and R. D. Maurer, 1967. Glass. In *Handbook of Physics*, 2nd ed, E. U. Condon and H. Odishaw, Eds. McGraw-Hill, New York, pp. 8-85–8-96.

Stumm, W., 1977. Chemical interaction in particle separation. *Env. Sci. Tech.*, **11**, 1066–1069.

Stumm, W. and P. A. Brauner, 1975. Chemical speciation. In *Chemical Oceanography*, 2nd ed, Vol. 1, J. P. Riley and G. Skirrow, Eds. Academic, London, England, pp. 173–239.

Stumm, W., C. P. Huang, and S. R. Jenkins, 1970. Specific chemical interaction affecting the stability of dispersed systems. *Croat. Chem. Acta*, **42**, 223–245.

Stumm, W. and J. J. Morgan, 1970. *Aquatic Chemistry*. Wiley-Interscience, New York.

Suess, H. E., 1955. Radiocarbon concentration in modern wood. *Science*, 122, 414–417.

Svensson, B. H. and R. Söderlund, Eds., 1976. *Nitrogen, Phosphorus and Sulphur—Global Cycles*. SCOPE Report 7, Ecol. Bull. 22, Stockholm, Sweden.

Sverdrup, H. U., M. W. Johnson and R. H. Fleming, 1942. *The Oceans: Their Physics, Chemistry, and General Biology*. Prentice-Hall, Englewood Cliffs, N. J.

Takahashi, H., 1959. Effect of dry grinding on kaolin minerals. *Bull. Chem. Soc. Japan*, **32**, 235–245.

Takahashi, T., 1975. Carbonate chemistry of seawater and the calcite compensation depth in the oceans. In *Dissolution of Deep-Sea Carbonates*, W. V. Sliter, A. W. H. Bé, and W. H. Berger, Eds. Cushman Foundation for Foraminiferal Research, Special Publication 13, Ithaca, N. Y., pp. 11–26.

Taylor, G. I., 1953. Dispersion of soluble matter in solvent flowing slowly through a tube. *Proc. R. Soc. Lond. A*, **219**, 186–203.

Taylor, G. I., 1954. The dispersion of matter in turbulent flow through a pipe. *Proc. R. Soc. Lond. A*, **223**, 446–468.

Taylor, S. R., 1964. Abundance of chemical elements in the continental crust: A new table. *Geochim. Cosmochim. Acta*, **28**, 1273–1285.

Taylor, T. D., 1968. Low Reynolds number flows. In *Basic Developments in Fluid Dynamics*, Vol. 2, M. Holt, Ed. Academic, New York, pp. 183–215.

Tchen, C.-M., 1954. Motion of small particles in skew shape suspended in a viscous liquid. *J. Appl. Phys.*, **25**, 463–473.

Tessenow, V. von, 1975. Lösungs-, Diffusion- und Sorptionsprozesse in der Oberschicht von Seesedimenten. *Arch. Hydrobiol. Suppl.*, **42**, 325–412.

Thorstenson, D. C. and F. T. Mackenzie, 1974. Time variability of pore water chemistry in recent carbonate sediments, Devil's Hole, Harrington Sound, Bermuda, *Geochim. Cosmochim. Acta*, **38**, 1–19.

Todd, D. K., 1964. Groundwater. In *Handbook of Applied Hydrology*, Ven Te Chow, Ed. McGraw-Hill, New York, pp. 12-1–13-55.

Toth, D. J., 1976. Organic and Inorganic Reactions in Near Shore and Deep-Sea Sediments. Unpublished doctoral dissertation. Northwestern University, Evanston, Ill.

Toth, D. J. and A. Lerman, 1977. Organic matter reactivity and sedimentation rates in the ocean. *Am. J. Sci.*, **277**, 465–485.

Truesdell, A. H. and C. L. Christ, 1968. Cation exchange in clays interpreted by regular solution theory. *Am. J. Sci.*, **266**, 402–412.

Truesdell, A. H. and B. F. Jones, 1974. WATEQ, a computer program for calculating chemical equilibria of natural waters. *U.S. Geol. Surv. J. Res.*, **2**, 233–248.

Turekian, K. K., 1965. Some aspects of the geochemistry of marine sediments. In *Chemical Oceanography*, Vol. 2, J. P. Riley and G. Skirrow, Eds. Academic, New York, pp. 81–126.

Turner, J. S., 1973. *Buoyancy Effects in Fluids*. Cambridge University Press, Cambridge, England.

Vanderborght, J.-P., R. Wollast, and G. Billen, 1977. Kinetic models of diagenesis in disturbed sediments. Part 1. Mass transfer properties and silica diagenesis. *Limnol. Oceanogr.*, **22**, 787–793.

Vanderborght, J.-P., R. Wollast, and G. Billen, 1977. Kinetic models of diagenesis in disturbed sediments. Part 2. Nitrogen diagenesis. *Limnol. Oceanogr.*, **22**, 794–803.

van Olphen, H., 1957. Surface conductance of various ion forms of bentonite in water and the electrical double layer. *J. Phys. Chem.*, **61**, 1276–1280.

van Olphen, H., *Clay Colloid Chemistry*. Wiley-Interscience, New York.

van Schaik, J. C. and W. K. Kemper, 1966. Chloride diffusion in clay-water systems. *Soil Sci. Soc. Am. Proc.*, **30**, 22–25.

Wagman, D. D., W. H. Evans, V. B. Parker, I. Halow, S. M. Bailey, and R. H. Schumm, 1968-1971. *Selected Values of Chemical Thermodynamic Properties*, National Bureau of Standards Tech. Note 270-3–270-6. U.S. Government Printing Office, Washington, D.C.

Walker, J. C. G., 1974. Stability of atmospheric oxygen. *Am. J. Sci.*, **274**, 193–214.

Wang, W. C., Y. L. Yung, A. A. Lacis, T. Mo, and J. E. Hansen, 1976. Greenhouse effects due to man-made perturbations of trace gases. *Science*, **194**, 685–690.

Wangersky, P. J., 1976. Particulate organic carbon in the Atlantic and Pacific Oceans. *Deep-Sea Res.*, **23**, 457–465.

Weast, R. C., Ed., 1974. *Handbook of Chemistry and Physics*. 55th ed. Chemical Rubber Co., Cleveland.

Weber, W. J., Jr., 1972. *Physicochemical Processes for Water Quality Control*. Wiley-Interscience, New York.

Weiler, R. R. and A. A. Mills, 1965. Surface properties and pore structure of marine sediments. *Deep-Sea Res.*, **12**, 511–52

Weiss, A., 1969. Organic derivatives of clay minerals, zeolites, and related minerals. In *Organic Geochemistry*, G. Eglinton and M. T. J. Murphy, Eds. Springer-Verlag, New York, pp. 737–781.

Weiss, H. V., M. Koide, and E. D. Goldberg, 1971. Mercury in a Greenland ice sheet. *Science*, **174**, 692–694.

Wendt, R. P. and M. Shamim, 1970. Isothermal diffusion in the system water-magnesium chloride-sodium chloride as studies with the rotating diaphragm cell. *J. Phys. Chem.* **74**, 2770–2783.

White, D. E., J. D. Hem, and G. A. Waring, 1963. Chemical composition of subsurface waters. *U.S. Geol. Surv. Prof. Pap.*, **440-F**.

Whittaker, R. H. and G. E. Likens, 1975. The biosphere and man. In *Primary Productivity of the Biosphere*, H. Lieth and R. H. Whittaker, Eds. Springer-Verlag, New York, pp. 306–328.

Whitworth, W. A., 1934. *Choice and Chance*, 5th ed. Stechert, New York.

Wild, A. and J. Keay, 1964. Cation-exchange equilibria with vermiculite. *J. Soil Sci.*, **15**, 136–144.

Wilke, C. R. and P. Chang, 1955. Correlation of diffusion coefficients in dilute solutions. *J. Am. Inst. Chem. Eng.*, **1**, 264–270.

Wimbush, M. and W. Munk, 1970. The benthic boundary layer. In *The Sea*, Vol. 4, E. D. Goldberg, Ed. Wiley-Interscience, New York, pp. 731–758.

Windom, H. L., 1969. Atmospheric dust records in permanent snowfields: Implications to marine sedimentation. *Geol. Soc. Am. Bull.*, **80**, 761–782.

Winsauer, W. O., H. M. Shearin, Jr., P. M. Masson, and M. Williams, 1952. Resistivity of brine-saturated sands in relation to pore geometry. *Am. Assoc. Petrol. Geol. Bull.*, **36**, 253–277.

Wolery, T. J. and N. H. Sleep, 1976. Hydrothermal circulation and geochemical flux at mid-ocean ridges. *J. Geol.*, **84**, 249–275.

Wollast, R., 1967. Kinetics of the alteration of K-feldspar in buffered solution at low temperature. *Geochim. Cosmochim. Acta*, **31**, 635–648.

Wollast, R., 1974. The silica problem. In *The Sea*, Vol. 5, E. D. Goldberg, Ed. Wiley-Interscience, New York, pp. 359–392.

Wollast, R., G. Billen, and F. T. Mackenzie, 1975. Behavior of mercury in natural systems and its global cycle. In *Ecological Toxicology Research*, A. D. McIntyre and C. F. Mills, Eds. Plenum, New York, pp. 145–166.

Wollast, R. and R. M. Garrels, 1971. Diffusion coefficient of silica in seawater. *Nature (Phys. Sci.)*, **229**, 94.

Wood, J. R., 1976. Thermodynamics of brine-salt equilibria II. The system $NaCl-KCl-H_2O$ from 0 to 200°C. *Geochim. Cosmochim. Acta*, **40**, 1211–1220.

Wyrtki, K., 1962. The oxygen minima in relation to ocean circulation. *Deep-Sea Res.*, **9**, 11–23.

Yaalon, D. H. and E. Ganor, 1968. Chemical composition of dew and dry fallout in Jerusalem, Israel. *Nature*, **217**, 1139–1140.

Yao, K.-M., M. T. Habibian, and C. R. O'Melia, 1971. Water and waste filtration: Concepts and application. *Environ. Sci. Techol.*, **5**, 1105–1112.

Yund, R. A. and T. F. Anderson, 1974. Oxygen isotope exchange between potassium feldspar and potassium chloride solution. In *Geochemical Transport and Kinetics*, A. W. Hofmann, B. J. Giletti, H. S. Yoder, Jr., and R. A. Yund, Eds. Carnegie Institution of Washington, Publication 634, Washington, D.C., pp. 99–105.

Zdanovskii, A. B., E. I. Lyakhovskaya, and R. E. Shleimovich, 1953. *Spravochnik Eksperimentalnykh Dannykh po Rastvorimosti Mnogo-Komponentnykh Vodno-Solevykh Sistem. Tom 1. Trekhkomponentnye Sistemy*. Goskhimizdat, Leningrad, USSR.

Zimen, K. E. and F. K. Altenhein, 1973. The future burden of industrial CO_2 on the atmosphere and the oceans. *Z. Naturforsch.*, **28a**, 1747–1752.

Zimen, K. E., P. Offerman, and G, Hartmann, 1977. Source functions of CO_2 and future CO_2 burden in the atmosphere. *Z. Naturforsch.*, **32a**, 1544–1554.

Zimmermann, P., 1961. Chemische und bakteriologische Untersuchungen im unteren Zürichsee während der Jahre 1948-1957. *Schweiz. Z. Hydrol.*, **23**, 342–397.

Symbols and Functions

Functions

erf   error function (Chapters 2, 5, Appendix A)

erfc   error function complement (Chapter 2, Appendix A)

exp( )   exponential, e raised to the power ( )

$f( )$ any function, used intermittently

$\mathcal{L}\{\ \}$ Laplace transformation (Appendix B)

$\mathcal{L}^{-1}\{\ \}$ Inverse of a Laplace transformation (Appendix B)

ln   logarithm to the base $e = 2.71828...$

log   logarithm to the base 10

sinh   hyperbolic sine (Chapter 8)

$\Gamma( )$ gamma function (Chapter 5)

$\Psi(a)$ function defined in Appendix B

Symbols

a thermodynamic activity (Chapters 3, 4, 7); constant parameter in Kelvin equation (Chapter 4); particle radius (Chapter 5, Appendix C); equatorial semiaxis of an ellipsoid of revolution (Chapter 6, Appendix C); constant defined in Appendix B

b constant in a vapor pressure equation (Chapter 4); parameter of particle-size distribution (Chapters 5, 6, 7)

c constant in permeability equations (Chapter 2); constant in size-distribution functions (Chapter 5); polar semiaxis of an ellipsoid of revolution (Chapter 6, Appendix C)

d differential

e base of natural logarithms (see ln under Functions)

f formation resistivity factor (Chapter 3); fugacity (Chapter 4); fraction (Chapters 5, 6)

g acceleration due to gravity

h a linear dimension, such as thickness of a layer,

used intermittently; height of a cylinder (Chapter 6, Appendix C); Planck's constant (Appendix D)

k   the Boltzmann constant (Chapters 3, 6, Appendix D)

k a rate parameter (or rate constant) of a flux or of a first-order chemical reaction (dimension time^{-1}); subscripts indicate rate constants of other orders

$\boldsymbol{k}$ permeability of a porous bed (Chapter 2)

ℓ linear distance of dispersal (Chapter 2)

m a constant parameter in size-distribution and reaction rate equations; stoichiometric coefficient in chemical reactions

n a constant parameter used intermittently; stoichiometric coefficient; number of moles (Chapter 4)

p partial pressure of a gas; constant in size-distribution equations (Chapter 5); notation for a solid phase (Chapter 5); ratio of the semiaxes of an ellipsoid (Chapter 6, Appendix C); constant of the Laplace transformation (Appendix B)

q water discharge rate (Chapter 2); notation for a reactive phase (Chapter 5)

r radius; root of a differential equation (Appendix B)

s Soret coefficient (Chapter 3)

t time; temperature in deg C, when identified

v volume fraction of solvent (Chapter 3); volume of a particle

x horizontal distance coordinate; a variable used intermittently

y horizontal distance coordinate; a variable used intermittently

z distance coordinate (Chapter 2); ionic charge (Chapter 3); vertical distance coordinate

A constant parameter in size-distribution equations; a constant parameter (Appendix B); cross-sectional area (Chapter 2)

B parameter in gas-diffusion equation (Chapter 3); parameter in the Stokes settling equation (Chapter 6, Appendix C)

C concentration; denotes a dissolved species, unless identified by subscripts

D coefficient of molecular diffusion or dispersal

E activation energy for diffusion (Chapter 3); Young's modulus (Chapter 5); collision efficiency factor (Chapter 6)

F flux

F	force (Chapter 6, Appendix C)
$\mathscr{F}$	faraday (Chapters 3, 4, Appendix D)
G	gradient (Chapters 2, 6); Gibbs free energy (Chapters 4, 5)
G	concentration in weight percent (Chapter 4)
J	rate of particle coagulation or aggregation (Chapter 6)
J	rate of regeneration (Chapter 6); rate of chemical reactions (Chapters 7, 8)
J_*	rate of a chemical reaction involving decomposition of organic matter (Chapters 7, 8)
K	coefficient of turbulent or eddy diffusion
K	equilibrium constant of a chemical reaction; distribution or adsorption coefficient (Chapters 7, 8);constant in an equation (Appendix B)
$\mathscr{K}$	coefficient of hydraulic conductivity (Chapter 2)
L	distance or length (Chapter 2); Onsager's coefficients (Chapter 3); dimension of length, latent heat of evaporation (Chapter 4)
M	mass; dimension of mass; gram-formula weight (identified by subscripts); molal or molar concentration (Chapter 4); mass-concentration of suspended materials (Chapter 6)
N	age-distribution function (Chapter 1), Brunt-Väisälä stability frequency (Chapter 2); number or number-concentration of particles (Chapters 5, 6)
N	Avogadro's number (Appendix D)
P	pressure
$\mathscr{P}$	collision frequency function (Chapter 6)
Q	fuel reserves (Chapter 1); activity quotient (Chapter 4); river water discharge (Chapter 5)
R	gas constant; electrical resistance (Chapter 3); radial distance coordinate (Chapter 4); radius of fragments (Chapter 5); radius (Chapter 6, Appendix C); shape ratio parameter (Tables 5.5, 6.3, Appendix C)
$\mathscr{R}$	rate of a chemical reaction (Chapters 3, 7, 8)
S	area or surface area; supersaturation (Chapter 4)
T	temperature in deg K; dimension of time; a fixed time (Chapter 1)
T_f	tensile strength (Chapter 5)
U	velocity
V	volume
W	specific surface energy (Chapter 5)
X	a variable (Chapter 5); mole fraction (Chapters 4, 7)

α rate constant of fuel consumption (Chapter 1); coefficient of thermal expansion (Chapters 2, 5); thermal coefficient of diffusion (Chapter 3); Bunsen's gas solubility coefficient (Chapter 4); fragmentation multiplicity (Chapter 5); particle-shape factor (Chapter 6, Appendix C); stoichiometric coefficient in a reaction (Chapter 8)

β biota growth factor (Chapter 1); a parameter of dimensions (L^{-1}) (Chapters 6, 7, 8)

γ activity coefficient of an aqueous species in solution (Chapters 3, 4, 7); constant parameter in size-distribution equations (Chapter 5)

∂ partial differential

δ delta notation of isotope enrichment (Chapters 1, 6)

ε radial rate of dissolution of a sphere

ϵ energy dissipation rate (Chapter 6)

η viscosity

θ tortuosity of a porous medium (Chapter 3); potential temperature (Figure 6.6)

κ coefficients of thermal conductivity (Chapters 4, 5); rate constant of chemical reactions (Chapters 7, 8)

λ rate of removal (Chapter 1); ionic conductance in solution (Chapter 3); coefficient of gas solubility (Chapter 4); activity coefficient in a solid or adsorbed state (Chapter 7); radioactive decay constant (Chapter 7); free path of molecules in a gas (Appendix C)

μ chemical potential (Chapters 3, 4); mean of distribution (Chapter 5)

ν stoichiometric coefficient (Chapters 3, 4); kinematic viscosity

ξ buffer factor for CO_2 (Chapter 1)

π 3.14159...

ρ density

σ measure of dispersal (Chapter 2); interfacial tension (Chapters 4, 6); scavenging efficiency of rain (Chapter 4); standard deviation (Chapter 5); density function (Figure 6.6)

τ residence time

φ hydraulic head and potential (Chapter 2)

ϕ porosity or volume fraction occupied by water or gas

χ parameter in the equation of droplet evaporation (Chapter 4)

Δ increment or finite difference

Γ measure of adsorption (Chapter 7)

Λ coefficient of gas solubil- Φ function in the alkalinity
        ity (Chapter 4)   equation (Chapter 1)
Σ sum

Author Index

Aitchison, J., 193
Allen, T., 420
Aller, R. C., 83, 339, 392
Altenhein, F. K., 35, 39
Anderson, T. F., 115
Andren, A. W., 172
Anikouchine, W. A., 389
Aoyagi, M., 182
Archie, G. E., 90, 91
Arrhenius, G., 298
Aston, S. R., 348
Athy, L. F., 380
Auluck, F. C., 197
Austin, L. G., 197

Bacastow, R., 31, 38, 39
Bachmann, K., 168
Back, W., 48, 49
Bailey, S. M., 173, 226
Bainbridge, A. E., 71, 72
Baker, C. W., 170
Bakker, J. P., 203
Barneisel, R. I., 377
Barnes, R. O., 381
Bartlett, R. W., 115, 240, 241, 243
Bé, A. W. H., 228, 270
Bear, J., 46, 47, 52-55, 61, 62, 64, 90, 91
Beard, D. C., 52, 54
Beardsley, G. F., Jr., 194
Beck, R. E., 93
Benedek, G. B., 87
Bennett, J. G., 197
Ben-Yaakov, S., 322
Berger, W. H., 261, 269, 320, 322, 365,
   407
Berner, R. A., 83, 226, 229, 320-322, 339,
   347, 375, 381, 392
Bett, K. E., 87

Betzer, P. R., 197, 277, 280, 282, 287
Bickley, F., 254
Billen, G., 171, 172, 339
Biscaye, P. E., 279
Bischof, W., 38, 39
Black, W. R., 345
Bladel, R. van, 345
Blifford, I. H., 213
Boerboom, A. J. H., 96
Bolin, B., 8, 24, 35, 36, 38, 39
Bolt, G. H., 232, 345
Börnstein, R., see Landolt-Börnstein
Bowen, H. J. M., 33
Bradbury, J. P., 261
Braitsch, O., 226
Bramlette, M. N., 320
Brauner, P. A., 171
Bredig, M. A., 205
Brenner, H., 52, 268, 420, 426-430, 432
Brewer, L., 138
Brewer, P. G., 279, 300, 306, 348
Bricker, O. P., 252, 254, 256, 389
Broecker, S., 322
Broecker, W. S., 36, 38, 39, 71, 72, 96, 144,
   322, 331, 332
Brown, J. A. C., 193
Bruggenwert, M. G. M., 232
Brun-Cottan, J. C., 274, 277
Brunskill, G. J., 300
Bryant, W. R., 52
Bryson, R. A., 213
Bumpus, D. F., 303
Burnet, T., frontispiece
Burns, J. H., 205
Burton, J. D., 318
Busenberg, E., 115, 244, 246
Butler, J. C., 202
Byers, H. R., 128, 272

Cadle, R. D., 127
Cappi, J. B., 87
Carder, K. L., 194, 197, 277, 280, 282, 287
Carlson, T. N., 213
Carnahan, B., 415
Carroll, D., 348
Carslaw, H. S., 135, 142, 410, 415, 418
Carter, H. H., 67
Cerrai, F., 347
Champion, W. A., 262
Chang, P., 94
Chapman, C. R., 197
Charnley, T., 169
Chen, C.-T., 227
Chilingarian, G. V., 52, 380
Christ, C. L., 138, 173, 226, 345, 346
Christiansen, E. B., 425
Churchman, G. J., 213
Clayton, R. N., 213
Clelland, D. W., 206
Clemency, C. V., 115, 244, 246
Cline, M. G., 248, 253
Cochran, J. K., 83, 339, 392
Cook, P. J., 358
Correns, C. W., 228, 240, 244
Coull, J., 425
Craig, H., 145, 306, 315
Crank, J., 142, 410, 415
Criado, J. M., 206
Csanady, G. T., 66, 68, 304
Culberson, C., 320
Curran, P. F., 78, 79, 83, 98
Currie, J. A., 103
Cyberski, J., 262

Dacey, M. F., 300, 302, 430
Danckwerts, P. V., 62
Danielsen, E. F., 213
Dapples, E. C., 193, 194, 211
Darcy, H., 46
Davies, C. N., 422
Davies, T. A., 260
Dayhoff, M. O., 227
Deevey, E. S., Jr., 23
Defay, R., 128
Deflache, A. P., 52
Degens, E. T., 36
Delaney, J. M., 226
Delany, A. C., 213
Dell, C. I., 261

Delwiche, C. C., 23
Dempster, P. B., 206
Dendy, F. E., 262
Dolcater, D. L., 345
Drake, R. L., 289
Drever, J. I., 361
Drozdova, V. M., 149
Duce, R. A., 150
Dufour, L., 128
Duguid, H. A., 135
Dullien, F. A. L., 107
Duursma, E. K., 348

Eagleson, P. S., 123
Eaton, G. P., 213
Eberling, C., 178
Eck, R. V., 227
Eckart, C., 69-71
Edgar, N. T., 403
Edmond, J. M., 320, 363, 364
Edzwald, J. K., 290
Eisenberg, D. S., 120
Eittreim, S. L., 279
Ekdahl, C. A., Jr., 35, 39
Eliason, J. R., 345
Ellis, J. H., 377
Emery, K. O., 380
Engelhardt, W. von, 47, 240, 244, 380
Eriksson, E., 36
Eshleman, A. M., 168
Eslinger, E. V., 204, 205
Eugster, H. P., 259
Evans, R. B., 107
Evans, W. H., 173, 226

Fairbridge, R., 320
Fanning, K. A., 83, 350
Feely, H. W., 71, 72
Feely, R. A., 300
Fenn, R., 197
Ferrell, R. T., 94
Fiadeiro, M. E., 306, 318, 319
Fick, A., 73-77
Ficklin, W. H., 170
Fisher, J. R., 226
Fitts, D. D., 83, 85
Fitzgerald, W. F., 170
Fleischer, M., 168, 170
Fleming, R. H., 301
Foland, K. A., 115

Fournier, R. O., 248, 387
Fredericks, J., 279
Friedlander, S. K., 214, 215, 290, 295, 296
Frischat, G. H., 115, 117, 118

Ganor, E., 151
Garrels, R. M., 3, 4, 8, 18-22, 24, 26, 28-30, 32, 35, 83, 123, 138, 149, 150, 167, 173, 218, 219, 226, 244, 254, 331, 345, 346
Gast, R. G., 345
Gaudin, A. M., 197
Gautschi, W., 253, 256, 410
Geins, W., 178
Gibbs, R. J., 422
Gieskes, J. M., 259, 320, 363, 364, 386, 392-394, 403, 405, 407
Giletti, B. J., 115
Gillette, D. A., 213
Gjessing, E. T., 151
Glew, D. N., 169, 174
Glymph, L. M., 262
Goldberg, E. D., 127, 168, 213, 217, 261, 381
Goldhaber, M. B., 83, 339, 392
Goldich, S. S., 218
Goldsmith, J. R., 205
Golterman, H. L., 392
Gordon, A. R., 252
Gordon, J. E., 181
Gordon, L. I., 145
Granat, L., 149, 153, 164
Greene, G. M., 202
Gregor, C. B., 4, 217, 218
Gregory, S., 81, 83, 347
Griffin, J. J., 213, 261
Griffith, A. A., 181
Grodzian, D. J. M., 206, 207
Gugelmann, A., 35, 38, 39
Guinasso, N. L., Jr., 339
Gumtz, G. D., 197

Habibian, M. T., 276
Hahn, H. H., 289, 290
Halley, E., 123
Halow, I., 173, 226
Hames, D. A., 169, 174
Hamilton, E. L., 380
Hanley, A., 279
Hansen, J. E., 32
Hanshaw, B. B., 48, 49

Happel, J., 52, 268, 420, 427-430, 432
Hardie, L. A., 259
Harris, C. C., 194, 197, 198
Harris, J. E., 300
Harrison, S. M., 228
Hariss, R. C., 150, 173
Hartmann, G., 34, 37, 39
Haul, R. A. W., 115
Hay, R. L., 237
Hay, W. H., 260
Heald, W. R., 345
Heath, G. R., 318, 320, 407
Heimann, R. B., 224
Heiss, J. F., 425
Helgeson, H. C., 226, 227, 240
Hem, J. D., 237
Hemingway, B. S., 226
Henriksen, A., 151
Hidy, G. M., 139, 145, 147, 290, 293, 295, 296
Himmelblau, D. M., 94-97
Hobbs, P. V., 126
Hoffmann, B., 206
Holdren, G. R., Jr., 229
Holeman, J. N., 218
Holland, H. D., 28, 375
Holm-Hansen, O., 315
Honjo, S., 270, 273, 322, 328
Horne, R. A., 87
Howard, P. F., 244
Hower, J., 204, 205
Hower, M. E., 204, 205
Huang, C. P., 289
Hubbert, M. K., 47
Huene, R. von, 379
Hunt, C., 3, 123, 150, 167
Hunt, C. D., 170
Hurd, D. C., 389, 390

Imboden, D. M., 314, 367, 375

Jackson, M. L., 205, 213, 345
Jaeger, J. C., 135, 142, 410, 415, 418
Jaenicke, R., 216
Jamieson, J. C., 205, 206
Jaquet, J.-M., 261
Jassby, A., 71
Jenkins, S. R., 289
Jerlov, N. G., 304
Johannessen, M., 151

Johnson, D. S., 87
Johnson, M. W., 301
Johnson, N. L., 193, 195, 197
Jones, B. F., 227, 237, 259
Jost, W., 78, 82, 110, 115
Junge, C. E., 123, 126, 139, 141, 145, 148, 150, 197

Kabanova, J. G., 315
Kadar, S., 279
Kaifer, R., 150
Kanwisher, J., 144
Kaplan, I. R., 322
Katchalsky, A., 78, 79, 83, 98
Kauzmann, W., 120
Keay, J., 345
Keeling, C. D., 31, 35, 36, 38, 39, 41, 163
Kellogg, W. W., 127, 217
Kemp, A. L. W., 261
Kemper, W. K., 377
Kennedy, C. S., 206
Kennedy, G. C., 206
Kennedy, V. C., 248, 249
Ketchum, B. H., 23
Khodakovskii, I. L., 226
King, E. A., Jr., 202, 210
Kirkham, D. H., 227
Kleyn, G., 96
Klimpel, R. R., 197
Klinkenberg, L. J., 90
Klobe, W. D., 345
Klotz, I. M., 138
Koblentz-Mishke, O. I., 315
Koide, M., 168, 261
Konishi, S., 170
Kothari, D. S., 197
Kotz, S., 193, 195, 197
Kraus, E. B., 294
Krishnaswami, S., 261
Krumbein, W. C., 51, 52, 183, 186, 194, 202, 221, 264
Kubsch, H., 206
Kuenen, P. H., 221
Kullenberg, G., 71
Kunishi, H. M., 345
Kuznetsov, S. I., 300

Lacis, A. A., 32
Lal, D., 197, 198, 261, 270, 274, 277, 300, 302, 319, 327, 337, 430

Lamb, H., 420, 428, 430
Lamb, H. H., 212
Landölt, H. H., see Landölt-Börnstein
Landölt-Börnstein, 89, 99, 125, 139, 140, 175, 226
Langleben, M. P., 268
Lapple, C. E., 422
Lasaga, A. C., 375
Laudelout, H., 345
Lawrence, J. R., 403, 405, 407
Leak, W. B., 8
Learned, R. E., 170
Leopold, L. B., 202, 220, 222
Levich, V. G., 294
Levorsen, A. I., 33
Lewis, D., 205, 206
Lewis, G. N., 138
Lewis, R. J., 345
Li, Y.-H., 8, 36, 38, 39, 71, 81, 83, 331, 332, 347
Liesegang, R. E., 120, 121
Lietzke, T. A., 347, 402
Likens, G. E., 23, 24, 35, 149, 153, 314
Lin, I. J., 206, 207
Link, D. A., 422
Linke, W. F., 226
Lippincott, E. R., 227
Lisitzin, A. P., 261, 277, 328
Liss, P. S., 145, 318
Livingstone, D. A., 219, 237, 314
Lotse, E. G., 345
Luce, R. W., 115, 240, 241, 243
Luther, H. A., 415
Lyakhovskaya, E. I., 226

McCarthy, J. H., Jr., 170
McCave, I. N., 277
McClelland, J. E., 240
McDonald, J. E., 126
McKeague, J. D., 248, 253
MacDonald, G. J. F., 206
Machta, L., 38
Mackenzie, F. T., 3, 4, 8, 18-22, 24, 26, 28-30, 32, 35, 123, 149, 150, 167, 171, 172, 218, 219, 252, 254, 256, 331, 339, 389
MacKenzie, R. C., 205
Manheim, F. T., 91, 92, 392, 393, 403
Manning, A. B., 197
Manning, J. R., 82, 110
Margenau, H., 412

Martell, A. E., 226
Martens, C. S., 83, 150, 151, 339, 381, 392
Mason, B. J., 128, 268
Mason, E. A., 107
Masson, P. M., 91
Matsunaga, K., 170
Matthews, M. D., 422
Maurer, R. D., 181
Mel'nik, Yu. P., 226
Meloy, T. P., 197
Menard, H. W., 258, 318
Menzel, D. W., 315
Meuschke, J. L., 170
Mezzandri, M. G., 347
Miller, D. G., 83
Miller, J. P., 202, 220, 222
Miller, W. L., 252
Millero, F. J., 227
Milliman, J. D., 321, 322
Mills, A. A., 390
Mills, R., 88, 97
Milne, A. A., 205
Miranda, H. A., Jr., 197
Mitscherlich, G., 13
Mo, T., 32
Moberly, R., Jr., 407
Mohr, E. C. J., 204, 225
Monk, G. D., 51, 52
Moore, G. S. M., 206
Moore, W. S., 71, 72
Morey, G. W., 248
Morgan, J. J., 138, 162, 173, 226
Morse, J. W., 228, 320-322, 339
Munk, W. H., 71, 306, 315, 338, 430
Murphy, G. M., 412
Murray, J., 320
Murray, J. W., 348

Nace, R. L., 123, 258
Nadiv, S., 206, 207
Nagarajan, G., 227
Naumov, G. B., 226
Nesbitt, H. W., 226
Neukum, G., 197
Niehaus, F., 38-40
Niiler, P. P., 294
Nishimura, M., 170
Northwood, D. O., 205, 206
Noyes, A. A., 230

Oberbeck, V. R., 182
Oeschger, H., 35, 38, 39
Oestlund, H. G., 71
Offerman, P., 34, 37, 39
Okubo, A., 66, 67, 274
Olafsson, J., 170
O'Melia, C. R., 276, 290
Othmer, D. F., 94

Page, A. L., 345
Pak, H., 194
Parker, P. L., 36
Parker, V. B., 173, 226
Parkin, D. W., 213
Parks, G. A., 115, 240, 241, 243, 346
Parsons, T. R., 315
Payne, L. E., 431
Pell, W. H., 431
Peng, T.-H., 36, 38, 39, 96, 144
Perdok, W. G., 221
Perry, E. A., Jr., 204, 205, 403, 405, 407
Perry, J. H., 102, 105, 140
Peters-Kümmerly, B., 220, 221
Peterson, M. N. A., 320, 322, 323
Petrenchuk, O. P., 149
Petrović, R., 115
Pettijohn, F. J., 183, 221, 264
Pettyjohn, E. S., 425
Phillips, R. E., 377
Pilson, M. E. Q., 83
Piper, D. J. W., 379
Pitzer, K. S., 138
Plummer, L. N., 227, 230-232
Poldervaart, A., 4
Pollard, W. G., 107
Potter, P. E., 221
Powell, T., 71
Prakash, A., 279
Present, R. D., 107
Pritchard, D. W., 67
Prospero, J. M., 213
Purcell, E. M., 87
Pytkowicz, R. M., 320

Quay, P., 71

Rad, V. von, 407
Radke, L. F., 126
Rammler, E., 193, 197
Randall, M., 138

Redfield, A. C., 23
Reid, R. O., 67
Reimann, B. E. F., 213
Renard, A. F., 320
Renkin, E. M., 93
Rettig, S. L., 259
Revelle, R., 320
Rex, R. W., 213
Rheinheimer, G., 381
Rich, C. I., 345
Richards, F. A., 23
Rieke, H. H., 52, 380
Riley, G. A., 303, 315, 430
Ristvet, B. L., 351
Ritchie, P. D., 206
Rittenberg, S. C., 380
Robertson, E. C., 52
Robie, R. A., 206, 209, 226
Robinson, R. A., 78-82, 88, 89, 120, 132
Rodhe, H., 8
Rogers, R. R., 123, 128, 136, 143
Romankevich, E. A., 315
Rooth, C. G., 71
Rösch, H., 407
Rose, H. E., 206
Rosenfeld, J. K., 83, 339, 392
Rosin, P., 193, 197
Rowe, J. J., 248
Rubey, W. W., 422
Ruth, E., 322
Ryther, J. R., 315
Ryzhenko, B. N., 226

Sachs, P. L., 279
Sackett, W. M., 300
Saffman, P. G., 291, 296
Sarmiento, J. L., 71, 72
Satterfield, C. N., 95, 102, 103, 105, 107, 120
Saunders, J. B., 403
Sayles, F. L., 392, 393, 403
Schachtschabel, P., 348
Scheidegger, A. E., 61
Schink, D. R., 339
Schlich, R., 386
Schneider, S. H., 127, 217
Schotterer, V., 35, 38, 39
Schrader, H. J., 229, 391
Schrader, R., 206
Schultz, J. S., 93

Schumm, R. H., 173, 226
Schütz, L., 216
Schwarzbach, M., 212
Scott, D. S., 107
Seiler, W., 178
Seinfeld, J. H., 150, 152
Selezneva, E. S., 149
Semet, M. P., 115
Shamim, M., 83, 85
Shaw, H. R., 115
Shearin, H. M., Jr., 91
Sheldon, R. W., 277, 279
Sherry, H. S., 115
Shewmon, P. G., 110
Shleimovich, R. E., 226
Siegel, B. Z., 168
Siegel, S. M., 168
Siegenthaler, V., 35, 38, 39
Siever, R., 221, 249
Sillén, L. G., 226
Silverman, S. R., 36
Simpson, E. S. W., 386
Skinner, H. A., 169
Skinner, K. J., 31
Sleep, N. H., 258
Slemer, F., 178
Smayda, T. J., 269
Smith, C. L., 279
Smith, S. M., 318
Smoluchowski, M. von, 289
Söderlund, R., 150
Soutar, A., 261
Southam, J. R., 260
Spencer, D. W., 279, 300, 306
Sridhar, K., 213
Stampfer, J. F., Jr., 135
Stein, L. H., 115
Stephen, H., 132, 226
Stephen, T., 132, 226
Stöber, W., 233
Stokes, G. G., 263, 421
Stokes, R. H., 78-82, 88, 89, 97, 120, 132
Stommel, H., 303, 305
Stookey, S. D., 181
Strickland, J. D. H., 315
Stumm, W., 138, 162, 171, 173, 226, 289, 290
Suess, H. E., 33
Sutcliffe, W. H., Jr., 279
Svensson, B. H., 150

Sverdrup, H. U., 301
Svistov, P. F., 149
Syers, J. K., 213, 345

Takahashi, H., 205
Takahashi, M., 315
Takahashi, T., 322, 331, 332
Taylor, G. I., 61
Taylor, S. R., 3
Taylor, T. D., 420
Tchen, C.-M., 432
Tessenow, V. von, 353
Thakar, M. S., 94
Theyer, F., 389, 390
Thomas, H. C., 345
Thomas, R. L., 261
Thorstenson, D. C., 339
Tisdel, F. W., 194, 202
Todd, D. K., 46
Toth, D. J., 351, 359, 372, 383, 392, 393,
   397, 398
Trabant, P. K., 52
Trillo, J. M., 206
Triulzi, C., 347
Truesdell, A. H., 227, 345
Truog, E., 205
Tschopp, J., 342
Turekian, K. K., 261, 320, 331
Turner, J. S., 69, 291, 294, 296

Upchurch, J. B., 290

Van Baren, F. A., 204, 225
Vanderborght, J.-P., 339
van Olphen, H., 93, 309, 377
van Schaik, J. C., 377
Van Schuylenborgh, J., 204, 225
Volkovinsky, V. V., 315

Wagman, D. D., 173, 226
Waldbaum, D. R., 206, 209, 226
Walker, J. C. G., 28
Wang, W. C., 32
Wangersky, P. J., 315

Waring, G. A., 237
Waterman, L. S., 91, 392, 393, 403
Watson, G. M., 107
Weast, R. C., 3, 163
Weber, W. J., Jr., 290
Weiler, R. R., 347, 390
Weiss, A., 119
Weiss, H. V., 168
Wendt, R. P., 83, 85
Wesolowski, J. J., 150
Weyl, P. K., 52, 54
White, D. E., 237
Whitney, W. R., 230
Whittaker, R. H., 24, 35
Whitworth, W. A., 197
Wigley, T. M. L., 230
Wild, A., 345
Wilke, C. R., 94
Wilkes, J. O., 415
Williams, M., 91
Williams, P. M., 315
Wilson, R. B., 262
Wimbush, M., 338
Windom, H. L., 213, 261
Winsauer, W. O., 91
Wise, D. U., 197
Wissing, R., 206
Wolery, T. J., 258
Wollast, R., 83, 171, 172, 254, 339, 389
Wolman, M. G., 202, 220, 222
Wood, J. R., 227
Woodford, N., 249
Worsley, T. R., 260
Wright, R. F., 151
Wyrtki, K., 306

Yaalon, D. H., 151
Yao, K.-M., 276
Yund, R. A., 115
Yung, Y. L., 32

Zdanovskii, A. B., 226
Zimen, K. E., 34, 35, 37, 39
Zimmermann, P., 70

Subject Index

Abundance, elements, 3, 372
Activity:
  chemical potential, 78
  definition, 79
  equilibrium, 138, 342-347
  water in saturated solutions, 132
Activity coefficient, 78-79, 82, 344
Adsorption, 142, 238, 341-349, 351, 355-356, 360, 374-375, 393
Advection:
  definition, 44-45
  in diagenesis, 340-342, 378-380
  effectiveness, 58-60
  pore water, 379
Age:
  humans, 8
  lunar soil, 210
  oceanic cherts, 407-408
  sediments, 8, 406
  trees, 8, 11, 18
Age distributions, 8-12
Aggregation:
  soils, 215-216
  suspensions, 288, 308
Aitken particles, 126
Algae, calcareous, 319
Alkalinity:
  definition, 162
  ocean water, 41-42
  rain water, 161-166
Alps, 213
Amphibole, 213, 218, 228
Analcite, 114, 119, 389
Anatase, 206
Annite, 114
Apatite, 373, 393
Aragonite, 205-206, 209, 219, 319, 321-322, 362

Atmosphere:
  chemical composition, 3
  moles of gases, 150
  thickness, 2
  volume, 127
Atmospheric particles:
  production rate, 127, 146
  removal rates, 146, 147
Augite, 218, 224-225

Bacteria:
  coagulation, 298-299
  lake water, 300
  ocean water, 25, 301
  sediments, 380-381
Baikal, Lake, 300, 301
Baltic Sea, 71
Barbados, 213
Basalt, 237, 348, 371
Bentonite, 345, 389
Bermuda, 213
Biological productivity:
  cessation, 21-23, 31-32
  lakes, 300, 314
  land, 31, 35, 40, 183
  models, 39, 312-318
  ocean, 24, 26, 35, 315
  sediments, 380-383
Biota growth factor, 39
Biotite, 218
Bioturbation, 339
Black Sea, 300, 306
Boundary layer:
  gas-solution, 141
  ocean-atmosphere, 144, 177
  ocean-sediment, 338
  see also Interface; Surface layer
Bowen reaction series, 218

Brachiopods, 320
Brines, 132, 259, 308-312
Brownian motion, 290-291, 294-297
Brunt-Väisälä frequency, *see* Stability
    frequency
Buffer factor, 36, 38

Calcite, 25, 149, 151, 205-206, 209, 219,
    228, 232-233, 300, 319-323, 362-364
Carbon:
    abundance, 3
    cycle, 23-32
    fuels, 33
    isotopes, 33-34, 36
    lakes, 314
    ocean, 315
    organic matter, 23, 33, 380, 382
    regeneration, 313, 331-332
    sedimentation, 25, 30, 331
Carbonate sediments, 24-25, 40, 52, 319-
    320, 336-338, 365, 402-403
Carbon dioxide:
    carbon cycle, 23-32
    fuel burning, 32-40
    ocean, 332
    pH solutions, 149, 155, 230
    soils, 108-109
    water-gas exchange, 38, 41-42, 140, 155
Carborundum, 103
Cariaco Trench, 392, 397-398
Caribbean Sea, 350, 382, 392, 398, 403
Chabasite, 114, 119
Chalcedony, 387
Chalk, 403-404
Challenger expedition, 320
Chemical equilibria:
    calcite, 25, 363-364
    carbon dioxide, 41-42, 155
    cation exchange, 345
    gases, 150, 155-159
    mercury species, 169, 171, 174-176
    mineral polymorphs, 206-209
    primitive atmospheres, 227
    silica minerals, 345
    water, 124, 132
    see also Solubility
Chemical potential, 78
Chert, 403, 407-408
Chlorite, 213, 224-225, 261, 345
Chromite, 224

Clays:
    airborne, 213
    cation exchange, 345, 348
    dissolution of silica, 389
    oceanic sediments, 345, 348
    surface areas, 390
Clinoptilolite, 254, 372, 403
Cloud nuclei, 126
Coagulation, 288-299
Coccoliths, 270, 328, 390
Coesite, 206-207
Collision frequency function, 289, 295
Comminution of grains, 205-208
Compensation depth, 300-301
Compressibility, 182
Conductivity:
    electric, 90-93, 377-378
    hydraulic, 47-48, 50
    ionic, 80
Coral reefs, 319
Coulomb's experiments, 421
Coupled reactions, 383-385
Cristobalite, 233, 387, 389, 403, 407
Crust:
    chemical composition, 3, 370
    weathering, 217-218, 221, 237
Cunningham's factor, 263-264, 422

Darcy's law, 46-50
    deviations from, 54-56
Dawson's integral, 251-256
Denudation, 6, 204, 212, 221, 261-262
Dew, 151
Diagenesis, 369
Diagenetic equation, 375
Diamond, 206
Diatoms, 229, 269-271, 277, 353, 373, 390
Diffusion:
    definition, 56
    effectiveness, 58-60
    Fick's experiments, 73-77
    Knudsen, 57, 104-107
    in sediments, 339-341, 373-378
    thermal, 57, 96-100
    see also Eddy diffusion; Molecular dif-
        fusion
Diffusional velocity, 78-79
Diffusion coefficients:
    in air, 103, 105
    in aqueous solutions, 85, 89, 97

in environment, 57
intrinsic, 82
in pore water, 83, 92
in sea water, 83
in solids, 111-115, 118-120
in water, 81, 96, 120
Disaggregation, 202-204
Discharge, 48-49, 220-221
Dispersal:
  coefficients, 64-65
  flow, 60-64
  lakes and oceans, 67-69
  velocity, 68-69
Dissolution:
  aragonite, 321
  basalt, 371, 386
  biogenic silica, 336-337, 353
  calcite, 230, 232, 320-323, 336-337, 362-
    365
  clays, 254
  crustal rocks, 218, 236-237
  diatoms, 391
  feldspars, 244-248
  foraminifera, 321-322
  magnesian silicates, 239-248
  quartz, surface layer, 207
  salts, 130-134
  silica minerals, 248, 387, 389
  soil, 248-249, 253
  solid particles, 228, 238-239, 272-275
  volcanics, 372, 386, 407
  zeolites, 254, 389
Distribution factor, 347
Dolomite, 237, 373, 405
Doomsday, 20, 21-23, 31-32
Drag force, 262, 272, 421, 426-433
Dust, 126-127, 145-147, 212-217

East Indies, 212, 225
Echinoderms, 320
Eddy diffusion:
  atmosphere, 57, 123, 179
  water, 57, 65-72, 287, 304, 367-368
Electrical double layer, thickness, 232
Enstatite, 114, 240-241
Epilimnion, 313, 318
Equation:
  Archimedes, 222
  Arrhenius, 111
  Carman-Kozeny, 52

Darcy, 46
Debye-Hückel, 227, 344
Fick, 74, 75
Freundlich, 346
Kelvin, 128
Langmuir, 346
Maxwell, 134, 135
Nernst, 80, 82
Othmer-Thakar, 94
Smoluchowsky, 290-292
Stokes, 262, 421
Stokes-Cunningham, 422
Stokes-Einstein, 86
Wilke-Chang, 94
Equation of state:
  ideal gas, 124
  ideal solid, 182
  sea water, 227
Equivalent sphere, 265
Erosion, 28, 30, 180. See also Denudation
Error function, 62, 63, 409-411
Euphotic zone, 303, 313
Evaporation, 45, 123, 125, 129, 258-259
Evaporites, 217, 259
Exponential distribution, 8, 10, 12, 198

Faujasite, 119
Fecal pellets, 269-271, 328, 336
Feldspars, 111, 113, 213, 218, 228, 236,
    244-248, 254, 256, 348
Fick's laws, 73-76
Filtration, 276-277, 287
Flocculation, 288, 309-310
Flocs, 308-312
Florida aquifer, 49
Flux:
  definitions, 43, 45, 58, 74, 340, 374
  and sedimentation rate, 360-362
Foraminifera, 228, 269-271, 277, 322, 390
Formation factor, 90-91
Forsterite, 114, 240-241
Fourier, 74
Fragmentation, 197-209, 272
Free energy of formation:
  chloride ion, 173
  hydroxyl ion, 173
  literature sources, 225-227
  mercury species, 173
  mineral polymorphs, 206
  water, 173

see also Chemical equilibria
Fuels:
  burning rates, 34
  composition, 33

Galena, 422
Gases:
  diffusion, 101-110, 134-137, 141-145
  equation of state, 124
  solubility, 137-141
Gas solubility coefficients, 139
Gaussian distribution, 10, 12, 185, 190-191
Geochemical cycle:
  biota and water, 312-319, 336-338
  carbon and oxygen, 23-32
  mercury, 166-179
  models, 14-15, 316-317
  phosphorus, 17-23
Gibbs free energy, 208
Gibbsite, 224-225
Glass:
  silicate, 117-118, 248
  volcanic, 114, 118, 237, 372
Glauconite, 348, 389
Gneiss, 237
Goethite, 224-225, 308
Goldich weathering series, 218
Granite, 50, 211, 237
Graphite, 206
Griffith crack, 181
Ground water:
  flow, 45, 48-50, 258-259
  silica concentrations, 237
Gulf of Mexico, 205, 300, 301

Hailstones, 126
Halflife, 11, 366
Halloysite, 348
Hawaii, 213-214, 351
Heavy metals, 342
Hemipelagic environment, sedimentation in, 260
Heulandite, 114, 119
Histograms, 185-189
Hornblende, 218
Humus, 35
Hydraulic gradient, 46, 48, 55
Hydrogen chloride:
  atmospheric production, 151
  pH of solution, 156

Hydrothermal circulation, 258-259, 369, 408
Hyperbolic distribution, *see* Pareto distribution
Hypolimnion, 318

Illite, 205, 224, 261, 345, 389, 403
Incongruent dissolution, 240
Indian Ocean, 382, 386, 394, 397
Industrial perturbation, 20-21, 28-35
Interdiffusion, 82-86
Interface:
  air-land, 2, 40, 179, 182-183
  air-ocean, 2, 145, 177
  air-water, 141, 145
  density, solutions, 308-310
  gas-solution, 134-137, 141
  sediment-water, 338
  solid-solution, 118, 132, 228, 371
Ion-activity product, 320-321
Ion-exchange capacity, 205, 348
Ionic charge, 3
Ionic radius, 3
Isotopes:
  argon, 113
  carbon, 34, 36
  cesium, 347
  hydrogen, 80
  oxygen, 213
  radium, 347, 366-367
  radon, 72, 366-368
  strontium, 347

Junge distribution, 197

Kaneohe Bay, Hawaii, 351
Kaolinite, 103, 203, 213, 224-225, 236, 261, 310, 348, 351-352, 389, 403
Karst, 227
Kinematic viscosity, 55, 291, 294
Kinetics:
  diffusion controlled, 233
  first order, 230, 233, 384
  fractional order, 244, 255
  mixed order, 249-256
  parabolic rate, 240, 244, 255
  reaction controlled, 233
  second order, 232
  and thermodynamics, v
  zeroth order, 244, 385

Knudsen number, 422

Lakes:
  Caspian Sea, 259
  Castle, 71
  Green, 300, 307
  Greifensee, 342, 367
  Huron, 304
  Superior, 259
  Tahoe, 259
  Tiberias, 71
  Ursee, 353
  Washington, 71
  Zurich, 70, 71
Laplace transformations, 415-419
Leucogranite, 211
Liesegang rings, 120
Ligurian Sea, 347
Limestones, 24, 50, 227, 237
Litharge, 206
Logistic curve, 34
Lognormal distribution, 188-193, 215
Long Island Sound, 382
Ludwig-Soret effect, see Diffusion, thermal
Lysocline, 320-321, 330

Madagascar Basin, 386
Magnetite, 224
Man-made reservoirs, 261-262
Marl, 403-404
Massicot, 206
Mean age, 9
Mean of distribution, 9, 185, 190
Median of distribution, 190
Mercury, 167-179
Meteorite impacts:
  earth, 207
  moon, 210
Mica, 103, 114, 205, 213, 218, 236, 348
Mineralization, 314
Mississippi River sand, 222
Mixing:
  flow, 61
  reservoirs, 16
Mode of distribution, 185, 190
Molar volume:
  gases, 95
  solids, 206
Molecular diffusion:
  gases, in air, 101-102

  in pore space, 102-104, 107-110
  in small pores, 104-107
  in solutions, 94-97
  self-diffusion, 80-82
  solids, 110-121
  solutes, in pore space, 90-93
  in small pores, 93-94
  in water, 80-89
  tracer, 80-82
Molluscs, 320
Montmorillonite, 261, 348, 372, 389
Mordenite, 119
Muscovite, 218

Natrolite, 119
Nepheloid layer, 279
Neritic zone, 215-216
Nernst layer, see Surface layer
Nitrogen:
  abundance, 3
  atmosphere, 3, 149-153, 156-158
  fuels, 33
  organic matter, 23, 33, 380
  sediments, 347-348, 383, 391-394, 397-398
Nontronite, 351-352
NTP (normal temperature and pressure), 139

Obsidian, 114. See also Volcanic glass
Ocean:
  chemical composition, 3
  mean depth, 2, 318
  residence time:
    particles, 6, 285-287
    solutes, 6
    water, 6, 258, 317
  salinity, 69, 334
  volume, 258
Ohm, 74
Olivine, 218, 240
Onsager's coefficients, 83-86
Ooze, 261, 389-390, 403-404
Opal, 362, 386-389, 407-408
Organic matter:
  cation exchange, 348
  composition, 23, 33, 380
  decomposition, 25-27, 312-316, 380-383
  diagenesis, 358-362, 380-383
  from fuel CO_2, 40

production, 31, 39, 312-315
    sedimentation, 24, 29-30, 331
Oxidation potential, 172-173
Oxygen:
    abundance, 3
    atmosphere, 3, 35
    fuels, 33
    geochemical cycle, 23-32
    isotope ratios, 213
    water, 315

Pacific Ocean, 318, 322-323
Palagonite, 372
Palygorskite, 348, 372
Pareto distribution, 187-190, 195-198, 210-
    212, 214-217, 238-239, 274, 277,
    279-282, 365
Particle-number spectrum, 183
Particles:
    atmosphere, 126-127
    biogenic, 270-271
    dissolution, 218, 228, 234-239
    rounding, 221-224
    settling, 262-272
    see also Atmospheric particles; Suspended
        material
Particle-size distribution:
    aerosols, 215-216
    carbonates, 365
    sediments, 205
    soils, 203-204, 211, 216
    suspended materials, 277-282
    theory, 183-202, 234-239, 272-287
    weathered rocks, 211
Peclet number, 64-65
Pelagic oceans, sedimentation of, 260
Peridotite, 224
Permeability, 50-54
Persian Gulf, 381
pH:
    acidic gases, 156, 158-159
    ammonia solution, 156-157
    calcite equilibria, 230, 320
    carbon dioxide solution, 155
    cation exchange, 241, 348
    dew, 151
    effect on alkalinity, 162-166
    ocean, 31, 394
    rain, 149, 153
    rivers and lakes, 151

Phillipsite, 119, 372, 389
Phi size scale, 186, 211
Phlogopite, 114
Phosphorus:
    abundance, 3
    global cycle, 17-23
    organic matter, 23, 33, 380
    sediments, 347, 358-360, 383, 391-393,
        397-398
Photosynthesis, 31, 40, 313
Piston flow, 16, 61
Plagioclases, 114, 218, 246
Plankton, 6, 18-26, 31-32, 36, 268, 287,
    303, 312-317, 328
Polymorphs, mineral, 206, 233, 387
Pore water:
    advection, 45, 340, 378-380
    diagenetic reactions, 349-365, 375-376,
        380-407
    flow, 46-50, 55, 61-64, 2'9
Porosity:
    definition, 47
    effect on diffusion, 83, 92-94, 102-
        110
    effect on permeability, 51-54
    flocs, 310-311
    sediments, 341, 379
    zeolites, 119
Power-law distribution, 10, 12. See also
    Pareto distribution
Precipitation:
    atmospheric, 123, 145, 151, 168, 219
    cement, 320, 406-408
    chemical kinetics, 230-231, 248-256, 383-
        385
Prehnite, 389
Probability density function, 190-191, 193-
    195, 198
Pteropods, 269, 319, 390
Ptilolite, 119
Pumice, 103, 403-404
Pycnocline, 301, 308-312. See also
    Thermocline
Pyrite, 25, 30
Pyrophyllite, 348
Pyroxene, 218, 241

Quartile of distribution, 53
Quartz, 112, 206-207, 213, 218, 233,
    236, 248, 348, 387, 389, 422

Radiolaria, 269-271, 373, 389-390, 403
Radon:
  lake, 366-368
  ocean, 72
Rain:
  acidity, 148-161, 219-220
  alkalinity, 161-166
  chemical composition, 149
  droplet size, 126
  evaporation and growth, 134-137
  pH, 149, 152-155, 161, 219
Rainout, 145
Raoult's law, 130
Reaction order, 229
Reaction rate parameters:
  calcite, 232-233
  carbonate-silicate sediment, 405
  clays, 389
  feldspars, 246
  magnesium silicates, 241
  nitrogen, 383, 397
  organic matter, 314-315, 383, 397
  phosphorus, 359, 383, 397
  and sedimentation rates, 383
  silica, 249, 354, 386-390, 403, 405, 407-
    408
  sulfur, 383, 397
  zeolites, 389
Red clay, 390
Redfield reaction, 380
Redox potential, 172-173
Regeneration:
  biological nutrients, 17-23, 312-313, 360-
    362, 380-383, 391-398
  calcium carbonate, 327-331, 336-337
  definition, 312
  manganese, 306
  models, 14-15, 19-20, 313-318, 323-327,
    354-358
  organic carbon and CO_2, 24-30, 314-315,
    331-332
  silica, 318-319, 336-337, 349-354
Regolith, 210-212
Relative humidity, 124
Residence time:
  carbon-oxygen cycle, 26, 35
  equations, 4-14
  gases, 6
  humans, 8
  land plants, 6

  ocean floor, 44
  particles, air, 6, 127
    water, 6, 44, 285-287, 297-299
  phosphorus cycle, 19
  rain drops, 44, 143
  salts, 6
  water, 6, 123
Resistance:
  electrical, pore fluids, 90-91
  particle settling, 302-304
  see also Drag force
Reynolds number, 54-55, 265
Ribonuclease, 94
Richardson number, 70
Rivers:
  dissolved and suspended loads, 217, 220
  input to ocean, 6, 258, 261
  volume of flow, 217, 219, 258
Rosin-Rammler distribution, 188-190, 193-
    195, 198, 215
Rounding of particles, 221-224
Rutile, 206, 224

Saanich Inlet, 382
Sahara, 213, 216
St. Lawrence River, 261
Salt concentration:
  fresh waters, 69
  ocean water, 69, 334
  rain water, 149
  saturated solutions, 132
  vapor pressure effect, 130-134
Santa Barbara Basin, 382
Sargasso Sea, 322
Scale distance, 59, 68
Scavenging:
  by rain, 139, 141, 145-147, 154
  by settling, 292-295, 298-299
Schists, 211, 237
Sediment:
  ages, 8
  chemical composition, 3
  mass, 4
  thickness, 2
Sedimentation:
  aerosols, 146-147
  calcium carbonate, 331
  carbon, 30, 331
  in diagenetic models, 340-343, 349-362,
    373-383, 395-402

rates in oceans and lakes, 6, 45, 260-262,
    334, 342, 350, 359, 392, 397
Sepiolite, 372, 405
Serpentine, 114, 224-225, 240-241
Settling retardation, 301-305, 308-312
Settling velocity:
  biogenic particles, 268-271
  coagulation, 292-295, 298-299
  mean, 274, 284-287
  particles, various shapes, 262-267, 420-434
  rain droplets, 45, 143
  snow flakes, 268, 272
  spheres in air, 264, 422
  spheres in water, 264, 420-422
Shape factors, 187, 262, 266, 423-434
Sideromelane, 372
Silica (SiO$_2$):
  adsorption, 249, 351
  biogenic, 318-319, 336-338
  deep-sea sediments, 318, 349-351, 402-
    408
  diagenesis, 205, 222, 349-354, 385-391,
    402-408
  diffusion, 83, 112-115, 243
  dissolution, 233, 240-241, 244-247, 249,
    254, 386, 389, 391
  ground waters, 237
  lake sediments, 353-354
  minerals, 233, 387
  near-shore sediments, 351-352
  ocean, 318-319
  pore waters, 350, 351, 353, 386, 403
  precipitation, 386-387, 406-408
  solid phase transitions, 206-207
  solubility, 236, 387
  surface waters, 237, 353
Silicon:
  abundance, 3, 370
  diffusion, 114-115
SI units, 435-436
Smectite, 205, 213, 224-225, 261, 372, 403
SMOW, 213
Snow:
  dust deposition, 213
  flake sizes, 126
  scavenging efficiency, 145
  settling velocities, 268
Sodium chloride:
  density, 131
  diffusion, 85, 88-89, 92, 103, 111-112, 312

saturated solution, 132, 310
solubility, 131-132
Soil:
  diffusion, 103, 108-110
  disaggregation, 202-204
  erosion, 28-30
  lunar, 202, 210
  mean thickness, 6
  minerals, 213, 224-225
  particle sizes, 203-204, 211, 216
  phosphorus, 18-22
  silica release, 248-249
  weathering, 218, 225
Solubility:
  chemical kinetics, 227-239, 248-256, 380-
    385
  gases, 137-141
  literature sources, 225-227
  minerals, 232, 236, 239-248, 253-254,
    320-323, 362-364, 387
  see also Dissolution
Sommes Sound, 382
Soret coefficients, 98-99
Soret effect, see Diffusion, thermal
Sorting coefficient, 53-54
Sphalerite, 206
Spinel, 224
Stability frequency, 69-72
Stishovite, 233
Stokes law, 262-267, 421
Suess effect, 33-34
Sulfur:
  abundance, 3
  atmosphere, 148-154, 158-159
  fuels, 33
  mineral sinks, 27, 373
  ocean, 3, 24-26, 30-31
  organic matter, 23, 33, 380
  sediments, 360-361, 383, 392-393, 397-
    398
Supersaturation:
  aqueous solutions, 248-256
  calcite-sea water, 25, 321
  vapor pressure, 129-130
Surface area:
  dissolving minerals, 232, 241, 246
  effect on reaction rate, 231, 243, 254-
    255
  sedimentary materials, 390
Surface energy, 182

Surface layer:
  adsorbed, 371
  conductance, 93
  diffusion, 118, 228-229, 233, 242-244,
    377-378
  laminar or Nernst, 228, 231, 235-236
  reactive, 114, 207, 228-240
  thickness, 118, 207, 243-244, 247-248
  see also Boundary layer; Interface
Surface tension, 125, 310
Surinam, 203
Suspended material:
  coagulation, 294-301, 307-308
  concentrations, 217, 278-280, 314-315,
    328
  pycnoclines, 300
  rivers, 217
  settling velocities, 45, 269, 305
Switzerland:
  erosion rates, 221
  lakes, 197
  Saharan dust, 213
Syenite, 211

Talc, 103
Tensile strength, 182
Tephra, 237
Thermal conductivity:
  air, 125
  rocks, 183
  water, 312
Thermal expansion:
  solids, 182
  water, 70
Thermocline, 70-72, 300, 304, 312, 317-318
Time to steady state:
  carbon-oxygen cycle, 29-30
  definition, 15-17
  gas dissolution, 142-143
  phosphorus cycle, 21-22
  pore-water reactions, 375-376, 402
Timor Trough, 358
Tortuosity, 90, 91, 103, 104, 107
Tridymite, 233, 389, 407
Turbulent energy dissipation, 65, 291, 294

Upwelling, 45, 313, 317-318
Urea, 94

Vaterite, 206
Velocity gradients, 70, 291-292, 295
Venezuelan Basin, 393, 397-398, 403-
    405
Vermiculite, 103, 224, 345
Viscosity:
  air, 263
  sea water, 264, 285
  sodium-chloride solutions, 88
  water, 86, 263
Viscosity anomaly, 87
Vivianite, 205
Void ratio, 51
Volcanic ash, 212, 261, 377, 390
Volcanic debris in sediments, 403
Volcanic eruptions, 212
Volcanic glass, 114, 118, 237, 372, 403

Washout, 145
Water:
  atmospheric, 123-125
  cycle, 257-260
  kinematic viscosity, 55, 294
  molecular radius, 120
  molecular weight, 124
  physical properties, 125
  residence times, 6, 258
  self-diffusion, 120
  viscosity, 86, 263
  viscosity anomaly, 87
Weathering index, 225
Weathering series, 218
Weibull distribution, 189-190, 193-195
Wind:
  transport, 212-217, 261, 369
  tropospheric, 45
Wurtzite, 206

Young's modulus, 182

Zeolites, 114, 119-120, 254, 348, 372, 389,
    403